UNCERTAINTY

THE LIFE AND SCIENCE OF

WERNER
HEISENBERG

UNCERTAINTY

THE LIFE AND SCIENCE OF

WERNER HEISENBERG

DAVID C. CASSIDY

W. H. Freeman and Company
New York

All photographs not specifically credited
are courtesy of the Werner Heisenberg-Archiv,
Munich.

Library of Congress Cataloging-in-Publication Data

Cassidy, David C., 1945-
    Uncertainty : the life and science of Werner Heisenberg / by David C.
Cassidy.
      p. cm.
    Includes bibliographical reference and indexes.
    ISBN 0-7167-2243-7
    1. Heisenberg, Werner, 1901–1976. 2. Physics—History.
3. Quantum theory—History. 4. Physicists—Germany—Biography.
I. Title.
QC16.H35C37   1991
530'.092—dc20   91-13818
[B]   CIP

Printed in the United States of America

2 3 4 5 6 7 8 9 0 VB 9 9 8 7 6 5 4 3 2

For Janet

Contents

Preface

Werner Heisenberg, born at the dawn of the twentieth century, became one of its greatest physicists. He is also among its most controversial. While still in his early twenties, he was among the handful of bright young men who created quantum mechanics, the basic physics of the atom, and he became a leader of nuclear physics and elementary particle research. He is best known for the uncertainty principle, a component of the so-called Copenhagen interpretation of the meaning and uses of quantum mechanics.

Heisenberg was also a man who chose to reside in Germany throughout his life. Born into an academic German family, Heisenberg experienced all the upheavals of the cultural elite in Germany: two lost world wars, a soviet revolution, military occupation, two republics, and Hitler's Third Reich. As the leading non-Jewish theoretical nuclear physicist to remain in Germany after Hitler came to power in 1933, Heisenberg, although not a Nazi, played a prominent role in German nuclear research during World War II, traveled frequently to German occupied territories, and helped to establish West German science after the war. He died in 1976. This is the story of his sometimes difficult life and his often brilliant science.

In approaching the life of this man, several questions immediately present themselves. How did this child, born in 1901, climb so quickly to the top of his profession, attaining a full professorship in theoretical physics at the age of 25 and the Nobel prize at the age of 32? What impact did the most turbulent period of his life, the events surrounding the end of World War I—the lost war, soviet revolution, and the German youth movement—have upon his mature political and scientific views? What impact did his private and professional lives have upon his scientific achievements? Why did this brilliant young man, this product of the best that Germany could offer, hold a prominent professorship under Hitler's Third Reich and perform nuclear fission research for Germany throughout World War II, working feverishly right up to the end of the war?

Every biography, and especially this one, brings together three lives: the subject's, the author's, and the reader's. A biography is one person's life as seen, researched, and interpreted through the efforts and sensitivities of another person. And, like other historical writing, the less the biographer intrudes the better. While I have attempted to adhere to this ideal throughout, it has not always been possible, or appropriate. Some of the questions raised by Heisenberg's life and science are difficult and profound, touching the lives of us all. For some questions, there are no easy answers; for others, the available sources are simply inadequate.

Especially difficult and controversial is a retrospective evaluation of Heisenberg's activities during the Third Reich and particularly during World War II. Since the end of the war, an enormous range of views about this man and his behavior have been expressed, views that have been fervently, even passionately, held by a variety of individuals. It is as if, for some, the intense emotions unleashed by the unspeakable horrors of that war and regime have combined with the many ambiguities, dualities, and compromises of Heisenberg's life and actions to make Heisenberg himself subject to a type of uncertainty principle—a principle that encompasses ambiguities and dualities by allowing for mutually exclusive perceptions of events—whence the title of this book.

My aim throughout has been to explore this principle as far as Heisenberg, its originator, is concerned. This I have attempted to do through intensive research and reflection, but in the process I have also had to rely upon my own experiences, background, and positions. Although my professional background is that of a physicist turned historian of science, two personal influences are especially significant.

The first is my Armenian heritage. My mother, descended from a well-to-do family of culture and nobility, escaped to the United States as a child after nearly her entire family was wiped out during the brutal Armenian genocide of World War I. Comprehending the nature of genocide, dictatorship, and the hold of mass persuasion over people's minds—even the minds of great thinkers—has always been of foremost concern to me.

The second influence is my late father, a man of erudition, universal esteem, and Irish heritage. His influence on his four children was summed up in the dedication that one of my sisters wrote to him in her recent doctoral dissertation: "a love of history and a compassion for the underprivileged."

Putting all these factors together, my approach to Heisenberg is twofold: first, to understand Heisenberg historically and biographically, as the complicated, multidimensional human being that he was, to comprehend what Heisenberg did and why he did it in the context of his own life, motives, science, thinking, and surroundings; second, to learn from his successes and failures and to appreciate how and why they occurred. In the first instance, this is neither to criticize nor to apologize but to comprehend. I hope what we do with that knowledge will help us and future generations to gain a better grasp of the nature of scientific progress as well as a fuller appreciation of the ethical demands on the citizen and the scientist in an age of technological weaponry, nationalistic fervor, and mass manipulation.

A work of this magnitude could not have been researched, written, and published without the generous help and encouragement of numerous friends and colleagues, individuals, institutions, and archives. Over a decade of research, interviewing, and writing in Europe and the United States has gone into the making of this book, although other projects and employment have also demanded my attention. I am especially grateful to Mrs. Elisabeth Heisenberg and to Dr. Helmut Rechenberg, director of the Werner Heisenberg-Archiv in the Max Planck-Institut für Physik und Astrophysik, Munich, for permitting me to examine the extensive materials in the Heisenberg collection. Without their generosity and kind support, this work would not have been possible.

I am also very grateful to the following people for their generous assistance in locating and using archival materials: Finn Aaserud and Erik Rüdinger, Niels Bohr Archiv, Copenhagen; Beat Glaus, Eidgenössische Technische Hochschule, Zurich; Dr. Heinrich, Handschriftenabteilung, Deutsches Museum, Munich; Winfried Mogge, Archiv der Deutschen Jugendbewegung, Burg Ludwigstein; Rolf Neuhaus, Archiv zur Geschichte der Max Planck-Gesellschaft, Berlin; G. Schwendler, Universitätsarchiv, Leipzig; Spencer Weart and Joan Warnow, Center for History of Physics, American Institute of Physics, New York; Gerald Wiemers, Sächsische Akademie der Wissenschaften, Leipzig; and the directors, archivists, librarians, and staff of the many other archives, libraries, and institutions that I have visited during the course of my research.

In addition to the above named individuals, my sincere thanks to the following, among many others, for informative comments and discussions at one or more stages of this project: Mara Beller, Walter Blum, James T. Cushing, Max Dresden, Hans-Peter Dürr, Michael Eckert, Paul Forman, Elizabeth Garber, Stanley Goldberg, John L. Heilbron, Martin and Apollonia Heisenberg, John Hendry, Armin Hermann, Gerald Holton, Friedrich Hund, Johannes Juilfs, Andreas Kleinert, Karl von Meyenn, Arthur I. Miller, Boris T. Pash, Wolfgang Rüdel, Manfred Schroeder, S. S. Schweber, Gottfried Simmerding, Imre Toth, B. L. van der Waerden, Mark Walker, Victor Weisskopf, Carl Friedrich von Weizsäcker, Linda Wessels, and C. N. Yang.

I would like particularly to thank my three avid readers, Barbara Sullivan, Robert Ubell, and Jeremiah Lyons, along with Diana Siemens, Nancy Singer, and the staff at W. H. Freeman and Company, who helped me turn a massive manuscript into a book.

Timely and generous financial support has been provided by the Alexander von Humboldt-Stiftung, the National Science Foundation, the Albert Einstein Papers Project, and the Max Planck-Institut für Physik und Astrophysik, Munich, for which I am truly grateful.

Last, but not least, I would like to express my gratitude to my family for their many years of encouragement, and especially to my wife Janet, who through her love and understanding helped us both to suffer through the birth pangs.

UNCERTAINTY

THE LIFE AND SCIENCE OF
WERNER HEISENBERG

PART ONE

Young Werner

CHAPTER 1

Family Matters

On November 11, 1901, August Heisenberg, a secondary school teacher of classical languages, presented a formal lecture to the faculty of the University of Würzburg—the last step in his candidacy for qualification as a university lecturer. Three weeks later, his wife Annie gave birth to her second child, a boy. Like his older brother Erwin, born in Munich, the infant arrived in the Heisenberg home, at Heidingsfelderstrasse 10 in the elegant Würzburg suburb of Sanderau. His birth certificate lists not only his name and date of birth but also the exact time: Werner Karl Heisenberg, born on Thursday, December 5, 1901, at 4:45 P.M. Before the month was out, the Bavarian Interior Ministry added to the joys of the proud father its approval of his appointment as a Privatdozent (lecturer) at the university, a position he held in addition to his duties at the secondary school.

The coincidence between Werner's birth and his father's appointment hinted at three essential elements in the child's future development: his timing with respect to important events, the high academic and cultural level of the family into which he was born, and the rapid upward social and academic momenta the family had attained by the time of Werner's birth. August Heisenberg came from a family of middle-class craftsmen. Within a decade of Werner's birth he would reach the top of the social and academic ladders as Germany's only full professor of middle and modern Greek studies.

The Heisenberg family's social mobility is evident in a carefully constructed family tree preserved in Werner Heisenberg's private papers. The tree, rather a typed pedigree replete with certificates of birth and baptism, owes its origin to the search by Nazi authorities for a Jewish ancestor in the scientist's past. It traces the Heisenbergs back five generations to one Heissenberg in Heidenoldendorf, a village in the northern state of Westphalia. The eighteenth-century ancestor is succeeded by a brandy burner, a master cooper, and a locksmith. The locksmith, Wilhelm August Heisenberg (1831–1913), dropped the second *s* in his name and moved north to Osnabrück, then in the state of Hanover, where he raised three daughters and two sons, one of whom was Werner's father.

After Werner's grandfather learned the locksmith trade, he set out on a "wander year," a common rite of passage in those days.[1] He obviously did well: on his return he purchased his master's business, barn, and house. With business, property, and title (master locksmith), he easily rose to the rank of official Bürger of Osnabrück, a voting member of the town's middle class. In 1858 he ensured his status by marrying the daughter of a prosperous local farmer. The two complemented each other well. Wilhelm Heisenberg is remembered as a quiet, cerebral man, an impression confirmed in a surviving photograph. His wife, Anne Marie, is remembered for her strong will and keen intelligence.

Werner Heisenberg's middle name was that of his father's younger brother Karl, the black sheep of the family. The fifth of the five children, he became a Tunichtgut, a ne'er-do-well. Always in trouble, he once stole a sum of money from one of his sisters, whereupon his father handed him 200 marks more and packed him off on the next ship for America—in those days, the end of the world. The clever young man quickly fulfilled the American dream: he opened a factory for uniform buttons in Flushing, New York, and soon became the richest Heisenberg. The exile's dollars proved invaluable to his German relatives during the inflation after World War I, as did his American connections to Werner after World War II.

Werner's father, Kaspar Ernst August Heisenberg, was born in Osnabrück in 1869, less than two years before the unification of the German Reich under Kaiser Wilhelm I. Hanover, then a Prussian territory, was already subject to Wilhelmine rule when Wilhelm's chancellor, Bismarck, induced the recalcitrant southern states to join a united Germany, the "small German empire" (minus Austria and the border lands). A period of enormous industrial, commercial, and technological expansion ensued throughout the Reich, matched by the rising nationalism of the middle and upper classes and the increasing solidification of a social and political hierarchy centered on the Kaiser and his chancellor. Like many others of his generation, August (the name he used) came to maturity under the Bismarckian monarchy, and, like many other German academics, he came to idolize Bismarck and the empire. Evidence suggests that August joined other academics in his allegiance to the National Liberal Party, a party on which Bismarck had greatly relied in unifying the Reich. Liberals believed that the best route to civil reform and the advance of their own upper-middle-class status lay in national unity under Prussian leadership, the predominance of secularized Protestantism, and commercial expansion—ideals that August later tirelessly impressed on his children.

August recalled a happy childhood in Osnabrück among "numerous siblings."[2] At the age of 10 he entered a nine-year course of study at the local gymnasium, his first step toward higher education and an academic or professional career. Only graduates of a gymnasium, which brought its students to the equivalent of the early junior year of a modern American college, could pursue higher education leading to the professions. August's study at a gymnasium, instead of apprenticeship to a craftsman, required a fundamental family decision, since it constituted the first break in the family tradition of producing middle-class craftsmen. With the complete support of his family, August would attempt to reach the next social stratum via the uncertain route of an academic career. When Wilhelm died in 1913, August, a lately appointed professor, wrote of the "sincerest and most trusting relationship" that he had enjoyed with his father "continuously from childhood on, until the very last. . . . In everything that I have in life, he stood by me with his counsel, and whatever I succeeded in doing pleased me, because it pleased him."[3]

Compared with a modern American university teacher, a German professor of the Wilhelmine era enjoyed much more prestige and power within the stratified world of Bismarckian Germany. During the decades following the Napoleonic era, which ended for Germany in 1815, German administrators had attempted to build up German national culture as one of the pillars of German national strength. Administrators and scholars regarded scholarship as an essential component of the cultural pillar, and neohumanist studies of Greek works of the heroic age— exemplars for a heroic new Germany—became its crowning achievement. Because of this, a German university professor, especially one in classical Greek philology, ranked in status along with other nonpropertied "bearers" of the Bismarckian state—judges, army officers, industrialists, higher bureaucrats—among the upper-middle-class elite. Just above them stood the nobility and the propertied upper class; below stood the middle class of craftsmen, farmers, lower civil servants, and gymnasium teachers. At the bottom were unskilled industrial laborers.

August's talents and Germany's economics encouraged the family strategy of seeking social advance through academic achievement. By 1879 the German industrial revolution was in full swing. Master craftsmen were finding it increasingly difficult to compete with mechanized industries and their growing pools of cheap labor. Although August Heisenberg's attempt to compete in the academic rather than the economic world entailed a grave family risk—only the best students could attain a university chair—the Heisenberg family was not at all unusual in taking it. Middle-class families were the main source of professors during that era, especially for the most prestigious discipline of the day, August's field of Greek philology.

The expansion of industry and empire, starting as early as 1850, required more administrators, jurists, and professors than could be supplied by simple replication. Sons of middle-class and lower-class families had to be recruited. (Women were not considered for such positions.) According to one study, during the period of August's education fully two-thirds of Prussian philology students originated from the middle class.[4] More than a quarter of such students came from the families of skilled craftsmen, small businessmen, and innkeepers— the families who could best afford to finance the long years of study. As more of the middle class attained the coveted title *Herr Professor Doktor*, they viewed themselves more consciously—for protection and privilege—as a group apart, a Bildungsbürgertum (academic bourgeoisie), a new upper middle class defined and established not by title or inheritance but by education and culture.[5]

Two years out of gymnasium, August abruptly headed for Bavaria, attracted to the southern province by the imperial Wagnerian music of the Bavarian capital, Munich, and by its enthusiasm for the glories of ancient Greece. Even more attractive were the efforts of Bavarian state officials to raise the cultural level of the rural province through generous funding of education and the importation of famous Prussian scholars, the so-called northern lights. August was drawn to one of these beacons at the University of Munich, Karl Krumbacher, a lecturer who soon founded Germany's only chair for Byzantine studies (middle and modern Greek philology). Heisenberg immediately converted to the promising yet nearly untouched field, assured of bright career prospects in the rural southern province.

In 1893, August Heisenberg completed his doctorate under Krumbacher,[6] passed the difficult teacher-qualifying examination, and soon became a teacher trainee at the prestigious Maximilians-Gymnasium in Munich under its learned and powerful rector, Nikolaus Wecklein. Two years later he abruptly left for his required year of military training, which he performed not in Bavaria but back home in Osnabrück with an infantry regiment under Prussian command. An unfortunate romance with the elder of Wecklein's two daughters, Annie, apparently precipitated the sudden move. Her esteemed family flatly disapproved of her unestablished suitor.[7]

August returned to Munich as a reserve officer—still dedicated to national unity and Prussian predominance—and soon came under Wecklein's scrutiny in a required pedagogical seminar. Wecklein's doubts apparently dissipated, for August's romance with his daughter revived. Heisenberg remained in Munich only six months before taking off to the Bavarian hinterland, this time to a Latin school in Lindau on Lake Constance. But three days before the start of his appointment he telegraphed his father the good news: he and Annie were engaged.[8]

Little is known of Werner's mother, Annie. Neither she nor her sister received a university education: German universities were closed to women, as a rule, until 1895, and Munich did not admit female students until 1903. Nor are there informative state personnel files on which to rely: German civil careers were open only to men. Both Wecklein girls no doubt attended one of the segregated girls' middle schools, which typically offered training in the fundamentals—math, history, and literature—and prepared its pupils for their future roles as genteel wives and cultivated mothers of educated sons.

Like Wecklein, August's father Wilhelm conferred his blessing on the union. But the marriage was postponed for over two years while the groom attained the acceptable status of gymnasium teacher, with corresponding salary, and a transfer back to Munich. During those two years August obtained a state research grant to prepare for academic advance after marriage. While August rummaged for artifacts in Greece and Italy, Rector Wecklein arranged his promotion and transfer to the Luitpold-Gymnasium in Munich (which its most famous pupil, Albert Einstein, had left only recently), and Annie changed her religion from Roman Catholic to August's Evangelical Lutheran faith in order not to risk opposition by the Catholic Church, the most influential church in Bavaria. In January 1899 August, then in Rome, submitted his required official request to the Interior Ministry for permission to marry. The Interior Minister personally concurred, after assuring himself of the bride's moral reputation (a state employee could not bring dishonor upon his employer).[9] With all in readiness, August returned in the middle of May 1899, and within a week the happy couple was married at the Erlöserkirche in the Munich suburb of Schwabing.

Nikolaus Wecklein had also married within the higher social stratum to which he had risen through academic achievement. The son of a long line of farmers in the northern Bavarian province of Mittelfranken, near the towns of Gänheim and Binsbach, Wecklein was married in 1870 to Magdalene (Magda) Zeising, whose ancestors all served with royal titles at the court of the Duke of Bernburg in the Harz Mountains between Halle and Magdeburg. After the failed liberal revolution of 1848, Magda's father, Dr. Adolph Zeising, an educator and poet, had gone to Munich to study the aesthetics of the Bavarian king's Greek statues. He died there in 1872 of a painful illness he contracted shortly after Magda's marriage to Wecklein.[10]

Wecklein owed his rise from humble beginnings to the top of the Bavarian school system to his scholarly command of classical Greek and to the helpful encouragement of powerful superiors whom he had impressed as a student. With his long white beard and stern demeanor, he had the look in later years of a patriarch, and when he wore his visored cap, he looked a little like an old sea captain. After the marriage of his elder daughter Annie (his younger never married) and the birth of her two sons, Wecklein often did function as the patriarch of the family and overseer of the professional advance of his son-in-law and grandsons. The grandsons entered the Max-Gymnasium while it was still under Wecklein's direction. Photographs of the extended family during their frequent Sunday outings display the familiar pyramidal arrangement of the era. One photograph shows the patriarch appropriately perched on the craggy summit of a hill with the succeeding generations dispersed beneath him in chronological order.

Wecklein aimed early for a teaching chair in Greek philology. He wrote a dissertation on the Greek sophists, qualified as a university lecturer with a work on Greek grammar, and became a leading authority on Greek tragedy.[11] But the year of his marriage marked the end of his advance and the beginning of a profound disappointment: his failure to obtain a university teaching chair. The Munich faculty refused to appoint him to an opening after it judged his lectures on Aeschylus too dull. The dull lecturer found himself instead on a fast track into the educational hierarchy when the Bavarian Interior Ministry for Church and School Affairs, encouraged by Wecklein's professors and his Liberal Party credentials, appointed him to gymnasium administration. Wecklein eventually rose to influential positions on the Bavarian school board and academy of sciences, while rectoring the prestigious Max-Gymnasium.

Still, without the prestige and standing of a university chair, Wecklein never felt his life complete.[12] When he retired from the top of the school system in 1913 at the age of 70 with all sorts of titles and awards, his family barely managed to keep him from becoming a lowly Privatdozent once again. A Privatdozent, or lecturer, having formally qualified for a chair, lectured for student fees until appointed by the state to a professorship upon recommendation by a university. In his last years, Wecklein must have been comforted in the knowledge that his son-in-law had been granted a university chair, that his two grandsons had by then obtained doctorates, and that Werner had habilitated (qualified) as a professor and had even recently substituted for a physics professor (Max Born) in Göttingen. After paying a goodbye visit to Werner in 1926, Wecklein died before seeing him occupy a chair in Leipzig less than a year later.

Even without a chair, Wecklein's administrative status positioned him well by the turn of the century to play a crucial role in the family strategy for social advance. When Krumbacher proved unable to secure the habilitation (qualification) of Wecklein's new son-in-law in Munich, Wecklein's connections saved the day.[13] After the birth of his first son, Erwin, in March 1900, and a return to Italy to complete his habilitation treatise, August Heisenberg learned through Wecklein that Wecklein's old alma mater, the University of Würzburg, would consider him. Shortly after submitting the treatise to Würzburg, August also learned through Wecklein of a vacancy at Würzburg's Altes Gymnasium. Should the habilitation succeed, he could lecture at the university while teaching (with sufficient salary) at the gymnasium. A request for transfer received the immediate approval of Wecklein's school board, and, after Heisenberg completed his annual six weeks of military exercises in the summer of 1901, he and his family moved to Würzburg, about 400 kilometers (250 miles) northwest of Munich. August began teaching in September, while his wife prepared to give birth to her second child.[14]

Dr. Heisenberg's wish to habilitate caused a tizzy in Würzburg. The faculty referees wanted their approval of habilitation to serve not as a vehicle for gymnasium advance (which they could not control) but as the means of attaining a university chair (which they could). The faculty finally acceded only after the candidate assured it privately that he intended to leave the gymnasium for the university at the earliest opportunity.[15]

The two-part habilitation process required a treatise of quality superior to that of a doctoral dissertation and a lecture and oral examination before the entire faculty, during which the candidate demonstrated full command of his discipline. But, as in all such university matters, state officials had to confer final approval. The faculty committee's glowing report to the Interior Ministry on Dr. Heisenberg's treatise and his performance during the test lecture left little doubt about the out-come.[16] August received his habilitation before the year was out and just weeks after his wife gave birth to her second son, Werner.

In contrast to Munich and other German cities, Würzburg during the first decade of this century remained a remarkably quiet, rural, tradi-tional, provincial town. The family's two-story suburban house near the Main River, with the hilly Franken vineyards along its banks and the nearby fields and woods, was ideal for the two growing Heisenberg boys. Despite the massive immigration of rural families into Würzburg and the sudden jump in gymnasium pupils the year Dr. Heisenberg began teaching,[17] the social and economic structure of Würzburg had remained nearly untouched since before Bismarck. Civil servants, mer-chants, landed nobles, and especially university professors retained con-trol of the political hierarchy.[18]

A wide economic gulf separated the cultured upper classes from Würzburg workers. According to one estimate, in 1902 the average annual living expenses for a working-class family of four in Würzburg amounted to 1200 marks, while the average annual wage of a skilled worker was 1200 to 1600 marks.[19] If this is correct, the Heisenberg family had few financial worries, despite the higher standard of living required of a professorial family. Werner's father started at the Altes Gymnasium with an annual base salary of 2820 marks, to which were added his seminar fees and salary for university teaching. In 1902, his gymnasium salary increased to more than 3000 marks, and after promo-tion to gymnasium professor in 1906, he received 4410 marks—nearly three times the highest wage of a skilled worker.[20]

August Heisenberg is remembered by his family, superiors, and pupils as a rather stiff, tightly controlled, authoritarian figure. A former stu-dent recalled that the schoolmaster demanded "unbending fulfillment of duty, absolute self-control, and meticulous precision."[21] "He treats his pupils with propriety but tolerates no lazy boys in his class," his Lindau rector noted.[22] August must have applied the same standards in raising his own two boys, who grew up in a family structure typical of Bürger families at the turn of the century: father centered, authoritarian, hierarchical. In the male-dominated, monarchical society of Wilhelmine Germany, it is not surprising that families were organized in the same fashion, or that men considered it an obligation to preserve such an organization.

By the same token, a German woman of that era, no matter what her interests or talents, regarded as her obligation being an obedient wife and a self-sacrificing mother.[23] As the wife of a gymnasium teacher and the daughter of a gymnasium rector, Annie's duties were self-evident. When Annie married August, she knew that self-realization or recognition could be achieved only in ensuring the success of her husband and the well-being of their children. She excelled brilliantly. Indeed, she probably made it possible for her husband to discharge at an outstanding level the almost incredible workload he carried during the Würzburg years. Despite the exclusion of women from higher education, she had obviously sought and received advanced instruction, probably from her father, for it was she who graded the daily homework of her husband's pupils. She even learned Russian in order to translate research papers for her husband's use — all this, of course, in addition to caring, no doubt without help, for the two growing boys, who surely required much coaxing to keep them from disturbing their busy father.

As a gymnasium teacher, Dr. Heisenberg at first taught 14 hours per week of Latin, German, and geography to large classes (35 to 40 pupils) of 9- and 10-year-old boys in the first and second grades. After promotion to gymnasium professor, he took charge of the more demanding sixth grade of 14-year-olds.[24] Throughout that period, he also offered three two-hour courses per week at the university on Byzantine philological topics, was deeply involved in the political affairs of the local Gymnasium Teachers Association, and generated scholarly writings at a prolific rate. His bibliography contains an astonishing 56 titles published while in Würzburg.[25] Among them was a two-volume account in 1908 of his own archaeological research in Constantinople — the foundation for his appointment to university professor two years later.[26]

Dr. Heisenberg's capacity for teaching and research also astonished his Würzburg gymnasium rector, who consistently gave him the highest marks as both teacher and scholar. In his evaluation for 1902, he noted in the man "a restless drive to expand and deepen his narrower professional knowledge." Yet Heisenberg's teaching did not suffer: "The lessons are carefully and thoroughly completed; the needs of the class are closely followed; the interest and encouragement of the pupils are close to his heart. Because of this, he exerts a stimulating and permanent influence on the education of the mind and morale of his pupils."[27]

Yet August's academic life took a heavy toll on him and his family. With the tremendous pressures exerted by his work, his restless drive, and his rigid role as family provider, it is little wonder that the professor is also remembered for his stormy temperament and rapid oscillations from pleasantness to depression. His wife, the daughter of an equally authoritarian and probably equally bad-tempered schoolmaster, had learned to handle such behavior with a carefree, childlike disposition and a quiet, pleasant, even-tempered manner. One visitor to the Heisenberg home recalled her as "a small, dear woman, concerned, kind, but not very conspicuous." Her husband, on the contrary, whether present or not, served as the center of attention and authority in all family matters.[28]

Children, on the other hand, were at the bottom of the Heisenberg family hierarchy. As Werner grew into adolescence, what he saw and felt from that position must have increasingly distressed him. Like any other turn-of-the-century Bürger family, the Heisenbergs cherished the appearance of genteel respectability, social grace, and allegiance to nationalist trappings. This was especially so for German Bildungsbürger and their families, whose public deportment and professional demeanor were expected to reflect their superior social station and the virtues of their Christianity-dominated, monarchical state. Respectability — frugality, devotion to duty, and restraint of the passions — writes one historian, defined and maintained the precarious position of the upper bourgeoisie during the late prewar period.[29] Bavarian gymnasium rectors, the educators of the future upper classes, were accordingly required to report on the ethical deportment of their subordinates. August Heisenberg's rector always wrote "impeccable" and usually added: "His family circumstances are the best imaginable."[30]

Werner eventually saw the respectability as a facade. Children and adolescents easily perceive and condemn hypocrisy in their elders, and it is usually nowhere more evident than in religious matters. With the recently increased role accorded churches in Bismarckian state affairs, no sharp separation between church and state had developed as it had in other countries. Both Heisenberg boys were duly baptized and confirmed in the German Evangelical Lutheran Church, and the family adhered closely to prescribed religious practices and to the requirements of Christian ethics. This combination of social, religious, and cultural virtues in the context of the monarchical bureaucracy produced that paragon of human virtue, the upright Wilhelmine professor, the decent and dutiful civil servant, exemplified perhaps to an extreme by the Munich-educated Prussian physicist Max Planck. According to Planck's biographer, "Respect for law, trust in established institutions, observance of duty, and absolute honesty — indeed sometimes an excess of scruples — were the hallmarks of Planck's character."[31] Certainly August Heisenberg's character came out of the same mold.

But how solid was the religious foundation of such scrupulous rectitude? Both Heisenberg parents were probably unusual among intellectuals in admitting to their sons the lack of any private religious beliefs.

European culture reeked in those years of bourgeois hypocrisy and duplicity.[32] Dr. and Mrs. Heisenberg judiciously left the matter of belief to the boys' personal preferences — as long as those preferences did not conflict with public norms. Years later Werner wrote to his parents that, as he saw it, for them Christianity was "just an empty form" used merely for appearances.[33] He told one interviewer: "My parents were far away from the Christian religion as far as the dogmas were concerned, but they would always stick to the Christian ethics. They would accept the rules of how to behave and to live, and say that we can take them from the Christian religion, but we cannot accept literally all these old stories."[34]

The bourgeois ambivalence of Werner's childhood may have played a role in his own adult ambivalence toward the sweeping claims of every system of thought and belief, including science. At middle age and again near the end of his life, Werner declared science and religion to be "complementary" aspects of reality, each with its own language and symbolism and each with its own limited realm of validity. Different religiously or intuitively apprehended truths should be viewed as different sides of the same truth, while rational science — his own profession —should be viewed as just one among a variety of ways of perceiving reality.[35]

While Werner reveled in ambivalences about ultimate reality, his brother Erwin became a convinced follower of a religious-philosophical system, anthroposophy, that enjoyed considerable popularity in Germany during the early decades of the twentieth century. Werner once recalled a heated argument with his brother in which Erwin declared that he knew his soul existed but was not so sure about the existence of his body or matter, an essential anthroposophic position. Teenaged Werner was sure that his body existed but was not so sure about his soul.[36] "If someone were to say that I had not been a Christian, he would be wrong. But if someone were to say that I had been a Christian, he would be saying too much," Heisenberg remarked to his longtime colleague and confidant Carl Friedrich von Weizsacker shortly before his death.[37]

While the problems of ambivalence and hypocrisy became especially acute during Heisenberg's adolescence and early adult life, he emerged from his childhood already endowed with his family's recently acquired cultural station and infused with his father's tremendous drive for recognition and success in academic affairs. The indoctrination took place in the close context of the four-member family.[38] Erwin is remembered as his father's favorite son, the shy and retiring "Wernerle" his

mother's favorite. Werner's allergies may have encouraged the favorit-
ism. At 5, he nearly died of a lung infection, which must have increased
his mother's protectiveness immensely. Disabling allergies and illnesses
recurred throughout Werner's life, while quiet and even temperament
encouraged by his mother's loving care became a permanent fixture of
his personality.[39]

In snapshots of Werner as a young man, he always appears radiant,
confident, alert, and pleasant. But in photographs as a child, the slightly
built youngster, with his close-cropped blond hair, freckled face, and
typical Bavarian Lederhosen, invariably appears uncomfortable in posture and expression, even hurt and withdrawn. His brother, in the same
photographs, always has a mischievous look in his eye. The two boys
were in continual competition.

The competition early focused on gaining the attention of their busy
father, and to do so, the sons had to excel in academic and cultural
skills. Among these skills was the playing of a musical instrument,
another essential activity of cultured Germans. Like impeccable deportment, the enjoyment and playing of classical music served the cultured
classes as recognizable expressions of a common cultural outlook. All
cultured children were thus taught a musical instrument and learned to
play the great classical works on it. Music served Werner throughout his
life as a significant and sometimes crucial social vehicle.

August Heisenberg and his sons reportedly practiced together daily
without the boys' mother, who, avoiding the male competition, claimed
complete lack of musical talent.[40] The father, endowed with a fine
operatic voice, was accompanied by Erwin on the violin and Werner on
the cello. Later Werner played the piano. One can easily imagine the
spirited father filling the house with arias accompanied by the halting
notes of the budding musicians, while wife and mother retreated to her
husband's homework.

Already inspiring ambition by example, August strongly encouraged
the competition between his sons, especially in the academic realm and
within that in mathematics. Years later Heisenberg recalled: "Our father
used to play all kinds of games with [us]. . . . And since he was a good
teacher, he found that the games could be used for educating the
children. So when my brother had some mathematical problems in his
schoolwork . . . he tried to use these problems as a kind of game and
find out who could do them quickly, and so on. Somehow I discovered
that I could do that kind of mathematics rather quickly, so from that
time on I had a special interest in mathematics."[41]

Early and intense rivalry, deliberately stoked by August, coupled with August's "restless drive" and the family's upward momentum, must have engendered Werner's own enormous lifelong drive to excel in everything he did—mathematics, music, even table tennis. His teacher at the Max-Gymnasium in Munich often observed in his grade reports: "The pupil is also extraordinarily self-confident and always wants to excel."[42] But the rivalry also must have engendered the increasing dissonance and eventual disharmony between Werner and his brother. As boys, the two often fought fierce battles with each other. As they grew older, they fought even more frequently and intensely. Finally, after one particularly bloody fight—in which they beat each other with wooden chairs— they called a truce and went their separate ways. After that, they had little to do with each other, except for occasional family visits as adults.

After 1916, Erwin was hardly at home, anyway; he was away first on military duty, then later in Berlin where he studied, married, and settled. Werner, meanwhile, moved increasingly outside his family as he became involved in youth-movement activities. Werner never mentioned his brother to his youth-movement comrades, nor do they recall ever having met him. Erwin, who became a chemist, is conspicuously absent from most of Heisenberg's public recollections. Even after World War II their relationship remained cool, especially after Erwin, now an official of the anthroposophic movement, unsuccessfully tried a second time to convert Werner. Erwin died in 1965 and was buried on the grounds of the movement's headquarters in Basel, Switzerland.

For Werner, competitive achievement outside the family seems to have served more as a personal challenge than as a means of impressing or subordinating others. Although he was a poor loser, he often settled for acceptance and recognition by his peers. The personal challenges that Werner continually set himself and the hard work required to achieve them were clearly noted by Heisenberg's later youth-movement comrades. He was a Willensmensch, a man of will, one of them recalled. This was especially evident in athletics, a favorite pastime of the youth movement.[43] Not a natural skier, Werner nonetheless trained himself to ski excellently and over difficult terrain. Nor was he an exceptional runner, yet one former comrade recalls him running laps alone at the school with a stopwatch in hand to improve his time. Whenever there were long-distance running races—supreme challenges to determination—Heisenberg, endowed with incredible endurance, was always among the early finishers. Later he won a Sportabzeichen, a state badge for outstanding physical accomplishment, that he proudly saved.

Eventually, the driven youth found other challenges to his enormous talents in the beauties of classical piano and, after 1920, in the seemingly insoluble puzzles of quantum physics.

For Werner the child, the family world in which he lived was simple, ordered, predictable, and full of motivations to excel. Family roles and expectations of propriety were clearly defined, even if the basis for bourgeois values was not. Although control lay in the hands of his male elders, and his father was given to unpredictable outbursts, Werner was endowed with his own ability to control attention. He never lost a sense of fairness. In competing with his brother for his father's attention, he was sure of the rules and certain that the competition would be judged fairly. And when he lost, he could always find solace in the arms of his mother—a situation that must have contributed early to his unusual insecurity and the importance he attached to trust.

When Werner entered school in Würzburg at the age of 5 or 6, one of his teachers falsely accused him of some misdeed—at least the boy regarded the accusation as false—and rapped him sharply on the knuckles with a switch. More emotionally than physically hurt, the lad immediately withdrew into himself, broke off all further interaction with the teacher, and refused to cooperate for the rest of the year.[44] The pattern repeated itself to the end of his life. Whenever Heisenberg felt that his freely (even naively) offered trust had been betrayed, instead of confronting the offender he would sever relations irrevocably. He never irrevocably rejected his parents—although he later dismissed their social outlook—but his brother was apparently cut off to the end of his life.

The significance of trust, as well as of its counterpoint, commitment, was underscored when Heisenberg emerged at the end of World War I as the leader of a small band of younger schoolboys. Amidst the confusion of that era, they recalled, he was one of the few older boys whom they felt they could trust to be fair, honest, and self-sacrificing, and—unlike their elders—he never betrayed their trust. Werner's self-confidence, which first begins to radiate from photographs in this period, probably reflected not only his discovery that he could easily disarm with a smile but also the newly formed special relationships of the youth movement. The fearful and betrayed younger brother had become what his older brother had denied him: a trusted older brother to a band of younger boys who felt as hurt, betrayed, and bewildered by their world as he did. Throughout his life, his closest personal friends were always a few trusted younger male companions and colleagues who looked up to him in unhindered admiration.

Whenever Werner felt maligned or betrayed, he retreated into the orderly, secure world of his inner thoughts and dreams. Never very imaginative or fanciful, this inner world grew increasingly abstract, logical, and mathematical. Long hours of solitary piano practice, immersed in a world of order and harmony, must have contributed to the shaping of this inner world. His gymnasium teachers noted the distinction: "More developed toward the side of rationality than of fantasy and imagination"; "the pupil . . . appears . . . to be developed mainly toward the side of understanding; feeling for form and imaginative powers seem to be less developed, which is probably related to the fact that he is less outgoing than would be desirable."[45] Among Heisenberg's favorite childhood stories was a collection about Klaus Störtebeker, a fifteenth-century Hanseatic sea pirate and folk hero whose adventures were less fantastic than realistic.[46]

In science, however, Heisenberg's intuition and imaginative powers were nearly unparalleled, as his teachers also began to appreciate. His capabilities must have originated and developed early, perhaps from his father's games. The emotional importance of trust to Heisenberg may have contributed here, too. One can always rely on phenomena that can be repeated and controlled or numerical harmonies that obey specific laws of nature and mathematics. The refreshing security of science and mathematics is often attractive to an insecure, sensitive youth who, like Werner, encounters ambivalence and betrayal within his family and society.

Because of the geographic dispersal of their relatives, Werner and Erwin became familiar with long train rides and even with traveling alone at an early age. They frequently visited their father's family in Osnabrück. August's sisters were especially fond of their two nephews, and Werner was particularly enthralled by the younger one, Aunt Grete. "She led us around on invisible reins," he recalled.[47] Werner always felt at home among his close, warm, and predominantly female Osnabrück relatives, and he always enjoyed the company of his grandfather, the kind and encouraging Wilhelm Heisenberg. Years later Werner still remembered him for his fine, white hands and his lessons in such practical matters as nailing a lid on a box — lessons he surely was not taught at home.[48] Perhaps because of grandfather Wilhelm the Heisenberg boys became fond of building technical gadgets in their teens. Their masterpiece was a 1.5-meter electric battleship, equipped with remote-controlled steering and electrically fired canons; it was proudly displayed in the Heisenberg home for years. With electric lighting only two decades old in Bavaria, the ship was no small achievement.

Family interaction of quite a different sort occurred with the academic Weckleins. Munich and Würzburg were separated by a two-and-a-half-hour train ride, but that did not deter the two sides of the family from frequent Sunday strolls together. Grandfather Wecklein was an enthusiastic hiker — a common activity in those days. He devoted every Sunday to the great outdoors, and during holidays he disappeared on long tours through Bavaria, Austria, and Switzerland. He often wandered with a small group of school colleagues, who called themselves the Alte Herren-Riege, which included, among others, the assistant rector of a Landshut gymnasium, Gebhard Himmler, father of the infamous Heinrich Himmler.[49]

During the Sunday walks with his Würzburg relatives, Wecklein must have taught little Werner the joys of experiencing the romantic beauty of their German homeland. Strolling with the family schoolmasters, Werner also learned — along with probable drilling in Latin verbs — the academic and social situation of his family. Matters of school policy were often discussed on those outings, for Werner's father and grandfather were thoroughly involved at the time in efforts to preserve the social gains of gymnasium teachers and in continuing their own climb up the social ladder.[50] Although Werner was only a boy, he later claimed that not all these discussions passed over his head: "The problems of the gymnasium of that period are thus well known to me from my youth."[51]

The pressures and idylls of Werner's Würzburg life, and with it his early childhood, came to an abrupt end when his father was suddenly called to Munich to succeed his teacher, Karl Krumbacher. Krumbacher was only 54 years old when, weakened by diabetes and overwork, he collapsed and died on his way to a lecture in December 1909.[52] The search for a successor began immediately. As was common for teaching chairs, the entire program in middle and modern Greek studies rested on the shoulders of one man. The search did not last long. Within nine days of Krumbacher's death, the dean of the philosophical faculty notified the academic senate that, instead of proposing the usual list of three candidates, the faculty had already decided on a single one, August Heisenberg.[53]

The dean had noted the elder Heisenberg's unusual pedagogical and scholarly talents. Even greater accomplishments could be expected with a full-time university appointment. More in his favor, Heisenberg was actually the only university-qualified teacher of middle and modern Greek in Germany.[54] The rector accepted the faculty proposal and passed it to the interior minister. On December 30, 1909, the interior minister, no doubt eager to silence an articulate spokesman for gymnasium personnel, ordered the immediate appointment of the candidate with an annual salary of 6000 marks, to which were added seminar fees. The candidate accepted readily, and in January 1910 Professor Heisenberg began lecturing in Munich as the occupant of the only chair in Germany for Byzantine philology.[55] The rest of the family remained in Würzburg to finish out the school year.

CHAPTER 2

A Great War

The Heisenbergs moved in June of 1910 from their Würzburg home to a large apartment on the top floor of a three-story building at Hohenzollernstrasse 110 in the fashionable Munich suburb of Schwabing.[1] The house, built at the turn of the century, stood near the intersection of Hohenzollernstrasse with Isabellastrasse and Fallmerayerstrasse, almost opposite Joseph-Klar-Strasse — all in existence today. Werner and Erwin shared a bedroom at the rear of the traditionally furnished, dimly lit apartment, heated by a coal-burning stove. Their window looked out across a small back yard toward the buildings on Fallmerayerstrasse to the north, away from the noise and bustle of the busier Hohenzollernstrasse. Their large electric battleship, when not afloat, lay docked on a dresser in their room. The third-floor quarters of the Munich flat were probably crowded, since servants lived in during that era. Servants were as much a practical necessity as they were essential to the prestige of a professorial family.

The boys must have keenly regretted trading their semirural Würzburg residence for a Munich flat, but they could not have had a more sheltered or cultured setting in which to grow up. Munich, a city of more than a half million people at that time, 75 percent of whom were Roman Catholic, served as the secular and religious administrative center of Bavaria, and the Schwabing suburb served as its cultural apex. The Bavarian king's residence flying the blue and white checked flag of proud Bavaria; the massive royal ministry buildings and the medieval town hall all stretching from the northern side of the center of town; Marienplatz, with its famous cathedral, the twin-towered Frauenkirche

(Church of Our Lady); the ornate stone Siegestor (Victory Arch) at the Schwabing border — all attested visually to the administrative power of the "court and residence city" nestled along the Isar River. Yet Munich was also the capital of the rural province of Upper Bavaria, which stretched south to the foothills of the Alps, near the Austrian border. Farm produce arriving daily at the "victuals market," also located near Marienplatz; the frequently seen traditional dress of the residents; the annual agricultural Oktoberfest; the Föhn, the wind that swept down from the Alps along the Isar River basin — these were constant reminders of the provincial mountain setting.

The suburb of Schwabing presented quite a different aspect of the city. Schwabing lay on the northern outskirts of Munich, just north of the university, and at that time it was approaching the height of its reputation as a center for art, music, and literature. It had also become the preferred neighborhood of the social elite who sought to escape inner-city life. While the working and lower classes settled in the Sendling district to the south of town, near the stockyards, breweries, and machine shops, upper-class nobles and officers and upper-middle-class administrators and academics shared the Schwabing area with artists, as did the owners of the numerous shops, pubs, and cafés that served them.

With Professor Heisenberg in a university chair and Dr. Wecklein at the top of the Bavarian school system and head of the prestigious Maximilians-Gymnasium, only blocks from the Heisenberg home, the Heisenbergs naturally mixed with the upper social and cultural strata of their new neighborhood and city. Those strata had long since eclipsed the monarchy itself in power and status. With Bavarian monarchs prone to madness, Munich ministerial officials, bureaucrats, merchants, and professorial civil servants formed a "ministerial oligarchy," dominated —despite the particularist Roman Catholic majority — by pro-Prussian, Protestant liberals. This oligarchy, and the Heisenberg family's solid position within it, remained intact until well after the outbreak of war in 1914.

But changes and threats of change were already apparent. Like Würzburg, Munich experienced an enormous population growth in those years, due to a rising birth rate and the migration of rural families to the city. These families entered the lower classes of the city at a time when the industrial revolution and the rise of the once-forbidden workers' party, the Social Democrats, rendered Munich's middle and upper classes fearful of "proletarianization." As suffrage extended to more sectors of Bavarian society, discontent with the ruling oligarchy caused the fortunes of the National Liberal Party to decline in the decades following Werner's birth. To the forms of protective insulation from administrative meddling invented earlier by professionals—many of whose liberal members had, like the Heisenbergs, immigrated to Munich from the Prussian north—were now added social barriers from below in the form of consciously fostered distinctive dialect and rural custom. The gulf separating the "Preissen" from the "Baier," Schwabing from the rest of Munich, was more than merely geographic.[2]

Still, Munich, as the leading commercial and cultural center of the south, continued to attract upper-middle-class outsiders. While the entire city experienced an overall expansion in those years, Schwabing underwent the greatest development of any Munich suburb. Many of the trees in the area had already fallen victim to lot clearing for future housing, and much construction was then under way. A map of the city published in 1911 shows that the Heisenberg home then stood near the edge of the area under development, one of the last buildings on the north side of Hohenzollernstrasse. There were no buildings at all on the opposite side of the street after Isabellastrasse.[3] Barren lots, fields, and mounds of dirt must have made tempting playgrounds for neighborhood children.

The most conspicuous feature of Werner's new neighborhood was an electric tram that thundered west down Hohenzollernstrasse past the Heisenberg home, before turning south at Kurfürstenplatz.[4] Almost no other motorized traffic ran along the streets. An engineer named Müller owned one of the few automobiles in the area—a fire-engine red contraption that yielded only grudgingly to the crank—much to the amusement of the local boys.[5] Save for the tram and an occasional car, horse-drawn wagons and carriages delivered practically everything: mail, ice, beer, milk—and street cleaners. The milk was delivered to the door by a milkmaid, whom neighborhood boys teased with the refrain "Millimadl, Millimadl, mit'm dicken Wadl!" ("Milkmaid, milkmaid with the thick calves!") The Heisenbergs could also catch the sounds of the many wandering musicians, watch the gas-lamp lighter, and wave to the policeman stationed near their home on the corner of Hohenzollernstrasse and Isabellastrasse.

But the high point of neighborhood life was the Bavarian soldiers who regularly marched or rode down Hohenzollernstrasse to the rhythms of marching bands and drums, on their way to and from training at the army barracks on the Oberwiesenfeld at the west end of the street. Everyone threw open doors and windows to watch them parade, as neighborhood boys ran cheering after them. In the summer, the boys would often follow the soldiers to the exercises at the barracks.

The extent to which Werner participated in such neighborhood fun is not known. He was not a gregarious youngster, nor, when school was in session, did he devote much time to frivolous play. Instead, being able to finish his regular school lessons with ease, he undertook outside studies. In addition, his parents enrolled him in piano lessons with the well-known Munich pianist Peter Dorfinger, who demanded hours of intensive practice. By the time he was a teenager, Werner could play master piano compositions, and he participated in chamber presentations and frequent musical interludes at school ceremonies. No evidence survives of any childhood playmate outside the family.

In the fall of 1910, Werner enrolled — for his fourth and last year of primary education — in the Elisabethenschule, several blocks from his home.[6] Erwin, however, having completed primary school, took the entrance exams for his grandfather's gymnasium and spent the nine years of his gymnasium education in the B section of each class. Werner followed Erwin to the gymnasium the next year and consistently attended the A sections, which were usually reserved for the brighter pupils. This circumstance and the one-year separation in grades, despite their nearly two-year separation in ages, must have contributed enormously to the boys' intense competition.

While awaiting the completion of renovations and new construction, the Max-Gymnasium occupied temporary quarters in the middle wing of the Damenstift on Ludwigstrasse near the university to the south of the Heisenberg home. The school was still there when Werner entered gymnasium. Going to school each day, Werner would join his father and brother on their early morning walk to their classes. Not until Werner's second year did the gymnasium reopen in its present location on the corner of Morawitzskystrasse and Karl-Theodor-Strasse, several blocks northeast of the Heisenberg home.

When 9-year-old Werner entered the Max-Gymnasium in September 1911, the school had already attained a reputation under Wecklein as an academically and socially elite institution — a "plutocratic gymnasium," as one former administrator called it.[7] Rector Wecklein's efforts to ensure the academic excellence of his teachers made the institution increasingly attractive to the wealthy elite. Among its illustrious students was Heisenberg's most noted predecessor, Max Planck, who also briefly taught physics at the school. Since school was not free and pupils were not assigned to neighborhood schools, the elite naturally chose the best for their children. The arrival of the upper classes in Schwabing reinforced the social composition of the gymnasium. Although a modest fund existed for needy pupils, few took advantage of it. Of the 37 pupils who survived the 1911–1912 school year in Werner's first class, nineteen of their fathers held titles that began with *Königlicher* (royal). Eleven fathers were jurists or state officials. Eight are listed as professors and five as military officers. The remaining thirteen held middle-class positions: businessman, craftsman, factory owner, artist. With 576 boys enrolled and 44 men as faculty members, the Max-Gymnasium was second in size only to its brother school, the Luitpold-Gymnasium.[8]

The Interior Minister for Church and School Affairs decreed the course of study and the content of each course throughout the Bavarian school system. During the decades preceding Werner's gymnasium study, German gymnasium education had undergone considerable controversy and reform. Neohumanism, the dominant educational ideology, had fostered the notion that only the classics should serve as the foundation for any professional career requiring higher education. By the turn of the century, the more practical demands of commerce, industry, and technology rendered the humanistic gymnasium no longer the only route to higher education. With Werner's classicist grandfather on the school board, Bavaria remained one of the last bastions of classical humanism, but even after the promulgation of reforms in 1914, the overall aims of gymnasium education remained the same: moral education on the basis of Christian ethics, "education in the spirit of the fatherland," and preparation for independent academic research — in that order.[9] Gymnasium curricula continued to stress the classics. Of the 263 credit hours over nine years of study, 63 were devoted to Latin, 36 to Greek, and 31 each to mathematics and German. The remaining 133 hours were divided among history, religion, athletics, French, geography, and nature studies, in that order. Physics received the least attention, next to drawing, with 6 hours spread over three years.

During his first three years at the gymnasium, Werner's main subjects were Latin and mathematics, of which he received eight and four class hours per week, respectively. Beginning in his fourth year, he received six hours per week of written and spoken classical Greek. Three years later, he began two weekly hours of physics. In his last year, Werner read Horace and Tacitus in Latin and Homer, Sophocles, and Plato in Greek; he pondered elementary classical mechanics in a single physics text. One can imagine the primly attired Werner standing at attention next to his desk as he recited some Latin or Greek text from memory, while the teacher reigned ominously over the teenagers with a switch.[10]

The outbreak of war in 1914 disrupted the Interior Ministry's precise curricular prescriptions. Moreover, the ministry's official neglect of science and technology was countered by unofficial interest in these subjects among the pupils. Current scientific developments contributed to the lessening of the pupils' enthusiasm for the classics in favor of science. Such notable events as the discovery of X rays and radioactivity, the rapid advance of technological industries, and the new Kaiser's personal interest in promoting and financing scientific and technological research gave rise to myriad new career opportunities.

Of the 41 pupils in the graduating class at the end of Heisenberg's first year, 20 intended to enter careers in science, technology, or medicine. The next choices reflected other advancing fields of the day: seven hoped to go into banking and five into the military. Only one brave pupil opted for a classical subject, archaeology.[11]

The school's teachers and their newly renovated facilities made advanced science studies possible in that otherwise humanistic setting. The Max-Gymnasium was the most modern school in Munich, with the most extensive facilities for physics instruction in the city. Since gymnasium teachers qualified in at least two subjects, mathematics teachers taught physics; the few science teachers usually taught in all nine grades. Consequently, a precocious younger pupil could be kept stimulated with material from the physics teacher's upper classes, while the teacher, perhaps inspired by the school's facilities, would refer more often to physics in his lower classes. Both were true for Werner and his favorite and most influential teacher, Herr Christoph Wolff.

Little is known of Herr Wolff, whose personnel records were lost in World War II. Apparently Wolff did well on his qualifying examinations in the early years of the century, for board member Wecklein, who kept the best teachers for himself, immediately assigned the young man to his school. Although Herr Wolff never obtained a doctorate — not necessary for science teachers — he had quickly advanced by 1910 to the top of his profession, gymnasium professor with the title of Königlicher Studienrat (literally, royal study councilor). Wolff taught Werner mathematics during his first three years at the school and both math and physics in his last three years there.[12]

Werner's need for study beyond that in the prescribed curriculum is apparent. For an ambitious student accustomed to solving his older brother's math problems, his course work offered little difficulty. Until the fourth grade, Werner was taught only simple arithmetic. From grades three to seven he studied plane geometry from a single elementary textbook. Not until the seventh and eighth grades did he finally encounter plane and solid trigonometry.[13] By fourth grade he was restless. His ability to dash off his assignments, fueled by his eagerness for recognition, left him few remaining challenges. His first-grade teacher noted in his grade report: "The thought operations, namely in grammatical and arithmetic questions, are completed rapidly and in most cases without error. Spontaneous diligence, great interest that the subject is done thoroughly, and ambition." At the end of the second grade, his teacher wrote: "He has attained his excellent achievements with playful ease; they have cost him no expenditure of effort." By the fourth grade, the teacher complained: "However, with his ease of comprehension [he is] often careless in his homework assignments."[14] Werner would master his homework in brief bursts of concentration, then go on to other, more challenging (and enjoyable) activities, such as music — a pattern that persisted throughout his life.

Another factor was already in play. Heisenberg's fourth school year coincided with Germany's first war year, and his father absented himself for long periods of active duty. Werner, now challenged neither at home nor at school, challenged himself instead with his own studies of mathematics. By the end of the school year his diligence had brought results: he had moved to the very top of his class, where he remained.

While not gregarious, Heisenberg was acknowledged and well liked at school, not only for his pleasant manner, mathematical talent, and musical ability, but also as one of the rector's two grandsons — a combination that can make life miserable for a quiet boy. But his classmates' respect for authority and accomplishment apparently prevented serious jealousy.

Werner's family connection brought him a special honor in March 1913, when Bavarian Prince Regent Ludwig came to dedicate the new gymnasium building. Werner's mother, a descendant of literati, wrote a poem for the occasion and her favorite son, 11-year-old Werner, recited it to the prince during the convocation, which concluded with an enthusiastic singing of the "Regentenhymne."[15] The delighted prince duly thanked the proud lad with an official certificate and a pair of cufflinks engraved with the letter L — a memento that remained among Heisenberg's most cherished possessions.

Werner's encounter with the prince was a high point for him of an era that would soon fade. On September 1, 1913, Grandfather Wecklein, approaching 71 and having seen his grandsons safely on their school careers, finally retired. He had by then attained the prestigious title of Königlicher Geheimer Hofrat (royal privy court councilor). Two months later, Bavaria's Wittelsbach dynasty itself underwent a change.

On the death under suspicious circumstances of mad King Ludwig II in 1886, his son, Prince Luitpold, had served as regent for the rightful heir to the throne, Luitpold's deranged brother Otto. Luitpold presided over the decline in monarchical influence until his own death in 1912, whereupon his son Ludwig assumed the regency. A year later, Prince Regent Ludwig declared himself King Ludwig III, promptly alienating his tradition-minded subjects, who five years later made Ludwig the last of the Wittelsbachs to rule Bavaria. Both transitions, Wecklein's retirement and Ludwig's coronation, marked the end of an era in gymnasium education, in German history, and in Heisenberg's life.

Werner's sheltered life ended after the world erupted in war in the early days of August 1914. Although they would later face bitter disillusionment, all sides greeted the outbreak of war almost with jubilation —a feeling that emerged less from the snapping of international tensions than from an explosion of nationalist fervor, social unity, and the romanticizing of war. In those early days, few of fighting age had ever experienced a real war, and no one could have imagined the horrors of trench warfare to come or the psychological devastation that defeat would bring.

Munich broke out in a patriotic fever. Germany, sandwiched in central Europe between potential enemies on the east and west, had settled on instant mobilization and lightning offense as its secret weapons against possible encirclement and a seemingly hopeless two-front war. The strategy instantly galvanized Munich and the rest of the country at the outbreak of war.

"There was plenty of excitement at the Pasing train station," began a story in a local paper on the first day of war. "Almost every minute brought in a new train with reservists waving from the windows, whom people greeted with fluttering handkerchiefs and shouts of 'Hurrah.' . . . On the streets themselves, groups stood around everywhere in front of the notice boards; people surround officers and give them ovations. . . . In front of the Feldherrnhalle, during the changing of the watch, there were stormy outbursts of enthusiasm and constant cheering for the troops."[16] On Kurfürstenplatz, near the Heisenberg home, people gathered daily during the first weeks to celebrate the many early German victories with patriotic songs and to cheer on their boys, who, they were certain, would be home by Christmas.[17]

The war had an immediate effect on the Heisenberg family. The family were as much in favor of the war as their compatriots, but they had to face the fact that Professor Heisenberg, still a Prussian reserve officer, might have to fight in it. The grim response of Werner's parents in those early days contrasted sharply with the town's jubilation, and the contradiction bewildered young Werner.[18]

In fact, Professor Heisenberg was called to active duty within days. Wife and sons accompanied him to Osnabrück at the end of August. A goodbye photograph of Werner and Erwin on each side of their moustached father, dressed in an officer's uniform with a sword and the traditional pointed helmet, reveals both pride and concern in all three faces. The nearly 13-year-old Werner, dressed in an ill-fitting suit for the occasion, is standing in his frequent defensive posture with his arms crossed and his face set.

The family stayed with their relatives until October, when August's regiment marched into German-occupied Belgium. In Osnabrück, Werner's father was promoted to captain with command of a local infantry company. His primary duty was to station one of his men, armed with a machine gun, on the roof of the tallest building in Osnabrück, in order to guard the city against enemy planes. Werner visited the soldiers in their perch from time to time, probably a welcome relief for the fellows, who certainly had little to do. At that time, airplanes were used rarely in warfare, and only for reconnaissance. Bombers and fighters were a later invention.

During those months in Osnabrück, Werner must also have discussed and reflected on his father's often expressed political outlook and his many activities that resulted from it during the years leading up to the war. A domineering father who dedicates himself to a higher cause in both word and deed can hardly avoid eliciting admiration from and profoundly influencing an impressionable son eager to please.

Professor Heisenberg's views were typical of his generation and station. Some historians have found that, following the unification of the Reich in 1871, socially prominent German professors turned ostensibly apolitical in public affairs. This was intended both to insulate themselves from the mass politics of the lower, working classes — officials elected by the lower classes could not be entrusted to administer their profession — and to establish their status as keepers of eternal, objective German culture. Although they still engaged in politics, they regarded their activities as being outside the party system and in service to German culture (especially to themselves as "bearers of culture"), so they could believe themselves free of political taint.[19] Many academics joined self-interest pressure groups, such as the Gymnasium Teachers' Association, which promoted German cultural interests in the international arena. They were encouraged by the example of other professionals — bankers, industrialists, military men — who were exercising their own influence in state service.

By 1910, the unified empire had achieved a powerful economy and military, but its diplomacy was weak. A late entrant in the international competition for power and markets, the country had failed both to achieve a meaningful alliance with any of its competitors and to establish a large colonial market and source of raw materials for its massive and expanding industries. The German people and their political leaders felt themselves increasingly isolated internationally and surrounded by hostile forces. The Balkan peninsula to the southeast, the "powder keg of Europe," provided one outlet and a possible geographic wedge to Germany's ally, Turkey. German classical scholars, including the occupant of the chair for middle and modern Greek studies, provided a connection—a cultural one—to Greece, the southern part of the Balkan peninsula.

When Werner's father assumed his Munich chair in 1910, he immediately took up the apolitical cultural barrage his predecessor, Krumbacher, had instigated and aimed it at the Balkans—Greece in particular. Although he followed Krumbacher's example on nearly every academic issue, Heisenberg's very first publication as professor contradicted his colleague's stand on the intense controversy then raging over vulgar Greek.[20] Heisenberg and other German scholars lobbied against a proposal to replace classical Greek as the official language of Greece, seeking to bind Greece more closely to Germany through the German preservation of classical Greek language and culture.

The outbreak of war intensified the political efforts of academics on all sides. No nation saw itself as an aggressor; each sincerely believed that it was simply fighting to defend home and culture from those who would destroy them. For Germany, lightning offense as a defense justified the invasion of neutral Belgium. This action, together with the unfortunate destruction of Belgian art works, set the seal on other countries' perception of Germany as an aggressor and a destroyer of culture. Germans, on the contrary, convinced themselves of the interdependence of military, nation, and culture, and thus regarded military measures as essential for their own cultural preservation.

The prim rationalist veneer overlying such fierce nationalist emotions pervaded professional German society, including humanistic gymnasiums, and even infected supposedly objective scholars. This is nowhere more evident than in the infamous German academic manifesto "An die Kulturwelt" ("To the civilized world"), which overtly supported the German cause. More than 4000 "apolitical" professors—almost the entire German professorate—eventually signed the manifesto (to their later regret).[21] Of course, they were not alone in their folly—numerous equally unscholarly, nationalistic manifestos and countermanifestos were written and issued throughout the war in nearly every land.[22]

Professor Heisenberg's war duties prevented him from joining his like-minded colleagues in the manifesto war, but they did not hinder his enthusiastic participation in the propaganda war. As early as 1912, as

Balkan tensions were erupting into battles, Heisenberg and his Munich philhellenic colleagues formed the ostensibly studious German-Greek Society.[23] It unleashed an increasingly overt propaganda campaign to persuade Greece to remain pro-German, or at least neutral, in the coming war.[24] Unfortunately, Bavaria's mad king Otto had once ruled Greece, leaving little infatuation for Germany among the Greeks.

The German-Greek effort ended when, in 1917, England and the Entente invaded Greece, deposed King Constantine, and encouraged the new parliamentary government to join the war against Germany. At least Professor Heisenberg could console himself with the thought that "We Germans were well represented with the intellectual leaders of Greece through the activities of our academic circles."[25] Werner's probable esteem for his father's efforts must have increased in 1914, as Werner encountered Greek for the first time and his father put his words into action.

Unlike most other older academics, Werner's father was not content with waging a mere cultural battle. Although he was of military retirement age, the 45-year-old professor demonstrated the strength of his commitment to his country by taking up arms for her. When his battalion was ordered into occupied southern Belgium in October 1914, Captain Heisenberg led the charge over the border. His regimental commander ordered his company to guard a 26-kilometer stretch of railroad track near Manage près Mons. Captain Heisenberg personally patrolled the track on horseback every day. His only other duty was to issue travel permits to the local populace from the home of a Belgian family that (willingly or not) provided him quarters.[26] But the enthusiastic soldier soon chafed under such tame assignments, and on Christmas Eve 1914, he transferred to Landwehr Infantry Regiment 56, headed for the front.

To prepare for battle, Captain Heisenberg underwent a month of infantry combat training in Belgium before shipping out in command of a company for the trenches of the Argonne Forest near Servon, France. Before leaving, he penciled his last words to family, friends, and colleagues on postcards. Just two weeks later, at the end of January 1915, his exhausted company pulled back for a nine-day respite. The much-sobered captain wrote to his Munich colleague, Otto Crusius: "If anyone speaks to you about the poetry of trench warfare, then please object, for holding out a grenade attack is nothing but pain, misery and suffering."[27]

By April, Heisenberg had had enough. The courage of his convictions had soon evaporated under fire. On April 24 he wrote again to Crusius

—this time from Osnabrück: "Despite all of my bellicose inclinations, the longing for peaceful activities had become very alive once more. Therefore, when, after a very difficult and strenuous fight, I was rather at the end of my nerves, I asked for my discharge from the front and my transfer to Area Command I Munich, in order to do garrison duty once again in Munich."[28] Because of his age, his request was readily approved, and within a month the erstwhile warrior was back in Munich delivering Greek lessons to a seminar of four students. His cherished company that he had eagerly led into battle could not escape so easily the horrors of a war that bellicose professors had helped bring about. Werner, who later remembered that his father had returned from the front because of wounds,[29] must have suddenly seen his father in an entirely new light: either as a tragic hero to a cause that ultimately failed, or as a man now compromised. Either perception would have served as a setup for his later total disillusion.

Werner's gymnasium immediately felt the effects of war. Within days of the war's outbreak, the Bavarian Army Command requisitioned the gymnasium's new building for the quartering of newly mobilized troops. Although the garrison was intended to last only until the men went into battle a few weeks later, the building remained occupied until shortly before Heisenberg graduated in 1920. With Bavarian troops settled in for the duration, the Ludwig-Gymnasium near Marienplatz, the center of town, hosted the required courses of its brother institution with half-day sessions for each school. Electives in music, art, and languages were curtailed, and, to Werner's dismay, the new physics laboratory was closed. Lack of coal in early 1917 forced the school to cease operations almost entirely. For a month, the pupils simply picked up and turned in their daily homework.[30]

The Bavarian Army vacated the Max-Gymnasium at the end of the war, only to be replaced by a hospital for the wounded. Once the hospital left, the Bavarian regime then in power quartered a unit of local troops in the building. Two Free Corps companies, among Reich troops dispatched to Munich, replaced the Bavarian troops in the fall of 1919. After a much-needed disinfecting, the building finally returned to classroom use, but a lack of coal again forced pupils onto half-day sessions until March 1920. Werner graduated four months later. Despite these interruptions, or more likely because of them, he not only mastered his assigned schoolwork, but progressed far beyond it through independent study.

The war profoundly affected the gymnasium's faculty and pupils. Six reserve army officers among the faculty, including Lieutenant Christoph

Wolff, were called to duty during the first few months. All returned unharmed, including now-Captain Wolff, who, like Werner's father, suddenly headed for home after tasting the horrors of the front. But the many youngsters they sent into the trenches fared far less well. Eager pupils answered patriotic calls to the colors by school, university, and academy officials, who also urged a draft for boys over 17.[31] Of the 452 pupils enrolled in the Max-Gymnasium in the fall of 1914, 74, including the entire graduating class, had joined the army by the end of the year. Eleven never returned. A plaque with their names and those of 22 other pupils and one student teacher killed in the war now hangs in the gymnasium.

To foster identity with culture and the military and to instill Vaterlandsliebe (love of the fatherland) in its pupils, the Max-Gymnasium subjected them to heavy doses of patriotic and military indoctrination throughout the war. The celebration of battlefield victories, the birthday of the popular Marshal Paul von Hindenburg, and the anniversary of King Ludwig's coronation offered frequent opportunities for patriotic school convocations. Carefully edited letters from former pupils at the front appeared in gymnasium annual reports and were used for instruction. Werner's 1916 composition class wrote an essay on the topic "My participation in the war for Germany's world stature."

For at least two years, from 1916 to 1918, Werner belonged to the Max-Gymnasium's paramilitary Wehrkraftverein, the Military Preparedness Association, which institutionalized the pupils' military indoctrination. His membership in the association led to many of his postwar extracurricular activities. The association was founded in 1910 by a group of Munich army officers intent on providing early training to gymnasium students, the officers of the future. A year later, they convinced Prince Luitpold to grant the association his official sanction, and soon a Wehrkraft-Abteilung (Military Preparedness Department) was attached to every Bavarian gymnasium. Initially, the association held only occasional after-school drills for the boys, but with the outbreak of war, it offered more extensive training and indoctrination. At the Max-Gymnasium, Dr. Ernst Kemmer, head of the school's Military Department, set up a Jungsturmregiment (Young Storm Regiment), one of whose later members was Werner. He joined the regiment probably for fun at first and in anticipation of his own likely entry into service upon turning 17 in December 1918. Fortunately for Werner, the war ended a month before his birthday.

Throughout the war, Dr. Kemmer, who had headed Werner's third-year class in 1913–1914, wrote yearly accounts of his youth training in the gymnasium annual reports. After the outbreak of war, Kemmer's training expanded to twice weekly, on Saturday and Sunday afternoons, and once monthly for an entire day. In keeping with Bavarian War Ministry guidelines, Kemmer's training focused not on weaponry but on physical and mental preparation for combat. This included physical fitness and familiarization with military discipline; instruction in marching, map reading, and reconnaissance; and such later youth-movement activities as tent making, outdoor cooking, and "the joy of hiking, the love of nature and of homeland."[32]

Ninety pupils descended on Kemmer's unit during the first year of the war. So many youngsters clamored to join that Kemmer had to lower the minimum age to 13. Yet interest in paramilitary training ebbed as the war dragged on. Parents complained that their tired children had little energy left for schoolwork and that the school itself was fast turning into a mere Rekrutenvorschule, a prep school for recruits for an increasingly nasty war. Kemmer argued that his program proved its utility as each graduating class marched into the trenches. The school could not abandon its moral duty, he wrote. The war "made [the school] responsible for the future of German culture!"[33]

Disillusion with such belligerence quickly spread as German forces bogged down in the west, casualties mounted, and food and coal began to run low. In Munich alone, 13,000 inhabitants died in the war. In the early days, reports by supply experts had encouraged undue optimism. A Munich paper boldly proclaimed: "The danger of a cut-off of food supplies to large cities does not exist, say the experts."[34] The experts obviously had not reckoned with or did not care to inform the public of the possibility of a near-total British naval blockade. It proved unbreakable even after Germany unleashed an unlimited U-boat war, which brought the United States to arms.

In 1915, rapidly inflating food prices led to demonstrations and bread rationing in Munich. Workers at the Krupp armament works, which was practically Bavaria's only heavy industry, were, as elsewhere, growing dangerously restless. In August 1916, the victorious commanders of the eastern front, Hindenburg and Ludendorff, took command of the entire German war effort — soon alienating the already anti-Prussian Bavarians. To make the best of shortages, the commanders introduced food rationing and a centralized control system for all food production and distribution. In Bavaria, the rationing of milk, meat, and sugar was administered by weekly ration cards distributed by gymnasium pupils. But even with ration cards, one needed to know a friendly farmer or a black-market source to obtain all the necessities. In late 1916, an early frost wiped out most of the potato crop, a staple of the Bavarian diet. A coal shortage in the ensuing months made the terrible "turnip winter" of 1916–1917 (when turnips were the main staple) the worst of the war. Food and coal ran out in Munich, soup kitchens were set up, and teenagers no longer sported on Hohenzollernstrasse. "We boys mostly went hungry," one of them wrote.[35] When one of Wecklein's colleagues asked how he was, the old fighter responded with a stiff upper lip, "We're starving and freezing, but we can take it."[36]

Faced with a failed harvest, a hard winter, and a lack of workers, in early 1917 Hindenburg and Ludendorff ordered all men between the ages of 17 and 60 who were not in military service to register for agricultural and factory war-assistance work—the Hilfsdienst. Since Bavaria had little industry, nearly all Bavarian helpers went to the farms. The authorities pressured schoolboys under 17 into volunteering during the spring and summer months and gave them an early final examination before they left for the fields in April. Thirty-two pupils from the Max-Gymnasium entered the Hilfsdienst that year, including Werner's brother Erwin and eight of his own classmates.

Pupils who did not enter any service were put to work in vegetable gardens during the holidays. Werner's class worked in gardens at the Torstickwerk and at King Ludwig's residence at Leustetten under the direction of their Wehrkraft master, Dr. Kemmer. "The boys worked diligently and for the most part happily," he wrote.[37] The Hilfsdienst was less successful. Newspapers reported that the poorly organized pupils regarded their early country adventure as "a pleasant summer holiday," and farmers had little patience with a pack of wild city boys rampaging through their fields.

When Werner joined the Hilfsdienst the next time—while his brother entered the army—the program and pupils were better prepared and more tightly controlled. The continuing hardships had rendered the boys much more cooperative. Although the Bavarian agricultural officer reported only moderate success in his personal campaign to recruit volunteers among the Military Preparedness units,[38] Werner, then a 16-year-old Preparedness member, required little prompting. The food shortage had taken its toll on the growing boy, whose parents had yet to locate a farmer or a black-market source to supply the family. The adolescent Werner, his voice already changed, required larger amounts of now unavailable food. He soon grew so weak from hunger that he once fell off his bicycle into a ditch. He and his family decided that he would go to the farms that summer.

Learning from past experience, the military overseer of the project divided the nearly 4000 pupils, called Jungmannen, into squads of 10 to 30 boys, each under the close supervision of a Preparedness leader or a military officer, who was in turn under the supervision of the district army corps commander. To encourage better work, the army supplied each boy with a new pair of combat boots. Judging from photographs of Heisenberg's youth group, the footwear was later put to excellent use on outings. The boys were also warned that a report would be placed in their annual school grade report. Werner, a well-behaved pupil throughout his schooling, received what sounds like the standard commendation for good behavior: "As a Jungmann he earned the recognition of his leader . . . through his good behavior and his work accomplishments."[39] Soon afterward, Kemmer promoted him to Gruppenführer, group leader, in charge of a squad in the gymnasium's Military Preparedness unit.[40]

The commanders posted Werner's farm squad at a large dairy school near Miesbach in Upper Bavaria, south of Munich. There he worked, save for short leaves, from early May until September 5, 1918, when he returned to school for ten days before leaving again to help with the potato harvest.

That summer marked Werner's first extended time away from home. It also marked the start of his frequent correspondence with his parents, a correspondence that lasted until the death of his mother in 1945. (His father died in 1930.) In nearly every letter home that summer, Werner wrote about his family's main concern — food. The diet on the dairy farm, although strange, apparently sufficed for the hard-working teenager. Dinner usually consisted of pancakes and sauerkraut, supper of potatoes, butter, cheese, and milk. Breakfast was at 5:30 A.M., and the boys had two breaks during the day for bread and milk. Bread, however, required personal ration cards. Meat was simply not to be had.

From photographs of Heisenberg and his farm companions, one might suppose that they, like their predecessors, led an idyllic country existence that summer. In fact, life was far from easy, and its hardships contrasted sharply with the playful ease of his school experience. "There I learned to work," he would tell his children. "Taking it all together, I think that was one of my most important times, considering my education, because on a farm you really learn to work. You know it's not like in school where you think it's not so important."[41]

The boys worked from 6:00 in the morning until past sundown (as late as 10:00 P.M.). After he recovered his strength, Werner spent entire days during the first month chopping and sawing wood. This was hard labor, but his only complaint was "The work is just too boring for me."[42] In June and July he and his companions were sent into the hay fields; during the rainy August they helped around the cow barns.

Heisenberg spent his free time in the evenings playing games of chess, which he always won. He took some mathematics textbooks and a copy of one of Kant's *Critiques* with him to study, but, he wrote home, "For school work and things such as reading etc. we are mostly too tired, i.e., we simply aren't interested."[43] He also brought some sheet music, planning to practice a piece by Liszt to perform on his return home. There was a piano in the farm school's central building, but piano practice lapsed until after the hay harvest in late July.

The ambitious pupil's confinement of his mental activities to chess testifies to the intensity of his physical labor. He also came into more direct contact with the people and life of the laboring world than he had ever known in Osnabrück. Both performing and observing concentrated physical labor may have fostered a similar type of concentration and perseverance in his later mental efforts. Years later, he recalled rather romantically the effect his farm summer had on his postwar years: "Others, including myself, had been working two years earlier as farm hands on farms in the Bavarian Highlands. So the raw wind was no longer alien to us; and we were not afraid to form our own opinions on the most difficult problems."[44]

Two months after the laboring Werner returned to Munich and his studies, the war came to a long-awaited but, for most Germans, surprising end. Having won a favorable armistice in the east with Bolshevik Russia, in the spring of 1918 Hindenburg and Ludendorff launched an all-out offensive in the west. At first it seemed to succeed, but a counteroffensive in July by massively superior forces, bolstered by newly arrived American doughboys, forced the German army into retreat. The ever-confident commanders couldn't believe they were losing. Only in September did they admit to the Kaiser the hopelessness of their situation. By November, the Entente had forced a capitulation and the acceptance of President Wilson's terms for surrender. The German public, thoroughly unprepared for defeat, was appalled. Even the *Münchner Neueste Nachrichten*, the highbrow paper of Munich's cultured classes, had failed to comprehend or to report the situation accurately.[45]

The armistice on November 11, 1918, imposed with the entire German army still in position far beyond the Rhine and occurring at a time of growing labor unrest at home, gave rise to the infamous legend of a "stab in the back" propagated in years to come by fanatic nationalists. On November 8, 1918, as the Kaiser held tight to his throne in Berlin, Kurt Eisner, the leader of the Bavarian Independent Socialists, declared in Munich an end to the Bavarian monarchy and the establishment of a republic. The next day, a defeated Kaiser Wilhelm abdicated the German throne, and a prominent Social Democrat declared the republic in Berlin—a further "stab in the back" to fanatics, and an end to the imperial era for all.

CHAPTER 3

Gymnasiast

Heisenberg graduated from Munich's Maximilians-Gymnasium in the summer of 1920 and entered the University of Munich that fall. During his first two years of university studies, he published four physics research papers, submitting the first just eighteen months after graduating from the gymnasium.[1] Three of the papers dealt with atomic spectroscopy, one with hydrodynamics. Their publication thrust Heisenberg, at the age of 20, into the forefront of quantum atomic physics research. This extraordinary achievement was certainly due in large part to the marvelous training he received from his university mentor, physics professor Arnold Sommerfeld. But Heisenberg could not have absorbed so much so quickly if he were not already advanced far beyond the gymnasium curriculum. The disruptions of the war years had only encouraged his independent study and accelerated his education.

Werner impressed his gymnasium teachers from the very start. On his grade reports they consistently noted not only his spontaneous drive, which always brought "very commendable achievements," but also his sheer intelligence: "The pupil is *very highly gifted*," noted his fifth-grade instructor; "A *highly talented, capable individual*," remarked his eighth-grade professor; "[He] is among the best in the class"—a unanimous opinion.[2] His achievements in grade 8, 1918–1919, were all the more remarkable, his teacher noted, because military activities in Munich had caused him to miss the last crucial weeks of school before the final examinations.

An average of Werner's final grades, weighted according to the number of prescribed hours for the listings available (starting in grade 4), confirms his teachers' assessment. It yields a grade average of 1.22 on a scale of 1 (very good) to 4 (unsatisfactory). His best subjects, for which he always received a 1, were mathematics, physics, and religion. In the main subjects, Greek and Latin, he received only one 2, the rest 1s. His worst subjects were German and athletics. Athletics accounted for his only 3s (received twice), and both subjects prevented his receiving all 1s during his last three years of school. His literate family notwithstanding, Werner wrote with little felicity. His fourth-grade teacher noted, "In essay, clear but dry."

On his Abitur (graduation exam), Werner received his lowest grade in German, his only 2. The theme of his essay, "What Makes Tragedy a Significant Form of Poetic Art?" would have intrigued his grandfather, an authority on Greek tragedy. The polite examiner, a close friend and colleague of Grandfather Wecklein, judged Werner's essay "a comprehensive, fluently written work that, however, does not always succeed in its argumentation."[3] In mathematics and physics, on the other hand, the pupil was simply "outstanding." The astonished state examiner reported: "[W]ith his independent work in the mathematical-physical field he has come far beyond the demands of the school."[4] Three months later, in October 1920, the young man arrived at Sommerfeld's institute.

Although Heisenberg and his classmates were heavily saturated with classical studies and German literature (as shown by the hours devoted to these subjects and the readings assigned), the late Wilhelmine age, as earlier indicated, had fostered an enthusiasm for science and technology throughout German society. Werner, like many precocious youngsters, became fascinated with technical objects in early adolescence. By his fourth term, 1914–1915, the 13-year-old's preferences had become apparent to his teacher, who wrote: "His interest has turned in a decisive and impressive way to physical-technical things." It was in those years that the Heisenberg brothers built their electric battleship.

Reconstructing Werner's intellectual development beyond technical things during his gymnasium years requires considerable unraveling. The available school reports and curricula do not fully coincide with his scattered recollections, which were committed to paper only late in life. Together, however, later memories and contemporary records enable considerable insight. In recalling his budding interest in matters scientific, Heisenberg often referred to his early fascination with technical toys, encouraged by his mechanical grandfather, as the starting point.[5]

This fascination led him, so he claimed, along the path of the geometry of objects into the realm of theoretical physics, especially the mathematical analysis of physical objects and data. But the shy teenager, ill at ease in the socially superficial "real" world, grew ever more fond of mathematics itself—particularly the harmonious, orderly beauties of abstract number theory. He learned differential and integral calculus, essential for physics, in his spare time. In addition to these independent studies, he later claimed encounters at this time with classic philosophical problems concerning the relationship between mathematics, experimental data, and atomic science within the context of ancient Greek philosophy. His urge for mental stimulation rapidly advanced him far beyond the meager demands of the school's science and mathematics curricula.

Werner was smitten with mathematics even earlier than the recollected stimulus of technical toys. As noted earlier, his father had set him competing with his brother in arithmetic at an early age, and the effects of these skirmishes were already evident when he entered the gymnasium. His teacher, Herr Wolff, noted the boy's skill in "calculational problems." Werner's fortunate experiences with Herr Wolff during his first three years at the gymnasium further encouraged the budding prodigy.

"He tried to interest me and give special problems to me. He told me, 'Try to solve that and that.'" But when the class turned from arithmetic to geometry, Heisenberg claims he lost interest. "I felt this to be very dry stuff; triangles and rectangles do not kindle one's imagination as much as do flowers and poems."[6]

The otherwise unimaginative Werner recalled feeling a sudden spark of interest in geometry only when Herr Wolff explained that universally valid propositions can be drawn from geometry, and that these propositions correspond to the transitory "real" world of physical phenomena. The correspondence between mathematics and the physical world "struck me as remarkably strange and exciting," he wrote. He remembered then applying mathematics to his homemade gadgets, fascinated by the notion that mathematics could be used to such ends. "Because of all this, I remained far more interested in mathematics than in science or apparatus during most of my life at school."[7]

The discovery that, as Galileo said, "the book of nature is written in mathematical symbols" comes as a revelation to any mathematically inclined youngster intrigued by transcendent harmonies. But exactly when this occurred to Werner is uncertain. Although Heisenberg recalled studying geometry in his third year under Herr Wolff, school records make no mention of Herr Wolff's teaching the subject at that time. Geometry was a fourth-year subject, and in that year Lieutenant Wolff was away from school at the front. One possibility is that Wolff introduced his intelligent class to geometry before the end of the third year. The following year's school curriculum (1914) recommended this. Alternatively, Heisenberg's remembered independent study of mathematics and its applications to his gadgets may actually have taken place when his father and Herr Wolff were both at the front. It was in that year, his fourth year (1914–1915), that Werner's grade reports note an unusual interest in "physical-technical things."

That geometry can correspond to physical reality, yet transcend it, was likely driven home to Heisenberg most strongly not by his gadgets or even by Herr Wolff but by Einstein. Like many other science-minded youngsters of his day, Heisenberg had heard of Einstein's theory of relativity and of its celebrated difficulty: "That naturally especially fascinated me." Werner accordingly obtained a copy of Einstein's explication of relativity theory—both the special and the general theories— written expressly for gymnasium pupils.[8] The opening section spoke right to the point: "Physical Content of Geometric Propositions."

From his geometry textbooks, Heisenberg had already learned that "as the name indicates, geometry arose from practical needs," particularly the needs of ancient surveyors. Einstein postulated that pure geometry deals only with logical relationships between concepts, from which logically valid propositions emerge. Determining the "truth" of these concepts and propositions, however, requires a comparison with "real" objects and apparatus, at which point "geometry is then to be treated as a branch of physics." The implication is that, once the "truth" of geometric axioms and propositions is ascertained through physical references, then any propositions derived from them are also likely to be "true." "Freely constructed" mathematical and physical laws are restricted by empirical and logical constraints. If Herr Wolff later followed the ministerial decree for teaching physics, he would have grounded physics even more in "real" data. Physics instruction, the decree ordered, "is to take its starting point from observations and facts of experience and not from mathematical considerations."[9]

Heisenberg later recalled that mastering the mathematics in Einstein's book gave him no difficulty—not surprising since the only mathematics in it was the Lorentz transformation—but that, ironically, he did have trouble with Einstein's physics: "It was extraordinarily difficult for me to think my way into this problem."[10] Einstein's theory thus did not convert him to physics as it did others of his generation, notably Wolfgang Pauli. But Heisenberg soon did manage to think his way into relativity through Hermann Weyl's even more advanced text, *Raum-Zeit-Materie* (*Space-Time-Matter*).

In the third edition of his essay, published in 1918, Einstein recommended Weyl's newly published treatise to those seeking an advanced treatment of relativity. Heisenberg, however, did not look at Weyl's text until after he had left gymnasium two years later in 1920. "And that again interested me a great deal," he said, "so I tried to understand the Einstein relation and the Lorentz transformation and so on. Still I didn't think about [studying] physics." Instead, that fall he inquired about studying pure mathematics.[11]

In his earlier cited recollection, Heisenberg remembered that he had taken up calculus during the first year of World War I in order to comprehend the advanced physics of his homemade toys. If true, it was a gigantic step for a 13-year-old with no apparent interest in physics. But another of his recollections indicates that he did not actually become adept in calculus until as late as 1918, when his parents asked him to help a family friend, Paula Fries, prepare for the mathematics section of her doctoral examination in chemistry. (Women began to gain unhindered admission to Munich universities in 1903.) For three months the 16-year-old gymnasiast tutored the 24-year-old university student in calculus. "And in that time I don't know whether she had learned it, but I certainly had."[12]

Fräulein Fries learned enough to pass her examinations, while her tutor applied what he had learned to the principles of elementary physics. By then he was in his second year of gymnasium physics and was beginning to feel a mild interest in the subject. Although his physics textbook required only a knowledge of algebra, by the end of his studies Werner could apply calculus to derive and solve the equations it presented.

On his final orals, Werner volunteered a demonstration of his newly acquired skills. He amazed the examiner with his use of calculus to solve the Newtonian equations of motion: "Heisenberg solved the problem posed with playful ease. Above and beyond that he now treats out of his own volition the problem of free fall and vertical throw with air resistance taken into account, whereby he makes use of infinitesimal calculus and proves that he has already gone far beyond the goal of middle school mathematics."[13]

Heisenberg's father returned from the front in 1915, but he absented himself again a little over a year later. In August 1916 the professor, fluent in Greek, left to serve as liaison officer between the Prussian War Ministry and a corps of captured Greek troops encamped near Görlitz, southeast of Berlin, where he remained until May 1918.[14] The war ended six months later. August's Munich interlude markedly accelerated the pace of his son's independent studies. The momentum occasioned by Werner's anxiety to impress his father helped to carry him through the following years of hardship and disrupted schooling. His teacher noted the obvious effect of his father's presence on the boy's performance: "Attentive and stimulating education at home, with which a good relationship existed, made itself noticeable."[15]

Professor Heisenberg was naturally most concerned with his son's progress in Greek and Latin. Noting the boy's enthusiasm for mathematics and his requests for mathematics texts from the state library, where August did research, the father sought out mathematics works written in Latin. Since Professor Heisenberg had personally catalogued the classical manuscripts of the state library, he had little trouble locating math texts in Latin, which had been the formal language of mathematics in Germany until well into the nineteenth century. Knowing nothing of mathematics, he brought home everything he could find that seemed related to the subject.

Among the treasures Professor Heisenberg presented to Werner that year was a copy of Leopold Kronecker's 1845 doctoral dissertation, *De unitatibus complexis*. In it Kronecker had attempted, unsuccessfully, to prove Fermat's famous "last theorem" in number theory by employing complex number units in cyclotomy, the science of algebraic rings.[16] Werner devoted considerable time to the thesis, fascinated by Fermat's last theorem and Kronecker's failed proof. Like most budding mathematicians, Werner tried his own proof of the deceptively simple last theorem: that there exist no positive integers x, y, z that satisfy the equation $x^n + y^n = z^n$, for $n > 2$. Like Kronecker, Heisenberg failed. To this day, no proof of the theorem has been found.

Shortly after encountering Kronecker, Werner happened on what is known as "Pell's equation" in number theory, which certainly does have solutions. His gymnasium mathematics teachers occasionally handed out offprints of current research papers. One such paper addressed the solutions to Pell's equation, $y^2 - Dx^2 = 1$, which arises from the quadratic representation of integers. For $D > 0$, the equation has an infinity of solutions, one set being the so-called elliptic functions.

Kronecker is best known for having contributed long treatises on the subject and on the properties of elliptic functions.[17] Werner, familiar with Kronecker's work, certainly knew how to manipulate elliptic functions. This was shown by his solution of one of Herr Wolff's special homework problems on the diffraction of light in a vessel of water.

Werner executed a long derivation of the diffraction equations, which led him into the realm of elliptic functions. Unfortunately for Werner, Herr Wolff couldn't say whether it was right or wrong!

His interest in technical objects having waned, in 1916 Werner began to devote all of his energies to music and to number theory. "I was fascinated by the theory of numbers," he later recalled. "It gave me pleasure to learn their properties — to know if they are prime numbers or not and to try to see if they can be represented as sums of quadratic numbers, or finally to prove that there must be infinitely many prime numbers."[18] He liked number theory much better than calculus "because it's clear, everything is so that you can understand it to the bottom." Werner apparently understood the offprint on Pell's equation to the bottom, for he applied the solution to other cases, wrote a short paper on the subject, and tried to get the paper published in a mathematics journal. The journal rejected it but without discouraging its author. In his only surviving school notebook, from about 1917 when disruptions forced the pupils to study independently at home, one finds such typical school exercises (with checkmarks penned by his teacher) as a plot of the function $y = \sin x$ in radians and the graphical determination of the roots of $x^2 - 2x - 3 = 0$. But on the first pages of the book, one also finds a graph of Pell's equation $y = \sqrt{Dx^2 + 1}$ for $D = 3$, worked out by hand in steps of 5 from $x = 0$ to 50.[19]

Neither Heisenberg's recollected research paper nor the offprint on Pell have been found, but remnants do exist of the works from which he probably first learned number theory. They were the standard, multi-volume series of advanced texts by Paul Bachmann, entitled *Zahlentheorie* (*Number Theory*).[20] Crumbling copies of the first two volumes survive in Heisenberg's book collection. The first volume, *Die Elemente der Zahlentheorie*, begins with the number system and arithmetic, which were of special interest to Werner. It then leads into a long section on quadratic forms that pays particular attention to Pell's equation. In the second volume, *Die analytische Zahlentheorie*, Bachmann notes that solutions of Pell's equation lead to applications of the theory of elliptic functions, but he does not elaborate, referring the reader instead to Kronecker for details. According to a surviving acquisitions list in the Maximilians-Gymnasium library, its collection at that time included a copy of Bachmann's first volume.[21]

In his seventh-grade class, 1917–1918, Werner received a dose of trigonometry and his official introduction to physics. The sole textbook for the three years of physics (1917–1920) was surprisingly good, though rather elementary.[22] It covered, without calculus, such subjects as elementary mechanics, electricity, magnetism, heat, kinetic theory of gases, optics, and energy conservation. Save for mathematics, it was comparable to a sophomore physics text at a modern American college. Contemporary physics — the relativity and quantum theories — did not exist for the author of Werner's gymnasium text. But, heeding the urgings of the ministry, he did provide material on other physical sciences, such as meteorology, astronomy, and geography, and offered explanations of such technical devices as the steam engine, water pump, telescope, and telegraph. The 500-page book was crammed with nearly 700 carefully detailed realistic drawings. Yet Heisenberg, although supposedly enthralled with technical apparatus, insisted that he had little interest in physics until his last two years at the gymnasium, beginning in 1918. And even then, he maintained, his curiosity was piqued by his philosophical ponderings on the problem of atoms, rather than by any specific desire to study physics.

Werner's pondering, so he frequently claimed late in life, derived from two encounters with atoms at about that time. One involved a drawing of multiatomic gas molecules in his physics textbook. In it atoms were joined into molecules with little "hooks and eyes." Accustomed to the realistic drawings of technical devices elsewhere in the book, the demanding adolescent was disturbed to find molecules portrayed in what was to him such a superficial, utilitarian manner. "To my mind, hooks and eyes were quite arbitrary structures whose shape could be altered at will to adapt them to different technical tasks, whereas atoms and their combination into molecules were supposed to be governed by strict natural laws. This, I felt, left no room for such human inventions as hooks and eyes."[23]

Heisenberg's prescribed physics textbook did treat atoms in a rather superficial manner. With the influential antiatomist Ernst Mach deceased only in 1916 and the ministry urging empiricism, the author accepted the notion of combinations of atoms in compounds as useful in describing the properties of gases, but he did not pursue the notion beyond that. Nevertheless, a search through Heisenberg's prescribed physics text yields no such picture of hook-and-eye atoms. The drawing most similar to his description is a new figure inserted in the seventh edition (1916), showing water molecules constructed of what might be Ping-Pong balls joined together by toothpicks. Either Werner was using another physics book, perhaps in independent study, or he remembered the picture differently. Heisenberg remembered the picture as being toward the end of his book, but the end of his prescribed book is devoted to astronomy. In any case, Ping-Pong ball atoms would have had the same distressing effect on him.

Heisenberg's second remembered encounter with atoms occurred when he read Plato's *Timaeus* while freed from school by military duty in May and June 1919. The relevant passage involved a fictional attempt by Timaeus to explain to Socrates that the observable properties of the four elements—earth, air, fire, and water—can be attributed to the transcendent properties of ideal geometric "atoms." To each of the four elements, Plato assigned one of the so-called Platonic solids. Plato, or a member of his school, had proven that there exist in nature only five solid bodies composed of equal-sided, two-dimensional geometric shapes. Timaeus used the properties of four of these solids—cube, tetrahedron (pyramid), octahedron, and icosahedron—in assigning each to one of the elements:

> Let us assign the cube to earth; for it is the most immobile of the four bodies and the most retentive of shape, and these are characteristics that must belong to the figure with the most stable faces. . . . And again we assign the smallest figure to fire, the largest to water, the intermediate to air. . . . Logic and likelihood thus both require us to regard the pyramid as the solid figure that is the basic unit or seed of fire; and we may regard the second of the figures we constructed [octahedron] as the basic unit of air, the third [icosahedron] of water. We must, of course, think of the individual units of all four bodies as being far too small to be visible, and only becoming visible when massed together in large numbers.[24]

Purist Werner reacted to this passage with astonishment and dismay, as he had to the drawing in his textbook. How could the sagacious Plato believe that atoms are cubes and pyramids? More important, "The whole thing seemed to be wild speculation, pardonable perhaps on the ground that the Greeks lacked the necessary empirical knowledge."[25] Atoms were not to be so rudely treated as objects either of pure speculation or of superficial utility. Certainly, Plato's atoms bore no relevance to modern science. Or did they?

In his 1969 memoir *Der Teil und das Ganze* (English: *Physics and Beyond*), Heisenberg recalled turning to two close friends from the gymnasium Military Preparedness Association and the postwar youth movement, Kurt Pflügel and Robert Honsell. As recounted in his memoir, the three young men entered into a neo-Galilean *discorsi* on Plato's *Timaeus* soon after Werner's encounter with the puzzling passage. In

Heisenberg's scenario, Kurt, a budding engineer, is cast as the crude pragmatist and Werner the perplexed seeker of enlightenment, while Robert, the deep thinker, is given the role of Platonist — atoms are not things but mental constructs, mathematical ideals or forms as transcendent yet reality-bound as mathematics itself. Werner's two friends argue their positions as though in a chess match, until Robert finally wins, convincing Werner of the validity of Platonism and helping him to comprehend Plato's Timaean atoms.

In another account of this encounter with Plato's atoms, delivered in defense of classical studies to the old Max-Gymnasium in 1949, Heisenberg went so far as to claim that his reading enlightened him to basic notions of atomic physics and that from then on "I was gaining the growing conviction that one could hardly make progress in modern atomic physics without a knowledge of Greek natural philosophy." The illustrator of his physics text "would have done well," he felt, "to have made a careful study of Plato."[26]

Much later in life, when he wrote many of the foregoing recollections, Heisenberg believed that he had found striking similarities between modern elementary particle physics and Platonic idealism — so striking, in fact, that he believed that Platonism provided genuine clues for contemporary physics. However, little corroboration can be found for his retrospective portrait of himself in memoirs as a lifelong Platonist, either in the surviving evidence or in the testimony of Carl Friedrich von Weizsäcker, one of his closest students and colleagues. Indeed, an examination of Heisenberg's work and of his own statements from that period suggests that Heisenberg was in fact singularly devoid of any systematic personal philosophy relating to physics. Only the stimulus of his colleagues and the needs of his science encouraged a deeper concern with philosophical matters. Responding in 1925 to a "philosophical letter" from his friend and colleague Wolfgang Pauli, Heisenberg, by then notorious for his pragmatic, utilitarian physics, remarked: "Unfortunately my own private philosophy is far and away not so clear, but rather a mishmash of all possible moral and aesthetic calculation rules through which I myself often can not find my way."[27]

Heisenberg did entertain a modest, usual interest in philosophical issues born of scientific research. The copy of Kant's Critique that accompanied him during his farm labor service in 1918 and his later youth-movement debates attest to his concern. But colleague Weizsäcker, whom Heisenberg first met in 1926, reports that what interested Heisenberg most in philosophy — other than what he needed professionally — was not the substance of philosophical systems but rather their beauty or their literary poetry; the same beauty and poetry he found in music — and in mathematics and physics.[28] Heisenberg was an idealist in science and music, but not in a systematic sense. This orientation became much more pronounced in the context of postwar events and during his early years with Kurt and Robert in the youth movement.

Weizsäcker reports that Heisenberg found neither Kant nor Mach beautiful and regarded only a few passages in Plato as worthy of the label. These, it turns out, are the standard passages he learned in school, which were taught precisely because of their poetic beauty: the first half of the *Phaedo,* the *Apology,* and the banquet scene in the *Symposium* (see Appendix B). Among Heisenberg's colleagues, Weizsäcker, both philosopher and physicist, probably best understood his feelings regarding the lack of poetic beauty in philosophy. And Weizsäcker suspected that, for this reason, Heisenberg never actually read Kant, Mach, or even the *Timaeus* in any detail, nor did he study the intricacies of Platonic thought in any depth. However, he did immerse himself in the beauteous harmonies of physics and music, while relying on his friends and colleagues for philosophical stimulus.

Heisenberg completed his gymnasium education in 1920 with a two-part Abitur. A successful performance certified one for entrance into any university. The written exam was administered at the end of June to 31 pupils in the Max-Gymnasium's ninth grade. They were joined by two girls from a nearby girls' school.[29] All passed. Werner did so well on the written exam that he was exempted from the subsequent oral section. But he underwent the oral ordeal, despite his exemption, as a candidate for support from the prestigious Maximilianeum Foundation.

Each year, the Max-Gymnasium entered its best graduating pupils in a Bavaria-wide competition sponsored by the foundation that King Maximilian I had established in 1849. Twenty-six of Bavaria's best students were provided with room, board, and cultural stimulus in the foundation's elegant Munich quarters. Originally, the honor was reserved for those intending to enter state service after passing through the legal profession. Later, students headed for the sciences and other disciplines were allowed to compete — if their number did not exceed a sixth of the total. Naturally, according to foundation rules, the trainees were to be "of outstanding intellectual talent and impeccable moral behavior." As future state servants, they were also expected to be unblemished Christian men; women, Jews, and the handicapped were excluded.[30]

In 1920, the year Werner graduated, the foundation had eleven openings. The Max-Gymnasium entered its top two graduates in the competition — Werner Heisenberg and his classmate Anton Scherer, who planned to study linguistics. Acceptance brought as much prestige to a candidate's school as it did to the candidate himself.

The Bavarian Interior Ministry instructed Johannes Melber—the ministerial examiner assigned to the Max-Gymnasium's Abitur committee, and a close colleague of Dr. Wecklein—to deliver a comprehensive report on the qualifications and suitability of the two nominees. Unfortunately, state records on the 1920 foundation candidates went up in smoke during World War II. Also destroyed was a no doubt revealing special report by Herr Wolff on Heisenberg's progress in mathematics and physics. A copy of Melber's report on the two Max-Gymnasium candidates does survive in the Maximilianeum files.

The examining committee, of which Melber was a member, administered a two-and-a-quarter-hour oral examination to Heisenberg in the gymnasium's seminar room beginning on July 7 at 8:00 in the morning. In his report, Melber praised the candidate's display of scientific prowess even more enthusiastically than might have been expected from an admirer of Werner's grandfather: "The examination provided in mathematics and physics really shining examples of the extraordinary and rare ability of the pupil in this field."[31] But Melber was far less impressed with Werner's German essay on his grandfather's specialty, tragedy as poetic art—especially when he compared Heisenberg's paper with that written by Scherer, the future linguist. Nevertheless, Melber was convinced that in science, in any case, "he . . . will one day accomplish something first-rate." Melber recommended the nominee "earnestly" for acceptance by the Maximilianeum Foundation, but he still ranked him a clear second "behind his equally extraordinarily gifted fellow pupil Scherer." Both were selected. On the foundation's 1920 list of successful trainees, Scherer is in sixth place; Werner secured the eleventh, or last, position.[32]

Werner declined the foundation's offer of free room and board. He recalled that he preferred to live with his parents. In its records, the Max-Gymnasium notes: "Entitled to support in 1920, but because of his economic situation not supported."[33] With the country in the midst of a raging inflation, others were needier than Heisenberg. In any case, although Werner did occasionally drop by the foundation for polite evenings of tea and music, by then he much preferred the companionship of his postwar youth-movement comrades to the genteel environment of either state or family.[34]

4

Red and White

Heisenberg completed his gymnasium studies during one of the most turbulent periods in modern German history. Defeat in the world war, collapse of the monarchy, and revolution across the Reich ripped away the fragile facade of bourgeois propriety and patriotism, throwing the entire nation into turmoil. Munich and Bavaria experienced some of the worst of it. The disillusioned 17-year-old went through his own turmoil at the turn of events. The last two gymnasium years—from his farm labor work through the political upheavals to his participation in the postwar German youth movement—were decisive for his adult political orientations. His reactions and orientations were reflections of political transformations occurring throughout the Reich and within his own family.

The seeds of transformation had sprouted before the war. With legalization after Bismarck's demise in 1890, the political representative of the working class, the Social Democratic Party (SPD), experienced a steady increase in influence until, in the last prewar election, in 1912, it gained over a third of the seats in the Reichstag (parliament), the largest representation in that body. The Catholic-sponsored Center Party showed equally dramatic gains. Elections to the Bavarian Landtag (state parliament) had parallel results. And in both cases, but especially in Bavaria, where 75 percent of the population was Catholic, the gains occurred at the expense of the mostly Protestant National Liberal Party, to which the Heisenbergs owed closest allegiance. As the party of upper-middle-class professionals—industrialists, merchants, professors, bureaucrats — the liberals strongly supported national unity under Prussian leadership as conducive to commercial expansion. They lobbied for the eventual institution of individual civil liberties — but not at the expense of national ideals. Bismarck, chancellor of the Reich, easily gained liberal favor; August Heisenberg, for instance, adulated "the creator of Germany" during his visit to Munich.[1] Liberals readily agreed to Bismarck's suppression of the Social Democratic Party in protection of their own interests; August, as expected, actively supported antiproletarian political pressure groups on behalf of gymnasium teachers and university professors.

The National Liberal Party achieved perhaps its greatest prewar influence in Bavaria, the only state to force a king from his throne during the liberal revolt of 1848. Liberal Party pressure — including that of one-time party deputy Nikolaus Wecklein — had helped induce the fiercely independent Bavarians to submit to Prussian leadership in joining Bismarck's Reich. As Bavarian monarchs succumbed to insanity, liberals grasped control of the ministerial oligarchy that ruled the Bavarian state from Munich.

By 1912, liberal predominance in Bavaria had itself succumbed to the rising Center Party. In that year voters reduced liberal representation in the Landtag to that achieved by the socialists and gave the Center Party control of the government and of the ministerial apparatus, with nearly triple the Landtag seats accorded either socialists or liberals.[2] While both socialists and liberals longed to gain or regain control of the government and to press for constitutional reforms, liberal professionals, such as the Heisenbergs, feared the dilution of their recently achieved social status and cultural prestige should socialist "proletarianization" ever occur.

All such tensions and differences were put aside as Germany went to war in 1914. But as the war dragged on and the body count of worker-soldiers mounted, German socialists came to believe that they had been duped by imperialistic capitalists. Annual Reichstag votes for war financing precipitated a split on the left. A radical minority, the Independent Socialists (USPD), opposed war credits and broke with the Social Democrats (SPD) in 1917. Leaders of both parties remained loyal to the constitution, but the USPD harbored a revolutionary wing, the Spartacus League. Encouraged by the surprising success of their Russian counterparts, the Spartacists agitated for a German Bolshevik revolution: the establishment of a workers' and soldiers' council — *soviet* in Russian, *Rat* (plural *Räte*) in German — to implement revolutionary working-class demands. They achieved their greatest success in Munich.

In Bavaria, the worsening food and fuel crisis, combined with the authoritarian rule of Marshal Hindenburg and General Ludendorff over nearly every aspect of Bavarian life, awakened antiwar and anti-Prussian sentiments. By the terrible "turnip winter" of 1917–1918, Bavarian socialists could count on liberal support — including the support of the Heisenberg family, whose two sons were now engaged in farm labor and military service and undergoing heavy indoctrination for a now unpopular war.

Bavarian socialists also experienced a split in 1917, but it was one of personality more than politics. The founder of the Bavarian USPD was the very un-Bavarian Prussian, Kurt Eisner, a Jewish writer and intellectual who had gone to Bavaria in self-imposed exile from Berlin. Eisner's Bavarian USPD quickly gained support from the war-weary man in the street. During a nationwide strike in January 1918 for food and peace, several thousand workers demonstrated in Munich under Eisner, the first such political defiance in Bavaria since 1848. The demonstration provoked brutal suppression by the Bavarian army and landed Eisner in jail until October 1918.

October was chaotic. The German army was in hopeless retreat, Ludendorff had lost his command, Austria had collapsed, and the Entente threatened to push north into Bavaria. August Heisenberg's border-guard company headed south to meet the threat.

The Bavarian Landtag finally promulgated democratic reforms on November 2, 1918, but events had already overtaken it. That night, when a ship was ordered to sail into a hopeless battle against the Entente fleet, the sailors mutinied at the north German harbor at Kiel, took over the city, and established a sailors' and soldiers' council, igniting revolution throughout Germany.

In Munich, socialists of both stripes called for a peace demonstration on November 7 on the Theresienwiese, the site of the Oktoberfest. During the rally, attended by 50,000 citizens of all classes and stations, Eisner seized the podium, proclaimed a socialist republic, and called for the abdication of the king and the establishment of a workers' and soldiers' council. The next day armed soldiers and civilians seized the army barracks, train station, newspaper offices, and Landtag building. King Ludwig III, informed of his overthrow as he strolled in his garden, quietly gathered his family, placed them in his new Mercedes, and drove into the countryside, vacating the Wittelsbach throne forever.

One day later, during a similar demonstration for peace at the Berlin Reichstag building on November 9, Social Democrat Philip Scheidemann declared a republic just ahead of what he thought would be a similar Spartacus proclamation planned for later that day. He assuaged middle-class and industrial anxieties by professing that much of the old bureaucracy would be maintained and that every attempt of more radical socialists to gain political control would be suppressed with any means necessary. The German social democracy would be neither revolutionary nor socialist, but parliamentary, bourgeois, and liberal. It would look for support to the army and to the upper middle class, not to the revolutionary councils or their representatives.

Eisner offered the same promise for the Bavarian "soviet republic," despite its name, which at first enjoyed wide support among the populace, including the support of Social Democrats, former Liberal Party members, and the newly influential Bavarian Peasants Party. Eisner did not tamper with the social and political structure of Bavaria, nor did he institute proletarian rule or try to socialize industry. Although more far-reaching in its goals than the SPD, Eisner's party sought little more than the introduction of the constitutional reforms already contemplated before and during the war.³ But the new regime proved pitifully unequal to its task, completely unable to control the forces it had unleashed. The party recruited its lower officials from among Schwabing coffeehouse radicals, men long on theories and short on practical sense. As the economy declined, Eisner and his regime lost favor with their supporters. Radicals on the left and the right prepared to seize control.

New parties sprang up to replace the old imperial formations. Bavarian liberals, such as the Heisenberg family, gravitated toward either the SPD or the Bavarian faction of the new German Democratic Party (DDP), which entered into a national and local alliance with the SPD. Bavarian Center Party members joined the more conservative, Church-supported, and now, in ostensible reaction to Eisner, more overtly anti-Semitic Bavarian Peoples Party (BVP). In the first postwar elections, held in January 1919, universal suffrage became a reality in Bavaria, with women voting for the first time. The results were staggering for Eisner's party. The USPD received only 2.5 percent of the votes. The BVP achieved a plurality, with the SPD and the DDP close behind.⁴

As prime minister of Bavaria, Eisner had presided over a socialist government (both SPD and USPD) that relied on the Bavarian army for military support. The war-weary army, independent of the Reich during peacetime, had rallied to Eisner's revolution in the early days. But the support quickly evaporated when—under restrictions imposed by the victorious Entente and in line with his own antimilitaristic sentiments—Eisner forbade the formation of a peacetime army. A failed Spartacist putsch in Berlin and a failed left-wing Munich coup attempt in December 1918 convinced right-wing extremists of the need for a counter-revolutionary militia. Secret protofascist societies, such as the Thule Society, organized private armies to protect against Bolshevism. One aristocratic Thule Society member, Anton Graf von Arco-Valley, apparently eager to demonstrate his anti-Semitic fervor to his comrades (because of a Jewish ancestor), gunned down the Jewish Eisner in the street on February 21, 1919. Ironically, his victim had been on his way to the Landtag to submit his resignation after losing the election and the support of the middle class.

Chaos reigned in Munich. A gunfight erupted in the Landtag, killing two deputies. Street fighting broke out all over the city, while Werner and other liberal students, now contemptuous of Eisner, burst into jubilant celebration at the news of his demise. The SPD, backed by the DDP, peasants, and the USPD, soon gained control of the government and assumed command on March 18. The government was headed by Johannes Hoffmann, minister for education and culture under Eisner. As one of his first acts, Hoffmann abolished the nobility and its privileges, which alienated gymnasium pupils and the middle- and upper-class majority. Gustav Wyneken, a well-known reform educator, scolded the elite pupils and their teachers in schoolmasterly fashion for their cultural snobbery and resistance to social change. They used their new freedom of opinion, he wrote, "in order impudently, spitefully, and scornfully to turn against the revolution and its leading men . . . and to form something like a silent conspiracy of resistance against the new order of things."[5] Wyneken's chiding only instilled deeper resentment in the students.

The "silent conspiracy" grew louder as events grew even more chaotic. When Hoffmann, no Bismarckian diplomat, attempted to integrate the now rabidly federalist province into the SPD-controlled Weimar Republic that had replaced the German Reich, right-wing extremists gained new support from the Catholic hierarchy and its party, the BVP. At the same time, communist victories in Hungary and Austria further radicalized Bavarian leftists, instilling even greater fear and resentment in the right.

Like Eisner, Hoffmann proved unequal to his task. On April 7 a self-styled Revolutionary Central Council composed of radical USPD members seized control in Munich and proclaimed a new soviet republic to rule Bavaria. The new regime of "coffeehouse anarchists," led by the expressionist poet Ernst Toller, vainly attempted to socialize Bavaria's press and educational system.

A regime newspaper proclamation to gymnasium pupils on April 12 drove them directly into the hands of the increasingly violent opposition: "Pupils! You have experienced the political and economic collapse of Germany; now you will experience the last and greatest collapse, that of her culture." Genteel Kultur had been used too long by the upper classes to separate themselves from the uneducated masses. "The collapse of our culture has now become an historical necessity," the revolutionary council proclaimed.[6] The next day the city was in revolt.

On orders from Berlin, Hoffmann and his officials fled to Bamberg in friendly northern Bavaria. On Palm Sunday, April 13, his Munich followers unleashed a coup d'état, bringing down the Toller regime. But after a bloody street battle at the main train station and a one-day rule of the Munich garrison, Bolshevik forces gained the upper hand, declared Toller's "pseudosoviet republic" at an end, and proclaimed a genuine soviet republic. Lenin telegraphed congratulations from Moscow. August Heisenberg and his family now turned to the SPD as the only hope for protection against Bolshevism and for preservation of national unity.

Like its immediate predecessors, the soviet regime faced a failing economy, a hostile populace, and a severe coal shortage, exacerbated by a brutal cold wave. On April 1 half a meter of snow had lain on the ground. The soviet regime's survival owed solely to the presence of its "red army," inherited from Toller, and to false rumors that red armies from soviet republics in Hungary and Austria were marching up the Danube. The Munich forces, without uniforms but well equipped with weaponry from the demobilized Bavarian army, were also well paid, receiving the highest army wages in Germany — in advance. Munich had little trouble raising a rag-tag army of 10,000 to 20,000 men, mostly unemployed workers, front veterans, and former Russian prisoners of war.

Supporting this expensive army proved impossible. Munich's tottering economy collapsed and remained in chaos after a near-total general strike by most of the Munich population during Easter Week, April 14–22. The plight of the people, including the Heisenbergs on Hohenzollernstrasse, grew ever more desperate: Hoffmann's forces had set up a total blockade of the city. In this they had the cooperation of the Bavarian peasants, who, though their party had once allied with the

Independent Socialists, violently opposed the soviet rebellion. The peasants prevented food and fuel from entering the city. Rejecting the regime's paper money, they refused to sell their produce.

The Heisenberg family found itself again in trouble. The wartime blockade of Germany had earlier deprived the family of food; the new blockade threatened starvation again. But this time Mrs. Heisenberg managed to locate a sympathetic farmer in Garching, about 15 kilometers north of Munich on the Isar River. The farmer agreed to supply the Heisenbergs with food staples, but only if the family could run the blockade and pick up the food at the farm. Many years later, 60-year-old Werner recounted to his former youth-movement followers how he, his brother, and a friend, Kurt Pflügel, set out one night to collect their black-market provisions.[7] "Kurtei," one of the debaters of Platonic atoms, was two years behind Werner at the Max-Gymnasium and a fellow member of the Military Preparedness Association. He and Werner had become acquainted through the tutoring Werner provided at the request of Kurtei's father, a demobilized major in the Bavarian army and no doubt an acquaintance of Captain Heisenberg.

Warmly dressed against the bitter cold but without their uniforms, which would have raised questions, the three young men who tempted death on that freezing all-night journey had reason to be glad of their military training. Munich at that time was encircled by a massive "white army" ordered into Bavaria by Berlin and poised to invade the city. Following the most direct route to Garching, the teenagers slipped through the red army line at one of its strongest points, the Krupp munitions works at Freimann, near the English Garden and the icy Isar River. Since numerous students left the city at that spot to join the white forces, the red guards paid particularly close attention to movement from that direction. Somehow, the three boys made it safely across both lines and reached the farm. When they tried to return, however, the white forces detained them, fearing that the boys might reveal the army positions if they were captured.

Werner and his companions managed to escape the white army, thanks to their intimate familiarity with the local terrain. Passing the Krupp works, they went past Aumeister, a beer garden in summer, and over the broad, wind- and snow-swept field near the North Cemetery and into Schwabing. They arrived home safely with knapsacks full of flour, butter, venison, and, so Werner claimed, unbroken eggs. Years later, partly in memory of the exploit, Heisenberg, director of the Max Planck Institute for Physics and Astrophysics, erected a new building for his institute at the very spot near Aumeister (still a beer garden) where the boys had crossed the army lines that night.

The communist regime in Munich had meanwhile made itself thoroughly hated and feared by most of the population, especially the educated upper classes. The closing of the university and newspapers, the confiscation of food and weapons, and the imposition of a "military dictatorship of the proletariat" further traumatized the people, who began to speak of a "red terror" and to equate communism with thievery and disorder.[8] As Werner later put it, "Pillage and robbery, of which I myself once had direct experience, made the expression 'Räterepublik' appear to be a synonym for lawless conditions."[9]

The red terror reached its zenith as the white army closed in. Red guards rounded up politically suspect persons and seized hostages from among the leading bourgeois and noble families. Adolf Hitler, though no bourgeois, claimed in *Mein Kampf* that he was to have been interned but defended himself with a carbine. Werner's father, like many potential hostages, went into hiding.

Meanwhile, Social Democrat Hoffmann had lost a skirmish at Dachau, just north of Munich, to red army commander Toller. Hoffmann retreated with his wounded pride to Bamberg, called on Berlin for help, and exhorted the right-wing secret societies, his staunchest supporters in Munich, to mount guerrilla attacks and to prepare for a general uprising.

The societies in their turn recruited the sympathetic gymnasium students to the cause, many of whom were already organized into paramilitary units through the Military Preparedness Association. The schools had closed for Easter recess and remained closed until early May 1919, but the pupils relayed the news from friend to friend. Kurtei's father, Major Pflügel, enlisted the Preparedness company of the Max-Gymnasium and organized it into an assistance unit under his command. The assistance units were to act as guides for the invading troops and as an auxiliary force during and after the invasion. Among those recruited into Major Pflügel's schoolboy unit were his son Kurt, Werner Heisenberg, their schoolmate Werner Marwede, and probably Heisenberg's brother Erwin. All except Erwin later joined Werner's youth-movement group.

Berlin's socialist war minister, Gustav Noske, charged with stamping out radicalism, dispatched a massive force of regular army troops and Freikorps (free corps) units to Bavaria. Noske had organized these units, with financial backing from German industry, to accomplish his mission without having to rely on the unreliable and severely limited regular army. Because of the postwar turmoil, Noske's free corps units came to consist mainly of adventurous, often ruthless, mercenaries — former officers (usually with royal titles), restless front veterans, and students lusting for action but too young to have fought in the war. They became a fertile breeding ground for right-wing extremism. Many infamous Nazi careers had their start in one of these units.[10]

On April 23, 1919, Noske ordered the onslaught on Bavaria to begin two days later under the command of Lieutenant General von Oven.[11] In anticipation of the attack, Werner's unit was assigned to assist Cavalry Rifle Command 11, with 1500 men a component of Gruppe Deetjen, commanded by Colonel Deetjen. In his invasion plan, Noske ordered Deetjen and his units to penetrate Bavaria from the north and to secure the northeast sector of a circle around Munich. At Jena, Deetjen assembled the cavalry command, part of a regular army unit that had fought at the front—a move that caused riots and a general strike among the prosoviet populace.

Once assembled, the cavalry unit traveled through Bavaria by train on April 28 via Regensburg and Freising, put down minor resistance it encountered at Freising, and positioned itself in the vicinity of the Aumeister beer garden, for its final assault. Werner and his companions had first encountered the men they would now support during their midnight expedition to the Garching farm.

By April 30, the north and east sectors of the circle around Munich had been secured, precipitating the final act of red terror. When the invaders captured the main red army base at Dachau on that day, the more bloodthirsty Munich reds reacted by murdering ten of their hostages in the basement of soviet headquarters, the Luitpold-Gymnasium. Among the victims were eight aristocratic members of the Thule Society and two prisoners from Hussar Regiment 8, attached to Gruppe Deetjen. The final assault on the city had been planned for May 3, but after the killings there was no stopping the momentum of events.

University and gymnasium students slipped through the red lines that night to inform Captain Ehrhardt, commander of the Second Naval Brigade in Gruppe Deetjen (and one of the leaders of the right-wing Kapp putsch a year later), of the senseless murders. Early on the morning of May 1, as a light snow fell, the Munich underground spontaneously rose up, stormed the Residenz and the Feldherrnhalle, seized weapons from the army barracks, and captured the university and the Luitpold-Gymnasium.

Hearing the clamor of battle, troops in the north broke ranks and began pushing toward their targets. Gruppe Deetjen, spearheaded by inflamed Hussars, attacked the Krupp works, where they met and overcame heavy resistance, then crossed in the snow through Aumeister to the North Cemetery. Fighting their way from house to house down Leopoldstrasse and Schleissheimerstrasse, they battled units of red

guards at the Max-Gymnasium and near the Heisenberg home on Hohenzollernstrasse, crossed the Siegestor near the university, and by
nightfall had pressed all the way to the Feldherrnhalle near Odeonsplatz,
just north of Marienplatz in the center of town. Student units and secret
societies guided the troops through the unfamiliar streets, but Cavalry
Rifle Command 11 remained as a rear guard in the vicinity of Schleiss-
heim, just outside the city limits. As daylight faded, the forward troops
of Gruppe Deetjen withdrew for the night into northern Schwabing and
camped at the Max-Gymnasium.[12]

General von Oven decided not to wait for the entire southern flank of
the city to be surrounded and ordered a full-scale assault for the next
day. At dawn on May 2, Gruppe Deetjen smashed its way out of
Schwabing toward the Residenz and the inner city. Local citizens spontaneously joined the invaders in heavy street battles raging at the war
ministry, the Luitpold-Gymnasium, and the train station. Kurt Pflügel
carried ammunition to his father, who spent the day blasting with a
machine gun near the Wittelsbach Fountain. On the evening of May 2,
General von Oven established his headquarters in the city and declared
the red terror at an end. But the white terror was just beginning.

It flared on May 3. Following Oven's orders and incensed by the
hostage murders, the ruthless white troops stormed for days through the
city once held by the fiercely hated red guards, summarily shooting
anyone who carried a weapon.[13] During the rampage, every home was
subjected to a systematic search for weapons and red army members and
sympathizers. Minor battles flared as residents defended their homes.
Anyone captured as a red was usually shot after the sketchiest of court
martials. Gymnasium units, Werner's among them, guarded the many
prisoners awaiting trial and execution.

The frenzy continued unabated until the traumatized populace was
finally shocked into sense by the murders on May 7 of 21 Catholic
journeymen, shot, bayonetted, and beaten to death by drunken free
corps soldiers who mistook their meeting for a red conspiracy. As one
writer described it: "At the municipal cemetery where dead White
soldiers lay on mortuary slabs adorned with wreaths and white-blue [the
Bavarian colors] garlands the remains of the massacred Catholic journeymen were dumped among the workers' corpses on dirty ground in
lean-to sheds. Crying women slithered on blood-soaked sawdust trying
to identify their husbands by numbered cardboard tags tied to the
corpses' limbs."[14]

By May 8, more than a thousand had died in the white terror,
including all the red leaders except Toller. Toller, arrested in a lady's

boudoir disguised in women's clothing, escaped summary execution only because of the sport he afforded his captors. In comparison, the white forces, by their own count, lost 58 men in the battle for Munich.[15]

Werner and his unit were stationed at the headquarters of the cavalry command in the Gregoranium, a Catholic seminary on Ludwigstrasse directly across from the university. The boys, dressed in the boots and green uniforms of the Preparedness Association, carried loaded rifles while on duty. Under pressure from authorities, Heisenberg stayed with his unit for several weeks, even after the Max-Gymnasium reopened on May 9, 1919. He recalled his reading of Plato's *Timaeus* early one morning while relaxing on the seminary roof where "it was nice and warm."[16] Weather records do not report a break in the cold and snowy conditions until the end of the month.[17]

The Berlin forces remained in Munich until July 1 to reestablish "quiet and order" — that is, to suppress every social democratic tendency in Hoffmann's reinstated Social Democratic government. The legitimacy of the once-moderate regime now rested on the antidemocratic right, which made the raising of a loyal Bavarian army a top priority. Socialist war minister Noske designated as the nucleus of the new army one of the more ruthless of his free corps units, that led by Bavarian Colonel Franz Ritter von Epp. Eager to send the "Prussians" back north, and believing that Bavaria's troubles were caused by the lack of an army to protect against Bolshevism, Bavarian officials carried on an intensive recruitment campaign for the harmless-sounding Schützen-korps (rifle corps). Free corps students, many from Prussia, eagerly joined the Bavarian cause, while local gymnasium units remained on duty into the summer.

School and university officials aggressively recruited Munich students, who were the most sympathetic to the new counterleft militia. On the second day of the new term, the university prorector informed his students that their studies must take second place to the defense of Bavaria against Bolshevism.[18] To allay any doubts about the new unit, the prorector assured the students that "in Free Corps Epp the *demo-cratic spirit* reigns in a practical sense." Among the champions of "democracy" on Epp's staff were Rudolf Hess, later to serve as Hitler's deputy Führer, and Ernst Röhm, soon to command Hitler's storm troopers. Epp himself became Hitler's Reichsstatthalter (Reich governor) for Bavaria in 1933. Anti-Semitism came to play an even more open role in German politics.[19]

Erwin Heisenberg, already an army veteran, signed up for service before school reopened, took an early Abitur in April, and graduated into Epp's Bavarian rifle corps. Werner, lately of military age and also under pressure by his superiors, would have joined along with his brother; but, the prorector notwithstanding, his studies took priority — as they always had. When Berlin needed recruits a year later to suppress another soviet uprising, this time in the Ruhr, Erwin again enlisted — and Werner again remained with his books. Of the 33 pupils in Heisenberg's ninth grade, 14 signed up for the temporary duty and were dispatched to the Ruhr in April 1920. But for Werner, military adventures had paled beside the excitement of his approaching Abitur.

Whenever Heisenberg looked back on this period of violent political upheaval, it always seemed to him a puzzle: "Why all this happened is no longer quite clear to me."[20] And whenever he described his activities during the red and white terrors of 1919, they seemed little more than youthful fun when enjoyable, otherwise a crushing bore: "Well, I was, you know, a boy of 17, and I considered that a kind of adventure. It was like playing [cops and robbers], and so on. . . . I just had to write things for an officer, and sometimes I had to take the guns somewhere; this was nothing serious at all."[21] Like many teenagers, he and his chums made the most of the opportunity: "We were freed from school, as so many times before, and we wanted to use our freedom to get to know the world from different sides."[22]

Yet the reality of that world was much less lighthearted than Werner made it seem. His future wife later supplemented her husband's reminiscences with two more serious episodes he had recounted to her.[23] During Werner's military duty, one of his Military Preparedness comrades accidentally shot himself while cleaning his rifle and died screaming in agony. On another occasion, his commander ordered Werner to guard a prisoner overnight — a "red" who was to be tried and executed the next day. Face to face with the enemy, probably for the first time, the teenager asked the man for his life story, which was as ordinary as the history of most other red soldiers. The majority were much more in need of the high army wages than committed to any political ideology. By morning, Werner was convinced of the man's innocence and managed to have him released. Death, apparently, was more than a frivolous adventure to the 17-year-old.

Nor should one readily accept that Werner did not fully comprehend the stakes. The political upheavals of the period entailed more than mere shifts in political power. Matters of class, social standing, and cultural recognition were at stake for everyone, and these were of primary concern to the Heisenbergs. The plethora of political parties, each representing a specific economic or religious interest group, suggests the extremely close identification of class interests with political aspirations. The shifting political alliances — even among such diverse groups as peasants and professors — in concert with the shifting economic and political situation also attest to the relationship.

The Heisenberg family was part of the pattern. Before the war, the liberal, Protestant professionals in particularist, Catholic Bavaria naturally supported a unified Reich and readily hoped for the expansion of middle-class rights as well as the protection of their status against the rising influence of masses of industrial workers. The collapse of the old order and the threat and eventual reality of local proletarian regimes bent on destroying the carefully crafted cultural bases of bourgeois academic status pushed the bourgeois liberal family into the seeming irony of supporting the party of lower-class workers, the Social Democrats. But the Social Democrats actually ascribed to liberal objectives and to liberal defensive measures against serious working-class challenges and eventually against any threatened change of the old social order.

If the "socialist" Weimar Republic that succeeded the Wilhelmine empire now subjected itself to mass party politics and to shifts in cultural and economic policies at the seeming whim of lower-class voting majorities, then the Weimar academic would continue to maintain the earlier fiction of an apolitical stance — even to appear fiercely oblivious to political change. Objective scholars and scientists could not be tainted by subjective, self-serving political intrigue — even as they engaged in such intrigue. Most of the established physicists of the period — Arnold Sommerfeld, Max Born, Max Planck, Max von Laue — reacted in this way. The most obvious exception, of course, was Albert Einstein, whose outspoken defense of the Weimar democracy earned him the disapproval of his colleagues.

By the middle of the war, young Werner had no doubt internalized his family's social and class allegiances. To the end of his life he always vigorously opposed the federalist tendencies of Bavaria and other German states in favor of a centrally governed nation. By 1919 he had also committed himself to the family goal of gaining and preserving social standing through academic achievement, and he immediately identified with the upper-middle-class academic elite during the Weimar period. He, too, assumed an apolitical stance, but for different reasons. The ideological and emotional allegiances of his elders, born of their station in Wilhelmine society, lost most of their attraction for him on the collapse of Wilhelmine Germany in bitter military defeat, brutal domestic violence, and bickering political factions. The frivolity of his military adventures seems less an indication of adolescent obliviousness than a relief from the intense adult pressures of home and school.

Faced with defeat and revolution, Werner and many other young people reacted with a bitter sense of betrayal and exploitation. The heavy-handed indoctrination they had received into German war aims and the hollow facade of bourgeois gentility now juxtaposed with armed revolution and counterrevolution on the streets of Munich rendered them angry and mistrustful. "We therefore took the right to see for ourselves," he later wrote, "what in this world is valuable and what is worthless, and not to ask our parents and teachers about it."[24]

Werner the physicist would enter the apolitical, bourgeois world of the upper-middle-class academic, but Werner the man would perceive his place within it in terms derived from the emotional and ideological commitments espoused by what he and his friends were now calling a youth movement.

CHAPTER 5

Pathfinding

I t must have been in the spring of 1920. The end of the First World
War had thrown Germany's youth into great turmoil. The reins of
power had fallen from the hands of a deeply disillusioned older generation, and the younger one drew together in larger and smaller groups in
an attempt to blaze new paths, or at least to discover a new star to steer
by."[1]

With these opening words Heisenberg set the stage for his 1969
reminiscences, *Der Teil und das Ganze* (English title: *Physics and
Beyond*). He began not with childhood or adolescence but with the
period that most profoundly influenced him as both scientist and
citizen — the chaotic years immediately following World War I. And he
focused neither on family nor on formal education but rather on his
participation in the postwar German youth movement, the experience
that most directly affected the formation of his adult values.

The first chapter of *Physics and Beyond* refers to Werner's diverse,
often difficult and confusing experiences during the early postwar years.
Between neo-Socratic dialogues on the nature of atoms, Heisenberg
discusses his assistance in suppressing the Bavarian soviet republic, his
remembered reading of Plato, and his study of textbook atoms. He also
recalls debates with his comrades about the lost war, the meaning of
social order, the search for order within their own lives, and their
developing notions of nature and homeland. One theme emerges clearly
from this rather muddled account: the desire for order in all aspects of
thought and life. Heisenberg and his friends longed to regain a sense of

purpose and belonging—and they found it with each other in the youth movement.

For Heisenberg himself, there were added benefits. The youth movement became a vehicle for his adolescent rebellion, adventurous impulses, and budding leadership qualities. It spurred his intellectual independence, taught him how his primary interests—science and music—could transcend the chaos of daily life, and gave him close and secure friendships with his comrades, with whom he formed valuable lifelong relationships.

As Heisenberg wrote in the opening lines of *Physics and Beyond,* the postwar youth movement grew out of a profound sense of crisis that engendered a spirit of rebellion among bourgeois German youth after the collapse of the old order at the close of the world war. But the roots of rebellion reached back into the prewar decades. Young people increasingly detested the charades of bourgeois propriety and nationalistic sabre rattling and felt no desire to pattern their lives on them. By the same token, middle-class society throughout Europe provided little place at that time for adolescence, the crucial transition from childhood to adult roles.[2] Bourgeois youth, like the children seen in Renaissance paintings, were expected to behave like miniature adults, to prepare for their adult careers and future station in life, and to accept without question the values and ideals handed to them.

The rapid urbanization of Germany at the end of the nineteenth century brought with it the problem of what to do with young people in large cities. Where could they come together outside school? Where could they find the adventure, romance, and excitement of youth? Before the war, some urban youngsters literally headed for the hills, seeking to rediscover basic values in the romance of nature, music, dance, and Germanic ritual. Groups like the Wandervögel (Migratory Birds) and the Freideutsche Jugend (Free German Youth) embodied the spirit of prewar youthful rebellion in northern Germany, but neither survived the war intact. Of the 11,000 members of the Wandervögel, 7000 perished in the war; the Freideutsche Jugend fragmented into factions.

For those too young to fight, the state provided youth organizations, paramilitary training, and agricultural assistance work. Youngsters were aggressively indoctrinated with nationalistic values to prepare them for the task their elders set them: to fight and die in a brutal war.[3] Those not battling at the front struggled at home with bitter cold, desperate privations, and near starvation. How carefree can a teenager be when he grows so weak from hunger that he falls off his bicycle into a ditch?

The sudden, humiliating defeat of Germany, the loss of friends and relatives, the collapse of the old regime, the political chaos that ensued, and the forced democratization of their schools traumatized middle-class youngsters, leaving them angry and mistrustful. "A gaping hole opened up for us young people," recalls Wolfgang Rüdel, one of Heisenberg's comrades. Their response: "We're going to make something for ourselves instead, without an organization from above."[4]

The situation was particularly acute for bourgeois Bavarian youngsters, many of whom belonged to the only existing youth organization, the gymnasium's Military Preparedness Association. Few had any use for the North German, "Prussian" youth groups, including the tradition-minded Boy Scouts. The Boy Scouts had originated in England and had spread to Germany in 1909, where they were called Pfadfinder (Pathfinders). Like their English counterparts, German Pathfinders were paramilitary and puritanical, but unlike the English Scouts they focused less on international ideals and more on preparing their young members to fit into the existing German adult social structure.[5] Two years after the First Munich Pathfinder Troop was founded in 1909, it joined the state-supported Military Preparedness Association.[6]

At war's end, the adult-led Preparedness Association lost any raison d'être, and Pathfinder units began dropping out. In January 1919, a Pathfinder troop in Regensburg rebelled against the "decadent" adult values that, in their view, had failed to preserve the monarchy. At the same time, they rejected socialist attempts to dilute the cultural elite by democratizing their schools. The Regensburg troop quit the state's Preparedness Association and pushed for a renewal of all German Pathfinders, and ultimately society itself, through the ideals of the Wandervögel—a genuine Jugendbewegung (youth movement) that would replace adult Jugendpflege (youth care).

On Easter Sunday 1919, at the height of the soviet republic's power, the equally traumatized Munich troop followed Regensburg's example. A month later, during Hoffmann's socialist restoration, the Preparedness Association changed its name to the more youthful-sounding Jungbayernbund (Young Bavaria League), and in the last months of the school year, during the bloody mopping-up operations in Munich, a group of Preparedness boys at the Max-Gymnasium debated their future.

Wolfgang (Wolfi) Rüdel, then 13 years old, had belonged only briefly to the Max-Gymnasium's Preparedness unit before it changed its name and some of its activities. Under intense pressure from their elders to support the socialist restoration, he and his friends now resisted "youth care" under any name. Wolfi, his older brother Eberhard, and several other boys from their Preparedness unit gathered one day during recess at the old fountain in the courtyard of the Max-Gymnasium. They agreed to reject adult youth care but still wanted the guidance of an elder. They decided to seek an older boy of suitable character to replace teachers and adults as their leader. At Wolfi's suggestion they turned to a well-respected older group leader in the Young Bavaria League—Werner Heisenberg.

Werner satisfied every prerequisite: he was an older student, disillusioned with youth care, well liked and well regarded at the school for his mathematical and musical talents, and endowed with intellectual self-confidence, good looks, and leadership qualities. He was also known as "a very great friend of nature," familiar with the mountains and countryside—a perfect choice.[7] Werner, then 17, in the eighth gymnasium grade, and just finishing his military duties following the suppression of the soviet republic, readily accepted the boys' invitation. By the summer of 1919 he was guiding Wolfi and eight or nine of Wolfi's friends into the postwar world.

Gruppe Heisenberg, as it was known, belonged at first to the Regensburg reform movement within the Young Bavaria League, then became independent in 1921. It remained closely associated with the independent Regensburg faction and officially rejoined it in 1922. According to Gottfried Simmerding, one of Wolfi Rüdel's classmates who joined Gruppe Heisenberg in the fall of 1919, the group was then part of Troop B18 of the Young Bavaria League, then headed by Dr. Kemmer, the gymnasium's former Preparedness commander and one of Werner's former teachers. The troop consisted at the time of six or seven groups led by Hans Schlenk, Heisenberg's friend in grade 9B and a war veteran who later became a well-known actor. Most of the troop members had previously served in the agricultural assistance service and in Major Pflügel's schoolboy unit during the suppression of the soviet republic.[8]

Besides Werner, the group leaders in Troop B18 included Heisenberg's comrades Kurt Pflügel and Werner Marwede. Marwede's younger brother Heini (Heinrich) helped found Gruppe Heisenberg. Werner's group met regularly with the other boys in Troop B18 in several basement rooms provided by the Max-Gymnasium; after breaking with the Young Bavaria League, they met in the Heisenberg home.[9]

Just days after the formation of Gruppe Heisenberg, the Regensburg reformers, led by Franz Ludwig Habbel, a wounded war veteran, and

Ludwig Voggenreiter, a publisher's son, called a meeting of all Path-finder leaders interested in reform. Held on the weekend of August 1–3, 1919, the meeting took place in a medieval castle, Schloss Prunn, in the Altmühl River valley near Regensburg. Group leader Heisenberg was still in the throes of his own postwar and postsoviet confusion when he encountered a young man his age on Leopoldstrasse near the university who, as he recalled it, told him of the Schloss Prunn meeting in the passionate words of an inspired youth: "'All of us intend to be there, and we want you to come. Everyone should come. We want to find out for ourselves what sort of future we should build.' His voice had the kind of edge I had not heard before. So I decided to go to Schloss Prunn, and Kurt wanted to join me."[10]

On Friday, August 1, young Werner, with his knapsack and a guitar, took the train with Kurt to Kelheim at the end of the Altmühl Valley. There they joined a stream of boys hiking the remaining several kilometers to the castle. The valley and castle made an ideal romantic setting for the adolescent adventure. The narrow valley, a prehistoric Danube River bed, is lined by steep cliffs and jutting rocks. The castle, still in existence, perches precariously at the top of one of the cliffs, and above it lies a large wood where the boys pitched their tents.

About 250 Pathfinders found their way from all over Germany and from Vienna, Austria, to the meeting. Gathered in their castle in the sky, the boys were alone at last to debate the questions of the day that concerned them the most: Had the German soldiers fallen in vain, now that the war was lost? How should young people respond to the new political situation? How should they interpret Boy Scout ideals of internationalism, self-sacrifice, and tradition? But the crucial questions were those of any reform movement: How was the movement to define itself, and how was it to address the decadent mass society in which it existed? The answers were vital to Werner, who had hoped to discover his own order at the castle—a philosophical, social, even personal harmony. "I myself was much too unsure," he recalled, "to join in the debates, but I listened to them and thought about the concept of order myself."[11]

Incredibly, their discussions were recorded and a transcript later published in *Der Weisse Ritter* (*The White Knight*), the periodical of reform-movement leaders.[12] The meeting was intended to be of lasting significance. The transcript and related writings vividly display the German youth rebellion—indeed, the rebellion of German society at large—against the modernity of urban, industrial "civilization" and the bitter sense of loss of common purpose, of meaningful traditions, of well-grounded values with the passing of a seemingly simpler age.[13]

According to the transcript, the young men agreed that their society had declined into lifeless mechanism, capitalistic greed, urban anonymity, and personal hypocrisy. Young people had to cut the chains of material and moral decadence. A year earlier, Regensburg reformer Franz Ludwig Habbel had declared: "The first demand of our conviction is for *truth* and *uprightness*. We must find our way out of the lie and swindle of our time."[14] Two years after the Prunn meeting, Dr. Kemmer, converted in his old age from youth care to youth movement and largely responsible for the Munich youth rebellion, declared in *Der Weisse Ritter*: "The youth movement is a freedom movement. It has freed itself from the soulless mechanism and materialism of modern civilization and has victoriously defended its value and the right of young life against the limitations of tradition and authority."[15]

As their first order of business, the Schloss Prunn conferees aimed to reestablish truth and uprightness. For them—unnerved by modernity and by the failure of bourgeois ideals to protect the cultured classes from the leveling of mass "civilization"—this meant an embracing of values derived from a revival of German romanticism: escape from the dead city to the living, genuine, fundamental virtues of pristine nature. There a complete renewal of the total man, a mystical revival of each human soul in unison with other souls, would occur through sustained contact with nature, with the cycle of seasons, and especially with each other—"not just an idea or a thought, but rather something internal, something fundamental, a harmony of souls."[16] Transformed and revitalized by immersion in the genuineness of nature, the young men believed they could eventually reclaim decadent German society and establish themselves as the new and incorruptible elite. "We want a new humankind that lives in our lifestyle. We believe that there is something good in everyone, but that it is suppressed in the shabby struggle for daily bread."[17]

Conflict arose with the practical question of when and how to bring about the social renewal. "Radical" reformers at Schloss Prunn embraced political elitism, seeing society and its masses as hopelessly lost and unredeemable by a handful of enlightened youth. They called for a total retreat from "civilization" to await its collapse, out of which a new order would emerge.

The "conservative" reformers at Schloss Prunn rejected elitism and called for the immediate reformation of society. A reformed society would encourage reformed individuals. Their most articulate spokesman declared: "We can't build a world of ideas next to us; we must reckon with the material side. For we have to deal with age, school, party, food, work. We have to remain standing on the ground with both feet, that is, we have to work for the masses." Did it really help the masses to tramp around the countryside or to enjoy good music and literature? The conservatives believed that the group must deal with the world in order to change it, and that meant dealing with parents, dealing with women, and dealing with what some regarded as the lowest form of materialism — politics. "Political struggles are self-evident for anyone who has learned to take pleasure in development, pleasure in the material side, pleasure in numbers, however reactionary that may sound."[18]

It did sound reactionary to the Schloss Prunn radicals, for involvement with politics, or with social activities in general, seemed an embrace of the very faults of youth care that the youth had rejected. They had grown up and been indoctrinated with political ideals that were suddenly shown to be a lie and a swindle. The older generation had exploited those ideals to wage a war that it had lost, plunging society into a chaos of "materialistic" (pecuniary, urban, working-class) regimes, all vying to control upper-middle-class young people and to command their allegiance.

The first issue of *Der Weisse Ritter* championed the sentiments of the antipolitical elitists: "*The White Knight* stays away from all efforts to win young people for the party politics of one or the other direction. It sees in these attempts an irresponsible crime against the highest right of young people to decide for themselves their relationship to a state whose crumbling society they themselves will have to rebuild anew."[19]

Yet, for all of their professed hatred of hypocritical bourgeois values, the young rebels rejected not their own middle-class social and economic status and mores but what seemed to them the hollowness of the ideals that sustained them.[20] The antipolitical reformers actually rejected not politics but the party politics of their elders and of the new Weimar democracy. Their antipolitical stance was merely pursuit of the political aims and interests of the upper middle class by other means.[21] Ultimately, the young elitists did a great disservice to the society they so zealously sought to reform. In their frustration, they reacted to the immediate situation by removing themselves from the postwar social and political arena, where the future of their country was being decided. In doing so, they nullified the salutary effects that their rebellion should have brought.

These highly educated, upper-middle-class young men could have exerted a democratizing influence by challenging and encouraging their elders, instead of literally heading for the hills. Their elitism left the arena open to those aggressive and ruthless enough to fill the void. Large numbers of conspiratorial secret societies, paramilitary organizations, assassins, and future Nazis emerged in Bavaria in the years just after the war, while the bewildered older generation, too, largely removed itself from the political arena. As young people entered universities and embarked on careers, they were unprepared for and easily overwhelmed by social unrest and political extremism. Ironically, in the late 1920s, the apolitical stance of many university members made those institutions fertile breeding grounds for dictatorial demagoguery—the very behavior that had so repulsed the students and made them apolitical in the first place.[22]

Throughout the 1920s, Heisenberg, too, held himself aloof from politics. But his apolitical stance coincided more with that of moderate older academics of the Weimar era than with that of the radical youth movement. Still, his reasoning derived from the latter. For Heisenberg, science and politics simply did not mix. This was not because of overt rejection of democratic politics—he actually considered himself a social democrat—nor because he regarded political intrigue as criminal exploitation, but rather because he came to view his physics—along with nature and music—as belonging to a higher plane of existence and truth that somehow transcended the ephemeral dirty world of politics. As Bavaria slid deeper into political extremism in 1923, Heisenberg wrote Kurt from Göttingen: "I never thought that I could interest myself in politics, because it seemed to me to be a pure money-business."[23]

Nor was Heisenberg quite the extreme social elitist his comrades were. Simmerding recalls that his views were even regarded as a little taboo, an indication of just how extreme his friends' opinions were. He was "absolutely for the simple, poorer levels of society, certainly very social."[24] Indeed, he participated briefly during his early university days in the so-called Volkshochschule (adult school) movement, an activity that the Schloss Prunn purists would have scorned.

The Munich adult school movement had arisen from an idealistic student goodwill program begun by Father Carl Sonnenschein. As had the Social Democrats before the war, the adult school sought to integrate uneducated workers into the respectable German middle class through cultural education. Social Democrat Werner, a little guilty over his role in the suppression of the workers' revolt, wanted to make amends—now that the threat was past.[25] In 1920 and 1921, Heisenberg, together with a young woman, conducted evening classes on German opera for the benefit of the workers: "She sang arias and I accompanied her on the piano." Heisenberg also offered a workers' astronomy course. He often went into the fields at night to observe the stars with several hundred workers and their wives or accompanied them to the state observatory, where his friend Hans Kienle served as an assistant.[26] Heisenberg's enthusiasm for the uplifting of the masses was earnest, if naive.

At Schloss Prunn, the question of apolitical elitism became so divisive that on Saturday the radicals stalked out of their own conference. The two sides met separately that afternoon and made up their quarrel by evening, but not before the radicals—mainly the Regenburg circle, the Austrians, and their Munich followers—proclaimed an oath of allegiance and gave themselves a new name, the Neudeutsche Pfadfinderschaft, the New German Pathfinders. They now defined their views in terms of three fundamental concepts: Gemeinschaft (community), Führer (leader), and Reich (empire). The three notions also formed the foundation of their successor organization, the Bund der Neupfadfinder, the League of New Pathfinders, with which Werner and his group were affiliated.

It is difficult to determine the precise meaning that the three abstractions held for Werner and the Neupfadfinder. The mystical and emotional overtones of the group's basic outlook—they were, after all, naive teenagers—make any rational analysis difficult. For Werner's friends in the Neupfadfinder, the symbol of the white knight seemed to represent the triumph of moralistic, feudal, Christian crusaderism over hollow, modern, bourgeois values—a symbol of their own craving for a revival of the old feudal Reich, or at least the revival of an idealized Holy Roman Empire. For decades, chivalry had served as the standard romantic antidote to modern debauchery and hypocrisy. But the white knight also seemed to symbolize the new man—both themselves and the idealized leader who would emerge from among their ranks to lead the masses—a chivalrous moral crusader, fearless in the face of danger and selfless in the suppression of individual wishes, base desires, and personal doubts.[27] For one of the Neupfadfinder leaders, the white knight became the crusader for a new third Reich that would emerge to fill the void "after the German throne has become vacant with the collapse of the corrupt ruling stratum and with the inability of the corrupt masses to rule. . . . Him, the 'White Knight,' we set on the empty throne of our longing; we fight for him as his knights, build up his people, and found his Reich. That is the meaning and mission of our Bund."[28]

Perhaps in an effort to placate the Catholic Church, the Bavarian Neupfadfinder often equated the white knight with St. George the dragon slayer. As a constant reminder of their calling, a portrait of St. George as white knight slaying the evil dragon of decadent debauchery hung over the door of the Bavarian ski hut built by Gruppe Heisenberg in the early 1920s. It was still there when Niels Bohr visited the hut over a decade later.

Like other romantic youth groups of the period, the Neupfadfinder believed their own organization to be the nucleus of the new Reich and themselves to be the founders of a new spiritual aristocracy that would replace the corrupt state aristocracy. Because of this, they attempted to make their relationships with each other as genuine and unhypocritical as possible. This was to be achieved by a covenant, their solemn oath of allegiance to each other and to their cause. The Neupfadfinder covenant, the "Prunn oath," contained the essence of their movement. Habbel had considered adopting the famous manifesto of the Frei-deutsche Jugend, proclaimed on the Hoher Meissner mountain in 1913, but he rejected its principle of the self-determination of the individual.[29] In casting off the bonds of the decayed society they so despised, they replaced them with their own equally rigid bonds in order to instill new and sounder virtues in the members of their own society. Their oath accordingly replaced the 1913 dedication to individual freedom with devotion to personal submission to the group: "We Pathfinders want to be young and gay and to live our lives with purity and inner rectitude. We want to submit to our leaders, to whom we give our absolute trust."[30]

Because of this solemn allegiance and devotion, the covenanting adherents of the Bund (league) formed not just a group of friends but a close-knit Gemeinschaft, a self-contained community, headed by a revered and trusted Führer. According to two authors, the notions of Bund and Gemeinschaft apparently derived from the prewar Wandervögel and probably from historian Hans Blüher's 1912 and 1917 monographs on the role of Eros in a male society such as that of the Wandervögel.[31] For Blüher, the most closely knit bond between men is formed only in all-male groups, because only in such groups are the creative forces of Eros, the homosexual urges, sufficiently awakened to generate the creative male bond between their members. Blüher identified three characteristics of the workings of Eros in the male bond: charismatic leaders; emphasis on the ideals of male beauty, masculinity, and athletic prowess; and the exclusion of women.[32]

The Neupfadfinder displayed all three characteristics. For them, the Gemeinschaft, the small group of boys totally dedicated to each other and to their leader, formed the basic unit of the Neupfadfinder. Here the new spiritual aristocracy would ripen, emerging at last to forge the new society. Because of this, the Gemeinschaft made all-encompassing and lifelong demands on its members. These were not schoolmates or even close comrades, but men dedicated with all their beings to each other and to their group leader. Said one Regensburg reformer at Schloss Prunn: "The way to the ennoblement of mankind leads through the Gemeinschaft. Only in the Gemeinschaft can a very slow process of development lead to the heights. . . . A Führer can only lead and bring about the good with a very small Gemeinschaft that depends on him, that will go with him through fire, and that truly believes in him."[33]

It is difficult to determine to what extent such relationships are attributable to Eros, bridled or otherwise. Certainly Werner's relationships with his followers and comrades went far beyond mere friendship, even involving a type of love for one another, but it seems more the love of comrades in arms. Though perhaps not devoid of sexual overtones, it expressed itself most extensively as a nonphysical Platonic love of kindred souls that excludes the outside world while uplifting and strengthening those privy to such feelings.

Women, of course, could not participate in such bonding; in fact, Werner and his fellows never had much to do with women. Although women were now admitted to German gymnasiums and universities, none of the boys encountered any women students or teachers in gymnasium and only a rare few at the university.[34] Far from being a social aberration, the all-male Bund actually reflected the predominantly all-male upper-middle-class professional world in which they lived. But it also represented a defense against a growing threat to that world. A vociferous German women's suffrage movement had been gaining in strength ever since its legalization in 1908. Because of the SPD support of women's rights, women increasingly rallied to the SPD cause.[35]

With suffrage finally granted women at the end of the war, many men feared the possible competition for jobs and the transformation of family life and roles that this could entail. Women became for the Neupfadfinder a part of the mass "civilization" they so longed to escape. A women's unit could not join the Bund until 1925, and even then it was composed largely of the boys' sisters. Perhaps because of the exclusion of women and the absence of attention to mature male-female relationships, the boys in Gruppe Heisenberg who married at all did so quite late in life. Werner himself did not marry until he was 35 — and then to a woman 13 years his junior who had to make clear to her husband that she was not one of his youth comrades.

As a Neupfadfinder leader, Werner could expect to exercise absolute authority over his Gemeinschaft, and, because he was chosen by the group, he could also expect to enjoy their absolute trust and personal respect. The group Führer contrasted, naturally, with the Erzieher, the schoolteacher. The latter led only by force, with meaningless rules and laws and with no personal authority. A Führer led by virtue of his charisma, his natural authority, and his total concern for the group. He had earned the trust of his followers, and he never betrayed it. An apt example of an Erzieher, said Simmerding, was the despised drill sergeant Himmelstoss in *All Quiet on the Western Front*; an apt example of a Führer was Werner Heisenberg.

In a diary of Gruppe Heisenberg kept during their outings, Simmerding recorded the following episode in 1921. In that year, Simmerding had recruited some younger boys to join their group, among them the "rather thick-headed" son of Colonel Seisser, chief of the Bavarian state police.[36] During an outing into the hills of Upper Bavaria, led by Werner, the antiproletarians managed to eat up all their personal food prematurely, "as if Communism had been declared." With nothing to eat, they fell back on their emergency recipe for a dish they called cement: flour, water, and blueberries. Seisser had withheld some food of his own, but when he tried to steal off to eat it alone, Simmerding caught him, confiscated his hoard, and insisted that he eat what the others had concocted. As the group prepared to leave, Seisser remained behind sulking, impervious to all attempts to force him to come along. Then Werner stepped in. Telling the others to go on ahead, he sat down on a log near Seisser, pulled out a letter, and read it silently. He then stood up and, without a word, Seisser accompanied him as he rejoined the group. Trust, not force, carried the day. That, Simmerding exclaimed, was a leader.

Group leader Werner had at last rediscovered the trusting relationships he had known before the war. But now he was at the top of the power structure, where his ambition and talents found new scope and freedom, but where he also found new, lifelong responsibilities. In Heisenberg his young followers found the admirable father figure they needed, the sincere elder who would set rules and challenges to them and in whom they could place their absolute trust. "He demanded a lot of us but not more than we could take," recalls Simmerding.

Before each outing, Heisenberg conducted a Lumpenparade (rag parade). Each boy had to take everything out of his knapsack to be inspected and weighed, and, if the contents weighed more than 10 kilograms, he had to discard the excess. At mealtimes, either Heisenberg or one of the other leaders in the troop—now called a Stamm, or tribe—designated a space on the ground as the "table" and marked it off with pine branches. No one was permitted to put his feet on the table, except the cook, and no one could come to the table without being properly dressed. Before every meal, a quotation from Eichendorff was read, and proper table manners were always observed.

A strict moral code, essential for the new man, was firmly enforced in Heisenberg's tribe. As in a cult, everyone was subject to judgment by the group. On extended outings, the boys regularly held intensive criticism sessions, called by the Germanic name Thing.

Tobacco and alcohol were strictly forbidden; the strongest drink permitted was cocoa, even after the members attained adulthood. But Werner was a slight exception. When Niels Bohr visited the group's ski hut in the early 1930s, Heisenberg's followers were shocked to see the local tavern owner arrive with a case of beer ordered by Werner. But when, during a swimming party on the North Sea, some of the more exuberant members of the other all-male groups threw off their swimming suits and frolicked in the waves, Heisenberg strictly forbade such untamed Eros.

Shortly after the Schloss Prunn reformers had solemnly pledged their devotion to their cause, they continued the meeting. Further debates ensued, but by Saturday evening the two factions were friendly enough to join together for an evening of song, music, and playacting in the romantic castle courtyard. Heisenberg, however, watching the debates, had as yet reached no such accord within himself. Partially sympathetic to both sides, he recalled that each seemed to offer only a Teilordnung, a piece of the true order he sought. He could be swayed by either side. A link to some sort of "central order" or cohesion was still missing: "The lack of an effective middle ground became more painfully obvious to me the longer I listened. I suffered almost physically under this."[37]

As the young men gathered in the courtyard to sing in the twilight, the impressionable Werner suddenly experienced what he regarded as that mystical central order. "As the song came to an end," the recorded transcript of the meeting reports, "out of the stillness the sound of a violin flows downward longingly from a narrow, mysteriously lit tower window and upward to the eternal stars. . . . The music ends, no wayward sound breaks the solemn silence."[38]

In that moment of silent epiphany, under the summer stars in the courtyard of the medieval castle, Heisenberg writes, "All at once, and with utter certainty, I found my link with the center." Suddenly, it seemed, everything fell together for him — music, science, philosophy, religion: "The clear phrases of the Bach Chaconne touched me like a cool wind, breaking through the mist and revealing the towering structures beyond. There had always been a path to the central order in the language of music, in philosophy and in religion, today no less than in Plato's day and in Bach's. That I now knew from my own experience."[39] Science, music, philosophy, and religion all now contained for him a transcendent inner harmony, an eternal truth or validity that did not inhere in the transitory, discordant, everyday world of a Munich student. Somehow, he had at last perceived the stabilizing insight that he so desperately needed during those early chaotic postwar days, and he experienced it only within the context of this faction of the reformed romantic youth movement.

Following the Schloss Prunn experience, Werner traveled to Osnabrück, where he stayed for two weeks while his brother was on leave from antisoviet military service. On August 16, the transformed youth wrote to tell his father that he would not to return to Munich until September; he planned to wander the mountains with his boys for three days, then to tour for a while with Hans Schlenk. He hoped that his father would not be too angered by his refusal to stay away from home. "It is indeed probably not right of me," he wrote, "but I simply prefer to be among my young friends than to sit alone and forgotten, and what I have [among my friends] older people could never give me, no matter how well or how kindly they care for me."[40] In his friends and in his experience of the central order, he felt he had found at last his place in the world and the place of thought and music within his life.

Franz Habbel's Regensburg faction, meanwhile, continued to work for reform within the Young Bavaria League and the German Boy Scouts (DPB). At a DPB troop leaders' meeting in Naumburg on Pentecost 1920, Habbel and his band joined forces with the Jungdeutsche Pfadfinder (Young German Pathfinders), a north German group led by a Prussian from Potsdam, Protestant minister Martin Völkel. Together they formed a new subunit within the DPB, the Bund der Neupfadfinder. To cement the relationship, Habbel named Völkel Führer of their new Bund, and Völkel named Habbel Bundeskanzler (Bund chancellor). Meanwhile, in Munich, Karl Sonntag, a later friend of Heisenberg, began vigorous organizing and recruiting for the new Bund. Sonntag acted as head of the entire Munich Pathfinder contingent, now allied with the Habbel-Völkel faction, and almost immediately began his own rise to leadership. During the summer of 1920, he organized a new group and promptly merged it with those led by Heisenberg's comrades Werner Marwede, Kurt Pflügel, and Wolfgang Ott, to form the Third Munich Pathfinder Troop, which he headed.[41]

The DPB tolerated the new faction only briefly before expelling it in November 1920. The newly independent Neupfadfinder quickly reorganized themselves along more explicitly feudal lines. Völkel named himself Herzog (duke); Sonntag was elevated to Bavarian Gaugraf (district baron) and editor of *Der Weisse Ritter*, while remaining head of the Third Munich Tribe; Karl Seidelmann from Augsburg became Gaukanzler (district chancellor) and editor of *Die Spur in ein deutsches Jugendland* (*The Trail into a German Land of Youth*), the magazine for Bund members; Heini Marwede became treasurer after moving to Berlin; and Heisenberg's friend Wolfgang Hurt became Führer of the entire Munich contingent, consisting of the three Munich tribes.

Although the details are sketchy, Gruppe Heisenberg apparently left the Young Bavaria League along with the Neupfadfinder, which consisted of the entire Munich contingent of the League. However, as a group, Werner and his followers did not remain official members of the Neupfadfinder. They did continue to maintain close ties with Sonntag's Third Munich Tribe, to which many of their friends from the old Preparedness Association and its successor, the Young Bavaria League, still belonged. When Werner was preparing to leave Munich to continue his studies in Göttingen in the fall of 1922, his group joined their companions in the Third Munich Tribe in order to gain the security of a larger organization. Heisenberg opposed the move, as he had since they left the Bavaria League, but he acquiesced in the decision.[42] A year later, when he moved permanently to Göttingen after receiving his doctorate, Heisenberg transferred the leadership of the group to his second in command, Gottfried Simmerding.

Besides their general contempt of official organizations, Heisenberg and his boys had several reasons to resist joining the Völkel-Habbel Bund. Foremost among them, as Simmerding later recalled, was a strong strain of anti-Semitism in the Bund, even though the Bund was apparently one of the few youth groups on the romantic right that admitted assimilated Jews. Anti-Semitic Volkism, argues one historian, did not prevent such groups from admitting Jews.[43] Ludwig Voggenreiter, a comrade of Habbel's who headed the Weisser Ritter-Verlag (White Knight Publishers) in Regensburg and Berlin, published several anti-Semitic pieces during the early twenties. Among them was Eros dévoté Hans Blüher's 1922 anti-Semitic diatribe on Zionism, *Secessio Judaica*, which prompted a well-publicized Jewish boycott of the press.[44]

In his "Bekenntnis zu Adolf Hitler" ("Avowal of Adolf Hitler"), published in *Die Spur in ein deutsches Jugendland* in May 1933, the rabidly antipolitical Voggenreiter claimed that he had met the

Nazis in Munich as early as 1919 and had presented them with copies of *Der Weisse Ritter*. In 1924, when Hitler went to prison for his Beer Hall putsch of the previous year, he asked Voggenreiter for a copy of Blüher's *Secessio Judaica*. Voggenreiter gladly obliged and included a special issue of *Der Weisse Ritter*, "Sendung," containing Völkel's romantic "Hie Ritter und Reich" ("Yon Knights and Reich"). To Voggenreiter's dismay, Hitler made no reply. But on his release, Hitler forged a new Nazi movement by recruiting students, many of whom were already attracted to the ultraromantic notions of Führer, Reich, and Gemeinschaft exalted by such youth groups as the Neupfadfinder.[45]

Anti-Semitism had also infected the Munich contingent. In a letter to Völkel in 1921, Hurt reported that, to his distress, the Munich members had split into pro-Jewish and anti-Semitic factions, the former headed by Walter Tuchmann. Since some Pathfinders refused to work with Jews, wrote Hurt, "a sort of Jewish counterpathfinder group has come into existence, whose relationship to me through its leader Walter Tuchmann, a very fine person, is at the same time both sincere and painful."[46]

Tuchmann is the "Walter" who appears in Chapter 2 of Heisenberg's *Physics and Beyond*. Here Heisenberg recalls that he and Rolf von Leyden, the violinist of Schloss Prunn, often met at Walter's home to practice classical chamber music, Werner on Walter's Bechstein grand, Walter on the cello, and Rolf on the violin. Two other Jews, Alfred Neumeyer and Kurt Bloch, were also closely associated with the Third Munich Tribe, suggesting that Sonntag's tribe was not overtly anti-Semitic. Bloch, a group leader in the Second Munich Tribe who remained in Germany after 1933 and survived internment at the Dachau extermination camp, was considered an honorary member of the Third Tribe.[47]

Another reason for the reluctance of Gruppe Heisenberg to join the independent Bund officially was the Bund's growing reputation in Bavaria for romantic, anti-Church, and immoral Schwärmerei (dreamy nonsense). Its overt rejection of church traditions—if not of religion itself—resulted in its proscription by some schools and parishes. Even worse, in late 1921 the Bavarian culture minister attempted to ban the entire Bund as a pack of Bolsheviks—this after the Reverend Völkel publicly supported education reformer Gustav Wyneken, now attracted to the Bund by Blüher's books. Wyneken had apparently taken Blüher at his word in seeking "a truly free community in which Eros could emerge out of the confining darkness"[48]: Wyneken was caught in a homosexual affair with two of his pupils and sentenced to three years in prison.[49]

The Wyneken affair occurred just before Heisenberg's group finally joined the Bund. Bavarian baron Karl Sonntag was caught in the middle, increasing tensions between himself and Völkel. Sonntag finally had enough and retired in 1925, turning over the leadership of the Third Tribe to Werner Marwede and the district leadership to Heisenberg's philosophical crony Robert Honsell. After his retirement, Sonntag became the head of a small group of Altmannen, young men who had grown too old for the Boy Scouts—one of whom was Dr. Werner Heisenberg.[50]

The Neupfadfinder were notorious throughout Germany for their romantic Schwärmerei: Teutonic rituals, visions of the white knight, and dreams of a mystical third Reich, most of which derived from Völkel's overactive imagination. Although he participated in some of their ceremonies, all of this may have made the more rational Werner a little uncomfortable. Most of the Neupfadfinder ceremonies took place in conjunction with their outdoor camps, the Lager, which, together with field games, distinguished the activities of the Bund from those of the prewar youth groups. Where the Wandervögel preferred wandering and singing, the Neupfadfinder favored the stationary camp, and they spent long periods, sometimes weeks, living in tents far from city and civilization. There they communed with each other and with nature in song, poetry, and ceremony and held day-long war games and feudal sporting matches.[51] The most important of the annual encampments—in which Gruppe Heisenberg usually joined—were the August Bundesfest (anniversary of the Bund), the Christmas-New Year camp, the spring camp at Pentecost, and the Sonnenwendfeier (solstice celebration) in June.[52]

After moving out of the gymnasium, Heisenberg's boys usually met as an independent group in the Heisenberg home one evening each week. They spent weekends taking day trips into the countryside—either alone or with Sonntag's Third Tribe. Weekday meetings were devoted principally to culture: music, song, and poetry. Although they were regularly in the Heisenberg home, neither Rüdel nor Simmerding recall much of Werner's parents; Mrs. Heisenberg usually disappeared after providing milk and cookies. During these meetings Werner often fulfilled his paternal role, dispensing advice (informed or otherwise) on school problems and adolescent dilemmas. Traditional Boy Scout crafts, practical skills, and merit badges were excluded as being formal holdovers from Victorian scoutism and wartime Wilhelmine youth care. Neither were uniforms worn until the group finally joined Sonntag's Neupfadfinder tribe in 1922 and adopted its plain grey shirt, devoid of badges or insignia, and blue neckerchief. A photograph of a smiling Dr. Heisenberg in his uniform playing a spirited game of kickball with his boys in 1926 appeared in a post–World War II newsletter of Sonntag's revived Neupfadfinder tribe.

During the evening meetings, Heisenberg also read aloud to his boys from the works of romantic German poets and writers—Goethe, Hölderlin, Ernst von Wildenbruch. Heisenberg's personal favorite work was Goethe's oriental *Westöstlicher Divan (West-Eastern Divan)*. These authors were so familiar to the boys that they often recited sections from memory on outings, gathered around the campfire at night. After their literature lesson, they took up their music. Each of the boys played at least one instrument, and during meetings they would play in a small ensemble. They also experimented—sometimes hilariously—with new instruments and combinations of instruments. Plucking the right chords, Heisenberg once tried to play piano on his guitar! But the high point of musical enjoyment came when Heisenberg turned in earnest to his piano and his beloved Bach and Mozart. Rolf and Walter occasionally accompanied him. Boys from other groups and the Pathfinder tribe were often invited to these musical evenings, for which Gruppe Heisenberg was widely celebrated. Music, poetry, and nature occupied practically their entire thoughts and activities. "That was our world," says Rüdel.[53]

On weekends, the group headed into the hilly countryside south of Munich or to nearby Lake Starnberg, where as a treat Heisenberg rented a sailboat. Sometimes they went out to the English Garden, the large park running through Munich along the Isar River, where they held field games such as Deutschball (German ball) and Speerwurf (spear throw). The latter was Heisenberg's favorite game, but it was also the most dangerous. The group divided into two teams that tossed a spear back and forth; the spear had to be caught in flight by an opposing team member. A missed catch meant that the intended receiver was "out," technically if not physically.

During school vacations, Gruppe Heisenberg embarked on long, adventurous hiking tours. Sometimes they ventured abroad, to South Tyrol or to Finland, but most of the time they remained in Germany itself, whose beautiful countryside they delighted in discovering. Werner's appreciation of the beauties of his homeland, instilled by family and teachers, became a profound attachment during those early youth-movement years.

The competitive Werner always enjoyed a flirt with danger, even in his later years. According to Friedrich Hund, Heisenberg's later colleague and a prewar Wandervogel, Werner and his comrades gained a reputation for engaging in dangerous activities. On long outings, the boys often hiked the most rugged mountain terrain where other groups did not dare venture. During one tour, one of the boys fell into a glacier crevice. Only after considerable effort was Heisenberg able to pull him out.[54] On another occasion a youngster broke his leg skiing. Heisenberg eventually managed to get him down the mountain and restore him to his nearly frantic parents—who thereafter paid regular visits to the group's ski hut. Werner's mother once wrote to Kurt while the boys were in South Tyrol: "How nice that you have such beautiful weather —but now be more reasonable than Werner and hold him back a little from the high peaks."[55]

In the summer of 1921, after Heisenberg's first year at the university, the group left for an entire month's tour. Under Heisenberg's leadership, they hiked first from Munich to the Harz Mountains in northern Germany to attend a Bundesfest, the annual celebration of the Bund, then made their way back to Munich through the countryside via Jena, Bamberg, and the Thüringen Forest.[56] They used no other means of transportation on their trips and carried no money for food or supplies. With inflation in full swing, they had to fend for themselves, which was just as they wanted it. They led an enviably unfettered life—hiking and singing during the day, then sleeping under the stars or on the sweet-smelling hay in a farmer's barn at night. In the morning, they did chores for the farmer as payment for his hospitality. In the diaries kept during those carefree tours, one finds an occasional drawing of a beautiful nature scene. Art, explained Simmerding, was simply one of the accomplishments expected of them. Werner tried his hand at painting, too, with typical success: a book of quite good watercolors from that period is among his private papers.

Having just offered a brilliant performance on his final gymnasium examinations in the summer of 1920 and about to launch his scientific career, Heisenberg went off with his group on a two-week hiking tour through the Altmühl Valley near Schloss Prunn. One day towards evening they came upon another castle and asked to spend the night. The castle keep consented and showed the boys to mattresses already laid out in a large hall. Little did the boys know that the castle had recently served as a military typhoid ward. As it happened, Heisenberg was the only victim, returning home terribly sick and suffering from a high fever. His father, alone at home, was at a complete loss what to do. He promptly put Werner on a train to his mother in Osnabrück. Since all the beds were taken at the Osnabrück house, the feverish Werner was settled, deathly ill and highly contagious, on a couch in the living room.

An uncle, a Dr. Mutert, was hastily summoned. He prescribed fresh milk and eggs as the only remedy, but both were practically impossible to come by in those early postwar years. Dr. Mutert apparently saved the future physicist's life by daily fetching eggs and milk himself from one of his patients in the country. Fortunately, no one else contracted the disease. By the time university classes began in the fall, Heisenberg had made a complete recovery.[57] Ironically, Werner's father died of the same disease a decade later, contracted while in Greece.

During their long excursions, when not singing or reciting, Heisenberg and his young friends often engaged in philosophical debates or played chess. Werner's skill at the game was legendary. Besides playing during leisure hours on the farm in 1918, Werner had often held chess matches under his desk at school. Even without a queen, he could still win. "That is indeed an accomplishment!" he bragged to his father.[58] Werner and Kurt often played chess in their heads while hiking, and one night when an actual board was upset an hour into play, Werner managed to reconstruct the entire game from memory. After he entered the university, his obsession with chess apparently became so obvious that Professor Sommerfeld finally had to forbid him to play, claiming it was a waste of his time and talents. But in one way it might have had a beneficial effect. As a physicist, Heisenberg possessed an unusual ability to perceive the physical result arising from a mathematical formulation after many complicated moves. Perhaps this skill, vital to any physicist, was enhanced for Werner by his many hours of complicated chess matches.

Although Heisenberg and his friends came together to rejoice in their youth, to be young and gay, they had their serious moments as well, as their solemn bonding, their strict moralism, and the intensity of their philosophical debates attest. During his youth-movement days, Heisenberg's favorite partners for serious discussion were the three young men closest to his own age and development: Kurt Pflügel, Karl Sonntag, and Robert Honsell. The first two planned to study engineering and often debated the merits of science with Werner—but it was Honsell who apparently proved most influential.

When Heisenberg later identified the three people who had most affected his intellectual development, he named Honsell second, between Niels Bohr and Carl Friedrich von Weizsäcker.[59] Honsell and Weizsäcker were slightly younger than he, and all three first came into his life during the postwar years. Weizsäcker suggests that all three gave Heisenberg, the pragmatist, a much-needed push toward philosophical reflection about his science and an equally necessary critical perspective on his scientific views. Honsell, apparently a well-read young man, may have provided Heisenberg with an education in the Western philosophical tradition that Werner would not have accepted at that time from a member of the older generation—namely, his father.

Honsell, a group leader in the Third Munich Tribe, is remembered as "a very deep man." Some say he once wrote a book on philosophy but lacked the confidence to publish it. Nor did he display much personal ambition. He was so scholastically advanced for his age that his teachers at the Luitpold-Gymnasium once urged him to skip a grade. He is said to have refused with a remark worthy of a bündischer youngster: "No, I want to enjoy my youth."[60] He later became a district judge.

In keeping with Neupfadfinder ideology, the discussions among these youthful thinkers focused almost exclusively on such otherworldly topics as theology and idealist philosophy. Economics and politics received hardly any attention. Their inquiries ranged from the reality of atoms to the bases of religious belief and the existence of the mystical central order. Heisenberg frequently discussed such questions with the theology students in his group. In 1926, Altmann Werner gave a campfire lecture to the Jungstamm, the youngsters of the tribe, entitled "God and the World." Karl Sonntag recalled a long debate—naive but spirited—on the difference between the concepts *unendlich* (unending) and *endlos* (endless) and another on the abstruse subject of the numbers 3 and 4 as dimensional "forms of the space-time world."[61]

Heisenberg clearly treasured his relationships with his young friends and the many journeys he made either with them or on his own. Once he started his university studies, these activities provided a necessary respite from intensive research, enabling him to return refreshed to another round of work. Indeed, the almost incredible intensity of his work during the early 1920s was possible only because he could relax completely during these outings. He had few friends or even acquaintances outside his youth-movement circle.

The deeper impact of the youth movement—as emotional, intellectual, and political phenomenon—on Heisenberg as both physicist and adult can be seen in a variety of ways and on a variety of levels. The intensive questioning and rediscovery of traditional values naturally included the value of science, physics in particular. Some blamed science for the "mechanistic materialism" of the age. "In addition to many other values, we also discovered science anew," he wrote.[62] In a little-known article entitled "Old Values in New Forms," Heisenberg notes that the youth movement led him to search for something new, to try "new ways of music cultivation," for example. "Even in science our interests concentrated on those fields in which it was not simply a question of the further development of what is already known."[63] In the end, the search reinforced his own intellectual interests and professional ambitions.

Heisenberg soon realized, thanks in part to his fellow physics student Wolfgang Pauli, that he could achieve something new and creative in atomic physics. Classical music, with all its beauties, and even other fields of physics seemed less promising areas in which to invest his talents, despite the urging of his youth-movement friends. In *Physics and Beyond*, Heisenberg recalls a long discussion with Rolf, Walter, and Walter's mother about his decision, soon after entering the university, to make a career of atomic physics rather than classical music. As Heisenberg tells it, all three strongly disapproved, Walter's mother giving as her reason typical youth-movement hyperbole: "The future of the world will be decided by you young people. If youth chooses beauty, then there will be more beauty; if it chooses utility, then there will be more useful things." Werner remained unmoved. The newness of atomic physics was too tempting: "Here, I believe, we are on *terra incognita*, and it will probably take several generations of physicists to find the definitive answers. It seems to me very tempting to play some part in all this."[64] Werner had, of course, already discovered that he had an exceptional talent for this work. Still, his early efforts reveal a taste for risk taking and a pragmatic toughness that were probably as much products of his youth-movement training as they were fruits of his innate intellectual self-confidence and ambition.

Although his early science was free of direct philosophical influence, especially the influence of Platonic idealism, he lived immersed in the romantic, otherworldly notions of the Neupfadfinder. Despite his choice of what seemed utility over beauty, Heisenberg was not unaffected by Neupfadfinder antiscientific romanticism. For instance, in his middle years he tried to link modern physics to Goethe's poetic world view and, later in life, to Platonism. During his last years, he seems to have supported efforts by some thinkers to connect contemporary quantum physics, including his own contributions, to certain irrational elements in Taoist philosophy.[65] The English translation of his memoirs, *Physics and Beyond*, appeared in 1971, with Heisenberg's consent, as Volume 42 of a series of mystical and religious monographs entitled *World Perspectives*. Heisenberg's name appears on the board of editors of this series. The general editor's epilogue on the series could have come straight from the antiscientific Neupfadfinder: "It is the thesis of *World Perspectives* that man is in the process of developing a new consciousness which, in spite of his apparent spiritual and moral captivity, can eventually lift the human race above and beyond the fear, ignorance, and isolation which beset it today. . . . *World Perspectives* endeavors to show that the conception of wholeness, unity, organism is a higher and more concrete conception than that of matter and energy."[66]

What are we to make of the obvious political dimensions of the Neupfadfinder? Did all the talk of a Führer and a coming third Reich have any impact on Werner and his group when Führer Hitler arrived proclaiming a Third Reich encompassing a racially conceived Gemeinschaft? Such questions are extremely difficult to answer. Nevertheless, looking to the future, the seeds of Heisenberg's reactions to the first years of Hitler's Reich already seem to have been planted within the formative environment of the youth movement. Heisenberg had already come to regard his science as above and beyond politics, but he had also come to see himself as party to a less transcendent special relationship with his young charges, for whom he had accepted full, all-encompassing, lifelong responsibility. In 1933, the established, successful, and famous physicist would feel the same way toward his students and his younger colleagues — even though the situation and his responsibilities as a German professor were now quite different from those of a youth-movement leader.

For Heisenberg, physics and physicists were meant to exist above the mere "money-business" of political intrigue, but, as a good Pathfinder leader should, he entered into that world of intrigue for the sake of those for whom he felt responsible. Perhaps even by gathering a small Gemeinschaft of his closest students and colleagues around him, he could rescue them from the outside world while awaiting a brighter world to come. Thus arises in part a divergence of perceptions that is usually encountered when examining that period. To Heisenberg himself his actions and motives largely coincided with the lofty ideals of duty and responsibility that he, as an exemplary Neupfadfinder leader, had internalized during his fundamental formative experiences in the youth movement; to others, viewing his responses to the Third Reich, he was so infused with the Volkish politics of the Neupfadfinder that he capitulated all too easily to a brutal, antiscientific, antihuman dictatorial regime with which he sympathized.

Heisenberg, like many other German intellectuals, did indeed sympathize with what he perceived to be the nationalist ideals of Hitler's Reich, but the realities of Hitler and his Reich fell far short of what he and the Neupfadfinder had envisioned. This became clear even during

Hitler's rise to power, as the Nazi movement rejected elitism and entered into mass party politics. As Karl Seidelmann told the Jungstamm in 1931, nationalism itself was admirable, but Nazi nationalism was not. "On the contrary, the Hitler party is a shocking example of what bad leadership can make of a good thing."[67]

For the Neupfadfinder, the coming third Reich was to be the culmination of centuries of German history, the final realization of the ideals of the first Reich, the Holy Roman Empire. Numerous petty princes and political parties would happily coexist within one apolitical empire, ruled by a single trustworthy, God-appointed Führer. He would ensure the peace and well-being of the German people — especially, of course, of the cultured upper middle class — in the same way a group Führer did for his small Gemeinschaft.

Such romanticism cast a spell on the Neupfadfinder, many of whose members had little trouble joining Nazi youth organizations when non-Nazi groups were banned in 1933. But among the members of Gruppe Heisenberg only one — or so it was remembered — joined the Nazi party.[68] Instead, it was the idealistic theology of their imagined third Reich that seems to have exerted the most immediate impact on Gruppe Heisenberg. As they conceived it, the coming third Reich bore a striking resemblance to the Christian concept of the coming kingdom of God, where all Christian believers will live together in peace and harmony under one God-given savior. Such ideals and the otherworldly orientation of their entire activities may account for the strong inclination toward Christian theology in Gruppe Heisenberg. Most of its younger members studied theology and later joined the clergy — both Rüdels became Lutheran ministers, Karl-Heinz Becker a Lutheran theologian, Gottfried Simmerding a Catholic priest, and Otto Heimeram a monk. The rest pursued other unworldly academic or scientific careers, save for Heini Marwede, the tribe treasurer. Much to Heisenberg's displeasure, Marwede went into banking.

If the transcendence of science and the Reich went hand in hand, then perhaps, when the bad leadership that corrupted the ideals of the coming Reich could be replaced, those ideals would flourish, bringing in the new moral age. Perhaps one needed only patience, one needed only to endure for a short while until a new leader of deeper vision arose. These were hopes that Heisenberg cherished in 1933, and they were hopes that he had apparently forged over a decade earlier during what he always remembered as "the most beautiful days of my life."[69]

PART TWO

Quantum Mechanics

CHAPTER 6

Sommerfeld's Seminar

Heisenberg entered the University of Munich in the fall of 1920. Between the start of his studies in Munich and his appointment as professor of theoretical physics in Leipzig seven years later, quantum physics underwent a profound transformation, from what Max Born called disorder in 1920 to the orderly system that Born and Heisenberg pronounced complete in 1927.[1]

While still a student and junior lecturer, Heisenberg was a prominent member of the small cast of talented physicists that performed this transformation. His role arose from the felicities of location, colleagues, and timing, in addition to his own extraordinary abilities. During those seven years, Heisenberg studied and lectured at three major quantum research centers — Munich, Göttingen, and Copenhagen — under three leading quantum theorists of the day: Arnold Sommerfeld, Max Born, and Niels Bohr. He also studied and worked with some of the brightest young physicists of the day: Pauli, Jordan, Dirac, Wentzel, Kramers.

The profound inadequacy of quantum theory in resolving the riddles of the atom was just coming to light in Munich when Heisenberg began his studies. Through his teachers and colleagues, Heisenberg became acquainted with particular problems and with various attempts to resolve them. This groundwork was vital in facilitating his own contributions over the coming years.

Those contributions were fundamental. They included the pivotal breakthrough to quantum mechanics in 1925, participation in the development of the complete quantum mechanics apparatus from 1925 to 1927, and perhaps Heisenberg's most famous achievement of all — his formulation of the principle of uncertainty, or indeterminacy, in 1927, Heisenberg's contribution to the "Copenhagen interpretation" of quantum mechanics. Although it remained controversial, quantum mechanics and the Copenhagen interpretation of its formalism completed the quantum revolution first ignited by Planck, Einstein, Ehrenfest, Bohr, and others during the first decade and a half of the century. It has formed the basis of research on the atomic scale to this day.

By the turn of the twentieth century, understanding of the workings of the physical world had reached a culmination in the classical mechanics of motion and force, derived from the work of Newton and his successors, and in the classical electrodynamics of electricity, magnetism, and light, based on the contributions of Maxwell, Hertz, and Lorentz at the end of the nineteenth century. In the same year that Einstein offered a revision of classical electrodynamics in his special theory of relativity (1905), he offered a profound and not readily accepted challenge to the classical electromagnetic wave theory of light with his hypothesis of "light quanta" — tiny, discrete bundles of light energy, each behaving like an independent particle and carrying identical amounts of energy.

Atoms, meanwhile, attracted increasing attention and — thanks again to Einstein — soon gained acceptance as real entities. Efforts to comprehend the structure of atoms and to account for their known properties led in 1913 to a quantum theory of the atom propounded by Bohr and extended thereafter by Bohr, Sommerfeld, and others. Bohr's model combined quantum elements, such as light quanta and discrete energies, with classical mechanics and electrodynamics. Although the resulting Bohr-Sommerfeld quantum theory of the atom enabled accounts of an astonishing array of atomic phenomena, the theory itself remained for many theorists a puzzling combination of classical and quantum notions that only somewhat seemed to work.

After World War I, improved experimental techniques and more sophisticated quantitative analyses began to illuminate more and more areas in which the theory seemed to work less and less satisfactorily. The limitations of the Bohr-Sommerfeld quantum theory of the atom and possible ways to overcome them or to revise the theory engaged quantum physicists throughout the early twenties. Heisenberg fully participated in these analyses and, after 1923, in the efforts to replace the by then inadequate quantum theory with a new quantum mechanics. The new mechanics and its Copenhagen interpretation, achieved by the end of 1927, were combined with other innovations — such as electron spin and the exclusion principle — to enable at last a consistent, self-contained theory of processes on the atomic scale. In opening up the realm of the atom, quantum mechanics enabled entirely new and profound advances in understanding all aspects of the physical world, from nuclei and quarks to the big bang—which had profound implications for the world in which we live, from philosophy to the technology of nuclear reactors, atomic bombs, semiconductors, and superconductivity. Heisenberg played a leading role in many of these developments from the very moment in the fall of 1920 that, as an 18-year-old student, he entered the University of Munich.

Heisenberg's work and studies continued to parallel the upheavals of Germany's social unrest. The military defeat in World War I and the collapse of the monarchy galvanized many Germans, especially young people. The Versailles peace treaty, imposed on Germany at the end of the war, blamed the defeated nation for the war and exacted oppressive reparations to the victors. At the same time, the new Weimar constitution imposed democracy on the already traumatized nation—all of which awakened violent nationalistic and antirepublican sentiments among many Germans, some of whom were now violently anti-Semitic as well. According to one count, 376 political murders were committed in Germany between 1919 and 1922, 354 of which were carried out by right-wing assassins. Between 1919 and 1924, 12 different federal cabinets governed in Berlin, of which only one came to power through normal electoral processes.[2] In the midst of such upheaval, inflation wracked the German economy, reaching astronomical proportions just as Heisenberg received his doctorate in 1923.

Politics in and around the University of Munich tilted precariously to the right during the early 1920s. The trauma of the Munich soviet experience led to the creation of secret civilian armies in Bavaria to resist what they regarded as "Bolshevism." The authorities placed the city under a permanent state of emergency to suppress every left-wing tendency. Right-wing extremism flourished unchecked in Munich, especially within the "nonpartisan" secret armies. It burst into Bavarian politics during the Kapp putsch. When Captain Ehrhardt's free corps brigade, which had earlier helped to suppress the Munich soviet, was ordered to disband as part of the German demobilization, his troops, now in Berlin, joined with other units and seized control of the government. A communist revolt erupted in the Ruhr in response, prompting the dispatch of Bavarian volunteers—among them Werner's brother, Erwin—to suppress it. Gustav Kahr, president of Upper Bavaria, took over the Bavarian government at the insistence of the secret armies. Hoffmann and his Social Democratic regime, whom the victors over the Munich soviet had restored to power in 1919, withdrew from office, never to return to power in Bavaria. Hoffmann's successor, Kahr, instituted an even more ruthless suppression of political opposition, preparing to declare Bavarian independence at his earliest opportunity.

When workers rallied to the socialist regime in Berlin, ending the putsch, many of the putschists headed for the fertile fields of separatist Bavaria, a magnet for right-wing fanatics throughout the Reich. When not soldiering, many of the zealots were university students, if only in name. The University of Munich quickly became a stronghold of right-wing extremism.[3]

The university administration openly encouraged its fanatical students. During the height of their power, the soviet leaders had aggressively attempted to subject the elite institution to proletarian rule. Armed student revolutionaries had seized control of the university, taken the rector hostage, and set up a revolutionary council to dictate university policy. The university senate promptly closed the institution and sent the faculty into hiding.[4] When the restored Hoffmann regime proclaimed Bavaria a social democratic republic in 1919, the move was denounced almost universally by the monarchical faculty. When in 1920 the courts sentenced Graf Arco to death for assassinating Eisner, the university rector himself joined his students in a demand of clemency for Arco that did not stop short of threatening violence. The threat succeeded. Max Weber, then a professor of sociology, attempted to defend a socialist student who opposed the demand, only to find his lectures constantly disrupted.[5]

The dire political and economic situation, combined with horrendous overcrowding at the university, contributed immensely to student fanaticism during the early 1920s. In the fall of 1920, when Werner entered the University of Munich, it had a student population of 6879, 62 percent of whom were Bavarian. A year later, 9659 students were enrolled, only 52 percent of whom were Bavarian.[6] Many students came from middle-class families who were among those hardest hit by inflation. During Werner's first semester, in the winter of 1920–1921, a counselor discovered that an astonishing 61 percent of the students existed on a monthly income that was below the minimum cost of living for a Munich student.[7] The lowest paid unskilled worker at the university earned more than the minimal living costs of a student—a circumstance that further embittered the already resentful students.

Although Heisenberg suffered a certain amount of economic difficulty, he was far better off than most. He was one of a small number (2.3 percent) of male Bavarian students who lived at home and avoided boarding expenses. His father's income, moreover, was not seriously threatened by inflation until 1923. Professor Heisenberg's salary was near the top of the civil-servant pay scale (step 12 of 13). It nearly tripled between July 1920 and October 1921, while the mark decreased in value relative to the U.S. dollar by only slightly more than that amount.[8] Even without the added income of seminar fees, the professor's salary in 1922 was still 4000 marks above the total expenses of an average five-member professorial family in the more expensive city of Berlin.[9]

With his elder son studying in Berlin between military stints, the professor found it necessary to put his younger son on a budget. In October 1921, August paid Werner's way to Jena so that he could attend his first physics conference, a conference at which he briefly met some of the great names in his profession for the first time—among them, Max Planck and Max von Laue, but, to Werner's dismay, not Einstein, who could not attend. During the conference, Werner decided to impose further on his father's finances and travel by train to Berlin to visit the Berlin contingent of his youth movement.[10] His parents maintained that they could not afford to give him more. On Werner's vehement insistence, his parents finally withdrew a sum from a gold-dollar account set up by Werner's rich American Uncle Karl to help his German relatives in emergencies. Independent minded and little concerned with parental dilemmas, Werner longed even more keenly for the day when he would not have to rely on family generosity.[11]

Weimar scientists, especially atomic scientists, reacted to the worsening economic conditions by developing increasingly innovative strategies to obtain research support for themselves and their students. However, in the political sphere, these same scientists, continuing to eschew politics, reacted to the upheavals of the era by becoming more insular. In keeping with the blame cast on Germany for unleashing the world war, international boycotts of German science followed the war. German scientists withdrew even further from political affairs, stubbornly refusing to allow outside events and pressures to contaminate them or their science. Richard Willstätter, a Munich professor of chemistry, deftly dodged machine-gun bullets on his way home for lunch during the battle of Munich—he refused to let his soup get cold.[12] During the Kapp putsch, Max Born encountered a heavy street battle in Frankfurt; yet, he recalled, "After things had settled a little we went about as if everything were normal."[13]

Heisenberg's future Munich mentor, Arnold Sommerfeld, managed to complete two of his most significant scientific papers in the midst of this turmoil. He submitted the Sommerfeld-Kossel ring model of the atom at the height of the soviet republic in April 1919; his spectroscopic magneto-optical splitting rule paper, the inspiration for Heisenberg's early work on spectroscopy, arrived at the publisher within four days of the Kapp putsch.

But even the ivory tower of Sommerfeld's institute could not protect its members from a collision with the university's anti-Semites. The issue, which arose during Heisenberg's second year of studies, was Einstein. Sommerfeld's students learned of the episode from Sommerfeld's correspondence with Einstein, which Sommerfeld often read aloud to his seminar.[14] Einstein, well known in Germany as a pacifist, openly supported the Weimar democracy and worked tirelessly in the cause of international understanding and accord. Since his newly confirmed theory of relativity seemed so profound yet so incomprehensible to most lay people, many saw in him a symbol of their own incomprehension of recent events and in the Jews a scapegoat for Germany's troubled situation. Outspoken, captivating, and Jewish, Einstein made an ideal target for both anti-Semitic and antiscientific hate. In 1920, Einstein's opponents unleashed an anti-Semitic campaign in Berlin against the man and his theory.[15] They were supported by several prominent experimentalists, including the Nobel laureate Philipp Lenard. Although the sources of their hatred were myriad, Lenard and his colleagues were especially furious at Einstein's sudden popularity and prestige as a theoretical physicist, which threatened to overshadow them and their field of experimental physics.[16]

During the nineteenth century, German physics and German physicists had established their power and prestige mainly in the field of experimental physics, the gathering and analysis of data. Experimental work also required mathematical methods and the framing of general hypotheses, but by the early years of the twentieth century a new professional discipline had emerged, especially in Germany, that focused in a new way on hypothesis, mathematical analysis, and empirically informed theories of natural phenomena — theoretical physics.[17] By the end of World War I, the startling results and successes of professional theorists such as Planck, Laue, and Einstein provided the new field with enormous popular appeal and prestige in Germany, but it still held a secondary professional status behind the established and supposedly more well-grounded discipline of experimental research. Many still held empirical data to be more fundamental than mathematically construed theories. Because of the lower status accorded theoretical work, Jewish physicists found more opportunities in theoretical physics, but — as everywhere else in German society — they also encountered anti-Semitism, both before and after World War I.[18]

Moderate Weimar physicists usually considered opposition to anti-Semitism to be not a moral or an ethical issue but primarily a political one. Such a view apparently arose from the politicization of anti-Semitism in Germany. Anti-Semitism had already become a plank in the platforms of several major political parties, and those who engaged in anti-Semitism often did so for obvious political ends. Because of this, scientists and academics regarded both blatant anti-Semitism and overt opposition to it as too political and therefore to be avoided. While determined opposition therefore failed to develop, implicit anti-Semitism flourished in German academe, at times infecting even Sommerfeld's institute.[19]

Einstein was a special case. While German repute abroad suffered because of the war, Einstein's international fame reflected favorably on the foreign image of German physics. The grotesque anti-Einstein campaign in Berlin threatened to tarnish that image. Sommerfeld and others, while carefully avoiding political involvement, realized that something had to be done. In 1920, Sommerfeld and several Munich professors formed an Einstein support committee. As their first order of business, they invited him to lecture in Munich. Sommerfeld obtained financing from a sympathetic philanthropist, and Einstein readily accepted, intending to arrive in Munich in November 1921. Heisenberg and his fellow student Pauli, disappointed at having missed Einstein in Jena, eagerly anticipated an encounter with the great man in Munich.[20]

The plan unraveled shortly before Einstein's appearance. An article in a leftist Berlin literary magazine reported an ominous meeting held nearly a year earlier between Sommerfeld and representatives of the Munich student government in the rector's office. Noting the disruptions of Einstein's lectures in Berlin, the rector — now suddenly worried about the reputation of his school — demanded assurances that no such disruptions would occur in Munich. The students responded with objections to Einstein's "person," prompting Sommerfeld to lecture them on the physicist's significance — but without success. During the next student government meeting, extremist representatives, especially members of the "Swastika Majority," refused any assurance against disruption. The article reporting these events appeared just before Einstein's arrival in Munich. Einstein immediately canceled his appearance.[21] Sommerfeld pleaded with his colleague to reconsider. His colleague absolutely refused. "There is just no other way," he told the non-Jewish Sommerfeld. "That you must feel yourself."[22] Heisenberg would have to wait three more years before meeting Germany's foremost theorist.

Sommerfeld and his students stubbornly maintained their insulation, both before and after the Einstein affair. These were talented and intense young scientists consumed by the demanding intricacies of their discipline. It was easy for them to relegate social issues to second place — and they welcomed the opportunity to do so. The arduous work and tantalizing promise of scientific research served as a convenient antidote to social upheaval — as it has for other scientists in similar situations throughout history.[23] Pauli, for instance, entered the university in the winter of 1918–1919, just before the end of the war. The years immediately following witnessed the soviet republic, civil wars in Munich and Vienna, the taking over and closing of the university, rampant inflation, and brutal violence. Yet he notes none of these events in his available correspondence nor in any of his published recollections. Years later, he wrote: "The war was over, with Sommerfeld I was in my right element. What then were the political and economic situations in Germany and Austria to me as a young man?"[24] Insular concentration fostered successful work, which in turn encouraged further insulation.

Heisenberg found his own escapes. "My first two years at Munich University were spent in two quite different worlds: among my friends of the youth movement and in the abstract realm of theoretical physics," he wrote. "Both worlds were so filled with intense activity that I was often in a state of great agitation, the more so as I found it rather difficult to shuttle between the two."[25] Heisenberg literally did shuttle between the two. During the warmer months of the summer semester, he usually camped out with his boys in the mountains at night, then hiked to the nearest train station early the next morning, arriving in Munich in time for Sommerfeld's 9:00 A.M. lecture. Constantly moving between the worlds of physics and youth conveniently left him little time for anything else.

When he left the Max-Gymnasium in the summer of 1920, Heisenberg still planned to study pure mathematics. Fresh from his brilliant Abitur, the ambitious young man intended to launch immediately into an advanced research seminar leading to a doctoral degree. Having passed the Abitur, students were automatically admitted to the lectures and exercise sessions of the German university of their choice. Most students attended local universities, trying for either a doctorate or a Diplom, the equivalent of an American master's degree (not available then in Munich). To obtain a degree, one had to be accepted by a professor into his seminar. There the student learned the fundamentals of research while working on an independent project. Rather than requiring the general education courses, examinations, and semester grades of American schools, the German university quickly introduced its students to focused independent research through early study with working specialists. Once in a seminar, the student completed a thesis project under the direction of a professor, who solicited faculty members in related fields for their written opinion on the result. Final approval of the thesis and a grade for the entire study were conferred at the final oral examination. The new graduate was now qualified to teach at a gymnasium. A university teaching career, however, required an even higher degree: the habilitation, or qualification, which entailed additional research, oral examination, and approval of the faculty.

Shortly after recovering from his bout with typhoid in the summer of 1920, Heisenberg had his father arrange an appointment for him with the mathematician Ferdinand von Lindemann, a colleague of the elder Heisenberg. Lindemann seemed an ideal candidate for the role of Werner's "doctor father," as advisors are called in German. He was well known in Werner's intended field of number theory for his proof of the transcendence of π and was codirector of the university's mathematical-physical seminar, composed of four professors and one assistant. It was not a seminar in the sense of a study group but rather an administrative unit, somewhat like an American department. The seminar was designed to train future gymnasium teachers of mathematics and physics in the fundamentals of their field and in basic research.[26] Presumably, a good researcher would make a good teacher. In 1920, Willy Wien, the newly arrived professor of experimental physics, codirected the seminar with Lindemann. Their two colleagues were Aurel Voss, professor of mathematics, and—in fourth place in the pecking order—the professor of theoretical physics, Arnold Sommerfeld.[27]

As a favor to August Heisenberg, Lindemann agreed to meet with Werner—but only as a favor. The old gentleman, a longtime chairman of the university's administrative committee, was two years from retirement. He had little patience with first-year students who intruded on him in his office—and none at all with audacious novices who demanded immediate admission to advanced research. The interview ended in disaster.

As Heisenberg recalled it later, Lindemann received him in a dimly lit office, seated behind a desk on which his pet poodle perched. When Werner began to speak, the poodle barked so loudly that the partially deaf professor could barely understand him. Finally, Lindemann asked his young visitor which textbooks he had studied. After mentioning Bachmann's *Zahlentheorie*, Heisenberg volunteered that he had just finished Weyl's *Raum-Zeit-Materie*. Lindemann, looking for an excuse and perhaps unsympathetic to Weyl's contamination of pure mathematics with physics, abruptly closed the interview with the remark: "In that case you are completely lost to mathematics."[28]

Stunned by his first rejection, the 18-year-old returned to his father to seek alternatives. They considered the three remaining seminar professors. Wien, now an experimentalist, would not do, and between Sommerfeld and Voss, the former was the more likely choice, since he and August were already well acquainted. The slightly built, balding, broadly moustached Sommerfeld, who always stood so erect that he looked, in Pauli's words, like a Hussar officer, had served as dean of the science faculty during the previous summer semester. During the coming year, he would serve as senator from his faculty.[29] These duties had already brought him into frequent contact with Professor Heisenberg, the university representative to the German Hochschullehrerbund (College Teachers Association). Father and son decided to try Senator Sommerfeld.

The physicist proved much more sympathetic than his elderly colleague. His office was well lit and devoid of poodles, and its less imperious occupant gladly received eager students of all levels. Unlike Lindemann, he was elated—and amazed—to learn that Heisenberg had read Weyl. "You are much too demanding," he told his visitor, with good reason. Obviously impressed, the perceptive Sommerfeld admitted Werner provisionally to his research seminar. "It may be that you know something; it may be that you know nothing. We shall see."[30] Heisenberg was on his way into theoretical physics.

Sommerfeld's approach to his science and his relations with his colleagues typified the state of German theoretical physics in the early years of the quantum revolution. Like most theorists of his generation (he was then 52), Sommerfeld began his career in mathematics. Coincidentally, he was born in Königsberg, East Prussia, the site of the first mathematical-physical seminar in Germany, which originated the Central European branch of theoretical physics. Sommerfeld had attended the local gymnasium with the cousins Willy and Max Wien, both of whom became physicists. In 1886, Sommerfeld began studying mathematics at his local university, attending the mathematical-physical seminar, directed by professor of mathematics Ferdinand von Lindemann. But, like many other mathematicians, he became intrigued with the mathematical physics of William Thomson (Lord Kelvin), as outlined in his attempted mechanical comprehension of Maxwell's electromagnetic field equations. Sommerfeld promptly switched from Lindemann's number theory to Kelvin's mathematical physics—the study of mathematical applications to physics — and wrote his doctoral dissertation on the subject under Paul Volkmann, professor of mathematical physics in Königsberg.[31] Doubtless Sommerfeld saw something of himself in Heisenberg.

Still a mathematician, in 1893 Sommerfeld headed for Göttingen, the capital of German mathematics. There he fell under the influence of the famous mathematician Felix Klein, a superb teacher and administrator who at the time pursued a program for mathematizing science and establishing institutes for applied mathematics.[32] Thirteen years later Sommerfeld began teaching theoretical physics in Munich.

Physics research in Munich derived from the university's instrument collection, its "physical cabinet," consisting of experimental apparatus and the professor who used it. In 1892 the cabinet moved into the new Physics Institute, which was headed until 1920 by Wilhelm Röntgen, the discoverer of X rays, who had previously taught at the University of Würzburg. Röntgen's successor at Würzburg was Willy Wien, Sommerfeld's old school chum. Wien again succeeded Röntgen at Munich, just as Heisenberg entered the university.

Professor Wien was proof of the respect experimental physics commanded in Munich. The Nobel prize-winning Wien, then 55 years old, would not leave Würzburg unless granted special concessions. He got everything he demanded, despite the grim economics of the day. He received a fat salary, four assistants, three technicians, and six-figure grants to expand and retool the institute.[33] By contrast, Sommerfeld's institute consisted of a lecture hall, three rooms, a modest laboratory, one assistant, and one technician. It was located on the ground floor and basement level of the university building, two floors directly beneath August Heisenberg's office. Aside from seminar fees, the institute received the modest sum of 2000 marks per year to purchase apparatus and to maintain a small library.[34]

Although Sommerfeld's chair and quarters were located in the university, they were administered as the state's scientific instrument collection, its mathematical-physical cabinet. Hence there were two professors of physics, two independent experimental laboratories, and two very different schools of thought as to how physics should be defined and taught. The university's physics professor handled experimental physics; the conservator of the state's cabinet pursued mathematical physics with a parallel university appointment. Boltzmann had occupied the position until 1894. In 1905, Röntgen, interested in electron theory, appointed Sommerfeld to the post, over the strenuous objections of Sommerfeld's former mentor, Ferdinand von Lindemann.[35]

On arriving at the university in 1906, Sommerfeld divested the cabinet of most of its outmoded apparatus and gave it a new name reflecting its new primary focus: Institute for Theoretical Physics. It quickly became a leading center of research in the new relativity and quantum theories. Sommerfeld was reportedly the first professor in the world to lecture regularly on both subjects, and he enjoyed world renown as one of the best and most stimulating teachers of the era. His institute produced a steady stream of first-rate theorists—the largest number of doctorates in the field until the 1930s. Einstein was amazed at its fruitfulness and, prompted by a report on Heisenberg, wrote to Sommerfeld in 1922: "What especially impresses me about you is that you have produced so much young talent, like stamping them out of the ground. That is something entirely unique. You must be able to activate and to cultivate the minds of your pupils."[36]

What particularly distinguished Sommerfeld as a teacher and researcher was not so much the brilliance of his physical insight but rather, as Born put it, his "logical and mathematical penetration of established or problematic theories and the derivation of consequences that might lead to their confirmation or rejection."[37] Sommerfeld combined this talent with an inspiring teaching style and a gradual selection process that served to weed out the weaker pupils. The institute's Munich location helped to ensure a steady supply of talent for the program.

As the state's conservator of apparatus, Sommerfeld was still obliged to allow at least some experimental research—work that he relegated to assistants whom he banished to the dark basement. Despite their lowly status, theorists Max von Laue and the outcasts made at least one major discovery: proof that X rays exhibit electromagnetic wave behavior. Laue received the 1914 Nobel prize for his discovery, and Sommerfeld's institute received generous grants to continue the research—thereafter in broad daylight.

By the time Heisenberg joined the institute, theory once again eclipsed experiment. Sommerfeld focused his theoretical interests on two topics: hydrodynamics and quantum spectroscopy. His interest in hydrodynamics arose with the financial support of the Isar Company in Munich, which had been contracted to channel the Isar River. Work on quantum spectroscopy grew out of Sommerfeld's concern with the modification of his quantum atomic model in the light of new and puzzling data. To Heisenberg's extraordinarily good fortune, Sommerfeld's institute was unique. It was one of only a handful of institutes for theoretical physics in Germany and one of only two or three that performed research on quantum theory. Moreover, it was the only one at that time concerned with theoretical quantum spectroscopy.

Heisenberg was also fortunate in the timing of his entry into Sommerfeld's teaching program. Like mathematics and other sciences, physics was part of the philosophical faculty, which required of doctoral candidates a minimum of only six semesters (three years) of study. Sommerfeld accordingly arranged the topics of his main lecture in a six-semester cycle, starting with classical mechanics. If a student entered in midcycle, he could either learn the material out of sequence or spend his first semesters on mathematics and required minors while waiting for the cycle to begin again. Heisenberg entered the program just at the start of a cycle, in the winter semester of 1920–1921. Sommerfeld had spent the previous year teaching a tiresome series of make-up semesters for war veterans and free corps volunteers.[38]

Sommerfeld designed his teaching program to satisfy a variety of needs. While Heisenberg studied under him, he offered five main lectures (one each semester for four hours a week) covering nearly all of classical theoretical physics. For advanced students, he taught contemporary nonclassical physics in a special lecture on current research. He also conducted the research seminar for doctoral candidates and gave an occasional public lecture on modern theories to raise money for the institute. The main lectures were attended by as many as 80 to 100 students from a variety of scientific fields. Students of chemistry and medicine who attended Sommerfeld's lectures on atomic models in 1916–1917 encouraged him to write his famous textbook *Atombau und Spektrallinien* (*Atomic Structure and Spectral Lines*), which was for a generation the "bible of the modern physicist."[39]

At each lecture, Sommerfeld assigned homework problems to be turned in during the weekly one-hour Übung (exercise) sessions. An assistant corrected the problems and discussed them with the students during the exercise, which Sommerfeld himself often attended. No grades were given; a student's work spoke for itself. Heisenberg recalled turning in such long and complicated solutions that Sommerfeld's assistant complained.[40]

The assistant was probably Peter Paul Ewald, who held the post when Heisenberg arrived. Ewald remembered Sommerfeld as a "true doctor father."[41] He took a personal interest in his charges, treated them with dignity, and gave them sympathetic fatherly counsel. He set an example for them as a hard-working, intensely active researcher — yet he was always accessible. Heisenberg was often in Sommerfeld's office for an hour or two each morning during his last semesters. On Sundays, Sommerfeld would invite institute members to accompany him on daylong outings in the countryside. Winter weekends were often spent with other physicists skiing at Willy Wien's country cottage in Mittenwald near the Austrian border. When students felt the pinch of economic inflation, Sommerfeld dipped into his own pocket to help them out. Werner, too, benefited from Sommerfeld's generosity, further increasing his admiration for his mentor.[42]

The stimulation and selection process began early in Sommerfeld's institute. It was his strategy to involve students at once with research and institute affairs, both to encourage and to test them. The professor, as Werner called him, gave his beginning pupils minor tasks, such as checking his calculations, analyzing newly received data, and correcting galley proofs of articles. Advanced students assisted with revisions of *Atombau* or with articles for Klein and Sommerfeld's multivolume *Encyclopedia of Mathematical Sciences*. It was in this work that Pauli's famous article on relativity theory, still considered one of the best summaries of the subject, first appeared.[43] Heisenberg recalled that Sommerfeld would often motivate a bright pupil by handing him a small problem with the remark, "Well, I can't solve this problem; now you try it." Based on performance of these tasks and in the exercise sessions, Sommerfeld assessed his pupils' suitability for admission to advanced training.

Sommerfeld offered his two-hour special lecture each semester on a topic that he was currently researching but had not yet fully grasped. When once asked how he could lecture on a subject he did not understand, Sommerfeld replied: "If I knew something about it, I wouldn't lecture on it!"[44] The object was to enable pupils and teacher to grapple with a current problem together and, in the process of searching for a solution (successfully or not), to arrive at a systematic comprehension of the subject. The communal effort made these sessions particularly stimulating. Sommerfeld prepared the special lecture in advance, but he usually tried to rederive the results at the chalkboard without referring to his notes. One can imagine the animated discussions that must have occurred when a derivation didn't work out. Throughout Heisenberg's studies in Munich, Sommerfeld devoted the special lecture each semester to the major atomic physics problem of the day: quantum spectroscopy. Young Werner was captivated.

Advanced study at the institute revolved around the research seminar, which was attended by all advanced students, assistants, lecturers, and the occasional precocious beginner. Heisenberg was one such beginner; before him, Pauli had also attended during his first semester. Both managed to survive the weeding out that the course entailed. Sommerfeld devoted each semester's seminar to a current field of research. Each attendee was given a small problem to solve or a large article to study, and the results were presented to the seminar for critical review. A successful performance was required for permission to write a dissertation on the subject. Heisenberg obviously did well, for his dissertation and several of his first papers grew out of his early seminar projects.

Sommerfeld put his precocious new pupil to the test from the very start. Within a few weeks of Heisenberg's arrival, Sommerfeld requested that he analyze some new data on the anomalous Zeeman effect in spectroscopy, and that he report to the seminar on a recent publication on quantum theory by Niels Bohr's assistant, H. A. Kramers. Both of Heisenberg's reports provoked long controversies at the institute. But in the quality of his work, Heisenberg proved himself worthy of the master's tutelage.

Heisenberg's first-semester registration form indicates that, despite his audacity and ambition in immediately entering Sommerfeld's seminar, his father had probably advised restraint: Werner had prudently protected himself in the event of failure by signing up for five hours of mathematics lectures and exercises conducted by Artur Rosenthal but for only one hour of theoretical physics — the exercise session following the main lecture. This meant that Heisenberg was in fact a guest auditor in Sommerfeld's seminar and main lecture and could withdraw promptly into mathematics should he prove unsuitable for physics. By the second semester, such caution was no longer necessary. Sommerfeld had admitted him without reservation to the program, and Heisenberg filled in his next registration form with all of Sommerfeld's offerings.[45]

Since students in the mathematical-physical seminar were required to take Wien's course in experimental physics, Heisenberg, like Pauli before him, also registered for the five-hour lectures in experimental physics (mechanics and optics) his first semester.[46] As a second-semester physics student, he registered, as required, for Wien's tortuous eight-hour beginner's practicum, or laboratory. Heisenberg continued to study mathematics with the aged Rosenthal and his colleagues Pringsheim and Voss — but he avoided Lindemann. Mathematics and astronomy were his two minor subjects, and in each he was expected to register for lectures, exercises, and one seminar. Already rejected by Lindemann, Heisenberg soon discovered that he had lost interest in Lindemann's abstract number theory but had gained interest in Rosenthal's "visualizable" geometry. The budding number theorist was ripening into a theoretical physicist.

Heisenberg's decision to study theoretical physics rather than mathematics caused his father no little concern.[47] Public interest in the relativity and quantum theories was certainly strong, as was demonstrated by

the large audiences at popular lectures such as Sommerfeld's. Nevertheless, employment opportunities seemed meager. Mathematics and experimental physics were well-established disciplines that could lead to
any number of jobs in industry and gymnasiums, but professional careers in theoretical physics were still restricted to university chairs, all of
which were already occupied.[48] Although academic positions would
increase during the next decade, Professor Heisenberg knew that his son
would have to do extremely well, particularly on the doctoral and
habilitation exams, to obtain a full professorship and thus achieve the
goal the family had set for him. And while Werner's abilities were keen,
other problems loomed. Personal and professional differences were
already arising between Sommerfeld and his new and more powerful
colleague, Wien. Both were required to sit on the doctoral committee
for physics students, and both had to agree on a single physics grade for
each candidate.

One historian has argued that, while German physicists vociferously
asserted their apolitical stance, they nevertheless polarized into professional camps corresponding roughly to their political temperaments.[49]
Such factionalism was rife in every area of professional activity: societies, journals, financial institutions, appointments, and research preferences. Although there were obvious exceptions, conservatives, such as
Wien, were usually experimentalists in senior positions who tended to
be skeptical about the quantum and relativity theories, usually lived
outside Berlin, and frequently contributed to the long-running *Annalen
der Physik (Annals of Physics)*, the leading German physics journal.
Moderates, like Sommerfeld, were mainly theoreticians and were usually
younger than experimentalists. They supported the quantum and relativity theories, either lived in Berlin or were closely allied with Berlin
physicists, and frequently wrote for the newly established *Zeitschrift für
Physik (Journal for Physics)*. Sommerfeld, who had helped to found the
Zeitschrift in 1920, differed from Wien, a longtime editor of the *Annalen*, on nearly every point.

Although Wien did not dismiss theoretical physics — he had, in fact,
once done theory himself — he simply regarded experimental work as
more fundamental. Any doctoral candidate in physics had to convince
Wien of his mastery of experimental techniques. Moreover, Wien insisted on a traditional, rigid program of study, leading gradually to
advanced work. This method was the very opposite of Sommerfeld's
habit of confronting his pupils early with research, while simultaneously
feeding them the fundamentals. By omitting courses outside their major
and minor fields, bright students, such as Pauli and Heisenberg, could

obtain doctorates under Sommerfeld in as few as three years. The rapid schedule might leave gaps in a student's knowledge, but, to Wien's horror, Sommerfeld assumed students could fill them in themselves. Wien soon discovered that Heisenberg's training did in fact leave him with serious gaps in his learning, but not apparently with any regrets. Years later, Heisenberg told a group of young people that, regardless of the many years now required to obtain a doctorate in physics, they should be doing original research by the age of 24.[50]

With such fundamental differences between the two Munich physicists, the final doctoral examinations could easily deteriorate into a pedagogical wrangle. To forestall problems, Sommerfeld ordered his pupils to enroll again in one of Wien's laboratory courses before the final orals. Pauli, who took the course in 1921, apparently did not encounter much difficulty with Wien, but Heisenberg—who suffered the course with ill-concealed scorn—did. His father's anxiety over potential trouble is apparent in a remark he made when Werner visited Göttingen for a semester in 1922, a year before his final orals: "How have Herr Professor Born and the other gentlemen received you? Please don't neglect the experimental physics!"[51]

In addition to his formal training, Heisenberg could credit his rapid advancement under Sommerfeld to the stimulus of an extraordinary group of colleagues and companions. Their names read like a *Who's Who* of their generation. When Heisenberg arrived, the principals included lecturers Karl Herzfeld and Wilhelm Lenz, assistant Peter Paul Ewald, and students Gregor Wentzel, Wolfgang Pauli, and Karl Bechert. Otto Laporte arrived in 1921 from Born's Frankfurt institute to continue his studies with Sommerfeld, and Adolf Kratzer habilitated in 1921 and served as lecturer thereafter. Outside the institute, Heisenberg met Hans Kienle, an assistant at the astronomical observatory who became a close associate, and the mathematician Robert Sauer, a fellow student in Rosenthal's lectures. As had Werner's brother Erwin, Sauer entered into a fierce competition with Heisenberg, and in vying with each other to solve the problems presented, they left the other students far behind.

Of the three nonlaboratory institute rooms, with their creaky wooden floors, high ceilings, and drab interiors, Sommerfeld used one for himself and designated another the seminar room, a forum where the select five to ten advanced students met daily to discuss and debate various problems and papers. Each student had his own desk. When the newly graduated Wentzel replaced Ewald as assistant in 1921, Sommerfeld appointed Pauli, a younger recent graduate, to the post of deputy assistant. Among his duties was the correction of Heisenberg's homework. Indicative of the position he would take in physics, Deputy Assistant Pauli's desk was perched on a small platform, from which he could oversee everyone's work.

On or off his perch, Pauli proved to be the most influential and vocal of the seminar members and especially so for Heisenberg. Young Pauli had come to Sommerfeld from Vienna even more advanced in the study of physics than was Heisenberg. He arrived in Munich with a paper on general relativity ready for the printer. Although only a year older than Heisenberg, Pauli was already in his fifth semester when Heisenberg first met him in 1920. Like Heisenberg, Pauli was the well-bred son of a university professor, and in their personalities they had much in common. Both were sensitive, naive, adolescent, personally insecure but academically confident, enormously ambitious, and thoroughly dedicated to theoretical physics. The similarities ended there. Outwardly, Heisenberg was quiet and friendly, at once retiring and almost recklessly daring — in life and in science — while Pauli was outspoken, aggressive, carefully systematic, and often devastatingly critical. The virtuous Werner loved the purity of the outdoors, youthful games, and the sunshine of long summer days. Pauli preferred city nightlife, risqué cabarets, and the pubs and coffeehouses of Weimar Schwabing. Heisenberg rose early in the morning, worked intensively throughout the day, and often sank into depression during long winter nights. Pauli haunted the cabarets in the evenings, worked feverishly until dawn, then slept until noon, missing his morning lectures. Sommerfeld tolerated Pauli's behavior since he was a mere deputy. But Pauli obviously annoyed Born, whom Pauli assisted in Göttingen beginning in 1921. Pauli left Göttingen for a new institute in Hamburg after only half a year, hungering for city nightlife. "He can't stand life in a small town," wrote Born.[52]

Although Heisenberg and Pauli were together in Munich for only two semesters, the two physicists — so opposite and yet so similar — formed a close professional friendship that lasted the rest of their lives. That association, recorded in their voluminous correspondence, is one of the most important in modern physics. Their admiration for each other as physicists persisted,[53] and each was significant, perhaps crucial, to the other's work. Although they never became close personal friends — they used the polite, collegial "Sie" ("you") form of address (the familiar form is "du," used in families and among close friends) until as late as 1927 — Pauli functioned for Heisenberg in ways remarkably similar to those of Werner's older brother. Pauli was more advanced in physics and offered Heisenberg brotherly advice on research. But as colleague and grader of homework, he could also issue ruthless criticism that pushed the insecure yet ambitious Werner to try even harder. Heisenberg once told an interviewer: "Pauli had a very strong influence on me. I mean Pauli was simply a very strong personality. . . . He was extremely critical. I don't know how frequently he told me, 'You are a complete fool,' and so on. That helped a lot."[54]

As noted earlier, Pauli apparently was also partly responsible for converting Heisenberg to the study of atoms. Having read Einstein and Weyl, Heisenberg at first considered work on relativity after abandoning number theory. During his first semester, he solicited Pauli's opinion of his prospects. Pauli was not optimistic. As author of the then definitive summary of relativity theory, he warned Heisenberg that research opportunities in the field would be meager. But Pauli was also the author of a dissertation on the quantum theory of the ionized hydrogen molecule that proved a failure in the agreement between theory and experiment. He could therefore assure his colleague that research in quantum atomic physics was wide open.[55]

If Pauli was Heisenberg's "brother" at the institute, Sommerfeld was his "father." Werner, rebellious toward his real father yet still searching for new authorities to replace the old, put his education and early career completely in Sommerfeld's hands. When Sommerfeld left to lecture in America for a semester in 1922–1923, he sent Heisenberg to Born in Göttingen. They had all agreed that Werner would return to Munich to complete his doctorate. During Heisenberg's visit, Born discovered that he needed a new assistant and hoped that Heisenberg might return to Göttingen to habilitate, after receiving his doctorate. When Born asked him about his future plans, Heisenberg responded: "I don't have to decide that! Sommerfeld decides that!" Born had to apply to Werner's guardian for permission to allow him to habilitate in Göttingen.[56] Werner had become by then a valuable commodity.

CHAPTER 7

Cutting to the Core

When Heisenberg arrived in Sommerfeld's office in 1920, the professor had been engaged for over a year in "the logical and mathematical penetration" of one of the most abstruse problems in contemporary quantum physics: an explanation for the behavior of optical spectra emitted by atoms. When white light is sent through a glass prism, it splits into a rainbow, or spectrum, of colors, each color corresponding to a different band of frequencies. If the atoms of one element are stimulated by heat or high voltage, they will emit not an entire spectrum of radiation but only certain colored lines corresponding to certain definite frequencies of light characteristic of that element.

Armed with clues from X-ray spectroscopy, Sommerfeld sought to explain the mysterious splitting of certain optical lines emitted by atoms. Each line seemed to split into doublets and triplets of lines, which are called multiplets. Even more of a mystery was the fact that, when the atom was immersed in a magnetic field, each multiplet line split in turn into a regular pattern of lines, a phenomenon known as the anomalous Zeeman effect.

Quantum theory called for explanations of these spectra and their various splittings on the basis of an imaginary mechanical model of the internal motions of the charged electrons within an atom. This model was expected to behave similarly to the quantum model of the hydrogen atom invented by Niels Bohr in 1913 and extended by Sommerfeld three years later. Their work formed the foundation of the prevailing quantum theory of the atom when Heisenberg arrived at Sommerfeld's institute.

Soon after his arrival, Heisenberg took up the search and was not long in finding an answer. Just a year after entering Sommerfeld's program, Heisenberg presented his atomic "core model" of complicated atoms that seemed to resolve every spectroscopic riddle at a stroke. But the model succeeded — it managed to "save the phenomena" — only because its youthful creator failed to follow the requirements of an acceptable theory as laid down by Bohr and Sommerfeld. Because they resisted any explanation other than Heisenberg's controversial core model, multiplet spectra and the Zeeman effect were among the most obvious and tantalizing phenomena regarded as irreducible to quantum physics — an intriguing challenge for any theorist. Heisenberg's model demonstrated his scientific attitudes and approach, as nurtured by Sommerfeld — a talent for leaping to solutions of seemingly insoluble problems that came to characterize Heisenberg's physics. In the end, he was often right. The characteristics of Heisenberg's core model found justification in the so-called quantum mechanics that Heisenberg and his colleagues presented nearly four years later. Heisenberg's core model for the Zeeman effect was an early step in that direction.[1]

Niels Bohr's quantum atomic theory offered an explication of both the structure and the properties of atoms, but it too had succeeded only by violating accepted physics.[2] In 1913, Bohr attempted to justify a nuclear atomic model — that proposed by Ernest Rutherford, his Cambridge mentor — in which the electrons orbit the nucleus like the planets of a miniature solar system. In his justification, he supplemented the so-called classical physics of orbiting electrons with two ad hoc quantum postulates.

The first postulate declared that orbiting electrons, obedient to Newtonian laws of motion in each orbit, seemed to ignore Newtonian physics when choosing possible orbits. Rather than a continuum of possibilities, only certain distinct stationary states were allowed. The states are stationary because the atomic electrons seemed to ignore classical Maxwellian electrodynamics by orbiting peaceably, without radiating as they should in the stable states. Each state, separated from the others by a gap in energy, could be identified by an integer, a principal quantum number, $n = 1, 2, 3, \ldots$, and a corresponding orbital energy, $E_1, E_2, E_3, \ldots$.

Bohr's second postulate stated that the emission and absorption of spectroscopic light rays (radiation) of a definite frequency v are associated with jumps, or leaps, of electrons between the stationary states, so that an Einsteinian light quantum of energy hv, h being Planck's constant, is absorbed (upward jump in energy) or emitted (downward jump in energy). For example, if the jump occurs downward from E_2 to E_1, the conservation of energy requires that the excess energy be radiated as a light quantum $h\nu$ that contributes to an observed spectral line, $E_2 - E_1 = h\nu$.

In Bohr's theory, the observed frequencies corresponded not as they should in classical physics to any motions within the atom but to differences between motions. The actual mechanism of emission and absorption was thus a riddle, while the ad hoc conjunction of quantum rules and classical mechanics seemed incomprehensible.[3] Bohr made no pretense of offering more than a hybrid system that somehow worked. "I am by no means trying to give what might ordinarily be described as an explanation," he declared, "nothing has been said here about how or why radiation is emitted."[4] Yet the utility of the system was so striking that it could not be ignored.

The utility increased with every account of additional complexities. Each member of the so-called Balmer series of hydrogen spectral lines, derived by Bohr, displayed an additional "fine structure," a mysterious splitting of each emitted line into a pair of lines (a doublet). In 1915 Bohr calculated a correction to the Balmer formula arising from the relativistic mass variation of the electron as it moved in its circular orbit. The result came close to the observed splitting of the Balmer lines, but Bohr hesitated to draw conclusions. Sommerfeld did not. Within months of Bohr's paper, he presented two papers on relativistic fine structure to the Bavarian Academy, and within a year he presented his famous three-part work, "On the Quantum Theory of Spectral Lines," generalizing the Bohr atom to the most abstruse of complexities.[5]

While Bohr considered only circular orbits, Sommerfeld, mindful of the solar system analogy, quantized every planetary orbit. He also included the relativistic rotation of the orbital perihelion for inner elliptic orbits, as Einstein had just done in applying general relativity theory to the planet Mercury. The combination of planetary orbits, quantum postulates, and relativity theory provided the precise, fine structure observed in the low-lying X-ray stationary states, along with all of Bohr's results in the limit of circular motions.

But Sommerfeld's discovery that even the orientation of an atom in space could be quantized (space quantization) if a third dimension (a z-axis) is defined by an external magnetic field went far beyond the Bohr atom. Planck and Bohr had earlier shown that, in quantum theory, a special relationship exists between each variable of motion of a periodic quantum system and the momentum corresponding to that variable. The total area covered by the system in a phase space formed by the variable and the momentum always equaled h times an integer, a quantum number, greater than or equal to zero. This quantum condition provided the connection between mechanical motions and quantum numbers.[6]

Using the quantum condition, Sommerfeld showed that not only is the orbital angular momentum of an atomic valence electron quantized — that is, it occurs only in discrete amounts — but also the angle of the plane of the orbit, the projection of the angular momentum on the z-axis (if defined), is quantized. Each of these quantizations gave rise to sets of integer quantum numbers, each set enumerating the possible quantum states within the atom for each variable. The myriad possible jumps between quantum states defined by these numbers yielded the myriad complexities of the observed atomic spectra. One such complexity was the Zeeman effect, to which Sommerfeld and Peter Debye, his Munich assistant, immediately turned.[7]

The Zeeman effect is closely entwined with the phenomenon of optical series spectra, which in turn produce the multiplets. The series spectra occur in multiple-electron atoms, such as sodium. They involve the separation of each of the lines observed in a simple hydrogen atom (one electron) into several series of lines originating from stationary states produced by various values of the orbital angular momentum quantum number k. For each principal quantum number n, there are n terms (stationary states), labeled s ($k = 1$), p ($k = 2$), d ($k = 3$), and so on. In 1916, following a suggestion by Bohr, Sommerfeld showed that the origin of the series spectra could be derived from the shielding of the nuclear charge by the inner electrons. Jumps of the outer valence electrons between states yielded the observed series spectra.

In addition to the series splitting of the hydrogen lines, each term, save for the normal (ground) state, splits further into doublets and triplets. (Higher multiplets were discovered in 1922.) In 1926, Heisenberg and Pascual Jordan attributed such multiplet line structure to the coupling, or interaction, between the orbital angular momentum of the electron and its own rotational, spin angular momentum. But the origin of multiplets remained a puzzle for the quantum theorists of the early twenties. The puzzle grew even more acute after the discovery that multiplet lines display in turn an anomalous Zeeman effect. That is, each spectral line of the series spectra, which occur only for heavier atoms, splits into multiplets (mainly doublets and triplets), each line of which splits again in the Zeeman effect.

The normal Zeeman effect, discovered by Pieter Zeeman in 1896, entails the splitting of an observed line into a further doublet or a triplet when the atom is placed in a magnetic field. The effect is explicable in Newtonian and Maxwellian theory by treating electrons as balls of charge bound to the nucleus by an elastic (spring) force. More precise spectroscopy revealed that Zeeman's normal effect actually holds only for singlet lines. Multiplets split into anomalous components, inexplicable in classical theory. The situation is even further complicated by the Paschen-Back effect. As the strength of the magnetic field increases, all of the anomalous Zeeman components coalesce continuously into a normal Zeeman triplet, suggesting some sort of highly complicated internal magnetic coupling or decoupling within the atom.

In 1916, Debye and Sommerfeld easily accounted for the normal Zeeman effect using their space quantization, the quantized projection of the angular momentum of the valence electron on the z-axis, the axis of the magnetic field. The orientation of the plane of the orbit is the crucial factor, because an electron orbiting in a closed loop acts like a current loop, and a current loop behaves like a little magnet. The direction of the north pole and the amount of the polarity are determined by the speed and orientation of the electron's motion, its angular momentum in the orbit. The potential energy of a magnet in a magnetic field is related to the angle formed by the direction of the field and the axis of the magnet. Debye and Sommerfeld discovered that this angle can assume only certain discrete values, represented by the quantum number m. The energy of a Zeeman state could be written $E = E_i + mh\nu_L$, where E_i is the energy of the undisturbed state, m is the projection quantum number, and ν_L is the Larmor frequency, the frequency of the rotation of the orbit about the magnetic field. From this equation it immediately followed that the normal Zeeman effect occurred if only certain jumps between values of the projection quantum number m were allowed: $\Delta m = 0$, 1, or -1, which yields the observed triplet of lines. The emitted light quantum can take on three energies: $E_2 - E_1 = h\nu$, $h\nu + h\nu_L$, or $h\nu - h\nu_L$.

That m did obey the selection rule $\Delta m = 0, \pm 1$, could be shown in two different ways using two different elements of quantum theory. The first derived from Bohr's 1918 version of the correspondence principle. According to this principle, which would play a crucial role in the subsequent development of quantum mechanics, quantum atomic orbits turn into classical mechanical orbits as n, the principle quantum number, grows very large. More precisely, the intensities and polarizations of spectroscopic lines are related for very large n to the amplitudes of an expansion of the orbital motion of an electron in a summation of harmonic (multiple Fourier) terms. The second way was provided by

Debye's successor as Sommerfeld's assistant, Adalbert Rubinowicz. Rubinowicz treated electromagnetic radiation emitted by an atom not as a light quantum but as a classical spherical wave. He equated the angular momentum carried by the emitted wave with the total change of quantized angular momentum within the atom, as controlled by the quantum number k. From this further ad hoc combination of classical wave theory and quantum atomic physics, Rubinowicz confirmed the selection rule for m. This in effect confirmed the Sommerfeld-Debye reduction of the normal Zeeman effect to the internal mechanics of quantum atomic models. But the anomalous effect remained unexplained by any theory.

After the disruption of postwar turmoil and heavy teaching loads, Sommerfeld rededicated himself to the arcana of multiplet lines and the Zeeman effect. In 1919, he tried a new tack. Instead of constructing intricate atomic mechanisms in the hope of reproducing observed data, he attacked the problem from the other end, by studying the remarkable regularities in the data in the hope of gleaning clues for an appropriate model. Working from the observed lines to the stationary states, one might, as Bohr had done, obtain a model interpretation. By 1920, this approach had yielded a new quantum number, a magneto-optical splitting rule, and what Sommerfeld himself, half in jest, termed a number mystery — all of which formed the foundation of quantum spectroscopy research in Sommerfeld's institute at the time of Werner's arrival.

Göttingen experimentalist Carl Runge had quantified the regularity of the anomalous Zeeman effect by expressing the separations between the anomalous lines by a ratio of two integers. Since in quantum theory the frequency of each observed spectroscopic line arises from the difference in energies of two unobserved quantum stationary states, Sommerfeld worked backward to the stationary states and assigned a Runge fraction to each state itself. When the difference is taken, the denominator of the observed line becomes the product of two integers — a magneto-optical splitting rule. The integers could be assigned according to a number mystery — that is, without theoretical explanation — for each term and type of multiplet.

Using these rules, Sommerfeld found that he could obtain all of the Runge fractions from the assigned quantum numbers of the multiplet terms. To him, the "musical beauty" of this "order and harmony" of integers bordered on the mystical. In the introduction to the first edition of *Atombau* (1919), Sommerfeld waxed Keplerian. As Kepler had solved the problems of planetary motion by speaking the "language of the spheres," one could solve the problems of quantum spectroscopy by learning to speak the "language of the spectra." This language, wrote Sommerfeld, is "an atomic music of the spheres, a harmonizing of whole number relationships."[8]

While the tonal harmonies of atomic music enchanted Sommerfeld and his pupils, atonal traditionalists such as Willy Wien stopped their ears. As late as June 1926, after the invention of two forms of quantum mechanics — wave mechanics and matrix mechanics — Wien was still complaining. In an address as rector of the University of Munich, the old physicist lamented the passing of "the greatest era in physics," when "mechanics was established and it had been possible to predict through it the movement of heavenly bodies with the greatest accuracy." Nowadays physics was plagued with an obscure quantum theory involving the combining of integers according to certain incomprehensible rules bordering on number mysticism. "Where these rules come from remains completely in the dark, and if you should have trouble following my remarks, you may be comforted to know that physicists do not understand them much better."[9]

Despite Keplerian verbiage and Wienian protestations, Sommerfeld did not really believe that atomic music could ever replace the unmusical mechanics of planetary atomic models. He even called a model interpretation of the Zeeman effect one of the two "unsettled questions of atomic physics" in 1920. He described two new avenues to a model, what he called pursuing the "techniques of quanta."[10] One avenue involved a return to direct inventions of models, the other the search for a possible hidden quantum number lurking behind his number mystery. Examining the atomic decay schemes yielding the series spectra, the technician noted that the nonappearance of some combinations of pairs of stationary states suggested a new selection rule for some unknown number. Obviously the number could not be the orbital quantum number k, which designated an external rotation; rather, the splitting of each of these terms into multiplets required a new number, "an inner quantum number, perhaps corresponding to a hidden rotation."[11]

For doublets, Sommerfeld surmised that $j = k$ or $j = k - 1$. He had, in effect, consigned the whole problem to unknown internal mechanisms. Sommerfeld published his assignment of inner quantum numbers, j, just after Werner's arrival at his institute. Since only differences between numbers are ever observed, the assignment of these numbers is nonarbitrary only up to an additive constant, a circumstance that Sommerfeld hardly appreciated but that Heisenberg would soon exploit.

Atomic mechanics, abstruse line splittings, and mystical number harmonies constituted the highly rarefied atmosphere that Heisenberg inhaled in Munich as a budding theorist. The atmosphere was rarefied both in quality and in quantity. Students pursuing science studies were clearly in the minority at that time, even in Germany; among theorists, the overwhelming majority did not concern themselves at all with the abstractions of quantum theory or atomic spectroscopy. In 1920, only a handful (8 percent) of all German students were studying science of any kind. Of the 337 doctorates awarded at the University of Munich at the end of Heisenberg's first semester, only 19 were in the sciences.[12] That quantum physics was of relatively little overall concern among physicists is suggested by a study of those publishing in the major physics journals of the time. The study showed that of the German physicists born in Heisenberg's generation, 1895–1909, only a little over a quarter devoted themselves to the quantum.[13] During the 1960s, the project Sources for History of Quantum Physics, sponsored by the American Philosophical Society, gathered interviews and archival materials from and about the main participants in the development of quantum physics through 1930. The project sought to preserve the historical record of quantum mechanics, one of the greatest intellectual achievements of the twentieth century. Worldwide, the project found that it could limit the information pool to a mere 200 individuals.[14]

While quantum atomic physics was statistically a rare and rarefied discipline, Heisenberg was statistically common within it. Perusal of the personal characteristics of the main contributors, including Heisenberg, to the matrix formulation of quantum mechanics in 1925 indicates that, like Heisenberg, nearly all stemmed from upper-middle-class academic families; most received their degrees from and were closely associated with the Munich-Göttingen-Copenhagen triad of research centers; all worked in quantum spectroscopy; the overwhelming majority were German; and, excluding their mentors, their average age in 1925 was 24 years.[15]

As indicated earlier, by the turn of the century Germany was leading other nations in numbers of theoreticians. This was the result both of international competition in cultural achievement and of the internal dynamics of the mathematical physics profession in Germany. Spurred by public fascination with science, technology, and the new discoveries in quantum theory and atomic science, by 1920 Germany had exploited its advantage in theoretical science by directing the efforts of its theoreticians into the abstruse yet significant realm of atomic physics.[16]

In the years immediately following World War I, German physicists were already adept at quantum atomic theory. Outstanding experimental research, especially in the field of spectroscopy, inspired the theorists to new heights. The striking contrast between Germany's scientific successes and its military failures did not go unnoticed by the German public; international triumph in atomic physics and science in general could garner at least a portion of the international regard that Germans sought but that their military and economic leaders had failed to deliver.

New and seemingly insurmountable difficulties in atomic research were becoming evident when Heisenberg joined the ranks of German quantum atomists in the fall of 1920. This was a critical moment in a field that had yielded impressive early successes. It was also an advantageous time for individual research. Postwar quantum theoreticians — similar in more ways than just their science — formed a close-knit, small, elite band in which each individual had the opportunity to express his unique talents within the whole.

To test and challenge the newcomer to the club, Sommerfeld early initiated Heisenberg into the mysteries of Zeeman spectroscopy. Tübingen experimentalists Paschen and Back, who regularly furnished Sommerfeld with spectroscopic data, forwarded a set of Zeeman data in the fall of 1920 that Sommerfeld was unable to reduce to his newly invented inner quantum numbers.[17] Four weeks after Heisenberg began attending his seminar, Sommerfeld suggested that he try his hand at analyzing the data. The eager novice promptly immersed himself in the intricacies of Zeeman spectroscopy and pored over Sommerfeld's latest papers on the subject. Toward the back of Heisenberg's only surviving gymnasium notebook is a neatly drawn scheme of the Zeeman effect for all doublet and triplet combinations, with intensities and polarizations of each component carefully indicated in standard fashion.[18]

The Tübingen data must have been for doublets, probably for sodium, for Sommerfeld's precocious student soon reported that the Zeeman lines could be easily obtained from stationary states by assigning not integers but half-integer inner quantum numbers to each state: $1/2$, $3/2$, $5/2$, and so on. Sommerfeld was shocked. "That is absolutely impossible," the number mystic retorted. "The only fact we know about quantum theory is that we have integral numbers, and not half numbers."[19]

Sommerfeld's seminar backed him up. The most striking feature of quantum theory was the existence of quanta — single, identical parcels of energy that could not be further divided. Such a notion was completely foreign to the classical mechanics of Newton and the electrodynamics of Maxwell and others. But on the atomic level it manifested itself in Einstein's 1905 hypothesis of light quanta — packets of light energy — in the hypothesized stationary states of the Bohr-Sommerfeld atom and in Planck's fundamental conditions for quantizing variables such as angular momentum. Integral quantum numbers derived from the counting of integral numbers of quanta in each case. Half-integer numbers simply had no physical meaning or place in quantum theory. Pauli noted that once halves were introduced, then fourths, eighths, sixteenths, and so on would inevitably follow.[20]

Yet the professor indulged his pupil's heresy long enough for Werner to discover a series of applicable rules for the Zeeman effect that apparently coincided with those discovered independently by Paschen's and Back's colleague, the theorist and Frankfurt lecturer Alfred Landé. When the 33-year-old Landé forwarded a set of rules involving half-integral combinations to Munich in March 1921, Sommerfeld replied: "Your new representation agrees well with what has been found by one of my students (in the first semester) but which has *not* been published."[21] Heisenberg would not let Landé get ahead of him again.

Landé published his empirical rules a month later. These rules, as Landé expressed them, formed the foundation for all subsequent work on the Zeeman effect, including Heisenberg's. Following Sommerfeld and Debye, Landé analyzed each observed anomalous Zeeman line into two unobserved stationary states, each possessing a quantum number m corresponding to the space quantization of the state. The anomalous Zeeman effect entailed a further, anomalous shift in the energy of each state. Landé introduced an empirical proportionality factor g, a rational fraction corresponding to this shift.[22] The previous energy E of any Zeeman stationary state could thus be rewritten with the extra factor g:

$$E = E_i + gmh\nu_L$$

in which all the anomalousness of the Zeeman effect is contained in the rational fraction g. If g is 1, the obedient normal Zeeman effect occurs.

Landé then employed the working hypothesis that Sommerfeld's inner quantum number j, not the external quantum number k, is the total angular momentum quantum number of the atom and that m is its space quantization number. This worked for singlets and triplets, which yielded the proper odd number of m-values $(2j + 1)$ in space quantization. But for doublets an even number of m-values was required for each value of j. Landé just as readily dropped the momentum interpretation of j and assigned without physical explanation half-integer numbers to m.[23]

With m assigned in this way and with empirically derived g values, Landé managed to recover the entire anomalous Zeeman effect, including Sommerfeld's number mystery and all of Runge's rational fractions. Although Landé's work was still number juggling, half integers were still incomprehensible, and a mechanical model for all of this was still missing, Landé's achievement and with it Heisenberg's was clear. Half integers were here to stay.

After Landé's coup, Sommerfeld consoled Werner by showing him the next step "into the still dark region of the complicated Zeeman effects": a quantum-theoretical reinterpretation of Woldemar Voigt's classical Zeeman theory. Sommerfeld hoped to use Voigt's theory to derive Landé's rules from a model.[24] In a 1920 address to the German Physical Society, meeting in Berlin, Bohr had suggested that multiplets and their Zeeman effects might arise from some mechanical interaction inside atoms between inner and outer electrons, a possibility that Sommerfeld and Debye had not considered. Woldemar Voigt, a Göttingen theorist, had earlier constructed a classical theory for the prominent sodium doublet (D-lines) in which the valence electrons were attached to each other and to a fixed core of electrons by what amounted to little elastic springs. Amazingly, the contraption worked, yielding all of the then-observed Zeeman data. It was, of course, superseded by Bohr's atom, but Bohr suggested that Voigt's model might still offer a clue to the "real Zeeman model."[25]

Within a year, Sommerfeld had taken Bohr's suggestion to its conclusion. He published his quantum-theoretical reinterpretation of Voigt's theory in the *Zeitschrift für Physik* immediately preceding Heisenberg's paper—his first—on the same subject. In his paper, Sommerfeld set out—much as Heisenberg would do four years later when he provided the breakthrough to matrix quantum mechanics—"to translate these [Voigt] equations from oscillation-theoretical to quantum-theoretical language."[26] Quantum principles provided the grammatical rules. Guided by these principles, Sommerfeld worked backward from Voigt's equations for the observed frequencies to the unobserved atomic stationary states, whose quantum numbers had been assigned by earlier analyses. The resulting quantum formula for the energy of any Zeeman state in any size magnetic field contained all the essential data: optical doublet splitting, Runge's fractions, Landé's g factors for small fields, the normal Zeeman effect for large fields, and the continuous transition between the two (the Paschen-Back effect)—but all without half-integral numbers, for which no physical basis could yet be found. Extra numerical factors could compensate.

Sommerfeld had derived his Voigt formula and composed a paper on the subject by the early fall of 1921, but he did not publish until

December. He still lacked the all-important model. But he did not lack a capable student. The 19-year-old Werner, unencumbered by integers, rewrote Sommerfeld's formula with half-integral numbers and thus obtained Landé's rules directly in the limit of a small magnetic field.[27] He now sought to derive this formula by translating not from classical to quantum language but directly from a quantized model involving half-integral angular momenta. As Werner constructed a model to yield the formula, the professor revised his manuscript at least twice to keep pace. The collaboration between professor and pupil proved vital to both: Heisenberg stimulated Sommerfeld to rethink and revise his theory; Sommerfeld tolerated his pupil's fracturing of accepted physics.

Heisenberg brought Sommerfeld's Voigt formula and his own early notions of a model for it to his first conference, the Jena meeting of the German Physical Society held in September 1921. Aside from meeting the great researchers of his profession, the awestruck beginner also met his competitor, Alfred Landé, for the first time. Heisenberg began an intensive — almost daily — correspondence with the more senior Landé, now an associate professor (Extraordinarius) in Tübingen, after returning from a visit to the Berlin faction of his youth movement and a mountain tour with his boys. Werner's frequent contact with Landé was probably intended in part to forestall further surprises from Landé, but it was also motivated by Heisenberg's problems with his developing model, the Atomrumpf, or atomic core, model.

Judging from his postconference correspondence, Heisenberg had already considered interpreting optical (as opposed to X-ray) doublets and triplets as a type of internal Zeeman effect — the magnetic field generated by one set of orbiting electrons would affect the "magnet" set up by another set within the atom. Heisenberg's core model involved the following scheme. He envisioned an atom displaying doublet or triplet lines to consist of one or two valence electrons orbiting an atomic core. The core consisted of the nucleus surrounded by the internal electrons orbiting in closed shells. This rigid "core" of nucleus and closed shells acted like a large spinning ball. Because it carried a net positive charge, the core had a magnetic polarity — it too behaved like a little magnet. It therefore assumed two or three space-quantized positions in the magnetic field generated by the orbiting valence electrons. In Heisenberg's conception of this model, these two or three orientations, which were the source of the doublet or triplet spectral lines, came about because, on the average, the core could somehow borrow a half unit of angular momentum from the valence electrons. Werner was attributing the half-integer quantum numbers to borrowed half-integer

momentum. The core could then align itself according to space quantization in the magnetic field produced by the orbiting valence electron.

Werner's ingenious model accounted for practically everything then
known about multiplets and the Zeeman effect: doublet and triplet line
structure, all of Landé's Zeeman rules, the Sommerfeld-Voigt equation
for doublets, and the Paschen-Back effect.[28] In retrospect, the essential
but problematic mechanical behavior of the core in Heisenberg's theory
was physically equivalent to the later almost equally problematic behavior of a spinning electron, and many of the mechanical interactions were
physically equivalent to such later interactions as spin-orbit coupling
and decoupling in magnetic fields. Even Heisenberg's version of the
Sommerfeld-Voigt equation finds its counterpart in the later matrix
mechanics of the Zeeman effect.[29] Heisenberg's model was an ingenious
and incredibly daring achievement, far ahead of its time. But in its time
it was far from acceptable according to established standards.

Heisenberg presented his new ideas to Landé and Pauli in Jena. After
one of the lecture sessions, the three physicists retired to consider the
Zeeman effect. While Pauli and Landé argued over whether k or j
should be the total angular momentum quantum number,[30] Pauli and
Heisenberg again disagreed over the existence of half-integral quanta
(but less so over half-integral numbers, which Landé had used to such
effect). To Pauli, half-integral quanta still made no physical sense. To
Heisenberg, physical sense was less important than achieving success.
Werner took particular pride in reporting to his mother his apparent
victory over his opponents: "That was now a three-way battle, in which
each had to defend himself against the other two. Naturally we did not
come to any conclusion. However, in the evening I got hold of the
professor [Sommerfeld] and he had a letter from Paschen in which it
turned out that once again I was completely in the right. Especially Pauli
was with that completely defeated."[31] Two days later Heisenberg even
dared to challenge "the professor," telling him that a newly written
section of the next edition of his textbook was all wrong. "Now that
too has been gotten rid of," he wrote home.[32] Werner's arguments were
so persuasive that Sommerfeld requested his assistance in completely
rewriting the chapter in question, while holding back his own Voigt
equation to see what developed.

Heisenberg was holding back, too. He had his core model by the end
of October 1921 but delayed publishing on objections from Landé. As
indicated earlier, in Heisenberg's model for doublets the valence electron shares an average half unit of its angular momentum k with the
core, retaining $k - 1/2$ units. This explained the origin of doublets as an

internal Zeeman effect, Sommerfeld's j as the total angular momentum number formed by the core and electron ($j = k - 1/2 \pm 1/2 = k$ or $k - 1$), and Landé's half-integral m-values as the magnetic quantum number. A similar though more involved sharing of half-integral units of momentum occurred for triplets, which are emitted by double-valence atoms. These were certainly remarkable results, but Heisenberg could offer Landé no explanation of why such a sharing took place, why it was precisely half-integral on the average, nor what half-integral angular momenta could mean physically in a quantum theory inherently consistent only with integral quanta.

Even more remarkable properties emerged when Heisenberg immersed his atom in a magnetic field, but they came about only through an even more deliberate fracturing of the rules. To obtain the continuous transition from the anomalous to the normal Zeeman effect with increasing magnetic field intensity (Paschen-Back effect), Heisenberg ignored the quantization of the core, permitting it to remain aligned along the continuously changing resultant of the internal and external magnetic fields. This "inefficacy" of the core directly violated Sommerfeld's space quantization, Larmor's precession theorem, Rubinowicz's selection rules, and their semiclassical derivation. But it did yield the Sommerfeld-Voigt equation for doublets and, in the case of triplets, Landé's g factors.

For small fields in the doublet core model, the Zeeman effect arises from a joining together of the magnetic interactions between the valence electron and the external field and between the valence electron and the core (internal Zeeman effect). As the external field increases, the electron-field interaction grows, the internal Zeeman effect is gradually overcome, and core and electron interact individually only with the external field, not with each other, thus producing the Paschen-Back effect. All this is expressed mathematically by the Sommerfeld-Voigt equation, which Heisenberg derived from his model, but only by ignoring the quantization of the core.

The implications were far-reaching and equally disturbing. As the first — and for the next four years the only — model for multiplets and the anomalous Zeeman effect, it had to be taken seriously, yet it violated nearly every quantum principle in sight. Attempts to come to terms with this dilemma occupied quantum spectroscopists for the next four years. Sommerfeld described the situation in a letter to Einstein in January 1922. Informing Einstein of new and "wonderful numerical laws of line combinations," he wrote: "A pupil of mine (Heisenberg, third semester!) has even interpreted these laws and those of the anomalous Zeeman effect using a model (*Zeitschrift für Physik*, in press). Everything works out but remains however in the deepest sense unclear. I can only promote the technology of quanta; you must make your philosophy."[33]

In addition to collaborating with Sommerfeld, Werner debated his model by letter with Pauli, who was now assisting Max Born in Göttingen. He also spent nearly two months haggling over details with Landé, who was his academic senior. Each of these interactions helped to establish and to solidify Heisenberg's approach to his work. He was both naive enough and overconfident enough to posit instant solutions to problems hitherto unsolved by the acknowledged leaders of his profession. Landé challenged Heisenberg on his improprieties, Pauli applauded his pragmatic acceptance of them, although he himself remained skeptical, and Sommerfeld condoned and finally sanctioned their use and publication. This ability to adopt a serviceable solution regardless of accepted wisdom was a great part of Heisenberg's genius. Born of brilliance, ambition, and youthful ignorance and independence, boldness remained with Heisenberg throughout his career and distinguished his audacious, intuitive style of physics from the more cautious, traditional, and rational approach of most of his colleagues.

Landé was willing to overlook half-integral momenta and even the transgression of Larmor's theorem, but he drew the line at any tinkering with Rubinowicz's principle; he had relied on it too heavily in his own work. Heisenberg sought support for his views in Bohr's latest work on the correspondence principle. Soon after returning to Munich from Jena, Berlin, and the mountain tour, Heisenberg forwarded to Landé a Fourier analysis, similar to Bohr's, of the conditionally periodic orbits represented by the quantum numbers n, k, j, and m. The analysis, which did not require a detailed model, yielded the correct Zeeman intensities and selection rules. Sommerfeld was enthusiastic, and Heisenberg managed to use the calculations and his inklings of atomic couplings to obtain the details of his core model. He was so excited that he wanted to place Bohr's correspondence principle above all else. He wrote to Landé: "The fact that the entire model interpretation of a process can be calculated from purely empirical material is another brilliant achievement of the Bohr correspondence principle, which I am beginning to want to consider as important as the entire quantum theory."[34]

Landé demurred. Although he had introduced g factors to obtain the fractional energies of the Zeeman effect, he had maintained Rubinowicz's integral selection rule, $\Delta m = 0$, $+1$, or -1, by equating quantum jumps by integral units of energy and momentum with the energy and momentum of the radiated wave, just as Rubinowicz had done. Heisenberg, on the other hand, sought to derive the observed fractional radiated energies from nonintegral quantum jumps. This could be achieved only by the peculiar behavior of the core—which, Landé complained to Bohr, led in turn to a choice between violations of Rubinowicz's widely accepted assumptions: "For the individual radiation process, *either* fractional momenta are given off as spherical waves in the ether, *or* one must give up the momentum conservation law between atom and ether."[35]

Heisenberg, then barely a third-semester pupil, seemed oblivious to the problem until Landé complained. Not until December did Sommerfeld lecture on "Maxwell Theory and Electron Theory." Uninfluenced by Rubinowicz, who had long since left Munich, Heisenberg relied solely on Bohr, the originator of quantum atoms, for selection rules. Landé was still complaining about doublets when Heisenberg forwarded his similar model for triplets and their *g* factors. "In the case of the triplets there really does seem to be a contradiction of Rubinowicz," Heisenberg told Landé. But apparently aware of Landé's alternatives, he asked, "Are momentum considerations really so sound now that every other minute they are no longer valid in the quantum theory? Do you really believe in the spherical wave?"[36]

Heisenberg seemed willing to write off anything necessary to save his model, and at that very moment Pauli the critic came to his aid. Pauli had received his doctorate in Munich the previous summer and had gone to Göttingen after Jena, where he revised his doctoral dissertation, on the failed quantum theory of the hydrogen molecule ion, for publication.[37] Pauli's calculation of the ionization states and the stability of the quantum orbits did not agree with experimental results, and a checking of the calculations confirmed this finding, clearly indicating a failure of quantum atomic theory for this simple case. Pauli's skepticism of classical mechanics in the construction of quantum atomic models now exceeded even Heisenberg's. When Pauli informed Heisenberg of his definitive results that fall, he included a motto (now lost) describing his attitude. Heisenberg responded with his core model and its deviations. By then, thanks to Landé, the "shady sides" of the theory were no secret, but Heisenberg now applied his own pragmatic motto to the situation: "Success sanctifies the means"[38]—at least until a complete calculation of a model, such as Bohr's work on the lithium atom, could be completed.

Heisenberg could be cavalier with his young friend and colleague, but dare he proclaim the same motto to Professor Landé, a published authority on Zeeman spectroscopy and one totally committed to Rubinowicz's rule? Heisenberg took the matter to his mentor. After brief deliberation, Sommerfeld sanctioned Heisenberg's means. The professor was now sympathetic to fractional selection rules and to the deviations they required, since, he noted to Werner and later argued in *Atombau*, the photoelectric effect indicated a breakdown of classical radiation theory or energy-momentum conservation or both.[39] Although a product of Sommerfeld's own institute, Rubinowicz's rule could be overturned if it blocked the way to a model. Werner left his mentor's office with permission to publish.

Armed with Sommerfeld's approval, Heisenberg responded to Landé's complaints with an ever stronger tone of self-righteousness. In his letter to Landé reporting Sommerfeld's blessing, Heisenberg openly dispensed with Rubinowicz: "One thus cannot regard the result of Rubinowicz as an absolute criterion for the correctness of a theory," he told Landé. "Selection rules and intensities (which amount to the same) must be calculated by means of the correspondence principle." Heisenberg had also picked his deviation: he would follow Sommerfeld in preserving momentum conservation, while violating classical wave theory not only with light quanta but with light quanta carrying fractional angular momenta. Heisenberg's model and approach and their implications for quantum physics were finally settled, and, as if to forestall any further objection from Landé, he informed the physicist that "Sommerfeld, too, regards my theoretical considerations as completely assured."[40] On December 17, 1921, two weeks after Heisenberg received Sommerfeld's blessing, the *Zeitschrift für Physik* received his first paper, the core model, for publication. Sommerfeld's paper on the Voigt equation had arrived at the same journal five days earlier.[41]

Heisenberg's core model put physicists in a difficult spot. The Bohr-Sommerfeld quantum theory of the atom, though inconsistent in its own way, was inconsistent for good reason and had accounted well for an astonishing array of spectroscopic phenomena and atomic properties. So far, only the anomalous Zeeman effect — the splitting of spectroscopic lines into multitudes of lines in a magnetic field — seemed irreducible to a quantum atomic model of some sort without such gross violations of the Bohr-Sommerfeld theory as half-integer quantum numbers, half-integer angular momenta, and improper mechanical motions within the atom. Many of these features were eventually transferred from the core to the electron and attributed to half-integral spin, enabling a complete theory of the Zeeman effect using quantum mechanics in 1926, but in 1922 Heisenberg's model posed a prodigious puzzle.

Physicists reacted by distancing themselves from the unruly model, but in different ways and to different degrees. Pauli and Sommerfeld, who had each encouraged Heisenberg, were both uncomfortable with what he had wrought. Although Sommerfeld publicly championed the model in his book and privately condoned it in his institute, he tried to make amends in his correspondence. Having expressed his reservations to Einstein, Sommerfeld wrote apologetically to Bohr—the physicist most committed to consistency—for failing to "bridle" his pupil: "I found his results so important that I consented to their publication, even though the form of the derivation is probably not yet the definitive one."[42] Apparently concluding that his pupil should fill the gaps in his knowledge of classical physics as displayed in his model, Sommerfeld urged Heisenberg to write his doctoral thesis not on quantum spectroscopy but in the more traditional and less controversial field of hydrodynamics—another subject in which Heisenberg had demonstrated an ability in Sommerfeld's seminars.

Bohr made no secret of his displeasure with the model after Heisenberg sent him a copy of his manuscript in early 1922. Not only had Heisenberg publicly recognized only two of his deviations while supposedly relying on Bohr's work, but also, Bohr complained to Landé, "The entire mode of quantization (half-integral quantum numbers, etc.) does not appear reconcilable with the basic principles of the quantum theory, especially not in the form in which these principles are used in my work on atomic structure."[43] For Bohr, the source of the anomalous Zeeman effect lay, as it had since 1913, in a failure of classical electrodynamics, not in a failure of quantum physics. Bohr insisted that only a program of consistent applications of quantum rules and procedures, joined by explicitly recognized and well-supported deviations from the rules, offered "a hope in the future of a consistent theory."[44] Heisenberg's half-integer model contradicted that program on every score.

Bohr told Heisenberg so that summer. Denmark's leading physicist and already recognized as one of the world authorities on the atom, Niels Bohr, then only 37 years old, would be regarded, together with Einstein, as one of the two leading physicists of the twentieth century (often followed, in third place, by Heisenberg). In June 1922, Bohr, who would receive the Nobel prize later that year for his work on the atom, delivered a series of comprehensive lectures on quantum atomic physics to German theorists and their students assembled in Göttingen—an event known affectionately thereafter as the Bohr festival. The festival marked Werner's first meeting with the master. It was the start of a lifelong, sometimes difficult, collaboration and friendship that was as important for Werner as his relationship with Pauli.

Bohr's lecture festival was also something of a political statement. Germany, condemned by its conquerors as the wartime aggressor, was at the time still suffering from an international cultural boycott.[45] German scientists therefore turned for scientific information and stimulus to each other and to scientists from neutral countries such as Denmark and the Netherlands. Bohr was unsympathetic to the boycott, impressed with German atomic theory, and grateful to Sommerfeld for supporting grant proposals for his institute.[46] He readily accepted invitations to lecture to the German Physical Society in Berlin in 1920 and, a year later, to deliver the first postwar Wolfskehl lectures in Göttingen, with which he had had a longstanding relationship through his brother Harald, a mathematician.[47] The lectures had to be postponed for over a year due to Bohr's heavy workload. By April 1922, he was relishing the prospect of lecturing to the Germans on quantum physics.[48] He had good reason. Although the third and latest edition of Sommerfeld's *Atombau* seemed more favorable toward his correspondence principle, the new papers coming out of Munich increasingly disregarded the content and methods of his research program.

Bohr delivered seven lectures over two weeks in June 1922 to packed audiences in the main lecture hall of the Göttingen physics institute. The smells of the garden roses and an occasional honeybee floated through the open windows overlooking the rear of the institute. Nearly fluent in German, Bohr presented in his characteristically soft and convoluted speech a careful and systematic account of the quantum theory of atomic structure, its problems, and how they might be resolved.[49] For many in the audience, Bohr's festival lectures were their first systematic exposure to the subject and served as a basis for most of their research over the next several years.

Mindful of Bohr's complaints and himself convinced that Heisenberg should meet other theorists, Sommerfeld paid Werner's way to the quaint university town of Göttingen in the northern state of Hanover. Inflation had forced the Heisenbergs to curtail Werner's travel and to rely more and more on "gold uncle" Karl for support. With Sommerfeld's help and his own initiative, Heisenberg easily gained access to the inner circles in Göttingen. During his stay, Werner quartered on the couch of a local mathematician, probably Courant. He delivered a private lecture on hydrodynamics to Ludwig Prandtl, a leading hydrodynamicist of the day, and eagerly joined the endless discussion rounds in apartments and coffeehouses and on walking tours. "This afternoon everyone is meeting in a café," he wrote to Uncle Karl and Aunt Helen. "Thus I must simply be there. I never get to bed before 1:00 A.M."[50]

Unfortunately, no contemporary record remains of Heisenberg's oft-recalled first encounter with Bohr, which probably occurred on June 14. On that day Bohr presented to his audience a favorable account of a calculation by his assistant, H. A. Kramers, of the relativistic (quadratic) Stark effect — the splitting of spectral lines in an electric field.[51] Heisenberg had already carefully studied Kramers's paper and had criticized it in the Munich seminar perhaps as early as his first semester. The audience listened approvingly to the speaker's summary of the paper and expected little comment from Bohr's peers during the discussion following the lecture. When Werner, a mere student already known for his candor, rose from his seat, an astonished silence fell over the audience. Seemingly without qualm, Heisenberg contradicted the master with a criticism of Kramers's calculation.[52] Bohr responded a little uneasily and afterward invited his critic for a walk to get a closer look at him.

Bohr had, of course, heard of Heisenberg and his disturbing core-model paper and had probably already made his acquaintance in Göttingen. As the tall, distinguished, well-dressed Professor Bohr walked alongside the slightly built, 20-year-old youth, their walk led them to a hill, the Hainberg, overlooking the town. Their discussions during the walk elevated with the terrain and ranged far beyond physics. As Heisenberg recalled years later (with much romantic admiration for his companion), they delved especially into philosophical questions concerning atoms, the epistemological use of familiar conceptions, and the meaning of consistent "understanding" in physics.[53] Such philosophical and methodological issues were important elements of their collaboration, both during this period and especially later in Copenhagen. Bohr's philosophical inclinations surely impressed young Werner, who had previously known only Sommerfeld, the quantum technologist.

The inevitable confrontation on the core model occurred early the next morning, on June 15. There were no lectures that day, and after breakfast Bohr played host to Heisenberg and Sommerfeld in his sumptuous lodgings at a local guest house. Later that day, Werner could brag once again to his family of his success. Always mindful of his family's high expectations, his letters redound with boyish pride. They show the same driven, almost reckless attitude toward his work that can be seen in his letters from Jena. He had to be successful — for his own obsessions and for his own survival in his new and competitive field. Werner carried the burden of his background into his profession.

Sommerfeld opened the breakfast-table debate with a brief lecture on Munich physics, including descriptions of the core model and a new helium model possessing half-integral momenta. Bohr responded briefly,

Heisenberg reported, "and then there developed a rather extensive discussion between Bohr, Sommerfeld, and me over my early paper. That was interesting. One can easily come to terms with Bohr."[54] The discussion was vindicating for Heisenberg: "In any case it was determined that until now a proof against my views is not to be found anywhere; at most only generalities and matters of taste speak against them."[55]

Bohr's reserved and diplomatic manner must have misled Werner as to how easily they had "come to terms"—neither Bohr nor his first three systematic lectures apparently impressed on Werner the profound nature of their differences. In his fifth lecture five days later, Bohr inserted an explicit complaint about Heisenberg's "very interesting paper" in the strongest words he would ever use in criticism: "It is difficult to justify Heisenberg's assumptions."[56] For Bohr, physics was more than the mere technical achievement of an end at any price. Understanding had to occur within the context of a complete theoretical apparatus. Tradition and structures imposed by old or new authorities had to be respected. Though always rebellious, intuitive, and unsystematic, Werner would gradually learn to appreciate these lessons during the next few years. Half integers and angular momenta, however, remained a principal point of contention between Bohr and the German physicists.

Bohr's diplomatic handling of Heisenberg obviously captivated the physics student. Werner wrote glowingly to his parents of the man who would come to exert a profound influence on him in every respect: "Bohr is the first scientist who also makes an impression as a human being. Always exercising only positive criticism . . . he is not just a physicist but much more. With me he was always especially nice. He always comes to me when he sees me anywhere, and he has invited me to see him once again next week."[57] Bohr would see much more of the young man in the months and years ahead.

8

Fair-Haired Farm Boy

B ohr's festival of lectures marked Werner's first encounter with the Copenhagen physicist and two other significant events. The festival served as the unofficial inauguration of Göttingen as a major center of theoretical atomic physics, the third point of a quantum triangle formed by Munich, Göttingen, and Copenhagen, and it inaugurated Heisenberg's long association with the Göttingen school. Werner's transfer to Göttingen for the coming school year was supremely fortunate. It coincided with the ascendance of the Göttingen school and enabled his own ascendance into the upper echelon of quantum research.

Heisenberg transferred to Göttingen after Sommerfeld accepted a guest professorship at the University of Wisconsin for the 1922–1923 school year. He did not want to miss the opportunity of promoting German culture in the former enemy nation.[1] During the Bohr festival, Sommerfeld arranged for his advanced pupils to study in Göttingen, and, on Pauli's recommendation, Born openly considered Heisenberg Pauli's successor as his privately funded research assistant. Sommerfeld agreed to the plan but only on condition that Heisenberg return to Munich the following summer to complete his doctorate: Sommerfeld did not want to lose his favorite pupil so quickly. After the Bohr festival, Heisenberg returned with Sommerfeld to Munich, where they hurriedly cowrote two papers before Sommerfeld left for Madison in August.[2]

The summer and fall of 1922 were busy seasons for Werner. In addition to writing the last-minute papers with Sommerfeld, Heisenberg delivered his first invited talk in September—probably arranged by

Prandtl — to an Innsbruck conference on hydrodynamics.[3] That same month, he attended the Leipzig meeting of the Gesellschaft Deutscher Naturforscher und Ärzté (Society of German Scientists and Physicians), or GDNA. Earlier that summer he had led his youth group on a month-long outing to South Tyrol.

The Tyrol trip served as the first of the "foreign policy" ventures instituted by the Neupfadfinder. As did many of their older academic counterparts in the Weimar era, the "apolitical" Neupfadfinder, ignoring the Weimar government, developed their own foreign policy. As youth leader F. L. Habbel expressed it, their aim was "to work successfully against the subjugation of German culture in the world."[4] Whether Werner knew of such an aim or not, it directly motivated his youth group's extended summer trips abroad. Werner joined the one to South Tyrol in the summer of 1922 and the trip to visit German-speaking Finnish nationals following the conferral of his doctorate in 1923. In 1924 his group journeyed without him to the German-speaking regions of Hungary and Poland.

In a 1927 retrospective entitled "The foreign policy of the German Pathfinder movement," Habbel, second in command of the Neupfadfinder, declared that the degradations of the Versailles Treaty, followed by the exclusion of German Boy Scouts from a 1920 international jamboree in London, impelled them to turn inward — a response similar to that of German academics to international boycotts of German scholarship. The Neupfadfinder concentrated on their own interests while taking personal responsibility for defending German culture abroad and in the "occupied territories" — those that Germany had lost to the Entente.

"The effects of the peace treaty," wrote Habbel, "forced us to defend and struggle against the suppression of German compatriots in the occupied territories, to support Germans in the separated boundary lands. The state as people, compared with the state as political accident and as a changing form of appearance, was the completely clear understanding and guide for all of our actions."[5]

For Werner and the Neupfadfinder, the current regime — beset as it was by transitory social democratic cabinets and coalitions and plagued by a faltering economy — seemed little more than an unpleasant interlude between two stable, monarchical or monarchiclike regimes. Heisenberg, for whom the war had brought an end to childhood comfort, wrote to his father in this vein in late 1922: "All of this was really only the fault of the war, which had destroyed what was earlier extraordinary and beautiful. Now we are at the point at which only one chapter is closed, only the end of the previous period is here, and the beginning of something new and 'solid' is not yet upon us."[6]

On declaring their independence from the German Pathfinders in 1920, the Neupfadfinder established their own cultural contacts and exchanges with Boy Scout units in German settlements abroad. By 1922, the plight of the southern Austrian province of Tyrol had captured the attention of all Germans. As a reward for supporting the Entente in the war, Italy had been granted South Tyrol all the way north to the Brenner Pass, just south of Innsbruck. This territory included not only the Italian-speaking region of Trentino but also the overwhelmingly German-speaking Bozen province, whose people wanted to remain Austrian. When the Fascists came to power under Mussolini in 1922, they began a systematic suppression of German culture and language in Bozen, now also called Bolzano.

Germans rushed to defend the culture of the German-speaking province. Scientists employed one of their favorite devices: meeting in sensitive locations. Prandtl held his September 1922 hydrodynamics conference at Innsbruck, the capital of all Tyrol, and the GDNA scheduled its 1924 meeting there, too. In a secret directive to Bavarian leaders in early 1922, Martin Völkel, head of the Neupfadfinder, ordered all his local units to travel abroad, especially south into Tyrol. "With this the separated Germans shall be greeted and at the same time a rigorous activity will be demanded of the groups."[7] In March, Bavarian Baron Karl Sonntag informed his subordinates and his own tribe of Völkel's orders, leaving no room for argument: "I expect from everyone unhesitating postponement of personal plans and wishes and faithful obedience."[8] Werner and his boys had just formally joined Sonntag's tribe. They headed for Innsbruck with Sonntag on July 15. Traveling farther into South Tyrol, they demonstrated their support of German Bozen by establishing camp for several weeks in this beautiful mountainous region. To make their point obvious, they journeyed all the way south to Venice before returning to Munich in mid-August of 1922.[9]

During the following summer, the Neupfadfinder exchanged visits with several hundred scouts in Hungary and Finland. In early August, the new Dr. Heisenberg led his group diagonally across Germany toward Finland. Stopping on the way in the Fichtel Mountains for a two-day Bundesfest celebrating the anniversary of the Bund, they eventually arrived at Stettin, near the northern Baltic coast of Germany.[10] From Stettin, Werner accompanied ten of the older Munich Neupfadfinder, among them Robert Honsell and Kurt Pflügel, on a visit with some families among their German-speaking Finnish counterparts. Their hosts were descendents of Austrian immigrants who had recently helped drive Soviet Russians out of neutral Finland.

The Finland trip was probably the most successful of their foreign policy gestures; the cultural contacts and new friendships lasted long afterward. Yet despite this and the political implications of the trip, the politically naive Werner and his followers regarded it as nothing more than an innocent adventure. They happily recounted their journey in a series of articles, published until as late as 1926 in *Die Spur in ein deutsches Jugendland (The Trail into a German Land of Youth)*, the magazine for Neupfadfinder followers. One of their first reports was an unsigned contribution, published in early 1924, entitled "The battle for the crossing."[11] At Werner's sixtieth-birthday celebration, he revealed with fond memories that he had written the piece himself.[12] In it, Werner recounted how he had applied charm, perseverance, and a bribe of some of Uncle Karl's dollars to convince the reluctant captain of a Finnish pleasure boat to take him and the older Neupfadfinder across the Baltic to Helsinki. The captain apparently had little use for inflated German marks. Only after a sympathetic Finnish passenger (probably of German extraction) added 500 Finnish marks to the offer did the captain grudgingly consent, and Werner and his companions clambered aboard just as the ship glided from the dock.

In another story, Kurt Pflügel recounted how he, Werner, and a friend named Wolfhard set out by boat on a hunting expedition in the Finnish lake region. They managed to shoot three ducks—but after getting wetter than the ducks, they recovered only one.[13] On their return to Munich, Werner and his comrades wrote enthusiastically to their Austrian-Finnish hosts to express their thanks.[14] In November 1923, Werner traveled from Göttingen to Berlin to receive a delegation of Finnish scouts paying a reciprocal visit.[15]

The foreign policy trip had proved a smashing success. Yet, apolitical policy aside, the most striking feature of Werner's foreign adventures is their juxtaposition with his other major pursuit during this time. He was a scientist of extraordinary abilities, already near the top of the profession that would ultimately produce quantum mechanics. He was immersed in research that was complex, sophisticated, and demanding. Yet he still engaged in adolescent romps and other uncritically naive activities as a German Boy Scout.[16] Apparently such immature behavior and the extended periods of outdoor fun provided a necessary counterbalance to and relief from the intense, technically abstruse physics he was creating.

Letters to his family and colleagues during and after long camping tours indicate that Heisenberg completely banished physics from his thoughts during his trips. For instance, after a month-long tour through Upper Bavaria in 1925 that occurred just after he had laid the foundations for the matrix form of quantum mechanics (an enormously exciting time for him professionally), the physicist warned Bohr: "Obviously I have not thought at all about physics during the entire last month and I don't know if I still understand anything of it."[17]

Heisenberg came face to face with the hard reality of current events soon after returning from his trip to Innsbruck in September 1922, which had followed the Tyrol excursion. He was home less than a week before setting out again, this time for Leipzig, the site of the biannual meeting of the GDNA. It was the centennial of the founding of the prestigious society, and Max Planck, then chairman, decided to use the occasion to promote a new sense of unity among German and Austrian scientists. He scheduled a general lecture for September 18, the first day of the conference. Albert Einstein, Germany's most renowned scientist, would speak on the theory of relativity, touted as one of Germany's most renowned achievements. Before leaving for Wisconsin, Sommerfeld encouraged Heisenberg to attend so that he might finally meet the great man. Werner's father generously provided the round-trip train fare plus 2000 marks, some of which Werner planned to use for another visit with the Berlin youth contingent after the meeting.[18]

Unfortunately, Planck's hoped-for demonstration of unity faltered in the face of the mounting discord over relativity and the mounting anti-Semitism aimed at Einstein. Still leading the attack on both fronts was the experimental physicist and Nobel prize laureate Philipp Lenard. After a debate with Einstein on relativity at the previous GDNA meeting in 1920, Lenard was unconvinced of relativity and unswayed from his own alternative, a classical ether theory of electrodynamics. He was already nurturing paranoid anti-Semitism.

As the 1922 meeting approached, Lenard published "A word of warning to German scientists" in his latest monograph on ether theory. In it, he dismissed relativity as a mere hypothesis and closed with an anti-Semitic diatribe against his critics.[19] Unbeknownst to Heisenberg and to Einstein's opponents, Einstein had withdrawn temporarily from public appearances after the shocking assassination in June 1922 of Walther Rathenau, the famous Jewish foreign minister. Einstein was replaced at Leipzig as the featured speaker on relativity by the man who would do so much for him a decade later—Max von Laue.

Heisenberg arrived in Leipzig on September 17 and checked into a cheap youth hostel in the poor quarter of town to conserve his funds for travel and the endless rounds of coffeehouse conversations.[20] As he approached the lecture hall the following evening, one of Lenard's disciples pressed a leaflet into his hand. The leaflet had been signed by 19 professors and doctors — with titles prominently displayed for effect — who proclaimed that they "not only regard the relativity theory as an unproved hypothesis, but even reject it as a basically failed and logically untenable fiction."[21]

Heisenberg was shocked. Unlike the earlier Munich episode, which involved fanatical local students — an incident that Werner himself did not witness and of which he learned nearly a year later — this was a direct confrontation with anti-Semites supported by learned professors, including the eminent Nobel laureate Lenard. The incident apparently shook him at last into brief recognition of the political implications. "I felt as if my world were collapsing," he wrote years later in his memoirs. He had always thought that science was above politics; indeed that was one reason he had chosen physics as a career. "And now I made the sad discovery that men of weak or pathological character can inject their twisted political passions even into scientific life." Heisenberg claimed that he even wondered at the time whether physics was "really worth bothering with" after all.[22] Such doubts, however, could not deter him from his richly promising future.

Despite whatever realizations — or remembered realizations — about the political susceptibility of physicists that Werner may have gained from the Leipzig encounter, they did not inspire him to take much action. If anything, the experience caused him to cling even more tightly to his apolitical illusions for science until forced to relax his hold a decade later. Nor did he display any increased interest in or concern for political affairs, even as a safeguard against their corrupting influence.

Werner returned to Munich the very next day in a depressed state — but not only because of politics. He had returned to his hostel after a lecture to find that he had been robbed of all his money and belongings.[23] Faced with showing up at the conference unwashed and unshaved, he returned to Munich, where he put in a stint as a woodcutter in order to earn back his money and enough to buy new belongings.[24] He did not know that the man who spoke that night in Leipzig was not Einstein but Laue.[25]

Heisenberg finally arrived in Göttingen in late October 1922 for the start of the winter semester. To Born, the slightly built sometime woodcutter looked "like a simple farm boy, with short, fair hair, clear bright eyes, and a charming expression."[26] Despite the farm boy's excellent references and despite his own intention to hire him, Born decided to see first if he would be getting his money's worth. Werner himself wanted to see what his new environment would offer him. What it offered was his first systematic introduction to mathematical atomic physics. Göttingen's mathematics tradition, Born's appointment, Bohr's lectures, and the generosity of various foundations had ensured Göttingen's place among the leading centers of atomic physics.

Göttingen mathematics boasted a long line of luminaries, among them Gauss, Riemann, and Felix Klein. Klein, who arrived in 1886 to head the mathematical-physical seminar, later founded a series of institutes and research programs for pure and applied mathematics that made Göttingen the leader in such research.[27] By the time Heisenberg arrived in 1922, Göttingen could boast of a Mathematics Institute headed by Courant, Hilbert, and Landau, an Institute for Applied Mathematics and Mechanics directed by Runge and Prandtl, and three separate Institutes for Physics. The last of these had been headed until 1920 by two full professors, Peter Debye and Woldemar Voigt, and an associate professor (Extraordinarius), Robert Pohl. When Debye left for Zurich in 1920, Pohl was promoted to full professor and head of the experimental physics section, and Max Born, the highly regarded former student of Hilbert and former assistant to the famed Hermann Minkowski, was called from Frankfurt to succeed Debye.

Born in Breslau of an academic Jewish family, Max Born had early devoted himself to mathematics and, like Sommerfeld, had turned to theoretical physics after encountering Felix Klein in Göttingen. Briefly assisting Minkowski until Minkowski's untimely death in 1909, Born continued Minkowski's work on relativistic electrodynamics, then turned to the quantum theory of crystals and molecular structure under the influence of Einstein's quantum theory of specific heats. He was best known by 1920 for his book on the dynamics of crystal lattices and for his work on the chemical consequences of a theory of ionic crystals that he had developed with Alfred Landé.

During World War I, Born, a noncommissioned officer, performed artillery research for the army in Berlin, while occupying an associate professorship at the university. While in Berlin, Born and his wife, Hedwig, a writer of romances, became close friends of Albert Einstein (then between marriages), with whom they frequently corresponded thereafter.[28] In 1919, Max von Laue, then in Frankfurt, suggested to

Born an exchange of positions, and Born readily agreed. It would mean promotion to full professor. Although he and Hedwig thoroughly enjoyed Frankfurt for its cultural offerings, they remained in the Goethe city barely two years. Born, then 38 years old, seemed an ideal candidate for the vacancy left by Debye in Göttingen.

Born hesitated. The shy and retiring theorist, plagued by hypochondria, was not attracted by big-science administration and had no desire to teach experimental physics in addition to theory. Meanwhile, the Frankfurt faculty was doing everything they could to keep him. Born went to the Prussian Culture Ministry in Berlin, which oversaw university appointments, to discuss the matter. He later recalled that, in reviewing the Göttingen files at the ministry, he discovered a notational error that provided for an extra associate professorship at the institute. Born easily convinced the Prussian ministry to make the extra position a full professorship, to be held by an experimentalist. To clinch the deal, the ministry doubled its salary offer.[29]

With Born's arrival in 1921, Göttingen physics was reorganized in typically German fashion. Three independent institutes were created, each headed by a full professor and all housed in one building, the box-shaped Physics Institute at Bunsenstrasse 9. Since experimental physics still enjoyed more prestige and more direct connections with the original institute, Pohl directed the First Physics Institute, which was devoted to experimental physics. Born selected his good friend James Franck, who with Gustav Hertz had lent experimental support of Nobel prize quality to the Bohr atom, to head the newly created Second Physics Institute — also devoted to experiment and to the beginner's laboratory. Born himself directed the Institute for Theoretical Physics, consisting of one small room, one assistant, one "private" assistant, and a half-time secretary.[30]

The bashful Born seemed overwhelmed by the number of students flocking to Göttingen — only the Berlin Technical College had a greater enrollment. "There are students here like hay," he wrote a colleague.[31] During the semester in which Werner studied in Göttingen, more than a third of all the students enrolled at the university studied mathematics or science, by far the most popular subjects in Göttingen.[32] Each of Born's lectures, like Sommerfeld's, drew about 80 students from all majors and degree programs, but many more physics doctoral candidates attended than in Munich. Born originally had nine advanced students in the winter of 1922–1923; Sommerfeld sent four more from Munich during his absence, one of whom was Heisenberg.[33]

Werner delighted at first in the picturesque walled town of Göttingen, with its "narrow alleys and strange [peasant] dialect" and with a university only a third the size of Munich's (about 3000 students).[34] A guide for English-speaking students described Göttingen as a small town where "life is comparatively quiet, and there are no noisy factories. In the outer town, peace and quiet are almost undisturbed."[35] The major industries manufactured instruments for the university laboratories, and town leaders were professors, army officers, and retired bureaucrats. Although it offered little public culture compared with Munich or Frankfurt, many of Göttingen's science professors were music devotees and often had Heisenberg to their homes for musical evenings. "Heisenberg is at least as [scientifically] talented as Pauli," Born reported to his friend Einstein, "but personally more pleasant and delightful. He also plays the piano very well."[36]

But as winter set in Werner sank into depressed loneliness. The many science and physics students around him were no replacement for his real friends, the Munich youth group. For the first time, he was away from both home and friends for an extended period. He soon felt trapped in the north German town and penned numerous mournful letters to family and friends. "In general there just aren't any people here, or I can't find them," he complained to his brother.[37] Werner tried to distract himself by cramming his weekdays with physics. On weekends, despite the cost, he traveled to Berlin to be with Heini Marwede and other expatriot Bavarians from his youth unit—but only rarely did he visit his brother, who was studying chemistry at the University of Berlin.

"You appear to give up your entire day only to physics," one of Werner's comrades noted.[38] By Christmas vacation, which Werner planned to spend skiing with his youth group in the Bavarian mountains, he had burned himself out. "If the holidays had started only 10 days later, I would go crazy with physics," he wrote to one of them. "In Munich I will not speak one word of it. In personal terms Göttingen remains a completely desolate hole."[39] Periods of intense and lonely work, followed by long and relaxing outings with the group, became the pattern of Heisenberg's life.

Like Munich, Göttingen had experienced a postwar soviet upheaval, but it was quickly suppressed after a battalion of government troops took up permanent residence in the town. Politically, Göttingen students tended to be right wing, nationalistic, and anti-Semitic, but they were not yet as brutal as their fellow students in Munich.

As everywhere in Germany, the economic plight of Göttingen students grew desperate after the war. But because of the central place of the university in Göttingen civic life and the dangers of student unrest, town and university officials paid more attention to the needs of students. Although all the universities were publicly funded, private contributions helped to establish perhaps the earliest student Mensa, a cafeteria serving subsidized meals, and to acquire a building to serve as a dormitory.[40]

With so many students flocking to the university town from all over Germany, the housing shortage became acute. Heisenberg began the search for a room early and soon found one through the efforts of Erich Hückel, Born's university assistant. Werner lodged in a spare second-floor bedroom of the large and stately Biedermaier home of the widowed Mrs. Ulrich at Walkemühlenweg 29, just a block from the Physics Institute.

Money was again a problem. When Mrs. Heisenberg stopped off in Göttingen on her way home from Osnabrück that November, shortly after the start of the winter semester, she gave her son 8000 marks for his monthly expenses. Inflation, though still in its early stages, quickly ate it up. For 1000 marks a month, Mrs. Ulrich supplied a bed, breakfast, afternoon tea, and evening potatoes. On top of this, Werner had to pay for heat, which cost 2600 marks; seminar fees for the semester were 718 marks; and once he had to buy Mrs. Ulrich a pound of butter for his fried potatoes, which cost him 750 marks. When in December Mrs. Ulrich gave him three eggs and a bouquet of flowers for his birthday, Werner fretted over whether he could afford to reciprocate with a Christmas gift.[41] The only bargain was lunch. Instead of eating at the Mensa, for 50 marks Werner and the other physics students attended a subsidized private table at a home across the street from the institute. This was the biggest meal of the day and even sometimes included a meat dish. The subsidy had probably been arranged by Born, who, like most science professors in the period, could rely on the generosity of a befriended industrial philanthropist.

By the end of November, August Heisenberg had to forward another 3000 marks to Werner and to raise his allowance for December to 10,000 marks. A thankful Werner offered: "In case you should go bankrupt, I can make out in an emergency with 1000 marks less per month."[42] By January his money problems were solved, at least temporarily. Professor Born had finally offered him the private assistant's position with a generous salary of 20,000 marks per month, supplied by the American philanthropist and financier Henry Goldman.

Born's offer came within days of Heisenberg's first triumph in the Physics Colloquium, the high court of Göttingen physics. There local and guest speakers submitted their latest work to careful scrutiny and themselves to withering cross-examination. "It was customary to interrupt the speaker and to criticize ruthlessly," Born recalled.[43] The Physics Colloquium was an example of the emphasis placed on oral and personal interaction in Göttingen. Born, a former Göttingen mathematics assistant who had handpicked one of his colleagues, worked much more closely with the local mathematicians and experimentalists than did Sommerfeld in Munich. Consequently, a variety of joint lectures, seminars, and colloquia were held, in addition to each professor's individual seminars.

Heisenberg registered for all of these symposia. In each, he was expected to defend his position — which was not difficult for him — and to employ proper oratorial and diplomatic skills — which was. "Thus I will no doubt learn how to lecture."[44] And learn he did. In January he reported to Sommerfeld: "The result so far is that Born and Hilbert are of the opinion that I can lecture very well."[45]

Heisenberg's new skill, combined with his intellect and demeanor, enabled his early acceptance by the Göttingen critics. Within weeks of his arrival, Heisenberg reported to the Hilbert-Born seminar on the recent Sommerfeld-Heisenberg papers, and in December he was called before the Physics Colloquium to report on his first and still only individually written paper, his controversial core model of the atom for the anomalous Zeeman effect. Usually only physicists attended the colloquium, but on this occasion Hilbert and Courant of the Mathematics Institute were in the audience, and the aged Carl Runge showed up to hear firsthand the author of the only atomic model that purported to account for the Runge fractions of the Zeeman effect.

The audience in that colloquium was already well aware of Heisenberg's core model and equally aware of Bohr's explicit criticism of it during the Bohr festival. As expected, the distinguished audience that gathered in the same lecture hall where Bohr had spoken the previous summer was already skeptical of Werner's model and prepared for ruthless criticism of the upstart 21-year-old. But Werner, who had demonstrated his daring before distinguished audiences in that same room the previous summer, was prepared for the challenge. He poured as much "verve and overassertion" as possible into his performance in an attempt to sway the experts.

"The result was resounding," he bragged to his brother. The audience did interrupt him several times — but only with applause. Even Runge conferred his blessing, no doubt because his fractions had at last found a raison d'être. "So now all of Göttingen is convinced of the theory," Werner proudly reported.[46] Within a month the newly polished lecturer was working for Born, and Born admiringly confided to Sommerfeld Göttingen's infatuation with the youth: "I have become *very* fond of Heisenberg; he is very well liked and highly regarded by us all. His talent is unbelievable, but his nice, shy nature, his good temper, his eagerness and his enthusiasm are especially pleasing."[47]

If Heisenberg impressed Göttingen physicists, they at first did not impress him. He found them "very strange" compared with his Munich colleagues. The former math disciple complained of too much concern for mathematics. "Even the physicists are actually interested much more in mathematics than in physics," he wrote to his father in November. "The result is that one has a somewhat bored impression of all the physics here; no one has the initiative to try something new; they pick out mathematically interesting topics that are in most cases exhausted as physics."[48] The quiet and subdued Born paled in comparison with the dynamic and forceful Sommerfeld, and Born's preference for rigid mathematical and physical consistency seemed much too tame for Werner.

Between the Bohr festival in June and Werner's arrival in October 1922, Born had settled on his own research program, which indeed contrasted sharply with Munich's: an even closer adherence to the elements of quantum atomic theory—both the quantum rules and the classical mechanics—than Bohr had just argued in Göttingen. "It was the time before the establishment of quantum mechanics," he later wrote, "and I was trying, with my collaborators, to find weak points and contradictions in Bohr's semiclassical theory of atoms."[49]

While Munich physicists unraveled spectroscopic number harmonies and constructed deviant models to explain them, Göttingen theorists, when studying atoms, constructed thoroughly consistent planetary atomic models, adopting the mathematical methods and mechanical techniques of planetary astronomy. Absolute consistency would illuminate the weaknesses and contradictions in Bohr's theory, they reasoned. Despite his skepticism, Werner quickly appreciated that: "For me personally Göttingen has the great advantage that for once I will learn correct mathematics and astronomy."[50]

By the time Heisenberg arrived, his two predecessors, Pauli and Ernst Brody, had helped to establish Born's program—and to appreciate its difficulties. Bohr's quantum theory of atoms and molecules worked well, according to the demands of mechanical stability and observed ionization potentials, but only when applied to two-body atoms, hydrogen (one electron orbiting a positively charged nucleus) and ionized helium (one electron orbiting a doubly charged nucleus). It failed when applied to anything more complicated, as Pauli had just demonstrated in his dissertation on the ionized hydrogen molecule (one electron orbiting two hydrogen nuclei).

Mathematically, however, Pauli's calculation was only a first approximation, since he assumed that the two nuclei are completely at rest. Allowing the nuclei to move about, as they do, required more detailed approximations and more sophisticated techniques. Born and Brody, his first assistant, developed such techniques by adapting the so-called perturbation theory of celestial mechanics to orbiting electrons.

In perturbation theory—a crucial technique used throughout and beyond the development of quantum mechanics—a very complicated motion is treated as an unending series of simple approximations to the real motion. The first approximation is usually the simplest and most basic—say, a ball whirled by hand on a string or a planet orbiting the sun. In the second approximation, a linear "perturbing" effect on the motion is taken into account—say, a lengthening of the string or the presence of a nearby planet. The effect is often added to the energy, multiplied by a linear factor. In the next approximation, quadratic perturbations are taken into account, such as an unsteady hand whirling the ball or an inhomogeneity in the gravitational field of the sun. Yet another term is added to the energy, this time with a quadratic factor, and so on. The energy, or the motion itself, is then expressed in the abstract mechanical formulation developed earlier by Hamilton and Jacobi. The general momentum and position variables for each degree of freedom are obtained using the Hamilton-Jacobi equations, and the motion is then quantized by substituting the position and momentum variables into the Bohr-Sommerfeld quantum conditions.[51] These conditions quantized the motion variables by requiring them to assume not a continuum of values, but only certain, discrete values, each associated with an integral quantum number.

Like the anomalous Zeeman effect, built up from a sequence of progressively refined splittings of the spectral lines of the original Bohr atom due to various perturbing effects within and outside the atom, complicated motions of planets or atoms or anything could be built up by progressive approximations. But unfortunately for Born and his assistants, quantized planetary mechanics could not be transferred directly to electrons orbiting within atoms because the quantization of electron orbits using the quantum conditions worked only for motions that were strictly periodic or conditionally periodic. Other types of motion required an approximation to periodicity at each order of perturbation.

Even worse, electrons in atoms repel each other with an electrical force that is almost equal to the attraction to the nucleus. Planets only attract each other and with gravitational forces that are much weaker than their attraction to the sun. Moreover, Bohr had recently hypothesized that each atom in a row of the periodic table, ending in a noble gas atom, is formed from the previous atom by increasing the charge of the nucleus by one and adding one electron to the solar system of electrons orbiting the nuclear sun. Each row corresponds with a shell of electron orbits, and the maximum number of electrons in each shell is governed by specific relationships, phase relations, between the orbital frequencies of the electrons in each shell. If correct, this meant that accidental degeneracies (the orbital frequencies are fractions of each other) occurred much more frequently in atoms than they did in planetary motions, further obscuring the already obscured planetary analogy. Finding an adaptation of planetary physics that would accommodate all these complexities occupied the Göttingen theorists throughout the early 1920s.

In the spring of 1922, Born and Pauli managed to develop more general perturbation techniques, applicable to both degenerate and nondegenerate cases.[52] With the Bohr model of the hydrogen atom well established, researchers turned to the second element of the periodic table, neutral helium (two electrons orbiting a doubly charged nucleus). But even this seemingly simple problem was further complicated by two different forms of helium, called parahelium and orthohelium and distinguished by their singlet and doublet spectra, respectively. The chemical stability of the noble gas helium suggested a high degree of symmetry within the orbits of the two electrons, which formed a closed shell. Yet every symmetric model but one proved a failure; the exception was a Sommerfeld-Heisenberg invention for parahelium. The ground state (lowest energy) of parahelium has zero net angular momentum, but since quantum theory left room only for integer quanta and quantum numbers, the lowest allowed total angular momentum with any motion at all had to be unity. In 1922 Sommerfeld borrowed the deviant half quanta from Werner's core model of the Zeeman effect to suggest that each of the two helium electrons is somehow endowed with a half unit of orbital angular momentum, their total yielding one as required. At the same time, zero net angular momentum in the ground state could be achieved if the two electrons move on orbits in the exact same plane but in opposite senses to each other. Their momenta would then be directed in opposite senses and thus result in zero sum momentum.

Along with the core model, Munich physicists brought their ingenious half-integral helium model to the Bohr festival, where Bohr summarily rejected it. He had already decided to blame another facet of the theory for helium's problems and to maintain the integers required by the quantum conditions. Although Born and Pauli had recently shown Bohr's own model for orthohelium (nonzero ground state momentum) to be unstable,[53] Bohr's assistant, H. A. Kramers, had been carrying out long calculations of each of the quantum energy levels for both forms of helium by expanding the energy in powers of the nuclear charge—one form of perturbation theory. Bohr reported the first results at the Bohr festival,[54] but by December 1922 Kramers's models had proved to be both unstable and too easily ionizable; the predicted energy required to pull one of the electrons from the atom was much lower than that experimentally observed. Assessing the situation, Bohr and Kramers decided to blame the use of classical mechanics in the calculation of the orbital motions rather than to lay the blame on the sacrosanct quantum rules. Nor did they fault (as did Pauli, their new Copenhagen colleague) the overall approach to atomic physics in which one reproduces the data by constructing mechanical atomic models.

Bohr proclaimed a new but consciously chosen violation of orthodoxy: Kramers's model really was correct, he declared, but the use of classical mechanics to analyze it was not.[55] Kramers echoed Bohr's sentiment nearly word for word. "Mechanics is not valid in this simple case," he wrote in December 1922. "As recently emphasized by Bohr, one must generally expect that in the stationary states these laws are different from those of the usual mechanics."[56] Born and Heisenberg would soon test that assertion.

Although Bohr had rejected their half-integral helium model along with Werner's core model, Sommerfeld encouraged Heisenberg to analyze it anyway, when he found time in Göttingen. Heisenberg had plenty of other work to fill his lonely hours. Before leaving for Wisconsin, Sommerfeld had assigned each of his pupils a research problem to keep them busy. Heisenberg received the difficult task of calculating the conditions for the onset of turbulent flow in hydrodynamics—with the promise that he could submit the results, if satisfactory, as his doctoral dissertation the following spring. Heisenberg delivered a short report on his preliminary results to the Innsbruck hydrodynamics conference in September 1922, and he sought further advice on the problem from

Professor Prandtl in Göttingen. He wrote Sommerfeld and Landé that fall that he was too busy with hydrodynamics to worry about atoms.[57] But in his last letter to Sommerfeld before leaving Munich for Göttingen at the end of October 1922, he reported that he had been unable to contain his curiosity: he had already studied the Born-Pauli paper and had just applied the Born-Pauli perturbation method to Sommerfeld's helium model.[58] A rough calculation yielded precisely the measured ionization energy!

Sommerfeld was ecstatic. After Heisenberg forwarded the detailed calculations, carried out with the help of a Göttingen student, Sommerfeld published the result in an American journal (with a thank-you to his German pupil).[59] Both Sommerfeld and Heisenberg were now convinced of the errant half-integral quantum numbers and momenta — prominent components of the core model and their helium atom. While Bohr and his Copenhagen colleagues contemplated the failure of mechanics in atoms, fractional angular momenta soon became a permanent option in Göttingen, Bohr's objections and the quantum conditions notwithstanding.

The Bohr festival had inspired a new dedication in Göttingen to consistent adherence to quantum theory. Five days after Bohr's last lecture in June 1922 — while Heisenberg busied himself that summer with other matters — Born announced his new approach: "The time is perhaps past when the imagination of the investigator was given free rein to devise atomic molecular models at will. Rather, we are now in a position to construct models with a certain, although still by no means complete, certainty through the application of quantum rules."[60]

To demonstrate his new devotion, Born freely constructed all imaginable models for the neutral hydrogen molecule, H_2 (two electrons orbiting two hydrogen nuclei), and then eliminated all but one, a crossed-orbit model, through the application of quantum conditions. A student, Lothar Nordheim, received the problem for his dissertation, while Born turned to a similar but mechanically more complicated problem: excited neutral helium.

Excited helium would serve as an explicit test case. By closely adhering to both quantum rules and classical mechanics, consistency would show if and exactly where the theory failed. But to handle helium, the Born-Pauli apparatus required extension to mechanical cases of accidental degeneracy. Hints could be gleaned from Poincaré's old advanced textbook on planetary celestial mechanics. Shortly after Heisenberg arrived, Born started a private Poincaré reading circle on Monday evenings in his home. Due to the housing shortage, the Born home, located near the institute, on a street named for another Planck, was a rented ground-floor apartment consisting of three enormous rooms and a kitchen. Born, his wife, and two young children subdivided two of the rooms into bedrooms and designated the third a combination parlor, study, and music room. A first-rate Steinway grand, which Werner often played, occupied one corner, the mechanics reading circle another.[61] Werner attended along with several other advanced students and assistants, and in seminar fashion each prepared a paper or talk on a section of the material, which the group then discussed with much critical enthusiasm. Continuing his hydrodynamics with Prandtl and his mathematics with Hilbert and Courant, Werner also studied Poincaré with Born "with every ounce of energy."[62]

By late December 1922, Born and Heisenberg had obtained the needed extension of Born and Pauli's method. They quickly used it to test Bohr's explanation for the closing of shells in the building up of atoms, confirming the hypothesis to their own satisfaction.[63] "Born is very enthusiastic about these results," Heisenberg wrote to Pauli, "because perhaps we now have a simple mathematical method for determining the symmetry relations and periodicities in the system of elements."[64] The entire periodic table could be constructed by building up each successive element through the addition of another electron to the outer electron shell, each of which closed as it should at a noble gas atom. Bohr aptly named this process the building-up principle.

At a time when problems in quantum theory were mounting, Göttingen had produced two positive results: the Sommerfeld-Heisenberg helium model and solid support for Bohr's building-up principle. But Bohr and Pauli (now in Copenhagen) both complained, rendering Werner more than a little uneasy: "I am somewhat unhappy over the fact that with all of these papers I constantly contradict Bohr and Pauli."[65] Half-integral momenta, detailed mechanical planetary models, and strict adherence to stationary-state mechanics all contradicted the latest Copenhagen line. Mechanics, not quanta, were to blame for the theoretical impasse.

But Heisenberg was not particularly attracted to the Copenhagen alternative. He responded with sharp criticism of Pauli's latest efforts to avoid the deviant core model by explaining the anomalous Zeeman effect with little reliance on any model at all. Pauli responded in January 1923 with a two-day stopover in Göttingen to argue the Copenhagen interpretation on his way back to Copenhagen from Vienna. Born, Heisenberg, and a handful of students tangled with the visitor. Although "he is besotted with Bohr [*verbohrt* (the German adjective for *stubborn* providentially contains Bohr's name)]," Heisenberg reported, "Pauli admitted that our standpoint is very consistent." But Pauli would make no concessions on Bohr's behalf: "Bohr in any case wants to allow mechanics to be no longer valid, i.e., only to a certain approximation. He does not yet believe in half quanta and helium."[66]

To resolve the debate over classical mechanics in quantum atoms, Born and Heisenberg turned at last to the excited helium atom as a careful, systematic test to prove or disprove the viability of quantum atomic theory. A rigorously consistent application of the most sophisticated planetary mechanics to the quantum orbits could decide the matter better than personal preferences. Heisenberg informed Bohr of the plan in early February 1923: "The other work of which I wanted to write you is a general investigation of all mechanically allowed orbits of excited helium. If in the end the experimentally found terms are not included, then one knows that the mechanics is wrong."[67]

Heisenberg, however inconsistent otherwise, kept his word, as did Born, in following the demands of theoretical orthodoxy to the letter.[68] The advantage of excited helium was that it could be treated almost exactly like a hydrogen atom—a perfect setup for perturbation theory. In an excited state, one of the two helium electrons moves in an orbit farther out from the nucleus than the other more tightly bound inner electron. Since the negative charge of the inner electron shields one of the positive charges of the nucleus, the entire effect of the inner electron and nucleus on the outer electron could be treated as a small perturbation of the motion of the single outer electron, hence as a small perturbation of the well-established hydrogen model (a single electron orbiting a single positive charge). This arrangement, the simplest possible three-body case after two-body hydrogen, should definitely prove whether or not the prevailing quantum theory of the atom was valid.

Bohr had earlier shown how the jumps between the quantum states of the orbiting hydrogen electron could yield the so-called Balmer series of spectral lines, if each state is represented by an integer. Bohr had also shown that a correction to the Balmer series, a Rydberg correction, could be obtained for heavier atoms if the effect of the nuclear charge were taken into account. Using their new quantum planetary mechanics, Born and Heisenberg derived an expression for the correction to the Balmer formula for the case of the outer helium electron. The exact value of the predicted correction depended on the mechanical orbit chosen. The physicists obtained four possible stable orbits. They dutifully discarded one of the four because it involved a nonintegral quantum number. A comparison between the observed value of the Rydberg correction for excited helium and the derived values for the remaining three possible orbits turned out, Born told Bohr, completely catastrophic.[69] None of the values agreed.

As early as February 19, 1923, Heisenberg was able to inform Pauli of the result and of his own conclusion: "This result appears to me to be . . . very bad for our present conceptions. One must probably introduce entirely new hypotheses — either new quantum conditions or new modification proposals for mechanics."[70] A month later the Göttingen conviction had grown even more radical: "All present He models are just as wrong as the entire atomic physics."[71]

Bohr, though vindicated in his skepticism of planetary mechanics, would not go as far as his Göttingen colleagues. Quantum rules and quantum theory should be preserved even after jettisoning the classical mechanics of atomic orbits. He wrote to Born that the helium calculation "is certainly very important as evidence for the inappropriateness of the present foundation of quantum theory, as far as systems with several electrons are concerned." But he still believed that somehow the failure of mechanics could be incorporated into quantum theory in such a way that the original quantum principles could remain untouched.[72]

Born, Heisenberg, and many of their colleagues were more willing to accept the inevitable. In July 1923 Born declared in a review article that the need for deliberate deviations in spectroscopy — that is, in Heisenberg's core model — and the obvious failure of all quantum models for simple atoms and molecules clearly demonstrated "that not only new assumptions in the usual sense of physical hypotheses will be necessary, but the entire system of concepts of physics must be rebuilt from the ground up."[73] With the core model held to be too controversial, Born and Heisenberg's strictly orthodox helium calculation, which was submitted in May 1923, gained widespread recognition as the beginning of the end for the earlier successful Bohr-Sommerfeld quantum theory of the atom.

Sommerfeld returned to Munich that month, and Heisenberg left Göttingen at the end of May to complete his dissertation during his sixth and final semester as a student. It was a great relief to be back in Munich. Not only did living at home mitigate the effects of the accelerating inflation, it also brought to an end Werner's painful separation from his comrades. Even as he looked forward to his return to Munich, on the eve of his departure from Göttingen Werner wrote a depressing letter to Eberhard Rüdel (Wolfgang's elder brother) in Erlangen, to which Eberhard replied: "From what you write in your last letter it appears to me that you were not in the best mood, that you did not know what you could do other than physics and music and [that you] nowhere really fit in, in other words that you missed your group."[74]

Heisenberg also returned eagerly to the circle of his Munich colleagues. Of those who were there a year earlier, Herzfeld and Kienle were still in Munich, Wentzel and Rüchardt had habilitated and were now lecturers, but Kratzer and Wagner had gone to chairs in Münster and Würzburg, respectively. Since the mathematicians Pringsheim and Rosenthal did not lecture that semester, and Perron and Voss offered only elementary topics, Heisenberg concentrated on his dissertation, Sommerfeld's lectures and seminars, and Wien's four-hour lab course. (Wien's eight-hour course conflicted with Sommerfeld's advanced lectures on spectroscopy.) Sommerfeld's main lecture that semester, on partial differential equations in physics, was probably most useful to Heisenberg in his work on his thesis problem: solving the horribly complicated equations for the stability and turbulence of flowing fluids.

Heisenberg had been working on the problem for over a year. It derived from the work of the English physicist Osborne Reynolds in the 1880s and from early experiments at Sommerfeld's institute, conducted at the request of the Isar Company, which had contracted to channel the Isar River. Heisenberg's puzzle concerned the determination of the transition from laminar (smooth) flow to turbulent flow for channeled liquids, a phenomenon that occurred in the greenish Isar River as it flowed north through Munich from the foothills of the Alps. Sommerfeld's student Ludwig Hopf had experimentally examined the problem over a decade earlier, but no one had yet discovered how to predict the precise transition to turbulent flow.

Reynolds had treated the problem on the basis of energy conservation and found that a dimensionless constant, which came to be known as the Reynolds number, governed the transition to turbulence. Heisenberg set out to derive Reynolds's results from the fundamental equations of hydrodynamics.[75] In his 59-page thesis submitted to Philosophical Faculty II (science) on July 10, 1923, Heisenberg divided the problem into two parts. In the first he examined the conditions under which laminar flow becomes unstable, while in the second he investigated the role of the Reynolds number.[76] Having already obtained, during his second semester, the solution for turbulent vortex motions between two parallel plates, and having spoken at Innsbruck on the derivation of the Reynolds number,[77] he found little difficulty in solving either part of his thesis using various approximation and simplification techniques. Willy Wien accepted his results for publication in 1924 in the *Annalen der Physik*. But after the mathematician Fritz Noether challenged the results, they remained in doubt for nearly a quarter of a century before

they were finally confirmed. Heisenberg did not publish again on hydrodynamics until 1946.

According to university regulations, a Munich dissertation was first
submitted to the dean of the subfaculty, who passed it along to the
student's advisor for critical assessment and a vote for or against acceptance. The work was then circulated among the entire subfaculty; if they
accepted it, the candidate was admitted to the final orals.[78] Sommerfeld
was worried about possible objections to the fact that Werner's solution
of the problem was only approximate and incomplete. While conceding
in his two-page, typed Votum that "the work still leaves much to do
with respect to the mathematics," Sommerfeld argued that the equations were so complicated that even approximate solutions were sufficient for a dissertation. But Sommerfeld's strongest argument in favor of
acceptance was founded on the talent of the candidate even more than
on the content of his work. "In the handling of the present problem,"
Sommerfeld concluded, "[Heisenberg] shows once again his extraordinary abilities: complete command of the mathematical apparatus and
daring physical insight. I would not have proposed a topic of this
difficulty as a dissertation to any of my other pupils. I therefore move
for acceptance of this work."[79]

More at home with classical hydrodynamics than with contemporary
quantum theory, Willy Wien seconded the motion, "even though
doubts may be raised from the mathematical side against the considerations presented." The rest of the science faculty signed without reservation, and Heisenberg's oral *examen rigorosum* was set for 5:00 P.M. on
Monday, July 23, 1923, in the seminar room of the Theoretical Physics
Institute.[80]

In Munich, a doctoral candidate's grades were based solely on his or
her dissertation and performance on the final orals. (Women were now
granted doctorates without prejudice in Germany.) Four passing grades
were possible: I (summa cum laude), II (magna cum laude), III (cum
laude), and IV (pass). At the completion of the orals, grades were given
for the major subject, for each of two required minors, and for overall
performance, the last being the most significant. The examining committee consisted of the professors in the candidate's two minor subjects
— in Heisenberg's case, Perron for mathematics and Seeliger for
astronomy — and the professor of his major subject. Since Munich
physics was split between Wien and Sommerfeld, both attended the
orals and both had to agree on a single grade.

Trouble was already brewing for Werner. Aside from the differences
between Wien and Sommerfeld, Heisenberg and Pauli had made the
mistake of working more on theory than on experiment when they first

took Wien's lab course, which Wien did not appreciate. Having taken a second lab course, Pauli managed to satisfy Wien in his orals and graduated with the overall grade of I.[81] Werner did not fare so well when he enrolled again in a lab course with Wien that summer. Wien presented him with a particularly difficult problem: Werner was to use the Fabry-Perot interferometer to measure the hyperfine structure in the anomalous Zeeman effect of mercury (an even finer splitting of lines later attributed to electron interactions with the nucleus).

Raging inflation made the acquisition and even the repair of equipment in Wien's laboratory nearly impossible, but Heisenberg did not know—or bother to find out—that he could use the institute workshop to construct his own equipment. Fresh from successes in the rarefied realms of Göttingen physics, the overconfident theorist had little use for laboratory exercises. Nor did he trouble himself to consider the theory of his instruments, despite its obvious relevance to his own work on the Zeeman effect. For his experiments, Werner simply threw together a slapdash contraption with cigar boxes and sealing wax. Such negligence only further incensed the already provoked professor, who proceeded to pounce on the offender in his orals.[82]

As the three aging professors met with their middle-aged colleague, Sommerfeld, late that summer day in the seminar room of Sommerfeld's institute, the object of their interest, the 21-year-old Werner, seemed confident. But that quickly changed. He easily handled Perron's mathematics questions and Sommerfeld's questions on theoretical physics, but he began to stumble over Seeliger's inquiries on astronomy—and fell flat on his face when confronted by Wien. Wien's ire mounted as Heisenberg proved unable to derive the resolving power of the Fabry-Perot interferometer or even that of the telescope or microscope, all of which Wien had discussed extensively in his lectures. Wien then asked Heisenberg to explain in detail how a storage battery works. The candidate was still lost—despite his earlier adolescent fascination with electrical gadgets. The vindictive Wien saw no reason to confer a degree on this ill-informed upstart, even if he were another in Sommerfeld's parade of prima donnas. A row promptly broke out over the relative importance of theory and experiment, resulting in Heisenberg's receiving the poor grade of III for physics, an average of Sommerfeld's I and Wien's V. Fortunately, he performed better in his other two subjects, receiving a I in mathematics and a II in astronomy. But he had to accept an overall grade of III (the equivalent of a C) for his doctorate—the second-lowest passing grade.[83]

Sommerfeld was shocked. Heisenberg was mortified. Accustomed to performing brilliantly in oral examinations, to presenting unassailable defenses of his work, and to having his lectures applauded by the leaders in his field, Heisenberg found a III for his doctorate hard to take. Sommerfeld held a small dinner party at his home later that evening for the new doctor, attended by the institute assistants and students, but Dr. Heisenberg excused himself early, packed a bag, and took the night train to Göttingen, showing up in Born's office the next day.

Born had earlier obtained Sommerfeld's approval for Heisenberg to habilitate in Göttingen, to qualify as a university lecturer through further examination and the publication of a major piece of original research. Born had already offered to continue his assistantship until Heisenberg completed his habilitation treatise. Because he had been so unhappy in Göttingen, Heisenberg had obtained Born's approval to remain in the south until the start of the winter semester. Born was astonished to see him in his office in midsummer and looking so depressed. Informing Born of the debacle of his orals, the youth asked sheepishly, "I wonder if you still want to have me." Born pressed for more details before answering. Together they went over Wien's questions and, once satisfied that "they were certainly rather tricky," Born let his offer of employment stand for the coming winter.[84] Born, who also had his difficulties with experiment, would hardly let Wien's objections hinder his employment of the Wunderkind. The Wunderkind left Göttingen a few days later for the trip to Finland with his youth group.

But Professor Heisenberg, knowing nothing of physics and determined to see his son succeed, was not so easily reassured. Nor was he so easily persuaded that physics was still the best route for Werner's academic career. By chance, August Heisenberg and Willy Wien were chosen deans of their respective philosophical subfaculties for the coming academic year, and both were members of the faculty senate the following year.[85] When the semester began in November, Professor Heisenberg received a disturbing firsthand report of his son. Wien no doubt maintained more than once that the boy did not know enough physics to survive in academe, the only place with jobs for theorists. At the end of November, Werner wrote to tell his father to stop worrying, since it didn't help either of them. He had now placed himself fully in Born's hands, and if his father wanted him to learn more experimental physics, he would have to take this up with Born. "For, as long as I am here in Göttingen I must do what Born wishes, just as in Munich I had to do what Sommerfeld wished."[86] The elder Heisenberg had long since lost any say in his son's affairs.

Even if Werner hid behind Born, it did not deter his father. In January 1924, he wrote directly to both Born and Born's experimentalist colleague James Franck, asking them what his son's chances were in physics. He asked Franck also if he would mind teaching the young man some experimental physics. Born attempted to calm the professor with his own report (now lost) of young Werner's extraordinary abilities, and Franck obligingly admitted Heisenberg to his laboratory course. Shortly thereafter, however, Heisenberg and Franck agreed that Heisenberg should leave the lab; the bored young man could make better use of his time doing theory.[87] Werner had made his choice: if he were to survive at all in academe, he would do so only in the field of theoretical physics.

CHAPTER 9

Principles and Politics

W hen Heisenberg received his doctorate in July 1923, Germany faced mounting challenges to its economic and political order, physics faced challenges to the quantum order of atomic structure, and Heisenberg's core model was under siege. These conflicts reached their peak in October and November 1923. By the end of the year, they had been confronted and temporarily defused.

The coincidence is extraordinary. Just as runaway inflation and a Nazi putsch gripped Germany, Heisenberg and his Göttingen colleagues responded to the challenge of new spectroscopic data with a new quantum principle and approach to atomic physics that again removed every difficulty in a single stroke — but again succeeded only by abandoning accepted procedures. A nationwide state of emergency was nearly two weeks old when the 21-year-old Heisenberg announced his new principle, a Zeeman principle, in Göttingen on October 9. As the authorities brought Hitler and the inflation under control in November, Heisenberg circulated a manuscript reporting his new principle. The paper would become his habilitation thesis eight months later.

In the wake of Göttingen's failed helium calculation in May 1923 and the discovery of new and puzzling spectroscopic data in 1922, proclamations of the "failure of mechanics" — and of quantum physics in general — abounded, ironically during the tenth anniversary celebration of the Bohr atom in the summer of 1923.[1] By the time Heisenberg returned to Göttingen in September, Landé had called for the creation of an Ersatzmechanik (substitute mechanics) and Born had proclaimed the necessity of rebuilding physics "from the ground up."[2] The "incomprehensible" union of quantum rules and classical mechanics that constituted the Bohr-Sommerfeld quantum theory had to be replaced by what Born was soon calling quantum mechanics. In attempting to take the first steps beyond the Bohr-Sommerfeld theory, physicists proliferated ad hoc principles and rules to be appended to the theory as a way of managing difficulties and gaining hints for the future. Werner's ingenious Zeeman principle was a product of the proliferation.

Heisenberg's abstract and abstruse physics again flourished in an atmosphere of extraordinary social upheaval. Yet again he hardly acknowledged the world around him. For Werner and his circle, physics served as both refuge and counterpoint. Heisenberg completed his doctoral thesis in Munich just as the German inflation began its frightful explosion. Inflation had already begun to gallop during the world war, as Germany poured its wealth into its war machine. During the last war years, only deficit spending kept the machine rolling, largely at the expense of the middle class. By January 1920, the mark had dropped to one-fifteenth of its prewar value relative to the U.S. dollar, the international standard. By then the Berlin regime had established political control, and the mark had nearly stabilized. The stability weakened in 1922 after the shocking assassination of prorepublican Foreign Minister Walther Rathenau on June 24—within days of Göttingen's Bohr festival. As investments slowed to a trickle, the mark fell to less than 1 percent of its prewar value. Unwilling to pay reparations to the war victors, Germany now found itself unable to pay. To cover demands for more money, the central Reichsbank merely cranked up the printing presses. Inflation brought windfall profits to German industries, and the politically powerful working class was temporarily content with commensurate wage increases. The less influential middle classes maintained appearances of prosperity as their savings evaporated.

Support for the mark faltered further in January 1923 when the bellicose French, demanding reparations, occupied the Ruhr Valley, the heartland of German industry. With Germany again humiliated and its economy deprived of heavy industry, the mark deflated to practically nothing. By November 15, 1923, the first day of the stabilized Rentenmark (RM), the mark had shriveled to an astonishing one-trillionth of its prewar value. The German people, already traumatized by recent political events, approached panic.

The Heisenberg family did not escape the economic crisis, nor could Werner totally ignore "the economic crisis and the misery all around me."[3] Still, Werner and his family fared better than most. The state continued to raise civil-servant salaries in pace with inflation. August Heisenberg's monthly salary jumped every few months, reaching 2 million marks as Werner received his doctorate in July 1923. August was also paid seminar fees, plus a local increment and a cost-of-living allowance of 1.2 million marks.[4] Yet without further emergency increases and the dollars sent by Uncle Karl, the Heisenbergs would have been in trouble.

In the month Werner graduated, the minimum monthly cost of living for a five-member working-class family in Munich was estimated at roughly equal to Professor Heisenberg's total salary, 3.2 million marks.[5] According to another estimate, the monthly expenses for an average "intellectual worker" in the Reich with two children was 9.4 million marks.[6] By the end of 1923, the Munich working-class family required 100 trillion marks. A kilogram of rye bread alone went for half a trillion marks in Munich.[7] Fortunately, the Heisenberg family managed to weather the storm, as did Professor Heisenberg's colleague Willy Wien. Wien claimed that his salary, slightly greater than Professor Heisenberg's, was just enough to pay for essentials. Wien received his salary every two days and handed it to his wife, who promptly converted it into groceries before prices rose again.[8]

Werner and his lower echelon academic and postdoctoral colleagues did not fare so well as the older academic generation. While professorial salaries regularly increased, government funding for apparatus, literature, and assistants nearly dried up. Young scientists managed to survive the inflation and the difficult years ahead only because of family support and prompt measures taken by established scientists. Many science professors astutely gathered private funds from embarrassingly rich German industrialists. They also befriended philanthropic foreigners, mainly Americans. Munich chemist Richard Willstätter found financial support through a New York brewer. Sommerfeld could count on the generosity of a Berlin industrialist, and Max Born tapped every source he could find "to feed my students."[9] Fortunately, Felix Klein had already established good relations with German industrialists, one of whom, Carl Still, funneled funds into Born's institute. When one of Born's friends left for the United States after the war to marry an American woman, Born asked him, half in jest, to find a rich American willing to provide research dollars. Shortly thereafter, Born made contact with New York financier Henry Goldman. Disturbed by the postwar mistreatment of Germany, Goldman, cofounder of Goldman, Sachs and Co., generously contributed to Born's institute. Born paid his private assistants—Brody, Pauli, and now Heisenberg—with these funds. His one university assistant, Friedrich Hund, received state funding.[10]

Other Americans, also dismayed at the plight of German culture—and eager to influence it to their own advantage—set up emergency committees. Their financial support (save for direct grants) was received and distributed by the Notgemeinschaft der Deutschen Wissenschaft (Emergency Association of German Scholarship). German academics and cultural administrators had established the association in 1920 in part to aid German research but also as a way of establishing their autonomy from the Weimar government.[11]

With its military and economy in disarray, only world-class German scholarship, German scholars argued, could sustain the country's international prestige. Culture could succeed where diplomacy failed. Extraordinary measures were thus required to bolster Germany's place at the forefront of world research. Still, German academics were reluctant to permit the democratic regime control over research funds—a prerogative that the prewar monarchy had enjoyed. In order to skirt government involvement, German scholars and scientists revived the notion of self-administration—and so invented the modern project-grant system of support. Under this system, neither governments nor foundations determine who should be blessed with research funds. Rather, in an innovative departure from previous practice, independent committees of leading scientists and policy administrators in each field, backed by panels of specialist referees, evaluate the merits of each grant application and make the appropriate awards. This unique and powerful system of research support—not adopted in the United States until after World War II—helps to account for the paradox of Germany's lead in atomic physics and other sciences during a period of extraordinary adversity.[12]

Thanks to Einstein and other German physicists, physics had become one of the most prestigious fields of German research by the early twenties. Accordingly, in its first report, published in 1922, the Notgemeinschaft announced that it had funneled the greatest support into physics. The physics disciplines most preferred were atomic physics, radiation, and the structure of matter—in particular, experimental research on the relativity and quantum theories.[13] While these specialized areas seemed to offer few practical applications, the generous funding was intended to maintain Germany's already impressive lead. Unless German physicists receive necessary support, the report declared, "financially better endowed physics in foreign nations will soon push us back."[14] Those who disagreed with the generous funding of relativity and quantum physics—notably Willy Wien and Johannes Stark—formed a separate agency to support their preferences: classical and technical physics.[15]

Coupled with its strategy of financing leading areas of research, the Notgemeinschaft created another influential novelty: research stipends to promising postdoctoral researchers in order to keep them in the field. One of the first stipends went to Heisenberg. Many of the other postdoctoral recipients would be, like Heisenberg, primary contributors to quantum mechanics.

Beginning in July 1923, a Notgemeinschaft subcommittee, the Electrophysics Committee, devoted itself almost entirely to atomic physics. Created with an annual sum of $12,500 from General Electric (and matched, in part, by German industries) ostensibly to help technical physics, its members—Planck, Laue, Franck, Haber, Max Wien, and Arnold Berliner—turned instead to the atom. Of the 140 grant applications the committee received between 1923 and 1925, it approved 71, most of which were in atomic physics. Of the 71 approved, 56 supported research assistants. Of these, 3 had been submitted by Born on behalf of Heisenberg.[16]

Heisenberg received his first stipend from the Electrophysics Committee as soon as the mark stabilized in November 1923. It amounted to only 50 RM a month, certainly not enough to live on.[17] It soon increased to 100 RM, then to 150 RM in March 1924—exactly half of an assistant's salary. The other half of Werner's support came from Born's private sources. Heisenberg's official Notgemeinschaft salary in Göttingen remained at this amount until the end of 1925.[18] Late in 1925, with Heisenberg and Born's breakthrough to quantum mechanics already in press, Born, supported by Courant, then dean of the science faculty, obtained a two-year lecturer's stipend—another invention of the Notgemeinschaft—for Heisenberg from the Prussian Cultural Ministry in the amount of 127.88 RM per month.[19]

These salaries, which supported Werner throughout this highly creative period of his career, did not add up to much more than the cost of living for an unmarried young man in the provincial town of Göttingen. Yet they served their purpose well: they kept talented young people like Heisenberg in science and especially in quantum atomic physics, where they helped to lay the foundations of quantum mechanics. In its 1926 report on the impact of its support during the previous five crisis years, the Electrophysics Committee congratulated itself for its perspicacity in financing Heisenberg and Born: "As is well known, quantum mechanics stands at the center of attention among physics circles of all nations.

The work of Heisenberg and Born, which the Electrophysics Committee has supported and without which the work would very probably not have been done in Germany but elsewhere, has shown the usefulness of the Electrophysics Committee in the development of physics in Germany."[20]

Soon after the poor showing on his doctoral examination and his postexamination visit with Born in July 1923, Werner headed for Finland with his youth group, where he remained until the end of August. Heisenberg returned to Göttingen in September and announced his new Zeeman principle to Pauli on October 9. He had begun putting it on paper in October and was well into it several weeks later when he wrote to his Finland traveling companion Kurt Pflügel of a new interruption: the latest news in Bavarian politics had "sent the blood to [his] face."[21]

Insane inflation, Bavarian separatism, and Germany's unnecessary humiliation at the hands of the Allied victors had reached crisis proportions when Chancellor Stresemann decided to end resistance to the French occupation of the Ruhr on September 26, 1923, igniting unrest throughout the country. At the same time that German President Ebert declared a nationwide state of emergency, Bavarian officials decided to declare their own state of emergency, aimed primarily against supposed left-wing agitation. They named Gustav von Kahr to the dictatorial post of General Commissar of Bavaria. Kahr, though close to militant Bavarian separatists, gained the temporary support of violent nationalists who sought to replace the democratic Weimar regime with a right-wing dictatorship. Among Kahr's nationalistic supporters were General von Lossow, commander of the Bavarian division of the Reich army, and the Deutscher Kampfbund (German Battle League), an alliance among the defeated General Ludendorff, Hitler's Nazis, and illegal nationalist military units made up largely of former Freikorps members.

Matters escalated on October 20 when Berlin authorities ordered Kahr and General von Lossow to close down the Nazi party newspaper, *Völkischer Beobachter,* for libeling Berlin officials. Lossow refused and was summarily dismissed by his Berlin commanders. Kahr immediately nationalized the Bavarian Reichswehr, placing it fully under Bavarian authority with Lossow back in command. The next day Heisenberg, with family ties to northern Germany, wrote angrily to Kurt of the Bavarian behavior. "I simply don't understand any of this: Lossow is a soldier, isn't he? If he doesn't obey, then a soldier is supposed to be shot, so it's written somewhere."[22] As did his family earlier, Werner openly sympathized with the Weimar regime over Bavarian zealots.

The more provincial Kurt, the son of a Bavarian officer, had meanwhile rejoined the Bavarian army after Kahr's action and now attempted

to explain to Werner the grounds for Bavaria's anti-Berlin attitude.[23]
Stresemann had just moved militarily against soviet regimes in Saxony
and Thuringia. The Bavarians demanded that he now move against
socialists of any stripe everywhere, while at the same time resuming
resistance, even if only passive, against the French occupation. In his
long reply on October 31, the more socialist-minded Werner, though
calling Bavarian motives noble, wondered how the government could
fight socialism when, in his view, a third of Germany was socialist and
had an equal right to national recognition.

Heisenberg felt that direct opposition to the French would ultimately
lead to the Heldentod (hero's death) of Germany as a nation, given its
present economic and military weaknesses. As during the Napoleonic
occupation, patience seemed the only practical alternative. "Out of all
of this I conclude that it is more correct (as our forefathers [did] a
hundred years ago!) to hold the enemy in check by cunning and deceit
and fright until a hope exists for weapons. That is precisely what the
Reich government appears to be doing."[24]

Hitler and his followers were not so patient. At a celebration com-
memorating the fifth anniversary of the Armistice on November 8,
1923, in the Bürgerbräu beer hall in Munich, the Bavarian government
and leading members of the upper classes, including professors, were
present. At the high point of celebration, just as Kahr unveiled his
authoritarian plans, Hitler and a band of ruffians burst into the hall,
fired several shots at the ceiling, and proclaimed the revolution: "To-
morrow we're all dead or we have a national government!"[25]

Inspired by Mussolini's march on Rome, Hitler and his followers
planned to march on Berlin with the Bavarian Reichswehr and the secret
army units that had been established to fight so-called Bolshevism in
defiance of the Allied limitations on German military units. Kahr, Lossow, and police chief Colonel von Seisser agreed, at gunpoint, to Hitler's
plan. But later that night the three slipped away, withdrew their support,
and ordered the Reichswehr to defend the government. The Hitler-
Ludendorff Kampfbund scheduled a march through Munich the next
day to force the issue over control of the Reichswehr. Gottfried Simmerding, a longtime member of "apolitical" Gruppe Heisenberg and
now its new leader, also belonged to an illegal machine-gun company led
by a Lieutenant Werner.

At 2:00 A.M., Simmerding recalled, he was awakened at home by a
friend from the secret company. He dressed quickly, stuck a pistol in his
pocket, and went to join the unit. As part of the German disarmament,
Bavaria was allowed only one machine-gun company, not enough in
Bavarian eyes to protect against Bolshevism. During the predawn hours,

after the Bavarian authorities had reversed their support of Hitler, they ordered Company Werner to the Feldherrnhalle in downtown Munich to help confine Hitler's march that day to the inner city. Apparently supportive of Hitler, Lieutenant Werner refused and his company disbanded without complaint.

As Simmerding made his way home later that day, after seeing to company finances, he heard machine-gun fire from the direction of the Feldherrnhalle. The Augsburg machine-gun company under the command of Kurt's indomitable father, Colonel Pflügel, had arrived in time to confront the march and helped to kill 16 of the marchers.[26] Hitler escaped unscathed, only to face prosecution and a brief stint in prison for his putsch.

Over the weekend, as loyal army units broke up repeated demonstrations by antirepublican students, Kurt, an army reservist and a student at the technical college, met with other Neupfadfinder to decide their political position, an action they would have shunned just a few weeks earlier. Kurt informed Heisenberg in quiet Göttingen that they, like most Bavarians, had decided to support the Berlin government for now and had even decided to assume what they regarded as responsibility for helping to alleviate social distress: "We ought not withdraw from the world and live our lives only in the woods and dream of a new Reich. If we are blessed enough to see the Reich, then it is our sin if we keep it for ourselves."[27]

Werner did not respond to his friend for nearly two weeks. In the meantime, most Germans had come to tolerate the republican regime (although the short-lived Stresemann cabinet had since fallen). Most still opposed dictatorial nationalists. Soon the economic basis for popular support was assured. On November 15, the government introduced the new Rentenmark (RM), or "mortgage mark," based on a mortgage against all German agricultural and industrial property. The Rentenmark, equal to a trillion paper marks, or one prewar mark, met with immediate acceptance.

On November 24, just as he received his first Rentenmark stipend from Born and was about to reveal his optimistic new quantum principle, Werner wrote Kurt of his pessimistic view of current affairs. While resistance to the French was still hopeless and Hitler's "November-Fasching" ludicrous, the Munich youth movement's turn to active social involvement was for Werner, as it had been earlier for the radical Neupfadfinder, unthinkable. "What role could we few little men play in a population of millions!"[28] Heisenberg had already settled on retreat to the pure worlds of physics, music, and youth. Youth-group follower

Wolfi Rüdel, soon after the upheavals settled in early 1924, wrote to Werner: "The time is now coming once again where we are the most strictly limited by external circumstances. But I can already imagine what your answer will be: physics and ever more physics."[29]

If Heisenberg's characteristic optimism in matters of theoretical physics did not prevail in the political arena, it was because he saw little hope of immediate success. Both in physics and in current events, the momentary situation was for Heisenberg only temporary, an unpleasant interlude between more permanent arrangements. But in the rarefied realms of theoretical physics, his abilities enabled him to achieve both influence and significant results of lasting consequence. Political entanglements were, to the contrary, merely superficial "money-business," a business in which he felt he would be just one of millions—a position that rendered him and others like him dangerously unprepared for what was soon to follow. "Therefore, Kurt," he wrote, "I believe that we should stay a while longer in the woods; there the air is cleaner. . . . We cannnot help the masses, and the few young people whom we can help will come to us all right 'in the woods'; a wonderful Gemeinschaft that at every hour is ready to forgo its life, the highs and lows of which it has come to know like few others. For, whoever has been at the peaks of life no longer needs to fear the end."[30] Werner held this view throughout the Weimar years.

Success in physics was one of the "peaks of life" for Werner, and when he wrote that statement he had just climbed again to the top by rescuing the core model, still his only independent contribution to quantum physics, with his new quantum principle. Returning to Göttingen, he now devoted full attention to the abstruse technicalities of Zeeman spectroscopy and the core model. His continuing struggles with this and other problems over the following year provided the basis and the setting for the breakthrough to quantum mechanics.

Heisenberg had invented the core model of the atom earlier in Munich to account for the puzzling division of certain spectroscopic lines emitted by atoms into doublets and triplets. Beginning with the Bohr quantum postulates—that electrons orbit in certain stationary states and that jumps between states result in the emission or absorption of an entire light quantum—Heisenberg had shocked his colleagues with an account of these doublets and triplets by inventing a model with half units of quanta and half-integral quantum numbers. In his model, outlined in Chapter 7, the one or two valence electrons orbit around an atomic core consisting of the nucleus and inner electrons, while sharing half units of angular momentum with the core. This completely violated quantum orthodoxy, since quanta appeared only in whole units and were thus numbered in whole integers. But to the further astonishment of his colleagues, Heisenberg showed that his renegade model, no matter how "deviant," provided the only explanation so far of the abstruse puzzles of the anomalous Zeeman effect—the further splitting of each member of the doublet and triplet lines into a subseries of lines.

Challenged on theoretical terms from the start, Heisenberg's core model encountered an experimental challenge within a year: the discovery that not just doublets and triplets appeared where only a single spectroscopic line should occur, but the single line split of its own accord into as many as nine closely spaced lines.[31] By December 1922 Ernst Back in Tübingen had amassed sufficient data on the Zeeman effect of the new multiplets of lines to enable his Tübingen colleague, theorist Alfred Landé, to reduce the entire phenomenon to a new number mystery—several numerical rules and formulas based on the assumption that every emitted line must arise from the jump of an electron from a higher energy quantum state to a lower energy state, each state being characterized by a variety of quantum numbers. Landé's formulas, which are still valid today, enabled his speculation on the properties of an atomic model that provided the basis for Heisenberg's extension of his earlier core model to handle such complications.[32] The aim remained, as before, to account for these lines and splittings using an imaginary quantized mechanical model of the atom.

Since each quantum number was supposed to correspond with a motion within the atom, each number provided a clue to the type of motion that might be involved. But first Landé had to discern the quantum numbers themselves through tedious analyses of Back's data. The analyses yielded three new numbers—each involving both integral and half-integral values—assigned to the stationary states: a quantum number R, equal to half the multiplicity exhibited by the atom (doublet, triplet, and so on); K equal to $k - 1/2$, k being the integral orbital momentum quantum number; and J, which behaved like a total angular momentum quantum number with half-integral values recently assigned by Sommerfeld.[33] Amazingly, with these three new numbers, Landé managed to reduce all of Back's enormously complicated Zeeman data to a single empirical formula for the value of g, the numerical constant that indicates the shift of each Zeeman line from the position given by the normal Zeeman effect.[34]

The task now was to find an underlying model of the atom that would account for these quantum numbers and formulas. Landé's speculations began with the core model, to which he had recently conceded support.[35] An electron orbiting a core, as a planet orbits the sun, acts like a little current loop, and a little current loop behaves like a little magnet. The strength of the magnet and the direction of its poles, or magnetic moment, are determined by the angular momentum vector of the orbiting electron (the direction and speed of the electron in its orbit around the nucleus). If an external magnetic field is applied to the atom, all the electron magnets orient themselves at definite angles to the external field, the angles being determined by Sommerfeld's space quantization.

Following Heisenberg and the usual procedure of associating magnetic quantum numbers with angular momenta, Landé assigned R to the rotation of the core, K to the orbital angular momentum of the valence electron, and J to their quantized vector summation. By fiddling with the orientations of the corresponding vectors in this vector model of the atom, Landé almost obtained his g-formula. To obtain the exact expression, he had to introduce two novelties — the treatment of the core as possessing twice the expected magnetic moment, and the replacement of J^2 by $J^2 - 1/4$ in the last step.

The effort succeeded but, as with Heisenberg's core model, only through untenable means. Neither Landé's two last-minute novelties nor indeed his introduction of half-integral quantum numbers bore any real physical justification, but they did reduce the data to numerical rules. Landé himself admitted that his derivation "serves as a demonstration of the fundamental difficulties rather than their resolution."[36] Pauli pronounced Copenhagen's official judgment: he and Bohr regarded the new relationships as purely "formal" and "hardly as a physical reality."[37]

Only Heisenberg, author of the model on which the scheme rested, seemed remotely enthusiastic. "I find Landé's result very beautiful," he wrote, "but his addition of the quantum numbers is arbitrary."[38] A month later, in March 1923, he declared: "There is still certainly much to improve for the triplets, but I believe the doublets are in order insofar as the logic is correct." But Heisenberg was still preoccupied at the time with helium and soon would turn to hydrodynamics for his dissertation. "At the moment I regard [helium] as the most important question — for the Zeeman effects it is still too early."[39] Heisenberg still suspected that the core model and perhaps even Landé's extension of it would be exactly calculable by applying the most sophisticated perturbation theory to the simpler atoms. But by the time Heisenberg returned to Zeeman spectroscopy in the fall of 1923, after his Finland interlude, planetary perturbation theory and every version of the core model had proved inadequate.

The further challenge to the core model came out of Copenhagen. Pauli had received an International Education Board (Rockefeller Foundation) fellowship to work in Bohr's institute for a year beginning in the fall of 1922. Ends and means became the issue. As indicated earlier, during the 1922–1923 winter semester Born had enlisted Heisenberg's help in an attempt to find the limits of the Bohr-Sommerfeld theory of the atom through a strict adherence to consistent means — the quantum rules (including integral quantum numbers) and the orbital mechanics of electron motions in atoms. During the same period Bohr, with Pauli's help, tried a different tack: not to explore the limits of the theory but to obtain a rational advance beyond those limits by allowing necessary breakdowns of orbital mechanics in the stationary states while adhering to quantum rules and principles.

While Born and Heisenberg scrutinized the properties of mechanical atoms in Göttingen, Bohr and Pauli scrutinized the challenges to quantum principles raised by Heisenberg and Landé. Half-integral numbers and momenta were still the prime culprits. Not only did they contradict the quantum conditions, threaten previous calculations using integral numbers, and undermine the adiabatic mechanical transformability of integrally numbered states (the slow transformation of one quantum state into another one of integral quantum number), but they undercut the correspondence and building-up principles, the most recent and acclaimed successes of the Bohr program.

Bohr's recently published building up of the periodic table of elements rested on a so-called building-up principle. One element of this principle required that all the quantum numbers of an atom remain constant as one very slowly (adiabatically) added another electron to the outer orbit of the atom to form the atom of the next element in the periodic table. Landé's new J number violated this requirement. Since in Landé's scheme J denoted the total angular momentum of the atom, it should be equivalent to the core rotation quantum number R of the next element. Instead, it appeared that J jumped by $1/2$ when it became the core R of the next atom. Not only did this cause inconsistencies in Landé's vector model, but even worse, the jumping of J destroyed the conservation of statistical weights, also essential to the building-up principle. The statistical weight of an orbital state is equivalent to the number of possible orientations (m-values) that the little magnet can take when an external magnetic field is applied. Since J jumped by $1/2$, the number of such states suddenly jumped by 1 as a new electron was added to form the next element.[40] If any version of the half-integer core model remained standing, Bohr's program of rational advance based on accepted quantum principles would soon collapse.

In early 1923, while Born and Heisenberg examined excited helium, Bohr and Pauli examined an elementary Ersatzmodell (substitute model) for the Zeeman effect, constructed in such a way that every quantum rule held. They hoped it would somehow yield some of the anomalous Zeeman data without half-integral quantum numbers. Heisenberg expressed his pleasure at the change of opinion in Copenhagen regarding detailed models,[41] but by early March this model, too, had failed. It had been, wrote Bohr, "a desperate attempt to remain true to the integral quantum numbers in that we hoped to perceive in the very paradoxes a clue to the ways by which we might search for the solution to the anomalous Zeeman effect."[42]

Further disheartened by the failure of Göttingen's helium model, Bohr took the most likely way out. In March 1923, as Göttingen's helium calculation went to press, Bohr reluctantly admitted one half-integral quantum number into his physics — Sommerfeld's inner quantum number j — but he steadfastly retained integrals n and k. Bohr also added to the growing popularity of ad hoc solutions. Exploiting his preference for allowing mechanics to fail in stationary states, Bohr came up with the remarkable idea of introducing an unspecified new mechanical force into the atom — what he called a Zwang, or constraint — that would account at a single stroke for all of the otherwise inexplicable features of quantized atoms.[43]

For Bohr, the new Zwang accounted somehow for the mechanical stability of otherwise unstable atomic models and the anomalous behavior of the core in the Heisenberg-Landé core model, and — in particular consequence — it rendered the proper, conserved statistical weight of an orbital state without the half-integral electron angular momentum number k. The constraint proved crucial to the future development of quantum theory. In the case of an atom such as single-valence sodium that displays a doublet of spectroscopic lines, the total statistical weight (number of magnetic states) is equal to $2(2k - 1)$. This expression was hitherto obtainable only by using half-integral k in Heisenberg's core model.[44] In his new scheme, Bohr allowed only integral k but hypothesized the presence of a vague Zwang. He first reduced the usual $2k + 1$ magnetic states by eliminating the orientation that is perpendicular to the magnetic field. The Zwang did the rest: mysteriously it reduced the allowed states even further to $2k - 1$, then forced a Zweideutigkeit (two-valuedness, or ambiguity) on the core such that it took up two positions under the action of the Zwang, rather than the two positions that would arise if the core shared a half unit of momentum with the electron, as in Heisenberg's model. The proper number of states, $2(2k - 1)$, resulted! The ad hoc nature of the argument was intended, but the rewards now justified the expense: Bohr had preserved the main elements of quantum atomic-model theory without the devastating heresy of half integers and with only a vaguely defined but rationally chosen modification of classical mechanics.

The temperamental Pauli wasn't buying; he had grown more radical than Bohr after the failure of the Ersatzmodell. Indeed, within a year Pauli declared himself "much more radical than the 'half-integer' atomic physicists."[45] After arguing with Bohr for days about the Zwang, Pauli "tortured himself" for weeks with the anomalous Zeeman effect. "But it would not and would not agree! . . . For a while I was completely discouraged."[46] What so discouraged Pauli that spring of 1923 was his failure to comprehend, using either mechanical interactions or quantum formalism, his own contribution to the approach of that period, a new permanence principle. This principle derived from a summation rule in Heisenberg's core-model paper from which, Pauli showed, he could derive Landé's g-factors for weak magnetic fields by transforming from the simple, well-understood expression in the case of strong fields (the Paschen-Back effect).[47]

Although "a satisfactory model interpretation of . . . the formal rule" could not be found, wrote Pauli, Bohr still urged publication. Pauli did so "with a tear in my eye."[48] But by the summer of 1923, Pauli was warning everyone, publicly and privately, against any attempt to interpret optical spectra using mechanical atomic models and current theory: "For I am convinced that there is no [satisfactory] model for the anomalous Zeeman effect and that we must create something fundamentally new."[49] Henceforth, for Pauli the core model was simply an unwelcome outgrowth of the failed quantum physics of atoms, while the core model's optical doublets were, like X-ray doublets, probably traceable to well-known, nonmagnetic relativistic effects.

As autumn tinged the rolling hills around Göttingen, Max Born explained his view of the source of the difficulties facing quantum theory after the failure of the helium atom and the blatant deviations of the core model. Since atomic electrons repel each other and valence electrons and core attract each other with much greater forces than the corresponding gravitational attraction of planets to each other and to the sun, the solar system analogy was severely limited and quantum celestial mechanics was inadequate to the task. A new quantum theory of coupling, or what he was calling in his lectures a quantum mechanics, was needed to replace the quantum theory of Bohr and Sommerfeld.[50] To achieve the new mechanics required, in Born's view, a "discretizing of atomic physics" such that quantum jumps between discrete stationary states would no longer enter as an ad hoc postulate of the theory, but would be incorporated into the foundations of the new physics through the replacement of differential equations for continuous processes by difference formulas for discontinuities. Behind this approach stood a new attitude toward all models, including the core model. Heisenberg summarized it for a sympathetic Pauli in his letter of October 9, 1923: "The model conceptions have only a symbolic sense, they are the classical analogy to the 'discrete' quantum theory."[51]

Born's discretizing program formed the background and basis for what Heisenberg called a new Göttingen theory of the anomalous Zeeman effect — his own Zeeman principle. Although the genesis of the new theory is unclear, its details were largely Heisenberg's, who outlined them for Pauli in the October 9 letter and for Bohr in a long letter of December 22.[52]

With stationary-state mechanics now subject to Bohr's Zwang, the basic notion of Heisenberg's theory derived from a study of Bohr's second quantum postulate of 1913. According to this fundamental postulate of quantum physics, if an electron jumps from a higher energy state E_2 to a lower energy state E_1, a light quantum is emitted with energy $h\nu$ equal to the energy loss of the atom: $h\nu = E_2 - E_1$, or $h\nu = \Delta E$. Here h is Planck's constant and ν is the frequency of the light observed in a spectroscope.

Heisenberg pointed out that this fundamental equation is actually an untenable mixture of classical and quantum notions and symbols. The energies E_1 and E_2 are calculated according to classical Newtonian mechanics, yet the observed frequency ν is related not to the energy of any one state, as it should be in classical Maxwellian electrodynamics, but to the difference between two states.

Suddenly smitten with consistency, Heisenberg replaced the classical energies — or, what amounted to the same thing, the classical Hamiltonian function H_{cl} — with a "quantum Hamiltonian," H_{qu}. He defined the quantum Hamiltonian as an average (an integral) of the classical energy, or Hamiltonian, over a supposed two-valuedness of each state. This two-valuedness was manifested in two values of J ($J + 1/2$ and $J - 1/2$) assigned to one state. This average, or integral, yielded a function $F(J)$, the differential of which yielded H_{cl}.[53] Following Born's discretizing plan, Heisenberg replaced this differential by a difference to obtain H_{qu}. H_{qu} could now replace the energies in Bohr's original formula, rendering the formula seemingly consistent in its exclusive use of quantum symbols: $h\nu = \Delta H_{qu}$.

Amazingly, it worked. "Now everything comes out of this," Werner wrote, and without Bohr's Zwang![54] Heisenberg demonstrated once again his incredible intuition and inventiveness. Applying the quantum

Hamiltonian to the now symbolic core model, he was able to derive all of Landé's results for line splittings and the Zeeman effect, restore Bohr's building-up principle, and even derive Pauli's puzzling permanence principle. With such success arising from consistency, consistency suddenly became a virtue: "The decisively positive aspect in this theory appears to me to lie in this," Heisenberg wrote to Pauli, "that the model and especially its angular momentum orientations are all entirely normal, just as one must consistently expect."[55]

The significance of Heisenberg's astonishingly facile removal of riddles and ad hoc inventions on the basis of a resurrected core model, even if the core model were only symbolic and heuristic, could hardly be overlooked. And this time Heisenberg even willingly admitted his main deviations. They were the usual important ones: half-integral momenta and a double core magnetic moment. The ad hoc averaging over what appeared to be two orbits per stationary state was also an oddity but one by then almost universally accepted. Landé, for instance, had suggested in March that the replacement of J^2 by $J^2 - 1/4$ in his g-formula was like replacing J by the geometric mean of $J + 1/2$ and $J - 1/2$: that is, by $(J + 1/2)(J - 1/2)$.[56] According to a diary kept by Born's university assistant, Friedrich Hund, months earlier Born had considered averaging the perturbations of an electron over the course of its orbit.[57]

Heisenberg no doubt discussed some of these matters and the new Göttingen approach with his colleagues when he arrived back in Göttingen in September 1923, and he announced his own discovery of how the Göttingen theory actually worked for the Zeeman effect on October 9. On that day, when he wrote Pauli of the theory, Hund wrote in his diary: "Heisenberg calculates the Zeeman effects with [his averaging formula]." Heisenberg soon began committing his ideas to paper — ideas that he no longer called a theory but a new quantum principle.[58]

Göttingen physicists were naturally enthusiastic. Heisenberg received the imprimatur of the ruthless Göttingen colloquium after delivering a well-received talk on the principle in November. Born urged immediate publication.[59] But Werner demurred under pressure from home. Having just learned about his son and his questionable physics from Willy Wien, Professor Heisenberg openly worried that the new theory, based on the controversial core model, might confirm his not yet habilitated son's ignorance of physics — and at the very height of frightful economic and political chaos. Sommerfeld liked what he knew of the new theory but urged caution, which August conveyed in the strongest terms to Werner at the end of November. Werner responded angrily that he was trying to be more consistent and self-critical than before and that this time he would seek advice and approval not just from Born but from the master of consistency, caution, and physical insight, Niels Bohr: "I realize ever more that Bohr is the only person who, in the philosophical sense, understands something of physics."[60]

True to his word, Heisenberg sent a copy of his manuscript—what he called a "Zeeman sausage with quantum sauce"—to Pauli in Copenhagen, asking his friend to scrutinize it before forwarding it to Bohr, then fund-raising in the United States, for his blessing. A copy also went to Landé and Back in Tübingen, who gave it their guarded approval.[61] But Pauli returned the manuscript with the complaint (or excuse) that it was handwritten in German Gothic script—which had not stopped Bohr before.[62] Heisenberg took the hint. On December 22, one day after a stormy encounter with Pauli in Sommerfeld's Munich seminar—they were both on Christmas vacation—Heisenberg wrote a long letter to Bohr (in Latin script) cautiously elaborating the theory in a way that would appeal to Bohr; the restoration of the building-up principle without need of a Zwang constituted "the main argument" for it: "I can hardly imagine another interpretation of the difficulties described by you in the Kayser issue of the Annalen [der Physik]."[63] Bohr, who had introduced the Zwang alternative in that issue, responded over a month later with an invitation to Copenhagen.[64] Heisenberg accepted for the semester break in March 1924. It would be his first visit to Copenhagen, and it would confirm and accelerate the already coalescing physics of the Göttingen and Copenhagen schools.

The critical and consistent Pauli, now a lecturer in Hamburg, didn't like any of the Göttingen theory. "I even regard it as ugly," he told Landé. "For, despite radical assumptions, it delivers no explanation of the half quantum numbers and the failure of the Larmor theorem (especially the doubling of the magnetic core anomaly)." For Pauli, the proper theory had to explain anomalies, not assume them. Even though Heisenberg had "hit the mark" in doubting the existence of realistic mechanical models and orbiting electrons, his latest, Pauli declared, "is not the theory for which I am hoping."[65]

Pauli and Heisenberg again discussed their differences during a two-day physics conference in Braunschweig in February 1924. This time Heisenberg had new support. Three of the four lecturers were Göttingen physicists—Born, Heisenberg, and Hund—and each reported new nonspectroscopic evidence for half-integral quantum numbers.[66] During the meeting Pauli had apparently raised objections against Heisenberg's new principle, and Heisenberg had responded with new diplomacy but even more outrageously radical assumptions.

Back in Hamburg, a disconcerted Pauli complained to Bohr of his frustration with Heisenberg. Pauli's letter to Bohr, dated February 11, 1924, provides probably the best intimate depiction of Heisenberg and his relationship with Pauli in this period. Pauli wrote: "I always feel very strange with him. . . . For he is very unphilosophical, he does not pay attention to the clear working out of the basic assumptions and their connection with the prevailing theories. However, if I speak with him, he pleases me very much, and I see that he has all sorts of new arguments—at least in his heart. Aside from the fact that he is also personally a very nice fellow, I regard him as very significant, even genial. . . . I was therefore very pleased that you have invited him to Copenhagen. . . . Hopefully then Heisenberg, too, will return home with a philosophical orientation to his thinking."[67]

Ten days after Pauli wrote to Bohr, he had had enough of Göttingen inconsistency with half integers. A new physics was needed, not, as displayed in Braunschweig, repeated calculations of the same effect using integral and half-integral numbers, then choosing the procedure that better fits the data.[68] On February 21, he informed Bohr of his withdrawal from the field: "I myself cannot obtain any taste for this sort of theoretical physics and withdraw from it to my thermal conductivity in solid bodies."[69] He did not publish again on Zeeman spectroscopy until he had found a way to destroy the core model.

Heisenberg arrived in Copenhagen by train and ferry on March 15, 1924, to an elaborate reception.[70] Danish hospitality aside, his reputation had obviously preceded him. He wrote home often during his two-week stay to tell of numerous luncheon invitations to sample Danish food; musical evenings at the home of the widowed Mrs. Maar, who rented rooms in her multistoried house to the institute visitors; a weekend excursion to Bohr's country cottage in Tisvilde, on the northern coast not far from Hamlet's castle; a meeting with Wickliffe Rose, head of the International Education Board, who happened to be in Copenhagen at the time; and a touring trip to Jutland with three visiting Americans. "I speak very well englisch," he assured his parents.[71]

But the main events were his conversations (in German) with Bohr. These increased in length to several hours each day. Bohr's three-story institute, a component of the University of Copenhagen, lay at that time on the outskirts of the small inner city near the large and secluded Fælledpark, with its stately trees, budding springtime flowers, and seemingly endless walking trails. Their conversations often occurred during strolls through the park or to the Copenhagen docks, where ships from all over the world unloaded their wares, and occasionally in the evenings in the Bohr residence on the third floor of the institute with "one (or more) glass of port wine." Within five days of his arrival, Werner had an invitation to Copenhagen for a year and a probable Rockefeller stipend provided by Rose.[72] "Everything has really gone better than I could ever have expected," he wrote home.[73]

Heisenberg's conversations with Bohr during the first few days were taken up not with the core model but with a common interest they may have discovered during their first meeting the previous summer — "philosophy," which meant to them everything nontechnical. "We have always talked about the most general questions and have picked apart their philosophical foundations (I see you scornfully smiling here, Papa)," he wrote home.[74] August, the positivistic philologist, made little secret of his abhorrence of philosophical matters.

Nevertheless, as Pauli had hoped, toward the end of his stay Heisenberg had begun to appreciate Bohr's systematic, or philosophical, approach to physics: "His manner of doing physics is really very 'practical,' he always attempts at first only progress in the details."[75] By then, Heisenberg also reported that he was spending his free time in the institute library reading physics textbooks, no doubt at Bohr's direct or indirect insistence, "in order to 'elevate' my general physics education." As the two physicists turned, after nearly a week, to discussion of specific problems—the core model and Werner's new principle — it was probably clear to the younger physicist that he required a deeper appreciation of physics.

One of the three Americans whom Heisenberg met that March in Copenhagen was recent Harvard Ph.D. John C. Slater.[76] During their outing to Jutland and in the institute with Bohr and Kramers, Werner learned of the physical and philosophical implications of the new Bohr-Kramers-Slater (BKS) theory. During Heisenberg's visit, Bohr's longtime assistant, Dutch physicist H. A. Kramers, submitted a daring extension of Rudolf Ladenburg's 1921 quantum formula for the dispersion of light by atoms, based largely on the BKS theory, then in press.[77]

The BKS theory itself had arisen in the context of the quantum theory of radiation, the study of the interaction of light with atoms. The new theory was designed to resolve one of the deepest mysteries of the period: the wave-particle duality. For over a century, light had been accepted to be a continuous wave. This interpretation had found decisive support in Maxwell's theory of electromagnetism—light being an electromagnetic wave. Only a wave could display such effects as interference, diffraction, and dispersion. In 1905, Einstein had revived the revolutionary hypothesis that light also behaves as a particle, a light quantum. In 1922, American experimentalist A. H. Compton had offered support for this interpretation by observing the collisions of particlelike light quanta with free electrons, the Compton effect. Faced with two mutually exclusive interpretations of light—the wave and particle interpretations—the cautious and consistent Bohr chose the well-established wave interpretation. He attempted to account for particle behavior without resorting to light quanta. The BKS theory was a product of this attempt. Later, in 1929, Heisenberg characterized this theory as a perspective that "contributed more than any other work at that time to a clarification of the situation in quantum theory."[78]

Quantum atomic physics as then practiced was based largely on calculations of stationary-state energies. The frequencies of emitted or absorbed light could be calculated from supposed quantum jumps between these states. Quantum radiation theory, on the other hand, concerned the scattering, dispersion, and absorption of light by an atom or groups of atoms, using Einstein's transition probabilities.[79] No one could yet calculate the probabilities for jumps between stationary states. So, on the basis of Bohr's correspondence principle, the harmonic (Fourier) components of the classical orbital motion were commonly used. However, this approach worked best only in the higher energy states and only for orbital motions that are periodic or conditionally periodic and nondegenerate (not overlapping).

Ladenburg had shown in 1921 that one could employ a more direct approach. It was also simpler.[80] The classical dispersion formula had been derived by treating atoms as little harmonic oscillators—charged balls on springs. The frequency of oscillation was the same as the emitted or absorbed frequency of radiation; the square of the amplitude of oscillation was related to the intensity. Ladenburg showed that the frequency and amplitude of the assumed oscillators could be matched with the quantum transition frequencies and intensities (intensities being related to Einstein's probabilities). Thus, in effect, for radiation problems the atom could be treated as an array of virtual harmonic oscillators—as if it were a series of charged balls all attached to a central point by springs. The frequency and amplitude of each oscillator would correspond to emitted or absorbed frequencies and intensities associated with each possible jump between two quantum states. In the very highest states (in the limit of large principal quantum number n), these oscillators would correspond exactly to the multiple Fourier components of the orbital motion, precisely what Bohr's correspondence principle had demanded.

As mechanical atomic models raised ever more problems for physicists, the utility of virtual oscillators became ever more attractive. For Pauli, they matched his preference for adherence to observed phenomena over unobserved models.[81] For Bohr, whose atoms were already infected by his invention of Zwang and two-valuedness, virtual oscillators became key elements in his new appreciation of correspondence as a tool. Since Maxwell and others had easily shown how harmonic oscillators emit and absorb light waves, oscillators gave Bohr the basis for adhering to the wave interpretation of light. Although the Compton effect seemed to support Einstein's light quanta, the support could be undercut if, as Bohr preferred, other established elements of physics were renounced.

Slater came from Harvard to Copenhagen in 1923 with the remarkable idea of a "virtual radiation field" or "ghost field" that carried no energy or momentum (hence "virtual") but that was continually emitted and absorbed by all virtual oscillators of an atom.[82] Making use of virtual oscillators and Slater's virtual field, Bohr and Kramers saw how to maintain a wave theory of radiation in interaction with matter, while ignoring both light quanta and the mechanics of stationary states (two elements that Bohr himself had introduced into atomic physics 11 years earlier). But this was no simple matter. It demanded the additional abandonment of some of the most cherished principles of physics — conservation of energy and momentum and the causal connection between events in distant atoms "so characteristic for the classical theories."[83]

In the BKS theory, a quantum jump, appearing as the emission of a light wave by the corresponding oscillator, could be induced by the virtual field even though no real energy or momentum had been transferred to the atom. Energy conservation required that the emission of light energy first involved the absorption of energy. The absorption arose in most cases through the absorption of a light quantum. At the same time, causality was violated because causes and effects were no longer connected to each other: absorption of light energy by an atom was not necessarily correlated with emission of light energy from another atom. Instead, the virtual field connected these processes, even though the field itself did not exist in the real world of energy and momentum. Conservation laws and causality were, for Bohr, Kramers, and Slater, no longer precise assertions but statistical concepts. Employing such radical notions, they managed to account verbally for numerous radiation phenomena on the basis of the wave theory of light alone.

Such ideas would resurface under different circumstances three years later.

Heisenberg's December 1923 letter to Bohr outlining his new Zeeman principle arrived in Copenhagen just as Bohr, Kramers, and Slater completed their radiation paper. Heisenberg suggested a possible connection between his principle and dispersion theory, but he did not then know of the BKS theory, nor did he think much of it before visiting the three authors. "Bohr's work on radiation is indeed very interesting, but I do not actually see it as an essential progress," he commented earlier. But by the end of his March 1924 visit to Copenhagen, Werner was a convert.[84]

While Werner seemed certain that Bohr's radiation theory was on the right track, he left Copenhagen uncertain about his own work and unsure "if and how much sense or nonsense is in my Zeeman salad."[85] Was there any way to merge Heisenberg's saladlike hodgepodge of the Zeeman principle with the recent work of the Copenhagenites? The Heisenberg and Copenhagen theories were based on observed frequencies, and both treated stationary-state models merely symbolically. Both required combinations of pairs of numbers or states. Still, Heisenberg's ideas depended much more on models than did Bohr's. Most important, as Pauli insisted, Werner's Zeeman salad, no matter how successful in reproducing results, could never satisfy Bohr, who demanded thorough justification for any radical innovation. To be sure, the BKS theory contained some of the most radical notions of the period—the abandonment of conservation laws and causality—yet it also satisfied Bohr's criteria for rational advance, which the core model and Werner's new principle could not. Once Heisenberg realized this difference, he seemed stunned. The newly philosophical physicist returned to Göttingen to rework his entire manuscript so "that I do not assert too much in the details."[86] In the title of his paper he no longer called his idea a "new quantum principle" but a "modification of the formal rules of quantum theory."

Heisenberg came down from the Bavarian hills in April 1924 to an incredibly intense round of work in Göttingen. He was so busy that spring that Born threatened to place a notice in the newspaper: "Assistent entlaufen" (assistant run off).[87] Added to his burden of thoroughly reworking the Zeeman principle and of preparing it (or another work) for his imminent habilitation under Born were the demands of completing papers on ion polarization and half-integral k, helping Born prepare his lectures for publication, contributing to Born's attempt to find the proper discretizing of atomic physics,[88] and laboring with Landé on a joint publication.

Heisenberg's collaboration with Landé in Tübingen began soon after Landé had received Werner's Zeeman salad late in 1923. Heisenberg's principle seemed to work for every atom he studied but one, neon, since the core of the neon atom possessed a larger angular momentum than expected.[89] Heisenberg went to the medieval university town of Tübingen in early January 1924. There, in a few days of intensive work, Heisenberg and Landé invented yet another new, and even more abstruse, principle, a branching principle. It allowed the total momentum quantum number J of the neon ion to "branch" into two values when one added another electron. The result again seemed to violate Bohr's building-up principle of the periodic table, but new dualities and alternative principles seemed worth multiplying.[90] As Heisenberg stated in their joint manuscript, finally sent to press in May 1924, the branching principle, too, would find its justification and any problems it caused for building up would find their resolution in his forthcoming publication on a new modification of the quantum rules. The new modification would obviate neon branching "almost of necessity."[91]

Three days before the publisher received the Landé-Heisenberg branching paper on May 18, 1924, Heisenberg prudently forwarded a copy of his latest modification manuscript to Bohr for prepublication approval. "I am still not entirely satisfied with much of it," Werner wrote in perhaps honest appraisal. "However, even if in the details much will turn out differently from what is here, I now actually believe that the basic idea is correct."[92] Bohr was so impressed by the obvious utility of the scheme—it restored the building-up principle without Bohr's Zwang—and by Max Born's latest work on discretizing atomic physics that Bohr, visiting Germany on financial business, stopped off in early June 1924 to discuss developments.[93] He had also gotten word of approval of a Rockefeller stipend for Werner and wanted to discuss Werner's visit to Copenhagen with the two physicists.[94]

Born's efforts had indeed borne fruit. After Werner's return from the March visit to Copenhagen, Born immediately used the BKS theory and Kramers's dispersion formula as clues for discretizing physics—a procedure that would hopefully precipitate the new quantum mechanics.[95] In the scattering of light by an atom, it is the electric moment of the atom (the charge polarization) that controls the scattering and dispersion (absorption and reemission) of the incident light. Ladenburg's 1921 formula for the atomic electric moment for dispersion had included only absorption of the light. Kramers, invoking the virtual oscillators of BKS, generalized Ladenburg's formula to include all possible transitions (absorption and emission) from any atomic state.[96]

Born perceived in Kramers's formula a hint of a new principle. In the published version of his new principle, received in June 1924, Born derived an expression for the electric moment from classical perturbation theory. This expression contained, in place of the quantum transition frequencies, a sum over the classical harmonic frequencies of each type of motion. Born used the Bohr-Sommerfeld quantum conditions to obtain a connection with the quantum frequencies corresponding to each summation over the harmonic frequencies. He discovered that "the actual (quantum-theoretical) frequency of the resonator is the 'linear' average of the corresponding (classical) frequency."[97]

This result led Born to at least two significant conclusions. First, it yielded a quantum mechanical discretizing rule: "We are as good as forced to adopt the rule that we have to replace a classically calculated quantity, wherever it is of [a certain] form . . . by the linear average or difference quotient."[98] This rule, applied to Born's expression for polarization, immediately yielded Kramers's dispersion formula, for which it was in fact constructed.

During or just before Bohr's one-day visit to Göttingen, Born discovered a second consequence of the new rule, "a very beautiful connection of [Heisenberg's scheme] with the Kramers dispersion theory."[99] He showed that Heisenberg's average over the classical Hamiltonian H_d arose just as did Born's average over classical frequencies, as long as the system was not too complicated or degenerate. Of course Heisenberg's Zeeman problems involved the most complicated degeneracies of the worst sort, but Born could conclude nonetheless "that averaging procedures over quantum integrals, such as were carried out by Heisenberg, look quite natural and lack artificiality when considered from the point of view of general quantum mechanics."[100] Bohr agreed. During Bohr's visit, Werner invited Bohr, Born, and the Danish scientist Svein Rosseland, Bohr's traveling companion, to his rooming house. While Werner's landlady, the hospitable Mrs. Ulrich, provided tea in the parlor, Bohr approved of the new Göttingen principles, and Born decided that Werner's salad now had enough tossing—and professional approval— to constitute a habilitation thesis.[101]

Everything seemed to come together in Born's new quantum mechanics. Although this was not yet the quantum mechanics that would replace classical mechanics in atoms or the semiclassical Bohr-Sommerfeld quantum theory, Born's new rule was a giant step in that direction and would prove essential to the formulation of the actual quantum mechanics a year later. While lending support to Heisenberg's principle, it established a connection between Kramers's dispersion theory and the future quantum mechanics, based on the virtual oscillators of BKS.

But the complete connection with BKS, still rather ill defined, became merely "intuitive" after another visitor passed through Göttingen within days of Bohr's visit. Albert Einstein was on his way to northern Germany, probably to Kiel for a health cure, and stopped off in Göttingen for a few days to visit friends and colleagues. Göttingen seemed overwhelmed by the succession of great men. Only Rutherford was missing, Mrs. Franck exclaimed.[102]

Heisenberg was elated. Einstein's visit constituted his first meeting with the great physicist. Einstein, then 45 years old, already rotund and with slightly greying hair, a moustache, and a frequent pipe, walked from the train station to the Göttingen physics institute for brief discussions with the assembled physicists and for longer discussions with Born and Heisenberg, less than half Einstein's age, on the BKS theory. Einstein honored Werner with a fifteen-minute stroll through the surrounding neighborhood, during which they further discussed the theory and Werner's latest work. The next day, a disappointed Werner reported to Pauli and his parents that "Einstein has a hundred objections [to BKS]."[103] Bohr's willingness to go as far as to relax the demands of such essential features of classical physics as causality and energy and momentum conservation was simply unacceptable to Einstein. He had already written Born before his visit that, if these elements had to be given up, "then I would rather be a shoemaker or an employee in a gambling casino than a physicist."[104] It was a position he would hold to the end of his life.

Born acquiesced without complaint. In his paper on discretizing, entitled "On quantum mechanics," Born announced that he would "make use of the intuitive ideas" of the BKS theory, especially the notion of virtual oscillators, but that "our line of reasoning will be independent of the critically important and still disputed conceptual framework of that theory, such as the statistical interpretation of energy and momentum transfer."[105] The now unified Göttingen physics promoted virtual oscillators, while ignoring the controversial (and soon disproved) BKS theory.

Heisenberg would soon return to Bohr's institute, where his work would at first build on but eventually depart from that of his Göttingen colleagues. On July 6, 1924, Heisenberg wrote Landé of the new Göttingen optimism. The new quantum mechanics seemed suddenly in sight: "The beautiful thing about the new Bohr and Kramers dispersion theory is precisely that one now knows (or suspects), especially on the basis of Born's calculations, *how* the quantum mechanics will look."[106] Within days of Bohr's and Einstein's visits, Heisenberg's modification and Born's quantum mechanics papers were received by the journal of choice in quantum theory, the *Zeitschrift für Physik*. They appeared together in the same issue later that year.

Early in July, Bohr, now back in Copenhagen, thanked Heisenberg for sending a copy of the final draft of his modification. He seemed happy with the paper—it coincided with the tendencies of BKS and Kramers's dispersion theory—and he informed the young man that the International Education Board had officially granted Werner a generous stipend of $1000 to work in Copenhagen for a year.[107] On July 28, 1924, after Heisenberg delivered a successful test lecture on his modification to the Göttingen science faculty, the science faculty voted to habilitate the author, conferring official certification of Heisenberg's suitability to lecture at all levels at German universities.[108] The next day, the 22-year-old left with his youth group for a three-week outing in the Bavarian hills.[109]

10

Quantum Multiplying

In the September 1925 issue of the *Zeitschrift für Physik,* Heisenberg published a 15-page article with the harmless-sounding title, "On a quantum-theoretical reinterpretation of kinematic and mechanical relations."[1] But the aim was ambitious — no less than "to establish a basis for theoretical quantum mechanics, founded exclusively on relationships between quantities which, in principle, are observable."[2] It dealt with observed frequencies and intensities of emitted and absorbed light, and, in so doing, it enabled a momentous breakthrough in physics, ensuring Heisenberg's place in modern science. Heisenberg's paper laid the foundation of a new theoretical "matrix mechanics," one form of the long-sought quantum mechanics — a new physics of the atom and its interactions that replaced the classical mechanics of Newton and Maxwell. Heisenberg, Born, and their colleagues brought the new physics to fruition during the months following Werner's initial breakthrough. Werner's breakthrough precipitated the culmination of the quantum revolution of the first decades of this century, a revolution that reached its conclusion two years later.

Heisenberg's path to matrix quantum mechanics was neither direct nor his alone. In the introduction to his paper, Heisenberg acknowledged those who had set the stage: "One can regard the [Bohr] frequency condition and the dispersion theory of Kramers, together with its extensions in recent papers, as the most important first steps toward quantum-theoretical mechanics."[3] He cited in particular Born's 1924 discretizing rule for differentials and the Kramers-Heisenberg dispersion theory of early 1925. To these important first steps, one must add

Pauli's relativistic destruction of Werner's core model in 1925, the Copenhagen sharpening of the correspondence principle, the demise of the Bohr-Kramers-Slater theory, and Heisenberg's revival of core-model physics in reciprocal dualities.

One must also acknowledge the hothouse "atmosphere of quantum theory" that pervaded Bohr's Copenhagen institute.[4] There, Bohr and his young exotics — Heisenberg, Pauli, and Kramers — struggled intensively and exhaustingly, with each other and with each other's idiosyncratic approaches, to cultivate their achievements.

Bohr's Institute for Theoretical Physics, like its inhabitants, rode the tail of a shooting star in 1924 and 1925. During the last years of World War I, in which Denmark remained neutral, Bohr, professor of theoretical physics at the University of Copenhagen since 1916, convinced the Danish authorities and the Carlsberg Brewery foundation to give him a three-story institute in place of his one-room office.[5] The sons of a famous Copenhagen university professor, Niels Bohr and his brother Harald, a mathematics professor, easily moved within the higher circles of Copenhagen social and cultural life. As with most of the young physicists and mathematicians who would come to work and study with the Bohrs, culture and breeding made an unspoken commonality of interest and outlook — a commonality expressed in such joint endeavors as musical evenings, horseback riding, hiking tours, and frequent trips to the local movie house to view the latest silent films.

In 1921, Bohr inaugurated his new building in the nearly rural outskirts of town. The institutional-looking rectangular building, with its grey-stucco facade, pitched red-tiled roof, and gabled third-floor windows, stood behind a wire fence only a few yards from the sidewalk at Blegdamsvej 15. Within a few years, flowers had sprouted by the front gate to beckon visitors, and collegiate ivy had grown to cover the entire first floor of the outer walls, reaching almost to the large letters embedded in the wall above the entryway: "Universitetets Institut for Teoretisk Fysik 1921."

With a permanent staff of eight and long-term visitors numbering more than nine, by 1924 the institute was overcrowded. Less than half the space was actually devoted to theoretical physics. The top two floors were residential; the basement was given over to experimental work, a typical feature of theoretical physics institutes at the time. Bohr, his wife Margarethe, and their two growing boys lived on the second floor; the family maid, lab demonstrator, and special guests occupied the third. Bohr often invited visitors to his rather elegantly furnished, three-bedroom apartment, and during his first trip to Copenhagen, in

March 1924, Werner had shared more than one bottle of wine there with the professor. "Often, the next day, what we had discussed actually turned out to be correct," Werner wrote his parents.[6] He occupied the guestroom then and on many occasions in the years to follow.

With the upper floors and basement occupied, only the ground floor was left for the lecture hall; library; functionally furnished offices for Bohr and his close assistant, Dutch physicist H. A. Kramers; and a drab study hall of wooden desks for the many visitors.[7]

Bohr returned from a visit to New York in November 1923 with a grant of $40,000 from the Rockefeller Foundation's International Education Board (IEB) to expand the building and living quarters. And, despite a raging inflation, the Danish Culture Ministry simultaneously increased the institute's funding to a remarkable 5 percent of the university's total budget — a clear indication of the esteem in which the Danes held the 1922 Nobel laureate and his institute.[8] New construction began soon after Heisenberg's arrival in 1924 for the start of his one-year IEB fellowship. The expansion was not completed until Werner returned again in 1926. By then, Bohr occupied a single-story house next to the institute.

The IEB, initiated by the Rockefeller Foundation in early 1923, funded one-year fellowships for postdoctoral scientists around the world to study abroad. According to one count, of the 135 fellowships awarded by 1930, 15 went to the 63 visitors to Bohr's institute.[9] During Heisenberg's first seven-month stay as an IEB fellow in Copenhagen, starting in September 1924, six of the nine long-term visitors were IEB fellows.

During his June 1924 visit to Göttingen, Bohr reached agreement with Born on Heisenberg's stay in Copenhagen. Born, who expected to be in the United States during the 1924–1925 winter semester, allowed his newly habilitated Privatdozent (lecturer) to absent himself from Göttingen during the summer and coming winter.[10] But Born wanted Werner back in Göttingen by May 1, 1925, for the start of the summer semester. Under prodding from Bohr, the IEB allowed Werner to complete the remainder of his one-year grant during vacations, an arrangement that, in the end, benefited the course of quantum mechanics.[11]

Kramers arranged for Heisenberg's room and board in Copenhagen in the turn-of-the-century home of Mrs. Maar, the recent widow of a university professor whom Heisenberg had met the previous spring. Mrs. Maar warmly befriended her young tenants, often inviting them for weekend outings at her family's country villa. Werner occupied a "small but nice" bedroom on the second floor of the Maar home, next to a room occupied by an American visitor, Dr. King, a chemist and violinist. The room, overcrowded with "many chests," faced west, but because the house was so close to its neighbor, little sunlight found its way through the lace-covered windows.

Meals, taken punctually with Mrs. Maar and Dr. King in the dining room at 8:30 A.M., noon, and 6:00 P.M., turned into language labs. Dr. King spoke only little German, Heisenberg only little English, and Mrs. Maar at least the three languages. They agreed, to Werner's advantage, to speak only English during meals. After breakfast, Mrs. Maar then helped improve their Danish by reading the newspaper aloud for a quarter hour before the two left for their respective institutes. During his stay in Copenhagen, Heisenberg apparently learned enough Danish to write and lecture (though haltingly) in the language—a distinct advantage when an opening came later at the institute.[12]

Soon after arriving in Copenhagen, Werner sent word home about his daily routine. During the day, he worked alone at his desk in the study room, visited Bohr's large ground-floor office, and discussed "all sorts of questions" with Bohr during their long walks in the quiet and verdant Fælledpark behind the institute.[13] But in none of his reports home did Heisenberg ever mention discussions with Bohr's assistant Kramers, with whom relations were already strained. By all accounts, the problem was envy. To the Göttingen Wunderkind, Kramers seemed superior in every respect.[14]

The tall, broad-cheeked Kramers, with his receding hairline and smoldering pipe, seemed older than his 28 years. The Dutch physicist was fluent in several languages, an excellent musician, and far more knowledgeable in the Copenhagen specialty of quantum radiation theory than Werner. Above all, he occupied the envied position of Bohr's personal assistant, confidant, and heir apparent. The moody and occasionally depressed Kramers exhibited a somewhat detached, condescending, and ironic attitude toward everyone, but his lack of deference toward Werner and his near disdain for Werner's core-model physics were worse. Kramers displayed obvious professional ambitions in an institute populated by a crowd of ambitious boy wonders—all of whom circled around the domineering, preeminent Bohr—the "Pope" of quantum physics—and his dutiful assistant—"His Eminence" the cardinal, as Heisenberg and Pauli sneered.

Heisenberg, in second place behind an older competitor, worried again whether he would be accepted into the inner circle. As in Werner's boyhood home, where he and his brother competed fiercely for their father's approval, the competitive relationship encouraged Werner's urge to excel and to achieve the status of the master's favorite, no matter how entrenched the principal disciple.[15] Even before he set foot in the institute, Werner was already in competition with Kramers. In the end, he succeeded. In May 1926 Kramers left Copenhagen for a teaching chair in Utrecht, whereupon Werner stepped into his rival's shoes as both Bohr's assistant and his confidant.

Heisenberg carefully timed his arrival in Copenhagen while Kramers was abroad at the 1924 Innsbruck meeting of the GDNA "so that I can then help you," he confided to Bohr.[16] After a stopover in Berlin to visit his brother Erwin and Erwin's new fiancée Marianne, Werner arrived in Copenhagen on September 17, 1924, at the very peak of mounting difficulties with the correspondence principle and virtual oscillators, essential features of Copenhagen physics.[17]

The extremely complicated properties of spectroscopic lines, emitted by atoms only at definite frequencies, still served as the primary clues to the internal workings of the atom. In the Bohr-Sommerfeld theory of the atom, electrons orbited around a nucleus in certain stationary states. A quantum jump between these discrete stationary states was accompanied by the emission or absorption of a light quantum—a particle of light energy bearing the frequency of one of the observed spectroscopic lines. The frequency of the emitted or absorbed light quantum was proportional to the energy difference between the two stationary states. Since the concept of light particles contradicted the well-established and experimentally verified electromagnetic wave theory of light, it could be avoided in practice only by treating each observed frequency and intensity of the emitted or absorbed radiation as if it arose from a little harmonic oscillator within the atom—a charged electron ball on a spring oscillating with the observed frequency and amplitude (related to intensity) of the emitted or absorbed light waves.

Bohr, Kramers, and Slater showed in their theory, the BKS theory, that the notion of atomic virtual oscillators worked well in concert with Bohr's correspondence principle. According to Bohr's 1918 formulation of this principle, the quantum features of atoms—discontinuities and jumps—are supposed to transform smoothly into the expected classical behavior as one climbs from the widely spaced lower quantum energy states of an atom at the bottom of the ladder to the near-continuum of highest state energies at the top. In the limit of infinite n (principal quantum number), the continuum is reached, and the properties of atomic electrons in those realms are purely classical (since they are essentially freed from the atom). The fiction of virtual oscillators became real in the limit of the correspondence principle, since Bohr required each virtual oscillator to correspond in the limit of infinite n with a real oscillator — one of the harmonic components in the (multiply periodic) Fourier expansion of the mechanical electron orbits contributing to the electric moment of the atom. The electric moment — in a sense the "antenna" of the atom — controlled the reaction of the atom to incoming and outgoing electromagnetic radiation. Virtual oscillators and their relationship to real oscillators in the correspondence principle proved a powerful tool for exploring the interactions between atoms and light.

In July 1924, soon after Max Born in Göttingen submitted his quantum-mechanical discretizing rule for publication, Kramers in Copenhagen submitted a brief outline of a new derivation of his dispersion formula based on virtual oscillators. He used the Göttingen prescription for replacing differentials by differences.[18] That summer, while Werner toured the Bavarian hills with his youth group, Kramers attempted a full derivation of general scattering and dispersion of light by atoms based on Born's rule and the virtual oscillators of BKS. But soon Kramers's approach looked questionable in the light of "Smekal scattering" — a form of incoherent scattering of light by atoms discovered by physicist Adolf Smekal.

Using oscillators, Kramers's theory amounted to a classical wave theory of the emission and absorption of radiation by atoms — he had scrupulously avoided light quanta. In Smekal scattering, incident light of a certain frequency is scattered — that is, partially absorbed and reemitted — at a different frequency from that of the incident light. Smekal interpreted the effect as arising from the atom's emission of light quanta at frequencies other than the frequency of the incident light.[19] Light of those frequencies arose not from virtual oscillators but from induced downward jumps of electrons to lower energy stationary states, with the emission of an appropriate light quantum. Successful rebuttal of Smekal using oscillators and light waves without light quanta would constitute, for Kramers, strong evidence not only for his theory and virtual oscillators but for the BKS theory itself. The BKS theory had become by then for Kramers a matter of personal belief.[20]

Heisenberg helped Kramers with this problem in the fall of 1924,[21] but devoted most of his attention to defending Bohr against new challenges to oscillator physics raised by spectroscopy. Every spectral line observed in a spectroscope has not only a frequency but also an intensity (brightness) and a polarization (direction of oscillation within the wave). These features of an emitted light wave can also be derived from the classical electrodynamics of oscillating charges. In view of this, Bohr and Kramers had speculated that even the intensities and polarizations of spectral lines emitted by atoms, in addition to their frequencies, should succumb to an interpretation involving virtu: ·oscillators instead of light quanta and quantum jumps.

In an essay submitted to the *Naturwissenschaften* in early November 1924, Bohr explained that, unfortunately, virtual oscillators did not quite work for intensities and polarizations of spectral lines. The calculation required reference to the behavior of the atom in the classical limit, using the correspondence principle. But for some states, more than one virtual oscillator corresponded with a single harmonic component. For such "degenerate" states, wrote Bohr, the correspondence is lost and the calculation of radiation properties is impossible.[22] Even Born's average of the classical harmonics to obtain a quantum frequency was of little help.

But things were far from hopeless. Heisenberg, together with Bohr, discovered that the correspondence principle could be "sharpened" to allow an analogy between degenerate quantum states and nondegenerate classical harmonics, yielding quantitative results. Bohr referred the readers of his November essay to a forthcoming paper on the subject by Heisenberg.[23]

In preparation for the fourth edition of *Atombau* in 1924, Sommerfeld had examined the ability of the correspondence principle to elucidate recent data on polarization intensities obtained by Dutch physicist Ornstein and coworkers in Leiden. They had found several rules for the total intensity of radiation emitted by an atom at all polarizations. In July 1924, Sommerfeld informed Kramers that Ornstein's rules could not be derived quantitatively from the correspondence principle and virtual oscillators. They were derivable only by the usual procedure of applying the quantum conditions to each degree of freedom.[24]

Heisenberg demonstrated his new enthusiasm for Bohr's virtual oscillator physics by immediately rising to the defense of correspondence. When Kramers returned from Innsbruck at the end of September 1924, Werner wrote Pauli: "With Bohr I have once again considered the question in exact terms and we have come to the conclusion that [Ornstein's] summation rules are not — as Sommerfeld says — unable to be understood using the correspondence principle, rather they are a *necessary consequence* of the correspondence principle. . . . We are pleased with this interpretation because now the attacks on the correspondence principle are completely refuted."[25]

Using correspondence to derive quantitative results, Heisenberg sharpened it for other applications. In mid-November, the new convert wrote to Sommerfeld of his complete admiration of Copenhagen physics: "My work moves very much along the tracks of the correspondence principle," he confessed, "which is clearly applicable in this case."[26]

Heisenberg explained the sharpening of correspondence and its uses in a letter to Pauli on October 8, 1924, and in the promised paper on the new approach, received by the *Zeitschrift für Physik* on November 30, 1924.[27] The essential feature was an "appropriate logical connection with the classical theory" made by tinkering with the virtual oscillators of the BKS theory.[28] Heisenberg simply hypothesized the existence of as yet unknown quantum laws that would somehow allow virtual oscillators in degenerate cases to correspond anyway with the harmonics of classical theory: "We believe that . . . virtual oscillators are subject to laws that make a narrow analogy between quantum theory and classical theory possible."[29] Heisenberg offered a number of sharpened applications of correspondence to the polarization and intensity of fluorescent light (the resonance absorption and reemission of light) and to the derivation of Ornstein's rules.[30] A forthcoming paper with Kramers would handle dispersion theory.

The Kramers-Heisenberg paper on the dispersion of light by atoms, submitted to the *Zeitschrift für Physik* on January 5, 1925, served as the apex of sharpened correspondence and virtual oscillator physics, the crowning achievement—in Kramers's view—of the BKS theory. In retrospect, their joint effort was the final touch needed for Heisenberg to fabricate quantum mechanics six months later—but without the BKS theory.[31]

Working in separate rooms—Kramers in his office and Heisenberg in the study hall—the two managed to produce a joint paper, written, however, mostly by Kramers. Following his July 1924 note on dispersion, Kramers turned to a complete quantum theory of dispersion. With orbital mechanics of little help, Kramers employed virtual oscillators, Born's discretizing rule, and a correspondence between Smekal's terms and classical atomic behavior. This led to the "surprising result"[32] that, if one neglects or treats statistically the atom's recoil, the entire interaction—including the Smekal terms—can be handled à la BKS theory: emission and absorption of classic spherical light waves, not light quanta.

Apparently, Kramers had arrived at most of his dispersion theory and calculations by the time he returned to Copenhagen at the end of September 1924. Heisenberg then joined in. While Kramers approached the problem physically, Werner calculated dispersion just as he would have in Göttingen: he attacked the problem with a complete mathematical apparatus.[33] Soon disputes erupted. Kramers and Heisenberg fought over the interpretation of the so-called false resonances, Smekal terms, virtual oscillators, and, crucially, the authorship of the paper. Kramers soon discovered that false resonances, which Heisenberg insisted on including, canceled out in the end. But the Smekal terms, oscillators, and authorship led to heated debate and ultimate disaffection.

Kramers preferred a close analogy between virtual oscillators and the orbital harmonics of nondegenerate systems, as demanded by the BKS theory and correspondence. Heisenberg preferred sharpened correspondence and only tenuous, symbolic connections between oscillators and harmonics. In the end they turned to Bohr. Only he could decide between analogy and symbol, and only he could rule whether Heisenberg's name should appear with Kramers's as coauthor. As Heisenberg recalled it, a long debate unfolded one chilly December afternoon in Bohr's office as the sun hung low on the Danish horizon.[34] The reserved Kramers was no match for the practiced Heisenberg, whom Bohr declared the winner. Bohr not only agreed with Werner's scientific conclusions, but he also decided that the men should appear as joint authors. Kramers drafted the paper and submitted it for publication during Heisenberg's absence from Copenhagen on Christmas vacation.[35] It appeared with both names.

An even more important upshot of the meeting than authorship was the giant step taken from the physics of mechanical atomic models to the behavior of observable quantities—the basis of Werner's breakthrough to quantum mechanics. Werner wrote: "The necessity for detachment from the intuitive models was for the first time stated emphatically and declared to be the guiding principle in all future work."[36]

Heisenberg may have overdramatized the oscillator debate in recollection. Pauli, for instance, still complained about Heisenberg's formalism—his mixing of quantum and classical notions. However, the Kramers-Heisenberg paper contains an even more remarkable statement on observables and detachment from intuitive models than that later asserted by Heisenberg. "In particular, we shall obtain, quite naturally," the authors claimed, "formulas that contain only the frequencies and amplitudes characteristic for transitions, while all the symbols referring to the mathematical theory of periodic systems will have disappeared."[37] In his subsequent "fabrication" of quantum mechanics, as he called it, Heisenberg relied only on the frequencies and amplitudes of emitted radiation, the secondary properties of the atom that could be observed and measured in a laboratory. By applying Born's discretizing rule directly to the differential quotients in the classical version of the electric moment of the atom, Kramers and Heisenberg obtained, as advertised, the general quantum-theoretical scattering moment for all forms of scattering and dispersion — and without elements of the unobserved motions of orbiting electrons.

Nevertheless, despite talk of emphasizing observables — the measured frequencies and intensities of radiation emitted and absorbed by atoms — Heisenberg was not yet ready to renounce the unobserved orbital mechanics of atomic models.[38] Especially, he would not abandon his symbolic core model or his ultimate aim of finding what he called "the real Zeeman model."[39] The transition from mechanical models of the atom, however symbolic, to laboratory observables as the foundation for a quantum mechanics was an enormously difficult one — even for the audacious and unphilosophical Werner. Heisenberg's extremely intense struggles with atomic models, accompanied by his gradual retreat to mere observables, became evident in his response to Pauli's newest assault on Copenhagen physics during the first half of 1925.

Early in the year, Pauli vehemently complained to Heisenberg that Copenhagen's "'virtualization' of physics" and Heisenberg's sharpening amounted to "imperialism of the correspondence principle."[40] Pauli himself had experienced two reversals in outlook in the fall of 1924. Thanks to Bohr, Pauli had abandoned his conservative "scientific conscience" and embraced the BKS theory in the spring of 1924. It was only a brief embrace. From the beginning, Einstein had opposed the BKS theory, with its radical notions of acausality and statistical conservation of energy and momentum, and had told Heisenberg so in Göttingen. At the same time that Heisenberg joined the Copenhagen program in September, Einstein was reviving Pauli's scruples at the Innsbruck meeting of the GDNA. Pauli went off in two new directions.

First, he explicitly rejected the "Copenhagen putsch" — the BKS theory. "Today I as a physicist stand in complete opposition to this conception of radiation phenomena," he wrote Bohr.[41] He also sent a letter to Heisenberg expressing his strong doubts about correspondence, sharp or dull.[42]

Second, under Einstein's influence Pauli began a close examination of the core model's relativistic effects. In several letters to Landé at the end of 1924, he presented his relativistic analysis. It decimated the core model. The crucial feature of the core model was the interaction between the orbiting valence electrons of an atom and the core — the nucleus and inner electrons — which bore an angular momentum due to its spin, or rotation. The core rotation, when combined with the orbital angular momentum of the outer electrons, contributed to the total angular momentum of the atom and thus to its behavior in magnetic fields. Pauli proved that this was impossible: "There can be no talk of a participation of the noble gas atomic core (the same for the other elements)."[43]

Pauli explained to Landé that if the nucleus and the electrons in a closed electron orbital shell form the core of an alkali atom, then the electrons in the core orbit around the nucleus so rapidly that they should exhibit a relativistic mass change. According to relativity theory, the mass of an object increases with its speed. Pauli found that this mass change should noticeably influence the rotation of the core and thus influence multiplet structure and the anomalous Zeeman effect. But such influence was not observed. From this, Pauli concluded: "The closed electron configurations should contribute nothing to the magnetic moment and to the angular momentum of the atom."[44] In other words, the all-important core of Heisenberg's core model was, in effect, physically inert.

The interaction between the core and the valence electrons served as the basic explanation for the observation of spectroscopic multiplet lines. However, the details of the multiplets had earlier led Bohr to attribute a two-valuedness to the orientations of the core under the influence of a vaguely defined force, a Zwang or constraint, within the atom. With the core rendered inert in 1925, Pauli transferred all the core's properties and behavior to the relativistic valence electron. This included the transfer of the inexplicable core two-valuedness to the now unmechanical electron—an action fom which Pauli did not shrink. "The valence electron is able in a puzzling unmechanical way to orbit in two states (with the same k) with different momenta."[45] Or, as he expressed it in his paper, received by the Zeitschrift für Physik on December 2, 1924, the electron possesses "a characteristic, classically nondescribable type of two-valuedness." And the origin of the two-valuedness was, as before, the mysterious "Bohr Zwang."[46] In effect, in transferring the two-valuedness of the core to the electron, Pauli had invented the behavior of the spinning electron long before the advent of spin.

Pauli sent his manuscript to press without informing Copenhagen, probably fearing a rupture. When Bohr inquired in December about the work, Pauli responded with a second manuscript and a long letter explaining "what I can actually accomplish positively with my notion."[47] He could, in fact, save Bohr's building-up principle for the periodic table—and without Heisenberg's core-model modification rule or the correspondence principle. Starting with a new numbering scheme introduced by E. C. Stoner, Pauli arrived at a general theory of the closing of electron groups in the atom, essential to Bohr's building up of atoms. The paper contained Pauli's most famous achievement, the exclusion principle. Every electron is assigned four quantum numbers, and no two electrons in a single atom are to have all four identical numbers. Although these quantum numbers were, for Pauli, no longer related to electron orbits within the atom, the exclusion of duplicate sets of quantum numbers accounted for the closing of groups and periods in the periodic table and replaced the now questionable phase relations between orbiting electrons argued earlier by Born and Heisenberg.[48] This principle, which lasts to this day, constituted the ground for Pauli's Nobel prize in 1945.

Yet exclusion required an obvious Unsinn—a nonsense—from which, again, Pauli did not shrink. Instead of invoking the core, Pauli attributed spectroscopic doublets to relativistic origins. This was incompatible with the successful orbital inclination theory of the core model, based on the correspondence principle. But this nonsense, he told Bohr, was no greater than the conjugate nonsense of the core model: "The physicist who finally succeeds in adding these two nonsenses will gain the truth!"[49] Bohr responded by inviting Pauli to Copenhagen.

Pauli probably had in mind the master of formal compromises, Heisenberg. Heisenberg, he was sure, would also be the most opposed to Pauli's devastating nonsense. He erred. As was usual, Bohr showed Pauli's letter and manuscript to his young collaborator. Werner fired off a postcard congratulating Pauli for bringing the so-called Schwindel (swindle) of current quantum theory to new heights. Pauli, Heisenberg, and Bohr were developing their own vocabulary to depict the difficult situation in which they found themselves. In their private language, Schwindel was derivative of Unsinn and a stepchild of the core model: an untenable formal or symbolic mixture of classical and quantum physics to obtain an otherwise unobtainable result.[50] Even Bohr indulged the linguists, informing Pauli of his enthusiasm for the "complete Wahnsinn [insanity]" of Pauli's paper, since "now the extent of the entire Schwindel has been so exhaustively characterized."[51]

Bohr nevertheless expressed hope of "perhaps coming essentially further with the help of the oscillator point of view"—virtual oscillators and correspondence.[52] But while Kramers drafted the Kramers-Heisenberg paper over Christmas recess, formalist Heisenberg began contemplating how to add nonsenses; that is, how to build a single theory out of the two available Zeeman schemes. One of the two available schemes was the Heisenberg-Landé core model, which involved magnetic interactions between orbiting electrons and—in Heisenberg's modification rule—a two-valuedness in the association of two values of the *J* quantum number with one electron orbit. The other scheme was Pauli's association of relativistic—not magnetic—effects with the appearance of spectroscopic doublets and the assignment of two values of the angular momentum number *k* to one electron in order to obtain the necessary four quantum numbers. In an apparent play on the Greek word *schema* ("scheme" in English) and the Yiddish word *Schummel* (swindle), Heisenberg called each of these Zeeman schemes a *Schimmel* (a German word whose literal meaning—"mold" or "white horse"—has no bearing on its use here.)[53] Since both Schimmel "are to be conceived of very symbolically or formally," he reasoned, a proper "addition" of the two should not be hard to find.[54]

No further progress occurred until Heisenberg returned to Copenhagen at the end of January 1925. He had been laid up in Munich for several weeks with a knee injury, suffered while skiing with his youth group during their winter encampment at the Bavarian ski hut. By February 18 he had hit upon an addition of Schimmel that arose from Kramers's latest work on fluorescent radiation on the basis of BKS. Kramers noted that incident radiation exerted its own Zwang on an atom, such that the reaction of the atom as controlled by the virtual oscillators involved its own two-valuedness: two energy levels associated with one reaction.[55] Each of the schemes supported by Heisenberg and Pauli contained its own two-valuedness. Kramers's work suggested to Heisenberg two ways of looking at two-valuedness. An interaction between two atoms involves a Zwang, such that either two state energies are associated with one interaction energy or two interaction energies are associated with one state energy. Transferred to an ion and its valence electron, the two Schemata—as he called them in his paper— were analogous to the two Schimmel of Pauli and Heisenberg: the Bohr Zwang causes a two-valuedness either of the valence electron in Pauli's scheme or of the core in Heisenberg's. Although the criteria never made it quite clear which of the two schemes was to be employed in a given situation, each was to be fully valid and each was to yield the same macroscopic observables, such as total energy. The two Schemata, wrote Heisenberg, "will form so to say a convergent calculation procedure for the determination of the properties of the stationary states of the atom."[56] Heisenberg was still searching for ways to calculate orbital energies, even while retreating from precise models of the atom.

Heisenberg eagerly reported to Pauli his discovery of the apparent identity of their two Schimmel on February 26 and even attempted to assign a definite momentum to the core of a noble gas atom. "Nevertheless, there is still stuff to argue over," he conceded. Two days later Pauli, repeating his objections, announced his visit to Copenhagen for mid-March 1925: "I hardly believe that we will be able to reach agreement there."[57]

Pauli's visit to Copenhagen coincided with the collapse of Copenhagen physics. Careful experimental studies of the scattering of free electrons by incident light (the Compton effect) in Germany and the United States were yielding no violation of conservation laws in individual collisions. Light really did seem to consist of light quanta, after all—and the BKS theory seemed on the wrong track.[58] Pauli's month-long stay in Copenhagen also marked a definite turning point in Heisenberg's route to matrix quantum mechanics.

During Pauli's stay, Heisenberg completed a draft of his paper on dual Schemata before leaving on April 7 for Göttingen and a two-week tour with his youth group through the Württemberg countryside.[59] Werner's younger comrade Wolfgang Rüdel was on an extended visit with Werner when Wolfgang Pauli arrived. Not until after Rüdel's departure in late March did the physicists get down to business, but with these three enormously gifted and active physicists together in one institute, the intensity of work—lasting daily from 9:00 A.M. to midnight—helped make up for lost time.[60] By April 7, Heisenberg had completed his duality paper, which Bohr and Pauli sent to the publisher after his departure. The paper served as a tacit admission of failure by everyone. For obtaining the intensities of spectroscopic lines, the correspondence principle worked only in the symbolic picture of Heisenberg's core-model Schimmel, where the electron orbits are unique.[61]

Equally devastating to what Bohr called "current conceptions" was the mere willingness to resort to "reciprocal dualities" in treating quantum phenomena. To be sure, dualism would become commonplace in quantum mechanics, but in 1925 it disconcerted everyone, including Heisenberg, who with Bohr had invented it. "One now has the strong impression with all models," Pauli had written Sommerfeld several months earlier, "that we are speaking a language that is not sufficiently adequate to the simplicity and beauty of the quantum world."[62] In August 1925, the American physicist J. H. Van Vleck, commenting on the demise of the BKS theory and the rise of dualities, declared: "Modern physics certainly is passing through contortions in its attempt to explain the simultaneous appearance of quantum and classical phenomena; but it is not surprising that paradoxical theories are required to explain paradoxical phenomena."[63]

Heisenberg returned home to Munich after two weeks of wandering among romantic medieval villages to find a depressing letter from Bohr. "There is much to report, for the most part negative, for I look at many things even more doubtfully than at the time when you were here."[64] Still uncertain how much of BKS to jettison, Bohr told of the "torture" that, with Pauli's comfort, he endured "in order to get used to the mysticism of nature. . . . I am attempting to prepare myself for all eventualities," he bravely maintained.[65]

Heisenberg remained calm at the news of the death of BKS. He claimed that he had never been fully committed to the theory's avoidance of light quanta anyway, although he had appreciated its pragmatic features. If waves and quanta demanded equal attention, then so be it: they could be treated just as he had treated the alternative Schemata — by a formal dualism. Responding to a report of confusion in Copenhagen over radiation, Heisenberg wrote prophetically in May 1925, "If I were there, I would, as in the case of the Zeeman effects, plead for a formal dualistic theory: everything must be describable both in terms of the wave theory and in terms of light quanta."[66]

Bohr, still unsympathetic to such duality, tortured himself with the consequences of admitting "coupling of quantum processes in distant atoms" — a direct correlation between the emission of a light quantum by one atom and its absorption by another.[67] The BKS theory had preserved a wave interpretation by renouncing a direct correlation, or coupling, and thus renouncing causality and conservation laws.

Faced with coupling and conservation, Bohr returned in April 1925 to his earlier position regarding internal atomic mechanics — allowing alterations of the mechanics rather than violations of the conservation laws. At the same time, Pauli reiterated his emphasis on observed data, expressed in December 1924: "I am convinced of an explanation (also for that of the coupling problem in general) from a physical analysis of the motion and force concepts in the sense of the quantum theory, for which one naturally will have to allow himself to be led by empirical points of reference."[68]

Heisenberg, meanwhile, was absent from Copenhagen during these crucial discussions. Away in the mountains, he gleefully wrote, "I have not thought a minute more about physics."[69] He returned to Göttingen on April 27, 1925, to begin his summer lectures and to follow what seemed a sure path to quantum mechanics — a careful, mechanical analysis of the simplest atom of all, hydrogen, a single electron orbiting a singly charged nucleus.[70] Without degeneracies or the devastating attractions between electrons, Born's discretizing rule and the loose linkage between virtual atomic oscillators and the orbital mechanics of the electron should enable him, he thought, to guess the correct quantum-mechanical model for hydrogen, just as Kramers had done for dispersion.[71]

Two guides existed for making the right guess. First, the planetary oribt of the electron as it revolved around the nucleus could be decomposed into a harmonic series of special functions, multiple Bessel functions. As in Copenhagen, each virtual oscillator of the atom should correspond in the limit of large quantum number n to a term in this series, and the series (for the real motion of the electron) could be adjusted to make the correspondence work. Second, the results should coincide with new spectroscopic line intensity ratios for hydrogen being obtained by Ralph Kronig, a German-American physicist who was visiting Copenhagen.

The plan seemed promising, especially after an encouraging letter from Kronig,[72] but by the middle of May, Heisenberg was bogged down in Bessel functions. "The present conditions together are still not entirely sufficient to get the intensities uniquely," Werner conceded to Bohr, "but I still want to try to make progress."[73] He worked at it a while longer but soon retreated to the much simpler problem of the anharmonic oscillator, observables, and quantum multiplying.

Pauli had come to a similar impasse at about the same time. He, too, had become interested through Kronig in the problem of obtaining multiplet line intensities by somehow modifying the Born rule for the virtual oscillators to obtain the correct amplitudes of vibration (which yield the intensities). But by the end of May 1925, Pauli, too, was enmeshed in difficulties and was now nearly ready to resign again from physics. "Physics is at the moment once again very wrong," Pauli exclaimed to Kronig in a now famous quotation. "For me in any case it is much too difficult, and I wish that I were a film comedian or something similar and had never heard of physics!"[74] Charlie Chaplin was all the rage in Copenhagen.

Right up until Heisenberg's breakthrough in July, Heisenberg and Pauli — when not watching Chaplin films — devoted their energies to empirical intensity effects rather than to quantum mechanics. Remarkably, both continued to keep their positions despite despair and rebellion. As late as June 24, Heisenberg explained their differences over the origin of the so-called Hanle effect for multiplet lines: "Our theoretical points of view differ insofar as you regard the line splitting as given and apply suitable oscillators to your representation; while I always attempt to hold on to the mechanics of the model."[75]

The more profound difference was that Heisenberg—unphilosophical, pragmatic, and formalistic—could advance on two fronts at the same time and respond quickly on each. In letters to Pauli and Kronig on atomic intensities, his new approach to anharmonic oscillators appeared almost as an afterthought. In the same letter to Pauli in which Werner defended the use of mechanical models, he made a leap to Pauli's side, declaring that observed phenomena, not models, would lead the way to quantum mechanics. Two weeks later, Heisenberg strayed further from orbital mechanics. He even chided Pauli for writing of electron orbits that fall into the nucleus. "My entire meager efforts go toward killing off and suitably replacing the concept of the orbital paths that one cannot observe," Werner declared.[76] He replaced the unobserved orbits with an ordered series of mathematical entities representing the observed radiation emitted or absorbed by an atom. Heisenberg, along with most other physicists, would soon learn that these entities were the elements of a little known mathematical object, a matrix.

Heisenberg's breakthough came because of his willingness to turn to a simpler problem (the anharmonic oscillator) and to try another approach (reliance on observed phenomena), even without being converted to it. It was Heisenberg at his best—simultaneously pursuing incompatible methods and employing inconsistent arguments intensely and brilliantly. Still, the work frustrated him. He could not advance beyond the simple problem to obtain a quantum mechanical explanation of even the simplest atom, and he dreaded his colleagues' criticism. "I have almost no desire to write about my own work," he told Pauli, "because everything appears to me still unclear and I can only surmise approximately how it will turn out."[77] Pauli held his tongue.

Heisenberg's fabrication of quantum mechanics may be reduced to three steps. First, he reinterpreted the equations of classical space-time kinematics as nonclassical formulas in quantum mechanics by making use of the observable properties of radiation emitted by virtual atomic oscillators. Then he raised the positivist criterion of observability of all quantities to a basic postulate of the theory. Finally, he "killed off" the mechanical orbits, replacing them with what became matrix elements.[78]

Reinterpretation and the appeal to positivism were, of course, not unique to Heisenberg. When the Kramers-Heisenberg paper—claiming concrete observability for all terms in the final equations—appeared in the *Zeitschrift für Physik* in April, Heisenberg had by then returned to

Göttingen as a lecturer in Born's institute. Born and his new private assistant, the 22-year-old German physicist Pascual Jordan, distinguished for his brilliance and an obvious lisp, were already at work on a quantum theory of aperiodic systems. Aperiodic systems, such as anharmonic oscillators, are nearly periodic. By May 1925, Born and Jordan found that the transition quantities, or the terms in their formulas roughly representing numbers of quantum jumps, always corresponded to squares of the amplitudes of classical harmonic vibrations, which in turn yielded the observed intensities of light corresponding to the quantum jumps in question.[79] Born and Jordan sought to reform radiation with these transition quantities. They raised their observability in terms of spectroscopic intensities nearly to a sine qua non.[80] In their published paper, completed by June 11, 1925, Born and Jordan appealed to Einstein's operational analysis of simultaneity in special relativity, announcing "a fundamental postulate of great significance and fruitfulness. . . . Only such terms enter into the true natural laws that are in principle observable and determinable."[81]

Heisenberg thought that the task his Göttingen colleagues had set for themselves was too difficult. Irritated, he complained to his parents that his own work was at a standstill and that "everyone here is doing something different and no one anything worthwhile." Leaving Born and Jordan to their transition quantities, Heisenberg tried to guess the quantum mechanics of anharmonic oscillators from the corresponding virtual harmonic oscillators.[82]

Even this simple problem required intensive thought and work, as revealed in his letters to Kronig and Pauli. By June 5, 1925, he had most of the final reinterpretation via observables, as he explained in a letter to Kronig—not Pauli—on that date.[83] Without justification for the approach, and mindful of Pauli's past criticism, uncertainty bred restraint.

Almost any periodic or quasiperiodic function $F(t)$ that varies with time can be expanded in an infinite addition, or series, of time-dependent harmonic terms—that is, it may be broken up into a summation of harmonic oscillators. Each term in the series—called a Fourier series—can be labeled by an integer n, n being 1, 2, 3, and so on. Each term consists of an amplitude of the oscillation, A_n, an angular frequency ω_n, and the time t. Each of these elements can be combined into a single term of the series by using the mathematical symbols e—the basis of the natural logarithms—and i—the square root of -1. The term may be expressed as $A_n e^{in\omega_n t}$. This represents a simple harmonic oscillator—a ball on a spring, or a swinging pendulum—oscillating with amplitude A_n

and frequency ω_n. A more complicated motion $F(t)$ may be represented by an infinite summation of such oscillators:

$$F(t) = A_1 e^{i\omega_1 t} + A_2 e^{i2\omega_2 t} + A_3 e^{i3\omega_3 t} + \ldots, \text{ or } F(t) = \sum_n A_n e^{in\omega_n t}$$

Heisenberg began his report to Kronig by looking at the observed electric field radiated by an unobserved charged, anharmonic oscillator. First he expanded the motion of the oscillator, represented by its position as a function of time, $x(t)$, in the preceding Fourier series. He then expanded the radiated electric field at a distant point off the axis of the anharmonic oscillator in a similar Fourier series. The square of the amplitudes of each term in this series yielded the observed intensities of the field at that point. Working backwards — as Pauli would have it — Heisenberg expressed these individual amplitudes of the electric field as functions of the amplitudes of a Fourier expansion of the position variable $x(t)$ of the charged anharmonic oscillator that emits the electric field as it vibrates.

So far, no quantum notions had entered into the calculation. But the calculation did suggest to Heisenberg that, in classical theory, all of the observed data concerning the emitted radiation could be expressed as relations between the Fourier amplitudes of the unobserved mechanical motion of the anharmonic oscillator itself. One can assume that this basic idea also holds in quantum theory, he wrote Kronig. The task was now reduced to reinterpreting the relationships between the Fourier amplitudes of the position coordinate of the oscillator so that they become quantum relationships between the observable properties of the emitted radiation.

Heisenberg's reinterpretation began with each individual harmonic oscillator. He regarded the nth term of a Fourier series, $A_n e^{in\omega_n t}$, as a virtual oscillator. Since each oscillator represents a quantum jump from stationary state n to stationary state $n - \tau$, the frequency and amplitude may be rewritten to reflect this combination of two states:

$$A(n, n - \tau)\, e^{i\omega(n, n - \tau)t}$$

Here n and $n - \tau$ are the integral quantum numbers of, respectively, the initial and final states in a quantum jump, $\omega(n, n - \tau)$ is the frequency of the radiation emitted or absorbed by the virtual oscillator, and $A(n, n - \tau)$ is the amplitude of the virtual oscillator (a real constant at this stage).

The amplitudes A_n of the classical motion are usually squared (multiplied by themselves) to obtain the intensity of the emitted radiation. Heisenberg sought the corresponding multiplication rule for the amplitudes of the quantum virtual oscillators, which yield the observed intensity relations of spectral lines. "The essential thing in this reinterpretation appears to me [to be] that the arguments [n and $n - \tau$] of the quantum amplitudes must be so chosen as to correspond to the connection of the [observed] frequencies."[84] The criterion of observability enters almost unnoticed.

In classical physics, the square of the first harmonic ($n = 1$) yields the second harmonic ($n = 2$), according to the relation $(a_1 e^{i\omega t})^2 = b_2 e^{2i\omega t}$, where $b_2 = a_1^2$. What, asked Heisenberg, is the corresponding quantum relation? In Bohr's atomic theory, the relation can correspond to two successive quantum jumps. The emitted frequency, $\omega(n, n - 2)$, is then equal to the addition of two individual frequencies, representing two successive downward jumps, from n to $n - 1$ and $n - 1$ to $n - 2$: $\omega(n, n - 2) = \omega(n, n - 1) + \omega(n - 1, n - 2)$. Hence, Heisenberg reasoned in his letter to Kronig, the quantum relation corresponding to the square of a harmonic is

$$b_2(n, n - 2)\, e^{i\omega(n,\, n - 2)t} =$$
$$a_1(n, n - 1) a_1(n - 1, n - 2)\, e^{i[\omega(n,\, n - 1) + \omega(n - 1,\, n - 2)]t}$$

The amplitude, which yields the observed intensity of the spectroscopic line of frequency ω, thus forms not a simple square of the term but the relation:

$$b_2(n, n - 2) = a_1(n, n - 1)\, a_1(n - 1, n - 2)$$

In his published paper, where Heisenberg reinterpreted the multiplication of two entire Fourier series, he drew on this simple result, and a similar expression in the Kramers-Heisenberg dispersion theory, to obtain a multiplication rule for all Fourier amplitudes, which were related to the intensities (or probabilities) of upward and downward jumps from a given state n. Since more than two successive jumps may be involved in the transition between the state n and the state $n - \beta$, β being an integer, the simplest and most natural assumption was to add up all of the terms representing jumps from and to all of the intermediate states $n - \alpha$:

$$C(n, n - \beta) = \sum_{n - \alpha} A(n, n - \alpha)\, B(n - \alpha, n - \beta)$$

where A, B, and C are now complex quantities, as in dispersion theory. This simple rule constituted the central feature of Heisenberg's quantum-theoretical reinterpretation of classical kinematic and mechanical relations — his breakthrough to quantum mechanics. While Heisenberg actually knew nothing of matrices, Born soon recognized this rule as the fundamental rule for multiplying two matrices together, the matrices $A(n, n - \alpha)$ and $B(n - \alpha, n - \beta)$. Various values for the two integers in the parentheses yield the various elements of the matrix. For example, if $n = 1, 2, 3$ and $n - \alpha$ can also equal 1, 2, and 3, then an array of values of A is formed, each element of the array corresponding to a particular n and $n - \alpha$. The array, or matrix, for $A(n, n - \alpha)$ looks like this:

$$\begin{pmatrix} A(1, 1) & A(1, 2) & A(1, 3) \\ A(2, 1) & A(2, 2) & A(2, 3) \\ A(3, 1) & A(3, 2) & A(3, 3) \end{pmatrix}$$

Heisenberg had his reinterpretation, but it was hardly a quantum mechanics. Indeed, Planck's constant was still absent.[85] Nor did Heisenberg seem satisfied with this result. Calculating the intensity of anharmonic oscillator radiation was still too complicated. He wrote Kronig, "The physical interpretation of the above scheme . . . results in a very peculiar point of view." Still uncertain, in a letter to Bohr three days later Heisenberg was silent about his work.[86]

At about this time, Heisenberg came down with a horrible attack of hay fever. Seeking relief, he left Göttingen for a barren rock in the North Sea, the tiny resort island of Helgoland off the German coast. Located in a guesthouse near the south shore, he was alone with the boulders and the sea and a strange multiplication rule for oscillator amplitudes.[87] What exactly happened on that barren, grassless island during the next ten days has been the subject of much speculation and no little romanticism. Years later, Heisenberg recalled suddenly realizing late one night that the total energy must be held constant, as recently reaffirmed with the defeat of the BKS theory. This apparently enabled him to derive the energies of the quantum stationary states by applying the new multiplication rule to the corresponding classical expressions. He hurriedly calculated the energies of the harmonic oscillator and the rigid rotator and obtained satisfactory agreement with known observations in time to watch the morning sun dawn over the eastern sea.[88] Heisenberg had his breakthrough.

On the return trip, when he stopped off in Hamburg to inform Pauli of his new physics, Heisenberg, now cured of hay fever, must also have had his version of the quantum-mechanical analogue of the quantum conditions, which were essential for obtaining the proper constants.[89] He immediately began to draft a paper on his results.

The quantum conditions had been perhaps *the* fundamental feature of the quantum theory of Bohr and Sommerfeld. For each variable of the motion, a momentum p and position q could be defined. The quantum conditions, applied to periodic motions, required that the total area of momentum-position space swept out by the system as it moved through a cycle should equal an integer n — a quantum number — multiplied by Planck's constant — that is, $\oint p\,dq = nh$. Heisenberg also reinterpreted this condition in terms of Fourier amplitudes and frequencies, such that only these observables remained in his equations. The new expression led directly to a form of the Kramers dispersion formula upon application of the Born discretizing rule.[90]

During the months following Heisenberg's submission of his paper presenting his abstract reinterpretation, Max Born and the Cambridge physicist and engineer Paul Adrien Maurice Dirac would independently interpret his new expression for the quantum conditions as the quantum-mechanical relation for the multiplication of the matrices for momentum and position, **p** and **q**, of any mechanical system (boldface type indicating a matrix). Usually, two quantities A and B multiplied together yield the same result when multiplied right to left, $A \times B$, and left to right, $B \times A$. Not so in quantum mechanics, Born and Dirac now discovered: **pq** is not the same as **qp**. Rather, $\mathbf{pq} - \mathbf{qp} = (h/2\pi i)\mathbf{I}$, where **I** is the unit matrix and h is Planck's constant. From this essential and fundamental equation and the properties of abstract matrix algebra the matrix form of quantum mechanics quickly emerged in the months following Heisenberg's breakthrough. This expression would later serve as the foundation for Heisenberg's most famous achievement: the uncertainty principle. But for Heisenberg in mid-June 1925, the primary accomplishment of what he called his new Schema lay in the reduction of interactions between the atom and the outside world to transition probabilities expressed as Fourier amplitudes.

Pauli was skeptical, if not openly critical. He still objected to Heisenberg's formalism and his use of half-integral numbers, this time for quantum energies.[91] By June 24, 1925, Heisenberg had evidently seen the Born-Jordan paper on radiation, completed and submitted during his absence on Helgoland. Heisenberg met Pauli's skepticism by raising the observability of all quantities in his equations to a postulate, and in nearly the same words used by Born and Jordan: "The basic postulate is: In the calculation of whatever quantities, for example energy, frequency, and so on, only relationships between quantities that are [observably] controllable in principle may appear. (Thus, for example, the Bohr theory for hydrogen appears to me much more formal than the Kramers dispersion theory.)"[92]

In the same letter to Pauli, Heisenberg continued to write of his own lack of joy about the paper and still wondered "what the equations of motion really signify if one conceives them as a relation between transition probabilities."[93] By July 9, he was eager to abandon orbits entirely, turning instead to positivism and mathematical manipulation — after all, it worked! He sent a copy of his manuscript to Pauli and handed his only other copy to Born, who apparently knew even less than Pauli about what his lecturer was up to. He had written his father shortly before: "My own works are at the moment not going especially well. I don't produce very much and don't know whether another [paper] will jump out of this at all in this semester."[94]

An uncertain Werner went to Born with two requests in early July 1925. First, he wanted permission to leave Göttingen before the end of the semester in order to accept an invitation to lecture in Cambridge, England. Second, he asked Born to look at his paper and to decide whether it was worth publishing. Born acceded to the first request and was fascinated by the second, but he was puzzled by the strange multiplication rule for two Fourier amplitudes. Somehow it looked familiar to him. Eventually, it reminded him of a rule he had encountered years earlier in a course on linear algebra that he had taken while a student in Göttingen. It was none other than the rule for the multiplication of two matrices.[95] At the end of the month, during Werner's absence in England, Born forwarded Heisenberg's paper to the editor of the Zeitschrift für Physik. The paper and its multiplication rule initiated the matrix formulation of quantum mechanics.

11

Quantum Mechanic

Heisenberg prepared two copies of his manuscript outlining a reinterpretation of classical mechanics. In early July 1925, he sent one to Pauli, who reacted "with jubilation."[1] The rapid evolution of this reinterpretation into a quantum mechanics over the following months gave the lately despairing Pauli "new hope, and a renewed enjoyment of life." As he wrote to Kronig in October, "Although it is not the solution to the riddle, I believe that it is now once again possible to move forward."[2]

Nearly two years later, Born, the recipient of the second copy, maintained that he still remembered the exact day and hour when he realized that Heisenberg's paper involved the old but little-known mathematics of matrix calculus.[3] While Heisenberg was lecturing on spectroscopy in Cambridge and finishing his IEB fellowship in Copenhagen that summer and fall, Born and his assistant, Pascual Jordan, fashioned Heisenberg's matrix calculation into a "systematic theory of quantum mechanics."[4] Their paper, sent to the *Zeitschrift für Physik* in September 1925, introduced most physicists, including Heisenberg, to the abstract methods of matrix calculation.

Applying these methods to quantum physics, Born and Jordan laid the foundations for a new quantum mechanics of matrices, a matrix mechanics. The new mechanics was (and is) nearly incomprehensible to the technically uninitiated. Since nearly every variable and function of classical mechanics was reinterpreted as a corresponding quantum matrix, the joining of matrix mathematics with quantum principles such that observable quantities could be derived involved at that time the most sophisticated properties and methods of matrix mathematics. The mathematics was made even more difficult by the circumstance that there are in principle an infinite number of possible stationary states in a quantized system. Fundamental variables, such as position and momentum, thus turned into infinitely large matrices containing infinite numbers of rows and columns.

Born and Jordan devoted most of their paper to explaining the abstract methods of matrix manipulation and adapting them to quantum physics. They discovered a matrix-mechanical analogue to many prior classical and quantum equations. Their efforts focused on the Hamilton-Jacobi formulation of classical mechanics, which Born and his assistants had used so extensively in their study of planetary atomic models. Briefly, according to this theory the motion of a mechanical system is controlled by a special function H, the Hamiltonian, which is usually equivalent to the total energy and is expressed as a function of the momentum p, the position q, and (if so chosen) the time t. Born and Jordan managed to construct the corresponding matrix Hamiltonian, **H,** expressed as a function of the corresponding infinite (and Hermitian) matrices for momentum and position, **p** and **q.** In addition, they showed that certain pairs of variables, such as p and q—known as conjugate variables in classical theory—display a special relationship in quantum theory. As noted in Chapter 10, Born discovered that Heisenberg's version of a summation rule in his July manuscript could be interpreted as a quantum commutation relation between the conjugate matrices **p** and **q:**

$$\mathbf{pq} - \mathbf{qp} = (h/2\pi i)\mathbf{I}$$

where **I** is the identity matrix. In other words, the multiplication together of two quantum matrices for conjugate variables differed when multiplied right to left or left to right (that is, for example, 3×4 does not equal 4×3). The difference is determined by Planck's constant, h.

Born and Jordan went on to demonstrate that the Hamiltonian matrix is diagonal (all elements are zero except for those on the diagonal) for certain (canonical) pairs of conjugate variables; that energy is conserved for such a matrix; that the diagonal elements of a diagonal Hamiltonian matrix are just the stationary-state energies of the system; and that jumps between states entail the emission or absorption of light quanta of energies equal to the energy differences between states, as Bohr had earlier postulated. The recovery of energy conservation, stationary states, and the Bohr frequency condition amidst this erudite and abstract matrix mathematics seemed "strong grounds" to Born and Jordan "to hope that this theory embraces truly deep-seated physical laws."[5]

P. A. M. Dirac, the lanky 23-year-old Cambridge student, soon confirmed Göttingen's hopes. The shy and retiring son of a secondary school French teacher, Dirac was just completing his bachelor's degree in electrical engineering and applied mathematics at Bristol University during the early 1920s when he learned of the British confirmation of Einstein's prediction of the bending of light in relativity theory. Fascinated by relativity through the popular works of cosmologist Arthur Eddington, Dirac headed for Cambridge to pursue a doctorate in physics. As a student of the notable R. H. Fowler, his interests in relativity expanded to include atomic physics, and in the summer of 1925 Dirac attended a lecture on atomic spectroscopy by one of Fowler's foreign guests — Werner Heisenberg.[6]

Dirac later recalled that he first learned of Heisenberg's reinterpretation only when he saw the galleys of his paper, long after Heisenberg had left Cambridge for Copenhagen.[7] Heisenberg, still unsure of what he had wrought, apparently did not mention it during his unrecorded Cambridge lecture. Working independently of Born, Jordan, and Heisenberg, Dirac exploited Heisenberg's multiplication rules to obtain an analogy between a classical bracket notation invented over a century earlier by Siméon Poisson and the quantum commutation relation between conjugate variables, such as p and q or energy and time.[8] This led to Dirac's famous distinction between so-called q-numbers and c-numbers, atomic quantities that do not commute $(AB \neq BA)$ and laboratory observables that do $(AB = BA)$.[9] At the same time, Göttingen further confirmed its own hopes by extending the Born-Jordan formalism to degenerate and perturbed systems of any number of degrees of freedom (numbers of particles). The result was the famous and fundamental Dreimännerarbeit (three-man paper) "On quantum mechanics II" submitted by Born, Heisenberg, and Jordan (in alphabetical order) in November 1925.[10]

Heisenberg contributed to the effort by letter from Copenhagen until Born ordered his return to Göttingen so that the three could finish the paper before Born left for a visiting professorship at the Massachusetts Institute of Technology that winter.[11] Heisenberg would help substitute for Born in his absence, so Born also wanted to familiarize Werner with institute affairs and teaching assignments.[12] Born, Heisenberg, and Jordan, experts on adaptations of Hamilton-Jacobi perturbation theory to degenerate quantum atoms, had little difficulty developing the matrix-mechanical analogue of their earlier work. Their method, however, raised the theory to new and (for many readers, then and now) intolerable heights of abstraction.[13]

Nevertheless, the three men proudly announced that this new quantum mechanics achieved the long-sought aim. It contained the basic postulates of quantum physics in its very foundations — the existence of discrete stationary energy states in atoms and jumps between states accompanied by the emission or absorption of light. And it allowed calculation in principle of any periodic system, such as atoms, and in close analogy with classical mechanics. The previously puzzling properties of atoms were now subject to a new mechanics. Pauli and Dirac immediately used the new mechanics — and their own ingenuity — to derive the well-known Balmer series of the hydrogen atom.[14]

But quantum mechanics became fully applicable to atomic phenomena only with the additional concept of electron spin — the hypothesis that electrons rotate on their axes like a spinning top or a miniature earth with a half unit of angular momentum. The invention and eventual acceptance of electron spin during the 1925–1926 academic year greatly facilitated practical applications of the new physics, resolved many of the lingering riddles of atomic spectroscopy, and solidified attitudes among leading matrix mechanicians regarding their physics and regarding the characteristics of nature they believed it contained. Matrix mechanics and spinning electrons provided the culmination of five years of intensive work in atomic physics.

One of the main riddles the new physics resolved was Heisenberg and Landé's old core model of the atom. This model had been designed to account for spectroscopic puzzles, but it contained a core that seemed to rotate with a forbidden half unit of angular momentum, $1/2\ h/2\pi$. All quanta were expected to appear only in integral units. Because of its half unit of momentum, quantum rules allowed the core to assume two and only two orientations (up or down) in the magnetic field produced by the orbiting valence electrons. This behavior accounted for many otherwise inexplicable spectroscopic data. (A rotating electric charge acts like a little magnet; with only a half unit of momentum, it aligns itself either up or down in a magnetic field.)

After rendering the core of the core model physically inert in December 1924 and January 1925 (it did not contribute to the observed behavior of the atom), Pauli had transferred all of the core behavior to the valence electron. But instead of endowing the electron with the core's half unit of angular momentum — what would later become electron spin — Pauli preferred the vaguer notion of a Zweideutigkeit (two-valuedness) in the orientations of the electron, caused not by the alignment of spin but by an even vaguer nonmechanical Zwang, or constraint. Two-valuedness was manifested in a mechanically undefined fourth quantum number that Pauli assigned to the electron in establishing the exclusion principle.

Ralph Kronig, who was visiting Copenhagen at the time of Pauli's papers, suggested to Pauli, Heisenberg, and others the idea of electron spin.[15] Although Heisenberg and Bohr interpreted Pauli's fourth quantum number as representing an additional degree of freedom of the electron, Pauli and Heisenberg rejected Kronig's idea as impossible — it required a point on the whirling equator of the electron to move faster than the speed of light. It was also unworkable, since it entailed the use of already failed pictorial atomic models in which atomic constituents were treated like little balls of matter — and it was simply too facile a solution of the horribly complicated problems that it supposedly would solve.

Even worse, allowing electrons to spin like toy tops would render electrons real particles, but electron particles were highly problematic, even in classical theory. A charged particle of finite size — a charged ball of matter — would be unstable, since the charge on one part of the ball would repel the like charge on another part, ultimately exploding the ball. If instead the electron possessed no size at all — a mathematically defined point — it would have infinite energy, since its charge and mass were squeezed into an infinitesimal space. Spin seemed to imply the actual rotation of a finite ball, but whether finite or infinitesimal, the electron charge, the ultimate source of electric and magnetic phenomena, remained, as Einstein had so aptly put it, "a stranger in the Maxwell-Lorentzian electrodynamics."[16]

When Dutch physicists S. A. Goudsmit and G. E. Uhlenbeck reinvented the idea of spin in the fall of 1925 to replace Zwang and two-valuedness, the situation in quantum atomic physics had entirely changed, but the "learned gentlemen" — as Bohr and Kronig called them — were still unreceptive and now had additional reasons to be so.[17] Both Heisenberg and Pauli expected all Zusatzregeln (ad hoc rules), including Zwang and two-valuedness, to arise naturally from the quantum-mechanical formalism. Spin, already problematic, seemed just another appendage.[18] But Heisenberg, ever open to pragmatic thinking and contradictory, so-called formal solutions, was easily tempted. "All difficulties would be removed with a magic touch," he declared, ". . . therefore I am not immediately opposed to the G[oudsmit] idea."[19] Only an extra factor of two in the calculated separation of doublet lines prevented practical results.

Pauli responded indignantly to Heisenberg's fascination with spin. Models of rotating and orbiting atomic particles — which looked even more than before like miniature planets orbiting the sun while rotating on their axes — were, to him, anathema. In a "philosophical" letter, now lost, Pauli probably complained to Heisenberg of what Bohr would soon tell the English-speaking world in an imprimatur attached to a note by Uhlenbeck and Goudsmit: "[Spin] opens up a very hopeful prospect of our being able to account more extensively for the properties of elements by means of mechanical models, at least in the qualitative way characteristic of applications of the correspondence principle."[20] Pauli's approach and the early matrix formulation rested on the renunciation of models and the banishing of orbiting electrons. Spin seemed to imply the return of not only planetary atomic models but the notion of actual particles of finite size. Pauli's philosophical objections and the extra factor of two notwithstanding, Heisenberg proceeded to run Goudsmit's model through the "matrix mill" to see what came out.[21]

After meeting in Leiden with Uhlenbeck, Goudsmit, and Ehrenfest, their mentor, an optimistic Bohr toured Europe in December 1925 as self-anointed "prophet of the magnetic electron gospel," which only encouraged believer Heisenberg's openly unphilosophical attitude toward spin.[22] By February 1926, Heisenberg and Jordan were assembling a manuscript on the subject in Göttingen, despite "the damned factor of two."[23] Heisenberg pleaded with Pauli to find a way to render spin acceptable. "I don't know what we should do with the Zeeman effects without Goudsmit."[24]

Pauli remained intransigent, even after L. H. Thomas, an English visitor to Bohr's institute, found a surprising error in the relativistic calculation of the relative motion of the electron as it orbited the nucleus. Pauli and Heisenberg had simply erred. Thomas's correction eliminated the obstinate two and with it any mathematical objection to the theory. Pauli, conceding the error but not electron spin, objected a while longer before "capitulating" with "heavy heart" on March 12, 1926.[25]

Four days later, Heisenberg and Jordan submitted a reworked manuscript from Göttingen, "The application of quantum mechanics to the problem of the anomalous Zeeman effect."[26] Göttingen's matrix mill had done its work, only now the nature of the enterprise had changed in the way Bohr had hoped. Heisenberg and Jordan applied electron spin and the full apparatus of quantum perturbation theory to what they called a calculation of the "quantum-mechanical behavior of the atomic model characterizing this [spin] hypothesis."[27] Electrons, formerly on the verge of virtualization, once again possessed actual mass, charge, and orbital angular momentum. But real spin was still problematic. When pressed, Heisenberg and Jordan placed their spinning ball in the more abstract nonclassical space-time of quantum mechanics. Such visualizable mechanical notions as electron particles, planetary orbits, and the resulting magnetic and spin-orbit interactions yielding a classical Hamiltonian all reappeared, but all had to be run through the mill of matrix perturbation theory (in nonclassical space-time) to achieve corresponding quantum-mechanical results.

The astonishing results served as both culmination and completion of years of old business — this time without the old unmechanical exotica of Zwang, two-valuedness, reciprocal Schimmel, and half-integral magnetic cores. Electron spin, however unpalatable, replaced them all. Among the fruits of the Heisenberg-Jordan combination of spin and matrices were derivations of the doublet fine structure, Landé's g-factors, and the old Sommerfeld-Voigt equations, now reinterpreted as off-diagonal matrix elements in the region of intermediate magnetic field strength. Further results followed during the coming years.

For Heisenberg and Jordan, such successes constituted important support not only for spin but for matrix mechanics itself. They also constituted fulfillment of the first of three program points for the new quantum mechanics of spinning electrons agreed upon by Bohr, Pauli, and probably Heisenberg in March 1926: derivations of the fine structure of spectroscopic lines emitted by atoms, the separation of the parahelium and orthohelium term systems, and the origins of the Pauli exclusion principle.[28] The three physicists expected such problems and puzzles, which had accumulated over the years, to find their natural resolution in the future quantum mechanics. Heisenberg's and Jordan's remarkable fulfillment of two-thirds of their program by late July 1926 using spin and matrix mechanics — they could not explain exclusion — both preceded and paralleled a surprising challenge to their approach raised by Zurich physicist Erwin Schrödinger, the author of an alternative contender as the new quantum mechanics — wave mechanics.

Just months earlier, physicists had despaired of ever finding a quantum mechanics; now they were suddenly confronted with two contenders for the title — each espousing entirely different aims, methods, and mathematical apparatus. Heisenberg's and Jordan's recent successes justified a strong defense of their candidate, matrix mechanics, and helped to justify their initial rejection of Schrödinger's sweeping claims for his physics. Heisenberg's prediction of allotropic forms of hydrogen based on his work on helium spectra in those months constituted the official grounds for awarding him the Nobel prize for physics seven years later.[29]

Erwin Schrödinger, a rakish 38-year-old Austrian, was professor of theoretical physics at the University of Zurich.[30] He approached the puzzles of quantum atomic physics from a very different direction and with very different aims from those defended by the inventors of quantum matrices. In a series of papers published during the first half of 1926, Schrödinger incorporated the opposite side of the wave-particle duality into the foundations of a new formulation of quantum mechanics. While matrix physicists tended to prefer quantum jumps between energy states and discrete spinning electron balls of charged matter, Schrödinger drew on the hypothesis of matter waves recently propounded by the French doctoral candidate Louis de Broglie and recently accorded Einstein's favorable notice.[31]

The son of a noble member of the French Assemblée Nationale, Louis de Broglie had set out for a civil service career, studying history and law in Paris. But after reading the works of Henri Poincaré, he was converted to physics by his elder brother, Maurice, a physicist who helped edit the papers for the first Solvay Congress in 1911. He learned of atomic physics firsthand in his brother's laboratory and published a series of papers leading up to his 1924 dissertation on matter waves at the Sorbonne.

If electromagnetic waves behave like particles—light quanta—then, de Broglie reasoned by symmetry and relativity, particles should behave like waves of matter—matter waves—under certain conditions.[32] If so, Schrödinger suggested, there should be a "wave equation" representing the evolution, or propagation, of the matter waves and a "wave function" that represented the waves themselves.

In the published version of his first paper, Schrödinger made use of Hamilton's old analogy between mechanics and wave optics and applied Hamilton's variational principle (small variations of a function or its integral yield only vanishingly small effects) to the mechanics of "matter-wave" electrons, described by a wave function ψ and moving in small spaces and on orbits with large curvature (small diameter). He arrived at the familiar equation for wave motion of any type: a second-order partial differential equation for, in this case, the wave function ψ and its "eigenvalues," E. The eigenvalues represented the energies of a standing wave in distinct modes of vibration.

Schrödinger interpreted these modes as stationary states. Without time dependence, this yielded one form of Schrödinger's famous wave equation. For the hydrogen atom, it may be written:

$$\Delta\psi + \frac{2m}{K^2}\left(E + \frac{e^2}{r}\right)\psi = 0$$

Here m and e are the mass and charge of the electron, r is the radius of its orbit in the hydrogen atom, E is its energy, Δ is the square of the gradient, and K is a constant $(h/2\pi)$. The energy eigenvalues, E, appeared as a series of distinct energies, which were soon related to integral numbers of waves oscillating on discrete orbits around the hydrogen nucleus. The orbits are discrete and distinct, as they were in Bohr's original quantum theory of the atom, because only certain orbits could accommodate integral numbers of waves. The constant K could be determined by equating the energy eigenvalues to the terms of the observed Balmer series of spectroscopic lines.

Introducing time as a variable in a subsequent publication, Schrödinger obtained the more general equation for propagation of the waves known as Schrödinger's equation. It emerged as the centerpiece of a new quantum wave mechanics, fully applicable to the same sorts of dynamic problems as Göttingen's erudite quantum matrix mechanics.

The power of Schrödinger's wave mechanics was awesome, its advantages obvious, its profound importance loudly proclaimed. While most physicists were already familiar with differential equations and eigenvalue problems of the sort contained in wave mechanics, many preferred the more familiar notions on which this mechanics rested. Schrödinger argued in his papers that, in his approach, matter waves replaced the problematic electron balls of discrete matter, and the harmonic vibration modes of the electron matter waves replaced the puzzling stationary states of Bohr's atomic theory. Continuous transitions from one harmonic vibration mode to another replaced the discontinuous quantum jumps between stationary states contained in matrix physics. If Schrödinger was right, he had just rendered irrelevant the formalism and fundamentals of matrix physics.

Recently presented with the abstractions of matrix algebra and q-number computations, most physicists, even matrix mechanicians, gladly welcomed Schrödinger's more familiar task of solving a mere partial differential equation, however difficult that might prove. Sommerfeld, for instance, who at first considered Schrödinger's approach "totally crazy," declared during a talk in Hamburg some months later that "although the truth of matrix mechanics is indubitable, its handling is extremely intricate and frighteningly abstract. Schrödinger has now come to our rescue."[33]

Schrödinger's facile derivation of the Balmer series in his very first communication obviously impressed the critical Pauli, who had just completed a tortuous derivation of the same result using matrices. He wrote Jordan in early April 1926: "I believe that this paper numbers among the most significant that have been written lately. Read it carefully and thoughtfully."[34] The possibilities afforded by Schrödinger's physics also encouraged Max Born, who was then working with Norbert Wiener at MIT on an extension of the matrix-mechanical method for nonperiodic processes, namely collisions.[35] Convinced that the logical structure of quantum mechanics was already closed by matrix mechanics, but mindful that wave mechanics did allow a facile treatment of collisions, Born voiced his enthusiasm for wave mechanics after returning to Germany: "I would regard [wave mechanics] as the deepest form of the quantum laws."[36]

Heisenberg, "not very pleased" with Born's apparent defection to wave mechanics,[37] at first exhibited far less enthusiasm for Schrödinger than did his colleagues. His skepticism derived as much from acquired conviction as it did from competitiveness. He called Schrödinger's paper "incredibly interesting,"[38] especially for its mathematical simplicity, but unlike Born, he refused to acknowledge that, as physics, Schrödinger's work had any advantage over his. Born wrote to Schrödinger a year later: "Heisenberg from the very beginning did not share my opinion that your wave mechanics is physically more significant than our quantum mechanics.[39]

During the formulation of the complete apparatus of matrix mechanics in the months preceding the publication of Schrödinger's work, and during the electron-spin period, which occurred simultaneously with Schrödinger's work, Heisenberg had acquired "very great confidence" in Göttingen's matrix mechanics approach.[40] But he already worried about the reception of the Born-Heisenberg-Jordan point of view regarding the characteristics of nature, a point of view that seemed obscured by the abstractions of matrix algebra. He had complained earlier to Pauli that Göttingen scientists, other than the three men, had fallen again into two camps: those who welcomed the unparalleled successes of matrix physics, regardless of its abstraction, and those who had given up hope of ever understanding matrices as physics.[41] Previously so motivated by success at any price, Heisenberg himself joined neither camp.

To Heisenberg, matrix mechanics was first and foremost a physics, not a mathematics, and it was, moreover, a physics to which he was long since committed. As Pauli put it, Heisenberg, like Pauli, hoped one day to strip the theory of its "formal learned words," laying bare its "physical nucleus."[42] Heisenberg was also not unsympathetic toward Schrödinger's attempt to link laboratory observables with atomic vibrations. After all, he claimed in a letter to Einstein in the same month as the exchange with Pauli, the formulation of matrix mechanics leaned heavily on the lessons of the BKS theory because of its focus on the amplitudes and frequencies of radiant waves, rather than light quanta, emitted by virtual atomic oscillators.[43] But in the end, he argued, one could not replace jumps and states with waves and modes.

In the series of papers presenting his theory, Schrödinger explicitly contrasted the origins and import of his wave mechanics with those of matrix mechanics. While the founders of matrix mechanics stressed the existence of quantum jumps and the elements of discontinuity, positivism, lack of visualization of atomic motions, and the substitution of matrices of discrete elements for classical continuous variables, Schrödinger attempted just the opposite: "a step from classical point mechanics towards a *continuum theory* [his emphasis]."[44] He argued that his theory was based on a continuous field, the ψ function, albeit existing in an abstract, multidimensional configuration space — not our everyday space of three or four dimensions. He believed that his theory returned physics almost to its classical visualizability.

"All philosophizing about 'principle observability' only glosses over our inability to guess the right pictures," Schrödinger told his colleague and supporter Willy Wien.[45] The future development of quantum physics would be best served, practically and intellectually, by adherence to nearly visualizable wave mechanics, he wrote, instead of "considering ourselves bound in atomic dynamics to suppress intuition and to operate only with abstract concepts such as transition probabilities, energy levels, and the like."[46]

In the end, the two alternative versions of quantum mechanics — wave mechanics and matrix mechanics — existed independently and for the most part peaceably for less than two months. In May 1926, Schrödinger published a proof that the two proposed formalisms — wave equations and matrix algebra — seemingly so different in form and content, were in fact mathematically equivalent! With that, Heisenberg and company began to deride Schrödinger's physics more openly, the debate becoming emotional and even vituperative.

Schrödinger himself was no help. In his equivalence paper he argued not for the equal consideration of the two intepretative contexts of the equivalent formalisms but "perhaps not wholly impartially" for the superiority of his own.[47] Moreover, in a famous footnote to his equivalence paper, Schrödinger declared: "My theory was inspired by L. de Broglie . . . and by short but incomplete remarks by A. Einstein. . . . No genetic relationship whatsoever with Heisenberg is known to me. I knew of his theory, of course, but felt discouraged, not to say repelled, by the methods of transcendental algebra, which appeared difficult to me, and by the lack of visualizability."[48]

Heisenberg expressed similar sentiments toward Schrödinger's theory in a letter to Pauli written shortly after the publication of Schrödinger's equivalence paper: "The more I think about the physical portion of the Schrödinger theory, the more repulsive I find it. . . . What Schrödinger writes about the visualizability of his theory 'is probably not quite right' [an echo of Bohr], in other words it's crap."[49] The only advantage to Schrödinger's method, he declared publicly, was that it enabled a simple calculation of the atomic transition probabilities for plugging into the matrices of quantum mechanics.[50]

It is important to note that it was not the mathematical equivalence itself that evoked such passion — Pauli had actually proven that earlier and without much ado — but rather what each side made of it.[51] Nor did the alternative yet equivalent formalisms actually cause much debate. Although matrix mechanics continued to be the most applicable to some problems, especially those involving spin, Schrödinger's wave equation soon won out as the formalism of choice by most physicists for solving most problems. Rather, the differences had to do with interpretation, with the two wholly different understandings of atoms that had given rise to the two formalisms that were now, in purely mathematical terms, suddenly recognized to be equivalent. With mathematical equivalence established in May 1926, attention focused on the physics that spawned each formalism — what the mathematics said about the characteristics of nature — and neither side was willing to concede.

The conflict between the respective defenders of matrix and wave physics formed the background of the interpretation of the quantum mechanics formalism that eventually was accepted — Heisenberg's uncertainty principle in early 1927 and Bohr's complementarity principle in the same period, the foundations of the so-called Copenhagen interpretation of quantum mechanics.

Much has been made of this conflict and its consequences. One view is that the fight was actually not about physics at all but over who would dominate the future of quantum mechanics. Some have pointed to Heisenberg's envy, ambition, and snobbery as the cause.[52] Heisenberg was, of course, no stranger to envy, ambition, or even snobbery — and neither were some of his close colleagues. But such traits seem less significant in accounting for the matrix mechanicians' attitudes toward the wave-mechanical challenge. Judged in a broader context, it is important to see that Heisenberg and his close colleagues had devoted their entire academic careers to struggling with the new and strange characteristics of nature contained in quantum theory — discontinuities, quantum jumps, discrete particles, and corpuscular light quanta. They believed profoundly that such characteristics actually existed in quantum phenomena and that they were incorporated into the matrix version of quantum mechanics. Schrödinger seemed to deny all of this.

During the period surrounding the publication of Schrödinger's equivalence paper, Heisenberg, Pauli, and Jordan had been astonishingly successful in exploiting spin in the context of matrix mechanics. This feat was at first not possible using the Schrödinger equation, which further prejudiced Heisenberg and others against Schrödinger's work and claims. Although during the year following the equivalence proof a more sophisticated unified formalism was developed that did include spin and that did allow Schrödinger's equation to reach wide acceptance, the quantum interpretation offered by Göttingen and Copenhagen began to dominate applications of that formalism.

While spin and matrices pervaded most of Heisenberg's thinking in 1926 (independently of anything Schrödinger did), something less elevated was also vying for attention — job hunting. Just as Schrödinger published his equivalence paper in 1926, several teaching chairs in theoretical physics became vacant. The lure of top jobs soon captured the professional attention of those, such as Heisenberg, Pauli, and Jordan, who had recently habilitated or soon would and contributed as well to their attitudes throughout the matrix-wave debate.

Late in 1925, as Born, Heisenberg, and Jordan put the finishing touches on matrix mechanics, Bohr's Copenhagen assistant, H. A. Kramers, accepted the teaching chair in Utrecht. This gave Bohr the chance to offer Heisenberg Kramers's dual position as university lecturer and institute assistant — a spot that Heisenberg had long coveted. Werner tentatively accepted Bohr's offer, but he made no secret of yet another goal.[53] As Bohr wrote Born: "No one in Germany need fear that [Heisenberg] will stay here very long." Born, still in the United States, agreed to allow Werner to go to Copenhagen, provided Heisenberg would stay in Göttingen until Born's return in April 1926.[54]

Heisenberg's desire to settle in his homeland paralleled the extraordinary popularity of theoretical physics in Germany and abroad. German work in relativity and quantum physics gained international recognition just as Germany itself suffered crises in other areas. Nearly every faculty in Europe and abroad sought to have a leading German representative of the new physics at its school. As Heisenberg's luck would have it, the deaths and retirements of several senior chair holders at German-speaking universities opened up room at the top.

In early 1926, the theoretician George Cecil Jaffé left an associate professorship in Leipzig for a full professorship in Giessen. On October 1, 1926, Max Planck retired from the University of Berlin (but remained active for another 20 years). A week later, Theodor Des Coudres, Leipzig professor of theoretical physics, suffered a fatal heart attack. In the same month, Karl Schmidt, theorist in nearby Halle, submitted his request for retirement. Three months later, in January 1927, Otto Wiener, Leipzig professor of experimental physics, also died suddenly of heart disease. "This damned Leipzig is a constant source of unrest!" Pauli exclaimed to Heisenberg (not to mention a constant source of heart problems).[55]

In the ensuing complicated "appointment klatsch," as Werner called it, Heisenberg and Pauli found themselves at the top of nearly everyone's list. In April 1926, Augustus Trowbridge, IEB European representative, inquired of Bohr what it would take to lure Heisenberg across the Atlantic.[56] At the same time, Heisenberg, Pauli, and Wentzel learned that they were all (in that order) on Leipzig's list to succeed Jaffé.

Heisenberg had already accepted Bohr's offer to replace Kramers and was scheduled to be in Copenhagen on May 1, 1926. Werner's startlingly late inquiry to Bohr about financial arrangements occasioned a worried telegram from Bohr announcing a salary increase for Werner. Werner was then in Dresden (the seat of the Saxon cultural ministry), where he was engaged in negotiations regarding the Leipzig position.[57] German academic tradition required a young person to accept the first offer of a professorial appointment, or he would be considered "impossible" for some time afterward. Werner's father, aware of this formality, strongly urged his son to accept the Leipzig offer—even if it meant backing out of the Copenhagen post.

An uncertain Werner turned to Göttingen for advice. Born, newly returned, and Courant, then dean of the science faculty, strongly urged Werner not to go to Leipzig in favor of the splendid opportunity of working with Bohr in Copenhagen.[58] Courant informed Bohr the next day: "I urged him to go to you *under all circumstances* and not to sacrifice the scientific and human advantages of his stay in Copenhagen to the superficial advantages of the call from Leipzig. It is my opinion that Heisenberg can calmly forgo this first opportunity [of a position] in Germany."[59]

The advice was repeated in Berlin. Earlier in April, Max von Laue, in charge that year of the famous physics colloquium at the University of Berlin, invited Werner to lecture on matrix mechanics. Heisenberg presented a grueling, comprehensive two-hour lecture on April 28.[60] He wrote to his parents from Berlin the next day that all of the "bosses of physics"—Einstein, Laue, Nernst, Ladenburg, Meitner—had assembled to hear him speak.[61] During their private conversations, the Berlin bosses, despite obvious doubts about Heisenberg's physics, unanimously urged him to take Bohr's post and to reject Leipzig's offer. They argued that Werner would certainly learn much more from Bohr; perhaps they even hoped that the Copenhagen connection might improve Werner's physics. They were certain that another chair would open for him later. "If I continue to produce good papers," he wrote, "I will always receive another call; otherwise, I don't deserve it."[62]

Werner accepted the unanimous advice and went to Copenhagen—but only on a temporary leave from Göttingen. He wanted to keep his options open to return to Germany if necessary. In early May 1926, Werner arrived in Copenhagen to begin his full-time duties as university lector and assistant to Niels Bohr.

Pauli, for his part, now promoted to full professor, refused to leave Hamburg. Wentzel, third on the list, accepted the Leipzig associate professorship, just as Schrödinger's equivalence paper appeared in print. If Werner were to receive another call to a German chair, the quality of his papers would have to be judged in terms of the new situation brought about by equivalence. This became clear when, just two weeks after Werner's arrival in Copenhagen, Grandfather Wecklein showed up, accompanied by his unmarried daughter.[63] The aged Wecklein never ceased to regret that he had settled for anything less than a university chair. His grandson's habilitation and recent substitution for Born had been the highlight of his last years. Now Werner's refusal to accept a German chair in favor of an assistantship abroad must have seemed incomprehensible to the ill and aged Greek scholar—as it did to his son-in-law, August Heisenberg. Three months later, Bohr received a worried letter from Werner's father about his son's future. Bohr responded with strong assurances that Werner was indeed on the right track toward a chair.[64] Before the year was out, Werner received news of his grandfather's death, made doubly sad by the fact that neither of the two grandsons — Erwin had not yet even received his doctorate — seemed close to achieving a chair. Werner would nevertheless realize his grandfather's dream within a year.

Heisenberg intended to room again in Copenhagen with Mrs. Maar, but when he arrived in May 1926, she was on holiday. Until her return at the end of the month, he occupied a cramped but nicely furnished third-floor service room beneath the roof of Bohr's Copenhagen institute. His window looked out on a splendid view of the Fælledpark behind the institute, where he and Bohr took their frequent walks.[65] The Bohrs, who had recently moved to the new house next to the institute, treated Werner as a family member, providing regular meals and access to their piano until Mrs. Maar's return. "I am already half at home with the Bohrs," he told his parents.[66]

Music and recreation — never neglected by those upper-class academics — consumed Werner's leisure time and that of his Copenhagen colleagues throughout this period. Apart from his usual daily walks and piano playing, Werner took numerous weekend tours to the countryside alone or with other institute members. He went on long sailing excursions and a hiking tour in Norway with Bohr, with whom he also took weekly horseback riding lessons. Such activities were never undertaken frivolously but "in order not to degenerate in physics." Still, as strenuous as these activities may have been, they could not counterbalance the intensity of his work.[67]

Heisenberg's duties consisted of helping Bohr direct the work of their many visitors. He corrected the research papers of numerous students and visiting fellows and delivered a weekly one-hour lecture in Danish. Heisenberg seems to have managed well with Danish, as he informed his parents, and as can be judged from lecture notes in Danish.[68] But Heisenberg's most demanding job was his work with Bohr to elucidate the interpretive situation in quantum mechanics following Schrödinger's challenging equivalence proof.

Schrödinger had publicly championed his continuum theory of visualizable waves over the fundamental discontinuity and quantum jumps of unvisualizable spinning particles forming abstract matrices. Heisenberg and his colleagues, while they accepted Schrödinger's wave method, rejected his claims. In the months following Schrödinger's publication of his equivalence proof and Werner's move to Copenhagen, they reaffirmed their belief in the existence of discontinuity and their desire to resolve the difficulties it presented.

Soon after reading Schrödinger's paper in May, Pauli told Schrödinger that a discontinuous element must be introduced into an understanding of quantum phenomena.[69] Continuum physics alone would not do, he repeated in November: "Do *not* think, however, that this conviction makes life easy; I have already tortured myself a great deal because of it and will probably have to do so again!"[70] After a face-to-face encounter with Schrödinger, Heisenberg complained to Pauli that Schrödinger's theory "did not fit the facts"; that is, it did not solve problems that seemed to require discontinuity. "Schrödinger simply throws overboard all 'quantum-theoretical effects,' such as photoelectric effect, Franck collisions, Stern-Gerlach effect etc.; then it is not hard to construct a theory."[71]

Heisenberg's encounter with Schrödinger occurred in Munich at the end of July 1926. Having heard Heisenberg's lectures on matrix mechanics, Berlin physicists were eager to hear Schrödinger on wave mechanics and invited him to speak in Berlin in July. On the return trip to Zurich, Schrödinger's avid supporter, Willy Wien, now rector of the University of Munich, and his colleague Arnold Sommerfeld invited the 39-year-old physicist to stop off in Munich to deliver two lectures on his new physics at the university—one to the Bavarian section of the German Physical Society on Friday, July 23, the other to quantum specialists the next day.[72] Heisenberg returned to Munich for the occasion—his first reappearance in Germany after moving to Denmark—and for a holiday with the young men of his youth group, whom, as usual, he missed terribly.

Heisenberg did not deliver a formal talk at the conference, but he attended both of Schrödinger's lectures presented in the crowded lecture hall of the physics institute. Heisenberg withheld his objections to Schrödinger's claims until the end of Schrödinger's second lecture, entitled "New results of wave mechanics." As the bespectacled Professor Schrödinger completed his discussion of applications that Saturday morning, the 24-year-old Werner rose from his seat in the crowded audience to argue that the new theory could not explain such basic quantum phenomena as Planck's radiation formula, the Compton effect, or Einstein's amplitudes for transitions between stationary states yielding spectroscopic intensities—all of which seemed to require discontinuity and quantum jumps. The audience clearly disagreed, and the aged Willy Wien, obviously annoyed, even motioned Werner to sit down and be quiet. Wien later tried to console Werner because his physics, "and with it all such nonsense as quantum jumps," was, in Wien's view, "finished."[73] Having failed to convince anyone of his views, even Sommerfeld, a despondent Werner soon headed for the hills with his youth group. But after the meeting and subsequent discussions with Bohr and Born during a conference in England, Wien wrote a worried letter to Schrödinger asking how indeed one could account for light quanta and the Compton effect without discontinuities. Schrödinger was, for the moment, equally uncertain.[74]

Heisenberg's earliest published response to Schrödinger had gone to the *Zeitschrift für Physik* over a month before the Munich encounter — in the same paper on resonance and many-body theory in which he derived the two distinct spectra of the two types of helium atoms, parahelium and orthohelium.[75] Although quantum electrons spin on their axes and behave like charged balls of matter, they are unlike so many identical marbles in a box in that they are not only identical but indistinguishable. In his derivation, Heisenberg introduced the idea that, since quantum electrons are indistinguishable, Platzwechsel (place exchange) — the possibility that two identical electrons can change places — must be incorporated into the energy function (Hamiltonian matrix). A single wave function describing a multitude of indistinguishable electrons should display certain symmetry properties as the order of the electrons is permuted. Heisenberg found that the wave function should be antisymmetric when the electrons undergo place exchange, since this was the only arrangement compatible with Pauli's exclusion principle for spinning electrons.

Having settled on the structure of the overall wave function, Heisenberg then constructed matrices of individual electron-wave functions, from which he obtained, using matrix mechanics, the hitherto inexplicable independent series of stationary states displayed by singlet parahelium and triplet orthohelium. Moreover, the Pauli exclusion principle was satisfied automatically by the structure of the overall wave function. The introduction soon afterward of so-called Fermi-Dirac statistics for spinning quantum electrons, which coincided with Heisenberg's antisymmetric wave functions, provided a way not of deriving Pauli's exclusion principle but of building it into the system of wave functions, which were in turn built into the matrix elements of quantum mechanics.[76] Extension of this procedure to hydrogen molecules led to the prediction of parahydrogen and orthohydrogen, discovered in 1929, for which Heisenberg received the Nobel prize in 1933. In 1926, the last of the Copenhagen program points for spin and matrices found its natural place — if not its natural resolution — in the new scheme, adding further cause to defend the theory.

Werner had exploited Schrödinger's wave functions to achieve the latest matrix success, but he had not yet responded in public to Schrödinger. Heisenberg prefaced his first helium paper (in June 1926) by defining his own approach explicitly in terms of corpuscular (particle) mechanics and arguing the limited practicality of wave mechanics. Spin was what made the difference. In at least one case, he declared, Schrödinger's "wave conception is more forced than visualizing the motions of particles in spaces in which a nonclassical kinematics holds" — the spinning electron. He showed that quantum mechanics — by which he meant matrix mechanics — does encompass the spinning electron, and it is based, he claimed, "on the corpuscular conception of matter."[77]

Heisenberg's June helium paper and his July encounter with Schrödinger heightened the controversy and shifted the focus from mathematical formalism to physical interpretations of the equivalent formalisms. The controversy reached its climax in the fall and winter of 1926 and its culmination in Heisenberg's formulation of uncertainty in February 1927 and Bohr's presentation of complementarity in September 1927.

After his Munich encounter with Schrödinger, Heisenberg was ready to take him on. "A few days ago I heard two lectures here by Schrödinger," he told Jordan, "and I am rock-solid convinced of the incorrectness of the physical interpretation of Q. M. presented by Schrödinger. That Schr.'s mathematics signifies a great progress, however, is clear."[78] Yet several months later, with most physicists enamored of wave methods and mechanics, as he observed firsthand in Munich, Heisenberg seemed much more defensive. In September 1926, he delivered a lecture in Düsseldorf to a joint meeting of the German mathematical and physical societies held during the large biannual conference of the Society of German Scientists and Physicians. His address to the audience, representing the most powerful bodies in German physics and mathematics, was published in the widely circulated *Naturwissenschaften* (*Sciences*). In it, he argued neither for a rejection of the rival wave conception (which would have been unlikely) nor for the superiority of the particle conception but for fuller appreciation of the particle conception on an equal footing with waves. This meant a return to a type of wave-particle dualism: "There exists in our intuitive interpretation of the physical phenomena and the mathematical formulas a dualism between wave theory and corpuscular theory of such a type that many phenomena can be explained most naturally by a wave theory . . . while other phenomena can only be explained on the basis of a corpuscular theory."[79]

Heisenberg then mustered every theoretical and empirical argument he could find for the reality and necessity of including an unvisualizable "typical, discontinuous element" in physics. He reiterated that Schrödinger's "purely continuous description of quantum-theoretical experiences in the sense of classical theory" simply contradicted accepted results, including, he now argued, the behavior of fluctuation phenomena—digressions from equilibrium caused by statistical effects of large numbers of particles.

Heisenberg's claims touched off a major debate at the Düsseldorf meeting, especially over the role of fluctuation phenomena. In response, Born, Heisenberg, and Jordan contemplated a new three-man paper to set the record straight. In a letter sent to Jordan, Born, and an opponent (Smekal), Heisenberg claimed he was withdrawing from polemics,[80] but Schrödinger's anticipated rejoinder to Heisenberg's remarks soon changed his mind.[81] Although Schrödinger apparently had no intention of responding to Heisenberg's assertions, Heisenberg submitted to the *Zeitschrift für Physik,* a journal usually reserved for noncontroversial research reports, a paper entitled "Fluctuation phenomena in quantum mechanics."[82] The aim of the paper is indicated by its tone and by Heisenberg's admission to Pauli that it was "only a pedagogical paper for the gentlemen of the continuum theory"—a polemical response to a perceived polemical situation.[83]

Heisenberg set out not to prove the fact of discontinuity but to show again "that the fact of discontinuity is contained in a natural way in the system of quantum mechanics."[84] Fluctuations again served his purpose. Einstein's early analyses of fluctuations in black-body radiation and Brownian motion were evidence for the existence of discrete light quanta and discrete particles of matter, atoms. Recently, Jordan had derived the first of these formulas in the three-man paper. Heisenberg now reiterated how quantum perturbation theory and an interpretation of the off-diagonal matrix elements as transition probabilities yielded fluctuations in light and matter—hence the existence of discontinuity—as natural consequences of matrix mechanics.

Heisenberg's fluctuation polemics also occurred in the wake of an unsettling visit by Schrödinger to Copenhagen in early October 1926, shortly after the Düsseldorf meeting.[85] Bohr apparently invited Schrödinger to Copenhagen after receiving a letter (now lost) from Werner after the Munich encounter the previous summer. Bohr was now determined to resolve the conflict in a way that accounted for, if it did not preserve, quantum jumps, discontinuous energy quanta, and rotating particles.

Schrödinger arrived in early October for the quantum summit meeting and was housed alone in an institute guestroom. Discussions began almost immediately. The usually gentle and congenial Bohr could, according to Werner, also "insist fanatically and with almost terrifying relentlessness on complete clarity" when it came to accounting for quantum phenomena.[86] Bohr's fanatical insistence soon rendered his unassuming visitor physically ill. Bohr pursued Schrödinger even to his sickbed — but without achieving the desired resolution. Heisenberg, still Bohr's lector and assistant, witnessed the struggle and no doubt joined in, but he recollected only its general features years later. A subsequent flurry of correspondence indicates that the main issue was, as expected, the existence of jumps and discontinuity — specifically, the existence of quantum jumps between discrete stationary states as revealed by various quantum phenomena and as recently interpreted by Max Born.[87]

In the second of his collision papers, received by the *Zeitschrift für Physik* in July 1926, Born had reinterpreted one of Schrödinger's early assumptions concerning his wave function ψ.[88] When multiplied with its complex conjugate, ψ^*, $\psi^*\psi$ acted like a density function — but the density of what? Schrödinger had assumed that, when multiplied by the electron charge, this density function represented the actual (three-dimensional) space density of the electric charge.[89] The assumption fell apart in Born's analysis. In order to derive such scattering phenomena as the collision of a matter-wave electron with an atom, Born interpreted $\psi^*\psi$ not as the density of particles but as the density of probability for scattering transitions — that is, as the probability that the waving electron will scatter into a certain quantum state after it hits the atom.

Probability and the closely related notion of statistical rather than exact results of calculations had actually entered physics years earlier. The second law of thermodynamics says that natural processes have a tendency to increase in entropy, or disorder, and most physicists viewed this as a statistical law. Radioactivity could also be treated only statistically and Einstein had shown in 1916 that the jump of an electron from one quantum state to another can be described by introducing transition probabilities. This feature had been incorporated into the BKS theory and the notion of virtual oscillators. The greater the probability of a specific jump, the greater the intensity of the radiation emitted by a gas of atoms at that frequency.

Born's intepretation went even further than prior assertions, providing one of the essential features of the current interpretation of quantum mechanics and an important challenge to Schrödinger's claims. The absolute square of Schrödinger's wave function, $|\psi|^2$ or $\psi^*\psi$, had nothing to do with the visualizable density of matter or charge in space, even in an abstract configuration space. Instead, it referred to the probability of finding a given matter wave in a given state, represented by $\psi (x,t)$, after a scattering process. For numerous particles, it thus represented the number of particles found in each state.

For Born, Schrödinger's equation did not describe the propagation of matter waves in space, as Schrödinger would have it; it described the propagation of probabilities to find particles at certain locations in space and time. "We free forces of their classical duty of determining directly the motion of particles," he told a British Association meeting at Oxford in August 1926, "and allow them instead to determine the probability of states. Whereas before it was our purpose to make these two definitions of force equivalent, this problem has now no longer, strictly speaking, any sense."[90] The implications of Born's probability interpretation of the wave function would soon grow even more radical and profound in the context of the Copenhagen interpretation as it developed in the months following Schrödinger's October 1926 visit to Copenhagen.

During Schrödinger's October visit, Bohr confronted his visitor with Born's interpretation and an argument in its favor. Born had suggested that, if one expands the perturbed wave function of electrons in an atom in an infinite series, the square of the coefficients of the series can be interpreted as probabilities for quantum jumps.[91] Bohr argued that the gradual decay of radiation of definite wavelengths emitted by a gas of excited atoms seemed to support the existence of individual, discrete states in atoms, and it supported Born's statistical interpetation of their decay as quantum jumps between states. Wave mechanics, on the other hand, gave equal treatment to all harmonic modes without notions of gradual decay and quantum jumps.[92]

Schrödinger would not budge. Bohr was equally unconvinced of Schrödinger's views. The upshot was the realization that no existing interpretation of the formalism was adequate — that is, no consistent correlation between the quantum-mechanical equations for atomic motions and the observed data in the physicist's laboratory had yet been established. Bohr concluded that his "'wishes' for a future physics" would be realized only after an adequate interpretation had been found.[93] Bohr's 24-year-old assistant, Werner Heisenberg, now concentrated all his energies on the search.

12

Certain of Uncertainty

On March 22, 1927, Werner Heisenberg submitted a paper to the *Zeitschrift für Physik* entitled "On the perceptual content of quantum theoretical kinematics and mechanics."[1] The 27-page paper, forwarded from Copenhagen, contained Heisenberg's most famous and far-ranging achievement in physics—his formulation of the uncertainty, or indeterminacy, principle in quantum mechanics. Together with Bohr's complementarity principle, enunciated later that year, and Born's statistical interpretation of Schrödinger's wave function, Heisenberg's uncertainty principle formed a fundamental component of the so-called Copenhagen interpretation of quantum mechanics—an explication of the uses and limitations of the mathematical apparatus of quantum mechanics that fundamentally altered our understanding of nature and our relation to it. Uncertainty and the Copenhagen interpretation, although controversial, marked the end of a profound transformation in physics that has not been equaled since; nor has any alternative theory proved so successful or so widely applicable to the phenomena of atomic scale as has quantum mechanics and its extensions and refinements since its completion in the Copenhagen intepretation of 1927.

Just two weeks after he submitted his paper enunciating the uncertainty principle, Heisenberg published the first of his many nontechnical summaries of the nature and significance of his work for nonphysicists.[2] In his insightful summary, published in a German periodical, Heisenberg suggested that the content of a physical theory may be easily recognized not by its mathematical formulation but by the new concepts to which it gives rise. Until the turn of the century, he wrote, Newtonian mechanics and Maxwellian electrodynamics had been seen as the foundations of all of physics. These theories involved the concepts of force, mass, absolute space and time, continuous processes, causality, and an objective reality existing more or less independently of the observer. Relativity theory changed our notions of space and time and showed that under certain conditions — that is, high speeds and large expanses of space and time — Newtonian mechanics had to be replaced by a new relativistic mechanics.

A similar transformation, Heisenberg continued, is required in the realm of small masses and short distances on the order of atoms and electrons. But here physicists encountered a difficulty. They cannot actually observe the internal workings of an atom; they can observe only the external properties of large numbers of atoms. The spectrum of light emitted by atoms and observed in the laboratory provided the most extensive information on the motions of the electrons inside the atom. Intensive analyses of spectroscopic light indicated the need to replace Newtonian mechanics for atomic electrons with a new quantum mechanics incorporating fundamental discontinuities of energy and momentum. Because of their extreme minuteness, these discontinuities, or quanta of energy and angular momentum, on the microscopic level of atoms are unrecognizable on the macroscopic level of the everyday laboratory, where Newton's and Maxwell's classical physics still held.

The same division between the atom and the laboratory also occurred for basic concepts in mechanics. The equations of the new mechanics indicated that atomic electrons cannot be described using everyday pictorial descriptions and concepts, such as position, velocity, and orbit. In the language of the day, the electrons were not "anschaulich" — literally, not "look-at-able." Again the physicist had to find a prescription — an interpretation of the mathematics — that would enable him or her to link the concepts and behavior of the atomic world to actual laboratory observations of atomic phenomena — and to do so in a way consistent with the mathematics of the theory. In other words, as Bohr would make much clearer in his work, the very procedures of laboratory observation and measurement, previously of only minor concern to the classical physicist, became a central concern of the quantum physicist — and a fundamental point for the creators of the Copenhagen interpretation.

Previously one could always describe the motion of an electron by noting its position and velocity at any given moment. Now, Heisenberg argued in his essay, such concepts are meaningful only when they are referred to or defined by the actual experimental operations used to measure them. The physicist cannot know any more than what he or she can actually measure. Here a puzzle arises. If one seeks to measure the exact position of an electron, he explained, one could use a microscope of very high resolving power, which would require the illumination of the electron with light of very short wavelengths. But the shorter the wavelength, the greater the energy of the light quantum (or the greater the pressure of the light wave) hitting the electron — thus the greater the recoil velocity of the electron. Because of this, Heisenberg noted, there seems to be a reciprocal relationship between the imprecisions, or uncertainties, with which one can simultaneously measure the velocity and the position of an electron at any given instant: "The more precisely we determine the position, the more imprecise is the determination of velocity in this instant, and vice versa." And this reciprocal relationship between uncertainties in measurement also holds for other conjugate pairs of variables, such as energy and time. This, in a few words, is Heisenberg's uncertainty principle.

The implications of this simple statement about the limitations of actual measurements of atomic properties were profound and far-reaching. Heisenberg already noted one profound implication in his original paper and in the nontechnical summary: "The above-mentioned boundary of precision, as determined by nature, has the important consequence that in a certain sense the law of causality becomes invalid"[3] — a revolutionary assertion that went beyond the technical minutiae of quantum equations. The law of causality — the association of a cause with every effect and the temporal sequence of cause and effect — was an underpinning of classical mechanics and, since Immanuel Kant's defense of the notion, an implicit or explicit assumption in practically every form of rational research. Laplace is credited with one of the simplest and most widely assumed definitions of causality in mechanics: if one knows the exact position and velocity of a particle at a given moment, one can calculate its exact position and velocity at any future moment. If one knows all of the forces and other external conditions, the motion of the electron can be fully determined by the mechanical equations for all time — all effects (the positions and velocities of the electron at all future moments) are linked directly and uniquely to a cause (the electron's position and velocity at the start, along with any outside forces).[4]

The uncertainty principle denies this, Heisenberg declared: "In the strict formulation of the causal law — if we know the present, we can calculate the future — it is not the conclusion that is wrong but the premise." Not knowing both the present position and the present velocity of an electron with absolute precision, one can calculate only a range of possibilities for the position and velocity of the electron at any future time, one of which will result from the actual motion of the electron. Because of this uncertainty about the future motion of an individual electron, Heisenberg continued, the laws and predictions of quantum mechanics "are in general only of a statistical type." One can never predict exactly the outcome of a single observation of any atomic process; only the probability of one outcome among a range of possibilities can be predicted. When joined with Bohr's notion of complementarity in the Copenhagen intepretation, this profound assertion would gain even greater impact in the months and years following Heisenberg's submission of his uncertainty paper in the spring of 1927.

Heisenberg came to his astounding interpretation of the quantum formalism — the uncertainty principle and its implications — during the months following his move to Copenhagen in May 1926 as Bohr's assistant. During those months, Schrödinger completed his formulation of wave mechanics, published the mathematical equivalence of the diverse wave and matrix formulations of quantum mechanics, and argued in print and in person (in Copenhagen and Munich) for a return to continuity and Anschaulichkeit (direct conceptualization) of moving electrons. Heisenberg, for his part, employed "unanschaulich" electron spin and matrix mechanics to account for atomic spectra and the behavior of systems containing many bodies, and he used fluctuation phenomena to argue the existence of fundamental discontinuities in nature. In the wake of the Bohr-Schrödinger encounter in Copenhagen in October 1926, the search for a physical interpretation of the quantum formalism — especially in the sense of incorporating discontinuity and quantum jumps into a prescription for linking the atom and the laboratory — became paramount in Bohr's institute.[5]

Two wider issues lent a sense of urgency to this task during the fall and winter of 1926–1927: the overwhelming popularity of Schrödinger's wave mechanics and the flurry of job openings throughout central Europe. Following Schrödinger's proof of equivalence, Heisenberg did not fail to notice the sudden drop in the number of publications employing orthodox matrix methods and the simultaneous increase in publications using the wave mechanics formalism.[6] He complained to Pauli nearly a year later, in May 1927, of a particularly threatening feature of such defections: some physicists had the audacity to rework old matrix mechanics papers, such as the Heisenberg-Jordan Zeeman work, in wave-mechanical terms. "I am always annoyed when even now physicists are writing a 'conjugate' wave paper to every matrix paper. I think that they should better learn both [wave and matrix mechanics]."[7]

By November 1926, the full extent of the teaching chair appointment klatsch had also become evident. Although Heisenberg continued to hold his research far above such mundane matters as a job, he could not ignore family pressures. Having turned down the Leipzig offer in April 1926, he already had one strike against him. His grandfather's disturbing visit to Copenhagen several months before his death in 1926, and his father's constant worry over Werner's future, did not allow Werner to neglect his career or to relax his production of "good papers," especially now that Schrödinger's more popular theory had been shown to be equivalent to Werner's most creative earlier work.

Because of his stature and his close familiarity with nearly every quantum theorist, Sommerfeld played a pivotal role in landing the new generation of quantum theorists in university teaching chairs. Soon after the death of Leipzig theorist Theodor Des Coudres in October 1926, the head of the Physics Institute and holder of the experimental physics chair, Otto Wiener, asked Sommerfeld to suggest a successor.[8] As a representative of the old school of theoreticians, Des Coudres had presided over a theoretical physics institute (a subsection of the Physics Institute) that also housed extensive experimental research laboratories. His research still covered turn-of-the-century topics: electrotechnology, high-pressure phenomena, theoretical studies of pre-Einsteinian electrodynamics.[9] Wiener, however, was not interested in having another old-school physicist appointed: "We wish rather an outstanding theoretician"—someone like Born or Schrödinger. "Obviously we are not thinking any more of Heisenberg and Pauli," Wiener wrote to Sommerfeld, "after they have already refused our offer."[10]

Sommerfeld replied that Born and Schrödinger were probably not to be had, that Heisenberg and Pauli were still the best prospects, and that several other physicists would be deserving.[11] On December 18, 1926, the Leipzig Philosophical Faculty unanimously accepted the search committee's recommendations for Des Coudres's chair: Peter Debye, Erwin Schrödinger, Max Born. Within a month, Wiener and his opposition to Heisenberg and Pauli died, and the search expanded for replacements for both chairs—Des Coudres's theoretical physics chair and Wiener's experimental physics chair. Wiener's favorite, Debye, more experimentalist than theorist, turned into the leading candidate for Wiener's chair. He accepted well before the July 27, 1927, faculty meeting. At that meeting the search committee announced a new list of candidates for Des Coudres's chair, as approved by Debye, the new head of Leipzig physics: Heisenberg, Wentzel, Pauli.[12] Two months earlier, Heisenberg's uncertainty principle paper and his nontechnical exposition of it had appeared in German journals.

As the appointment klatsch heated up in the fall of 1926, quantum physicists pondered their sudden surfeit of theories: two very different yet equivalent mathematical schemes for quantum mechanics — wave and matrix mechanics — and two very different interpretations of the fundamental properties of nature and the fundamental meanings of the symbols used in the two mathematical theories. Unlike most similar situations, in which scientists must make do without a theory, they now had one too many. Everyone recognized that a single formalism and a single interpretation of its meaning were now needed — a fusion of wave and matrix mechanics, as Born called it.[13] The search culminated for most physicists within months in a unified formalism — the Dirac-Jordan transformation theory — and, within a year, in a unified interpretation — the Copenhagen interpretation, of which the uncertainty principle was a part.

Heisenberg's intellectual route to uncertainty lay through the work of his closest colleagues — Born, Jordan, Pauli, Dirac, and Bohr.[14] As each struggled during the last months of 1926 with interpretations of the mathematical formalism, each informed Schrödinger of his opposition to Schrödinger's assertions that a theory involving continuous waves alone would suffice to account for phenomena that seemed to demand discontinuities, quantum jumps, and spinning particles. To Heisenberg and his colleagues, the particle side of the wave-particle duality seemed paramount. Pauli and Jordan even tried to throw the weight of majority opinion against Schrödinger. Wrote Pauli: "But I am convinced now as before (together with many other physicists) that the quantum phenomena cannot be encompassed with the conceptual resources of the continuum physics alone."[15] A continuum theory could not account for phenomena that seemed to require jumping, rotating, orbiting balls of charged matter — electrons.

Following his work on electron spin and his encounters with Schrödinger in Munich in July and in Copenhagen in October 1926, Heisenberg himself had grown even more committed to the existence of particles, jumps, and discontinuities. In this context, a prime impetus toward uncertainty came as usual from Pauli. Born had earlier interpreted Schrödinger's wave function, ψ, not as a matter wave, as Schrödinger would have it, but as a probability wave. The absolute square of $\psi(q)$, $|\psi(q)|^2$, represented the density of probability for a transition to occur to a certain quantum state in a collision between two electrons.

When multiplied by the number of colliding electrons (and integrated over all values of q), this yielded a statistical distribution, the predicted number of electrons that, statistically, should end up in each state.

Writing to Heisenberg from Hamburg on October 19, 1926, Pauli reinterpreted Born's interpretation of $|\psi(q)|^2 dq$ (dq being an infinitesimal increment of q) as representing not only the transition probability in a collision but also the probability of finding an electron in a definite atomic stationary state of the atom with the position coordinates q of the particles lying between q and $q + dq$. Such notions were far removed from Schrödinger's electron waves evolving continuously (rather than jumping) between harmonic vibration modes.

Pauli's reinterpretation and his quoted objection to Schrödinger coincided with a shift in his previous attitude toward particles, orbital stationary states, and atomic models. During the past years, he had renounced all three in favor of a theory based only on quantities observable in the laboratory, such as the frequencies and intensities of emitted spectroscopic radiation. The historical development, he said, had led to a connection between these observed properties of radiation and the matrix elements of matrix mechanics. In rational retrospect, Heisenberg had originally obtained matrices and the matrix multiplication rule by relying exclusively on such a connection. Pauli now went one step further: he believed that a direct connection could and should be established between the matrix elements, which give the probabilities, and the unobservable motions of electrons in stationary states by making use of the probability interpretation of the wave function. "I am now however convinced with all the fervor of my heart that *the matrix elements must be connected with principally observable kinematic (perhaps statistical) data of the particles in question in the stationary states* [his emphasis]."[16]

The direct connection between observed data and unobserved particles might be only statistical, but the return of electrons orbiting a nucleus like planets orbiting the sun came at an even higher price—as Bohr would say—than statistics. Pauli looked again at Born's work in which two electrons, treated as matter waves, collide with one another. He noted that when the electrons are far away from each other, the variables p and q, momentum and position, may be chosen without difficulty and behave classically—the electrons act like simple plane waves. But when the electrons approach each other in a collision or in an atom, their quantum behavior comes into play and manifests a "dark point": "The p's must be assumed to be *controlled* [by observation or calculation], the q's as *uncontrolled*. That means one can always only calculate the probabilities of definite changes of the p's for given initial values and *averaged over all possible values of the q's*." Thus one cannot speak of a definite "path" of the particle, nor, wrote Pauli, "may one inquire simultaneously about p and q." In other words, of the two conjugate variables p and q, a probable value for the change in momentum (or velocity, which is just momentum divided by mass) occurred simultaneously with an indeterminate value for the position.

Heisenberg was "*very* enthusiastic" about Pauli's letter, which "continually made the rounds" in Copenhagen, and especially about Pauli's dark point, to which Heisenberg often returned over the following months.[17] His enthusiasm culminated in a 14-page letter to Pauli on February 23, 1927, in which Heisenberg outlined nearly all the essential features of the paper he submitted a month later—his uncertainty, or indeterminacy, principle.[18]

Schrödinger had been arguing that a discontinuous particle physics, such as that espoused by the champions of matrix mechanics, is essentially irrational because it is "unanschaulich." Heisenberg countered this argument at the very outset of his uncertainty letter and paper by redefining the word "anschaulich" to refer to the physical or experimentally meaningful, rather than Schrödinger's supposed preference for the merely visualizable or pictorial. Heisenberg's interpretation of the quantum formalism would be empirical or operational—referring to actual measurement operations in the laboratory—a circumstance indicated by the very title of his paper (where "anschaulich" is translated as "perceptual").

After defining such basic concepts as the position, velocity, and path of a particle in operational terms, Heisenberg turned in his paper to the unified quantum formalism proposed by Dirac and Jordan and to the quantum "commutation relations" for multiplying matrices representing conjugate variables. For any two variables, such as p and q or energy and time, that do not "commute," Pauli's dark point had suggested that one or the other of the variables must be regarded as uncontrolled. Using the Dirac-Jordan formalism, Heisenberg discovered an extension of the dark point: not one or the other variable is uncontrolled, but *both* are uncontrolled in a reciprocal way.[19] A fundamental imprecision in the measurements of conjugate atomic variables is required by nature, no matter how precise the researcher's instruments. The uncertainties — or indeterminacies or imprecisions — in the measurement of the two conjugate variables form a reciprocal relationship that can be expressed in mathematical symbols with Planck's constant h playing the pivotal role —Heisenberg's famous uncertainty relations:

$$\Delta p \Delta q \geq h/2\pi$$
$$\Delta E \Delta t \geq h/2\pi$$

The first relationship expresses the circumstance that, if the position q and momentum p of a particle are measured simultaneously to an error in precision of Δq and Δp at a given instant, then the product of these two imprecisions, or uncertainties, must be at least equal to $h/2\pi$ (which is a very small number, h being 6.6×10^{-27} erg-sec). The same holds for energy E and time t.[20] For example, if the position q of a particle is measured so exactly that the experimental uncertainty in the value of q, Δq, is very small, then the uncertainty in p at that instant, Δp, must be very large, since $\Delta p \geq h/2\pi\Delta q$. In the extreme case of zero uncertainty in q, $\Delta q = 0$, then Δp becomes infinite, and vice versa. Heisenberg also showed that these uncertainty relations are not just mathematical abstractions, they are in fact consistent with actual experiments. This principle and these relations were firmly rooted in both the equations and the experiments of quantum physics. In other words, they were — as Heisenberg defined the word — "anschaulich."[21]

As Schrödinger had done in his papers the previous year, Heisenberg included in his uncertainty paper several sweeping and profound implicit or explicit assertions about physics and human inquiry. First, he asserted, the Dirac-Jordan formalism of quantum mechanics, on which his derivation of the uncertainty relations rested, could be regarded as complete and finished — thus ensuring the irrefutability of the uncertainty relations. Second, because the formalism was complete and its interpretation in uncertainty irrefutable, all past and future experimental research is and will be subsumed under quantum mechanics. Future research would never alter the fundamental validity of quantum mechanics, nor would it offer any hope of surmounting the limits imposed by the uncertainty relations. Neither of these bold statements could be justified by this paper alone, but such statements were repeated and eventually accepted after the assemblage of the entire Copenhagen interpretation.

Third and fourth, Heisenberg introduced his claims for the decimation of mechanical causality (in the Laplacean sense) and the closely associated notion of statistical laws. Quantum mechanics, he argued, contains a fundamental statistical element, but that element is not a property of nature itself — rather, it enters with the physicist's examination of nature. In the very act of measuring the properties of a quantum object at a given moment, the physicist selects one of the possibilities for that property. At the same time, he or she introduces a new range of possibilities for the result of any future measurement of that property, but without a causal determination of which possibility will actually be observed. The very act of measuring the position of an electron disturbs its velocity, because light is required to "see" the electron. The future course of the electron's motion, its sequence of positions in space, cannot be predicted with certainty, because its present position and velocity are not known with absolute certainty. One can only calculate the probabilities that it will be measured in a range of positions in the future. Moreover, it is "fruitless and senseless" to inquire any further about the actual motion of the electron. As Bohr would soon emphasize more explicitly, we can never know nature as it really is; we can know it only as it appears to us as we become part of the experiment itself. And because quantum mechanics is complete, no future theory will ever be able to improve this circumstance. Heisenberg's assertions were indeed revolutionary. They would find further definition and refinement in the context of the Copenhagen intepretation.

Heisenberg's sweeping claims for his interpretation and his actual derivation of the uncertainty relations evolved in late 1926 and early 1927 in especially close connection with the work of Cambridge graduate P. A. M. Dirac, then a visiting fellow in Copenhagen, and Pascual Jordan, still working with Born in Göttingen. Schrödinger's unsuccessful attempt to reduce particles to packets of waves (the packets would not stay together),[22] Pauli's "dark point" letter to Heisenberg, and Heisenberg's paper on fluctuations in the fall of 1926 had all encouraged Dirac to search for the proper mathematical procedure for relating the symbols in quantum equations — what he called q-numbers — to the symbols in classical equations — so-called c-numbers. Unlike c-numbers, q-numbers come in pairs of variables (conjugate) that do not commute, and this property is closely related to the existence of discontinuity. In his "Physical interpretation of quantum dynamics," received by the *Proceedings of the Royal Society of London* on December 2, 1926, Dirac built the discontinuities displayed by quantum jumps into an abstract "transformation theory" that at last united the different but equivalent mathematics of wave and matrix mechanics. The highly mathematical theory contained a prescription for transforming from one set, or definition, of conjugate variables to another — that is, it established a mathematical link between the symbols of the familiar classical world of the laboratory and the symbols of the strange quantum world of the atom.[23]

Dirac's work and Jordan's independent efforts along similar lines[24] convinced their Copenhagen colleague, Heisenberg, that the new transformation theory represented a "mathematically rather closed" formulation of quantum mechanics. Moreover, Dirac was arguing to Heisenberg as early as November 1926 that, with respect to the new transformation theory, "all physical applications of Q.M. that there are now and, as Dirac says, that there ever will be, fall under this."[25] Heisenberg's daring assertions about the completeness and universal applicability of quantum mechanics were, though still daring, neither entirely unreasonable nor quite his alone.[26]

Yet Dirac's physics raised a difficulty regarding the observed path of a supposed particle. Schrödinger had invented his wave packets to replace discrete particles and had claimed that the trajectory of the microscopic packet as calculated using wave mechanics would yield the observed macroscopic path of the particle. Heisenberg realized that neither Schrödinger's wave-mechanical packets (even if they did not dissipate) nor his own quantum-mechanical particles would suffice for obtaining the observed path of a particle in a laboratory, because Dirac's q-number world was essentially discontinuous. In an atomic world composed of Dirac's q-numbers, the motion of an electron can form only a series of discontinuous dots through space, not a continuous path.[27] Because of this, Heisenberg noted to Pauli, the velocity of a particle as used in Dirac's theory really had no meaning, since the velocity is defined as the change in position between any two arbitrarily close points divided by the change in time. Discontinuous dots cannot be arbitrarily close.[28] Instead, he wrote, "the solution can now, I believe, be expressed pregnantly by the statement: the path only comes into existence through this, that we observe it."[29]

This was empiricism of the most radical sort, even more radical than that in Heisenberg's original matrix mechanics paper or in his remarks about statistical laws. What he seemed to be saying is that not only is it fruitless and senseless to inquire about reality beyond our observations, but as far as we are concerned, there really is no more to reality than what we can observe of it. In a sense we create reality ourselves by becoming a part of it through the disturbance of our measuring instruments. Again, the full import and implications of such a radical notion would become most apparent only in the context of the fully formulated Copenhagen interpretation.

Eighteen days before Heisenberg came to his radical conclusion about the path of a particle in his uncertainty principle letter to Pauli in February 1927, Pascual Jordan published his Göttingen habilitation lecture, entitled "Causality and statistics in modern physics."[30] One historian has argued the crucial role of Jordan's paper in Heisenberg's thinking. First, Jordan asserted that causality is not a given, a priori law, but an experimentally defined, a posteriori concept. Kant's answer to Bishop Berkeley's argument that causality is only an appearance derived from a series of observations seemed endangered by quantum mechanics, and Jordan's raising of the problem served as the source of Heisenberg's interest in causality and statistics in his uncertainty paper. Second, the paper helped to induce Heisenberg's shift from abstract definitions of concepts such as position and velocity to empirical, operational definitions and assertions, from which emerged his radical notions about reality and causality.[31]

Jordan may well have provided the stimulus that resulted in such radicalism, but Heisenberg, while citing Jordan in his paper at this point, years later recalled only Einstein's influence. Before taking up his post in Copenhagen as Bohr's assistant in May 1926, Heisenberg had lectured on matrix mechanics before the Berlin physics colloquium on April 28, 1926. Following the lecture and a long discussion with the many skeptics in the audience, an intrigued (though skeptical) Einstein invited young Werner to accompany him on the walk home to his apartment.[32] Werner gladly accepted, and during the half-hour walk along the tree-lined streets of Berlin to his apartment in Haberlandstrasse, Einstein got to know the brilliant young man a little better. Werner had first met the great physicist two years earlier in Göttingen. But it had been only a brief encounter and had concentrated on Einstein's objections to the Bohr-Kramers-Slater theory. This time Werner was a principal author of a revolutionary, yet baffling, new mechanics, and the two had exchanged several letters on the subject during the previous months.[33] Einstein, then 47 years old, wanted first to know more about Werner's background, education, and research; Werner, half Einstein's age, wanted Einstein's opinion on whether or not he should refuse the Leipzig job offer in favor of working with Bohr. Einstein urged the young man to work with Bohr.

When the two men finally arrived at Einstein's elegantly furnished apartment—with its heavy oak Biedermaier furniture, glass-enclosed breakfront, overstuffed leather sofas, and built-in bookcases containing the complete works of Goethe, Schiller, and Humboldt—the conversation turned to the issue at hand: quantum mechanics. In a sense the conversation reflected Einstein's own role in quantum physics. From the very beginning of the quantum revolution at the turn of the century, Einstein had been a principal player but never a principal contributor to an encompassing quantum theory. His work, more than that of any other physicist, had indicated the very existence of quanta of energy and the necessity of radical revisions in physics to encompass them. He had argued the hypothesis of light quanta; he had presented a theory of specific heats of crystals in which the atoms appeared to oscillate like balls on springs but with only certain discrete amounts, or quanta, of energy; he was the first to introduce the radical notion of probability into quantum physics by suggesting that individual quantum jumps between stationary states could not be treated exactly but only in terms of probability amplitudes. Quantum oscillators and probability amplitudes formed the basis of virtual oscillators and a fundamental component of the Born-Heisenberg-Jordan formulation of matrix quantum mechanics.[34]

Einstein did not like it. He preferred instead the approach that led to wave mechanics. He shared with Schrödinger the conviction that the quantum had to be understood in traditional terms, not merely accepted or assumed. Thus, while Heisenberg, Bohr, and others struggled to obtain a new atomic theory that would somehow consistently encompass quanta, jumps, and discontinuities, Einstein's research and each of his fundamental papers provided a new and cogent argument for the appearance of energy quanta in nature. But each argument was always heuristic — nature behaved under certain circumstances only *as if* energy quanta really existed. The existence was only "for the hour," until energy quanta could be properly understood, more or less, on existing principles. Schrödinger's approach, based on the continuous wave nature of matter, which Einstein had encouraged, coincided with his own aims and seemed to hold promise of an understanding of quantum phenomena without relying on quanta, discontinuity, problematic particles, or unvisualizability. Just two days before Heisenberg's visit, Einstein wrote Schrödinger that he was convinced that Schrödinger's work represented "a decisive step forward, . . . just as I am convinced that the Heisenberg-Born approach is off the track."[35]

Heisenberg's much later recollection of the meeting focused on Einstein's objections to the empirical and positivistic elements of the Heisenberg-Born approach. Heisenberg, Born, and Jordan assumed the existence of electrons inside the atom but made no attempt to describe their actual orbital motions. Instead, Heisenberg had built his multiplication rule on equations that, he argued, involved only quantities that could be observed in the laboratory — primarily frequencies and intensities of emitted radiation — and, in his enthusiasm, he had elevated this approach to a prescription for the formulation of any cogent theory.

"But you don't seriously believe," Einstein objected, "that none but observable magnitudes must go into a physical theory?" Heisenberg attempted to raise Einstein's formulation of the special theory of relativity in his defense. Einstein had excluded such notions as absolute space and time because they could not be observed, and he had used an operational definition of the simultaneity of two events.

Muttering that a "good trick should not be tried twice," Heisenberg's recollected Einstein called such empirical reasoning nonsense. "In reality the very opposite happens," he declared. "It is the theory that decides what we can observe."[36] Confronted ten months later with the unified formalism of the Dirac-Jordan transformation theory, but without a satisfactory interpretation of its symbols, Heisenberg recalled suddenly remembering Einstein's statement just before writing his uncertainty paper, thus probably just after reading Jordan's paper. Operational definitions of fundamental concepts subject to quantum mechanics and the uncertainty relations quickly followed. The theory did indeed decide what could or could not be observed, or remembered.

Heisenberg first discussed his views on the implications of the uncertainty principle for statistics, causality, and our knowledge of reality only in the closing paragraph of his paper.[37] With a closed formalism tightly in hand, his sweeping pronouncements took on a degree of finality. But although we have no written clue that Heisenberg thought about causality before Jordan's paper, he must have considered the problem on several occasions during the previous years.[38] A relaxation of causality figured prominently in the BKS theory and in Bohr's concerns ever since the demise of that theory in 1925. Causality was also a problem of foremost importance in Göttingen after Born's statistical interpretation of Schrödinger's wave function and his own far-ranging remarks on its implications. "The motion of the particles follows probability laws," Born wrote, "but the probability itself spreads out in accordance with the causal law."[39]

Born and Jordan believed that physical laws on the atomic scale are objectively probabilistic — that is, they are probabilistic with or without a physicist to observe them — while quantum mechanics is a causal theory when applied to large numbers of particles. Heisenberg differed on both counts — and they would remain points of contention for years to come. Dirac's influence was again crucial. According to Heisenberg, Dirac was arguing in Copenhagen that, unlike the view held in Göttingen, "all statistics are first brought in through our experiments."[40] Nature itself is not probabilistic; the probability enters through the observer's interference. Second, Dirac pointed out in his transformation theory paper that "one cannot actually set up a one-one correspondence between the values of these coordinates and momenta initially and their values at a subsequent time" — a causal unfolding of probabilities, as Born and Jordan preferred, would be impossible.[41] From there it was only a small step to Heisenberg's claims for causality and to his formulation of the uncertainty principle in the paper in which he assembled and modified, as necessary, all of the disparate positions and points of view held among his closest colleagues in that period to form one uniquely Heisenbergian interpretation — the paper that, more than any other, established his position among the leading physicists of the twentieth century.

Heisenberg's paper nevertheless contained an error that even his pragmatism could not surmount. Heisenberg first derived the uncertainty relations themselves from the Dirac-Jordan quantum mechanical formalism.[42] Then he tried to show the consistency of these relations with actual experimental situations. The most basic situation involved the microscope "thought experiment" — the attempt to determine the position of an electron using an imaginary microscope — and this is where the error occurred.[43]

Since all concepts are supposed to be defined operationally, the position of an electron requires an account of the operations used to locate it. For example, Heisenberg argued, one would build a microscope using light of the shortest wavelengths available, gamma rays, to illuminate the electron and thus to locate its position. Collisions between light and free electrons had been studied several years earlier by A. H. Compton and had provided convincing evidence of the existence of light quanta, bundles of light energy acting like corpuscles (particles). Recently, Schrödinger had attempted a wave interpretation of the Compton effect.[44] But committed to corpuscles and unconvinced by Schrödinger's argument, Heisenberg assumed the gamma rays to behave strictly according to the opposite side of the wave-particle duality — as corpuscular light quanta.

Since, Heisenberg reasoned, the Compton scattering of a gamma ray into the microscope for detection entails an inconstant recoil of the electron — that is, a finite, discontinuous jump in its momentum — its exact position is indeterminate. The smaller the wavelength of the gamma rays, the greater the indeterminate recoil of the electron: "Thus the more exactly the position is determined, the more inexactly the momentum is known, and vice versa."[45] (An analysis of the Stern-Gerlach experiment yielded the other uncertainty relation, $\Delta E \Delta t \geq h/2\pi$.) Theory and thought experiment agreed; the uncertainty relations provided, in Heisenberg's opinion, a satisfactory and sufficient interpretation of the quantum formalism.

Bohr begged to differ. He had been discussing with his assistant and visitors the problems of interpretation and measurement ever since Schrödinger's unsettling visit to Copenhagen in October 1926. While Heisenberg grew ever more attached to particles and discontinuity — despite his earlier overture to wave-particle dualism at the Düsseldorf meeting in September 1926 — Bohr became ever more convinced since the refutation of BKS by the Compton effect of the need to accommodate both waves and particles equally in any interpretation of the quantum formalism. Heisenberg saw no reason for such generosity. By mid-February 1927, Bohr had had enough of arguing and left for a skiing trip in Norway — without Heisenberg.[46]

To clear his own thoughts, several days after Bohr's departure Heisenberg wrote his 14-page uncertainty principle letter to Pauli, which relied exclusively on particles, discontinuity, and the Dirac-Jordan theory. Bohr returned around the time Heisenberg's final manuscript espousing the uncertainty principle was received by the *Zeitschrift für Physik* on March 23, 1927. Two days later, Bohr sent out a cry for help to Pauli, offering to pay his way to Copenhagen.[47] He and Heisenberg were locked in a deepening dispute over the origin of uncertainty.

Bohr insisted and later demonstrated that Heisenberg's gamma-ray microscope was, as Heisenberg put it, "not quite right."[48] Earlier, this imaginary instrument had nearly cost him his doctorate — he had been unable to derive its resolving power during his final orals. Now it threatened to cost him his physics. Bohr informed the author of the uncertainty principle that the principle arose not, as he claimed, from the inconstant recoil of the electron under bombardment by light quanta in the Compton effect, but from the scattering of the waves making up the light quantum into the aperture of the microscope lens — an essential limitation on the resolving power of any microscope.[49] Not only was the finite aperture of the instrument essential to the analysis, Bohr argued, but most important, the analysis required a *wave* interpretation of the scattered light quantum, which carried momentum according to the quantum relation $p = h/\lambda$, λ being the wavelength of the light. The same held for the uncertainty relation between time and mechanical energy, which also required a wave interpretation of the light quantum. The light quantum carried energy $E = h\nu$, ν being the frequency of the light. Together, Bohr showed, these two formulas and the aperture of the microscope easily yielded the uncertainty relation for p and q.[50]

Bohr made an even more profound observation. The two simple formulas for momentum and energy of the light quantum actually manifested the wave-particle duality. They involved a mixing of wave — λ and ν — and particle — E and p — symbols. The simultaneous use of both wave and particle symbols and interpretations in any given experiment — not just particles alone, as Heisenberg would have it — constituted for Bohr an essential quality of any quantum experiment and an essential source of uncertainty.

Heisenberg at first refused to accept Bohr's argument, and especially his suggestion that Heisenberg should withdraw his paper, which was now in press. Not only did the paper encompass every aspect of nature and physical inquiry to which Heisenberg had become committed during the past year, but the uncertainty relations seemed to him a fully consistent consequence of the unified quantum mechanics formalism that, unlike any alternative Schrödinger could offer, contained those commitments in its very foundations. The basis, the innovations, the sweeping claims, the demonstration of Werner's abilities that this paper contained could not be withheld because of a mere error in a fundamental thought experiment or the mere neglect of one side of the wave-particle duality.

Bohr strongly insisted, Werner stubbornly refused, and Pauli greatly regretted that he could not come to Copenhagen as mediator. Werner soon learned what had driven Schrödinger to his sickbed. As the controversy raged through the offices and halls of Bohr's institute, Swedish theorist Oskar Klein, then on a visiting fellowship, entered the fray and, as Heisenberg reported it, supported Bohr out of friendship to Bohr and out of opposition to his assistant, who had apparently offended Klein with overweaning criticism of his work. The controversy soon degenerated into "gross personal misunderstandings."[51] In the heat of the battle, Heisenberg, who at one point burst into tears, managed to wound Bohr with his sharp tongue. Obviously much was at stake for Werner — his past commitments, his alternative insights, his future plans.

Apparently the prime cause for tears was neither the enormous pressure exerted by Bohr and Klein nor the microscope error, but frustration at Bohr's regarding Heisenberg's argument as "too special a case of the general rule" of what Bohr was now calling complementarity — Bohr's assertion that both wave and particle notions appear simultaneously in any given experiment, and that the experimenter's necessary choice of one or the other notion introduces a disturbance that unleashes the uncertainty relations.[52] The disturbance enters through the experiment, but in a way that differed from Heisenberg's notions — not in the act of fixing a variable in a measurement, but in the necessary act of choosing one side of the wave-particle duality in making the measurement.

In Bohr's view, complementarity provided the general framework, the raison d'être, for uncertainty, which in turn provided the epistemological grounding for the unified formalism of quantum mechanics. Together, uncertainty and complementarity represented the interpretive culmination of quantum mechanics, the basis for comprehending the dualities of waves and particles, wave functions and matrices, atoms and laboratories, causal and acausal descriptions—the foundations of the so-called Copenhagen interpretation of quantum mechanics. Heisenberg's paper, resting solely on particles and discontinuity, was, as Bohr now suggested, both too narrow and too premature.

The Copenhagen clash continued unresolved through Easter. Not until May 16, 1927, after further clashes, was Heisenberg willing to capitulate in writing. The scattering of light waves into the aperture of the gamma-ray microscope was, he conceded, the basis of the uncertainty relation for p and q. "Thus the relation $\Delta p \Delta q \approx h$ indeed comes out naturally, but not entirely as I had thought."[53] At Klein's urging, Heisenberg agreed to add a postscript to the galleys of his otherwise unaltered paper.

The postscript concerned several "essential points that I had overlooked" and that, he wrote, Bohr had since brought to his attention.[54] Foremost among them was the realization that the "uncertainty in the observation" arises not exclusively from discontinuous particles or continuous waves but also from the attempt to encompass simultaneously phenomena that arise from both wave and corpuscular origins.

Bohr, already at work on a paper under the rubric "there are waves *and* corpuscles,"[55] attempted to clarify the situation even further at the Lake Como celebration of the hundredth birthday of Alessandro Volta, the inventor of the voltaic cell. Nestled in a villa overlooking the deep-blue mountain lake in northern Italy, the celebrants listened as the soft-spoken Bohr raised the existence of discontinuity in nature to the status of *the* quantum postulate—an assertion that no doubt pleased one member of his audience, Heisenberg.

In the course of his rambling and wordy paper, an obvious reflection of his own imprecision at that point regarding his ideas, Bohr offered the perfect compromise between the opposite positions espoused so forcefully by Heisenberg and Schrödinger.[56] Wave physics and particle physics were not, as hitherto supposed and debated, antithetical— either wholly right or wholly wrong—they were complementary— mutually exclusive yet jointly essential. Rather than explaining away the wave-particle duality in favor of one or the other extreme, Bohr incorporated it into the very interpretation of quantum mechanics. The uncertainty relations between momentum and position ($\Delta p \Delta q \geq h/2\pi$) and between energy and time ($\Delta E \Delta t \geq h/2\pi$) provided the clue. Bohr pointed out that each relationship combines a symbol referring to wave notions and a symbol referring to particle notions. Particles are usually associated with energy (E) and momentum (p), and these variables are usually necessary for causal descriptions. Waves, on the other hand, are usually associated with space-time descriptions (q and t). The uncertainty relations, each of which combines a particle symbol with a wave symbol, are the price we pay for having to encompass in one situation the mutually exclusive notions of wave and particle. However, observation and measurement by the experimenter force a choice between one or the other set of complementary physical concepts and thus, by reducing the choices to one or the other side of the wave-particle dualism, disturb the system. Because of this disturbance, Bohr seemed to be saying, precise, simultaneous observations of complementary pictures or measurements of complementary variables are limited according to the demands of the uncertainty relations. "We are concerned with a demand of causality complementary to the space-time description, the adequate application of which is limited only by the restricted possibilities of definition and observation." Heisenberg rose from his seat in the audience that day to confer his public approval on Bohr's interpretation of their physics.[57] The Copenhagen interpretation was born.

Heisenberg's battle with Bohr reached its climax and diffusion in the spring of 1927, just as several faculty committees were deciding on candidates for vacant teaching chairs. Heisenberg's academic ambitions seemed close to realization. He had produced another good paper, and faculties were meeting to award appointments. The changing situation on the job front is reflected in part in his changing relationship to Bohr after the publication of the uncertainty principle — but only in part. Jobs aside, Werner always looked up to Bohr as a father figure and could not let a disagreement over waves and particles come between them for long.

Probably during his Easter vacation but certainly by June 1927, Heisenberg learned from Sommerfeld of Debye's intention to have him called to Leipzig as successor to Des Coudres. (In the German tradition, a candidate, even one who had not applied for the job, was not appointed but "called" to the position; the professing of knowledge was not a mere job but a calling.) Heisenberg received Leipzig's official call at the end of July. By then he had received two other calls to full professorships. "All good things come in threes," he gloated.[58] In June 1927, a search committee in Halle had recommended Heisenberg, Wentzel, or Hund as a replacement for Schmidt. A month later, the Saxon education ministry, which officially issued the call, requested Sommerfeld's confidential judgment of the three. But Heisenberg was already carefully weighing the alternatives. The call to Halle, coming at a time when he felt he had not yet made up for his injurious remarks to Bohr, caused Heisenberg considerable concern. In a personal letter to Bohr written in a very shaky hand, Heisenberg informed his mentor in apologetic terms of the Halle appointment. Diplomatically, Werner asked for advice: should he stay in Copenhagen or "seek my own future"?[59]

But Sommerfeld offered another alternative. Both Sommerfeld and Schrödinger were on the list to succeed Planck in Berlin. In gratitude for his turning down the Berlin appointment, the Bavarian education ministry increased Sommerfeld's salary and his institute budget significantly. They also promised to establish a long-requested Extraordinariat (associate professorship) in Sommerfeld's institute.[60] Sommerfeld hoped that Heisenberg would fill that post. Having heard from his father that Werner might want to remain in Copenhagen a while longer, Sommerfeld wrote to Werner offering to reserve the new position for him.[61] The post itself was actually secondary to its most important features: it was in Werner's beloved Munich, and it would give Werner "simultaneously the right after further years to become my successor in the Ordinariat [full professorship]." The implication was clear. With his sixty-fifth birthday approaching in 1934, Sommerfeld had decided on Heisenberg as his heir apparent. In the event that Heisenberg decided to leave Copenhagen, Sommerfeld also told him unofficially of Debye's intention to recommend Heisenberg rather than Schrödinger to succeed Des Coudres. Sommerfeld counseled acceptance: "I thus advise you with heavy heart, but at Debye's wish, to accept the Leipzig Ordinariat, which is indeed especially attractive because of Debye and Wentzel and colleagues."

Having turned down one call already to a German chair, Heisenberg could not afford to reject another and decided to accept either the Halle or the Leipzig offer. Bohr reluctantly conferred his blessing. Werner's only real preference was for Munich. If he could not be called immediately to a full professorship there, he told his parents, it did not matter to him where in Germany he resided until he could move to Munich.[62] He duly thanked Sommerfeld for his kind offer, which he wanted to reconsider later.[63] Expressed less diplomatically, Heisenberg had accepted the mantle as Sommerfeld's heir apparent, but he would occupy an Ordinariat elsewhere until Sommerfeld retired.

Dresden's official call to Heisenberg to occupy the Leipzig chair occasioned another shaky letter to Bohr.[64] Heisenberg wanted to reassure Bohr that he was not courting callers because of their earlier differences. Two more offers followed: one from Trowbridge of the IEB, on behalf of American universities eager for a German quantum theoretician, and another from the esteemed Zurich Polytechnic (ETH) as Debye's successor. The Polytechnic experimentalists wanted a full-time quantum theoretician as Debye's replacement in order to keep the institute abreast of "the enormous development of theoretical physics."[65] Neither of these offers had a chance with Heisenberg, although he did dangle the Polytechnic offer in his negotiations with the Saxon ministry. Negotiations finally concluded in late October 1927 with Heisenberg's acceptance of the Leipzig Ordinariat for theoretical physics, retroactive to October 1. The position came with a lucrative salary, an opportunity to work with Debye, and equally important, a place in Germany — none of which the Swiss could match.

A week later, Heisenberg, having recently returned from Lake Como, traveled to Brussels where he and Born would proclaim the completion of the quantum revolution to the prestigious Solvay Congress of selected physicists. There Heisenberg also tried to smooth over any lingering personal and scientific misunderstandings with Bohr. The undertaking did not quite succeed, however, apparently in part because Werner somehow overslept on the morning of Bohr's departure.[66] On delivery of his inaugural lecture before the Leipzig faculty on February 1, 1928, Heisenberg was celebrated as Germany's youngest full professor.[67] He had just turned 26.

With Heisenberg and Debye finally settled in Leipzig and Sommerfeld firmly established in Munich, the rest of the central European appointment klatsch quickly fell into place. Schrödinger succeeded Planck in Berlin; Wentzel followed Schrödinger at the University of Zurich; Hund took Wentzel's place in Leipzig; Pauli, second on the Swiss Polytechnic list, succeeded Debye; Jordan followed Pauli in Hamburg; and Bohr's ally Oskar Klein took over Heisenberg's post as Copenhagen lector and assistant. Within a year of the Como and Solvay meetings, a new generation of quantum theorists had come of age. With their new status as tenured professors and established leaders of their field, Professors Pauli, Heisenberg, and Bohr at last began to address each other with the familiar German pronoun "du."

13

Spreading the Spirit

Heisenberg's appointment to the Leipzig chair for theoretical physics in October 1927 and his attendance at the prestigious Solvay Congress in Brussels in the same month—the fifth of occasional meetings of leading physicists to consider fundamental developments—represented milestones in the completion and acceptance of quantum mechanics. During the six-day meeting, Heisenberg and other former matrix mechanicians rallied behind Bohr and the Copenhagen interpretation of quantum mechanics, refuting objections raised by Einstein and others more closely associated with Schrödinger's interpretation of wave mechanics.

The Bohr faction managed to rebuff its opponents, but it could not convince them for long. Despite this failure, acceptance and exploitation of the Copenhagen interpretation spread rapidly among most physicists. This was facilitated by two factors: first, the almost missionary zeal displayed by proponents in carrying the "Copenhagen spirit," as Heisenberg called it, around the globe, and second, the establishment of the new science in the seats of academic power, the university teaching chairs.[1] The settling of the European appointment klatsch redefined the aim and structure of the physics institute to accommodate the new science. To the already established centers of quantum physics were added new centers that employed the science to train future generations wholly within the new spirit. Heisenberg and his Leipzig institute are prime examples.

Yet, throughout most of 1927, Heisenberg seemed an unlikely candidate for such distinction. As noted in Chapter 12, Bohr and Heisenberg

had argued bitterly over the origin of the uncertainty relations—
Heisenberg emphasizing the primary role of discrete particles, quantum
jumps, and the disturbance of the measuring instrument, Bohr pointing
to the simultaneous appearance of waves and particles as the root cause.
Even after he admitted the errors in his tearful defense of uncertainty,
Heisenberg's lack of enthusiasm for Bohr's insistence on waves and
corpuscles was obvious. In May he wrote condescendingly to Pauli,
"Naturally, if one starts with that, one can do everything without fear of
contradiction."[2] A month earlier, Bohr had forwarded a copy of Heisenberg's uncertainty manuscript to Einstein, at Heisenberg's urging. In
his accompanying letter, Bohr presented his own very different views on
the matter and expressed his intention of preparing a paper on the
subject.[3] Five months later, when Bohr presented the promised paper at
Como, Italy, Heisenberg had dried his eyes and completely changed his
tune.

To the audience assembled in the auditorium of the Instituto Car-
ducci on the shores of Lake Como, Bohr provided his own contribution
to the Copenhagen interpretation—the principle of complementarity
—the notion, briefly, that quantum processes manifest contradictory
properties, such as wave and particle natures, that are nevertheless
jointly necessary for a complete description of the process. He showed
how Heisenberg's poorly argued uncertainty relations could be better
understood using the "complementary features of the description."[4]
Not only does the quantum-mathematical formalism demand redefini-
tions of the basic concepts of classical physics, as Heisenberg had argued
in his uncertainty paper, but, Bohr continued, it also limits classical
descriptions of phenomena based on causal space-time events. The
wave-particle duality forces a choice between complementary descriptions: one must choose to interpret an experiment in terms of what Bohr
regarded as either particlelike causal events or wavelike space-time
events. The experimenter's necessary choice of one side of the duality
introduces a disturbance that excludes the other side of the duality.
Applied to Heisenberg's much-debated gamma-ray microscope, Bohr
showed (though not without an error of his own) that the disturbance
arising from the complementary nature of the two descriptions can be
expressed quantitatively by reciprocal relations between the uncertainties in the measurements of energy and time or momentum and
location—that is, by Heisenberg's uncertainty relations:

$$\Delta p \Delta q \geq h/2\pi$$
$$\Delta E \Delta t \geq h/2\pi$$

The more sharply one can measure the particlelike variables, energy E and momentum p, the more indeterminate are the wavelike variables, position q and time t, and vice versa. Heisenberg's uncertainty relations became, for Bohr, not merely a manifestation of discontinuity, mathematical formalism, or the intrusion of instruments, but a direct consequence of "the general complementary character of the description" — a product of complementarity.[5]

In the published version of the audience discussion following Bohr's Como paper, Heisenberg, previously so confident of his own derivation of uncertainty, made the first of his generous acknowledgments to Bohr for clarifying the origins of uncertainty and for assembling the Copenhagen interpretation of the quantum formalism: "The physical interpretation of the inexactness relation $\Delta p \Delta q \approx h$ and its relationship with the general points of view raised by Bohr have been made entirely clear for the first time through Bohr's researches."[6] In his own version of the history of quantum mechanics, published just over a year later, Heisenberg, while gliding over wave mechanics, gave Bohr sole credit for proving "in every detail" that together complementarity and uncertainty yield a fully consistent interpretation of the quantum formalism.[7] And within months of the Como conference, he wrote to Bohr upon receiving an offprint of Bohr's paper: "During the last weeks especially, I have . . . seen once again how much I have learned from your work, and I was satisfied with your work in every respect. I now consider all questions concerning principles as completely answered."[8]

By reconciling uncertainty with complementarity, Bohr rescued the doctrines of the matrix and Copenhagen adherents from the challenges of their opponents — at least for the moment — and for that Heisenberg was clearly grateful. But he had other motives, as well, for lauding Bohr. Heisenberg had always admired and respected Bohr, and he wished to make up for his adolescent behavior in defending uncertainty. During the five years following the Como and Solvay meetings, the personal and professional relationship between the two men, as judged by the extraordinary number of mutual visits, personal letters, and joint outings, grew as close as it ever would.

Beyond the healing of old wounds lay a professional motive. Just a month before the Como conference, the Saxon Education Ministry informed Heisenberg of his call to Leipzig, and within weeks the new professor had begun lecturing.[9] Heisenberg's need to underscore his unique contributions to quantum mechanics suddenly evaporated, to be replaced by a new ambition that now required Bohr's help to realize. Heisenberg was determined that he would build a permanent, first-class

research program in Leipzig on the basis of the new, solid, and successful physics. Bohr's views not only buttressed Heisenberg's debatable
arguments for uncertainty by placing uncertainty within a solid, unified
framework, they also provided a rallying point for Bohr's followers.
Bohr's disciples were, like Heisenberg, eager for a completed physics
that they could propound from their chairs and podiums and exploit in
their papers. Even the ever critical Pauli declared himself "very much in
agreement" with Bohr's Como paper—"both with the overall tendency
and with most of the details."[10] After Como, Heisenberg and other
Copenhagenites no longer gave their allegiance to individual programs
and discoveries, matrix mechanics or uncertainty, but to the Copenhagen spirit—a spirit epitomized, they believed, by Bohr's Como presentations, personified by Bohr himself, and propagated by his ordained
ministers, namely themselves.

Max Born promptly announced the new consensus in his reply to
Bohr's Como lecture. "Herr Professor Bohr has presented the views that
we have formed about the basic concepts of quantum theory in such an
appropriate manner that there is nothing left for me to do but to add a
few remarks."[11] A month later Born and Heisenberg, his former assistant, surveyed those basic concepts for the fifth Solvay Congress. Should
any doubt remain, as it did for Schrödinger, discontinuity topped the list
of basic concepts—not continuous waves but discrete, discontinuous
particles, energy quanta, and quantum jumps were fundamental to understanding nature. Their opening sentence: "Quantum mechanics is
founded on this idea, that atomic physics is distinguished essentially
from classical physics by the existence of discontinuities."[12]

After first hailing Bohr for grounding uncertainty in complementarity,
Born and Heisenberg declared the completeness and sufficiency of
quantum mechanics in even bolder and more provocative terms than
Heisenberg had used in his uncertainty paper. "We regard *quantum
mechanics* as a complete theory for which the fundamental physical and
mathematical hypotheses are no longer susceptible of modification."
Nor would future developments ever alter any of the fundamental
features of the theory. Its implications for indeterminacy in particular
were fixed: "Our fundamental hypothesis of essential indeterminism is
in accord with experiment. The subsequent development of the theory
of radiation will change nothing in this state of affairs."[13] Like it or not,
quantum mechanics and the Copenhagen interpretation were etched in
stone.

The Copenhagen circle could hardly have been surprised, however, by
objections raised by those outside the "community of true believers," as

Pauli once called a similar coterie of Copenhagenites.[14] Some physicists, notably those who preferred Schrödinger's early interpretation of wave mechanics, were still unconvinced of the Copenhagen doctrine and less than eager to admit what seemed a fundamental failure of the physicist to do his job. Physics should enable one to comprehend nature so completely that exactly what occurs in any given experiment can be explained or predicted. Settling for anything short of that age-old goal was seen by some as resignation or even despair.

With such a divergence of attitudes and opinions, the many formal and informal discussions during the six-day Solvay Congress held in Brussels at the end of October 1927 became a running debate between imbibers of the Copenhagen spirit and their more sober opponents. Ever since physicists Walther Nernst and H. A. Lorentz convinced the wealthy Belgian industrialist Ernest Solvay in 1910 that his generosity could best be used in support of occasional conferences at which a select group of leading physicists would present and discuss their latest research on a basic problem of physics, the Solvay Congress had come to mark both the culmination of a phase of research and the stimulus toward new directions.[15] Lorentz, the organizer of the fifth Solvay Congress in 1927, settled on the topic of electrons and photons. Of the 29 invited participants, every major contributor to quantum physics — Heisenberg, Bohr, Schrödinger, Einstein, Born, Planck, de Broglie — was in attendance at this summit of quantum physicists. As they assembled in the sumptuous Hotel Métropole in downtown Brussels, their papers and discussions in the meeting hall, dining room, and hallways turned to the interpretation of the quantum-mechanical equations provided by the Copenhagen faction.

Both supporters and opponents of the Copenhagen doctrine had already reaffirmed their positions during the month since the Como conference. Their debate, which resumed with even greater force at the next Solvay meeting in 1930, ultimately pitted Einstein against Bohr.[16] The opposing arguments often revolved around a favorite device of the era: so-called thought experiments, imaginary experiments involving various ideal arrangements of photons and electrons in boxes or passing through slits in infinite, impermeable screens. During the scheduled sessions and even more during the informal evening discussions, Einstein, master of the thought experiment, contrived numerous such arrangements to demonstrate the inadequacy of quantum mechanics as a physical theory. By the following morning, Bohr, the master of Copenhagen dialectic, managed with the help of his supporters to refute each one.

What particularly bothered Einstein was the statistical nature of quantum mechanics. Because of the uncertainty relations and the disturbance introduced into the experiment by the observing physicist, the present and future results of any measurement of mechanical properties of an individual system, such as a single electron, cannot be predicted with absolute precision—the theory is essentially indeterministic, or acausal. The Copenhagen doctrine maintained that only the probabilities for a range of possibilities, or the statistical number of systems with a given property out of a large number of systems, can be obtained from the quantum equations. Although Einstein himself had introduced probability into quantum physics as early as 1916, such a notion was, for him, intended as only temporary. Every individual event in nature is governed by precise physical laws, and any physics that permanently settles for only probabilistic or statistical predictions about the outcome of a given experiment is ipso facto incomplete. Although Einstein conceded that quantum mechanics and the Copenhagen interpretation together formed a closed, logically consistent, and complete theory of statistical events, the theory is nevertheless incomplete when applied to individual events. Einstein had already expressed himself on this point in his now famous theological response to Born's 1926 probabilistic interpretation of Schrödinger's wave function: "The quantum mechanics is very worthy of regard. But an inner voice tells me that this is not the true Jacob. The theory yields a lot, but it hardly brings us any closer to the secret of the Old One. In any case I am convinced that *He* doesn't throw dice."[17]

Years later, Bohr recalled one of Einstein's thought experiments to prove his point to the 1927 Solvay conferees.[18] A beam of electrons impinges on a large screen containing a tiny aperture. Behind the screen is a photographic plate in the shape of a hemisphere of large radius. The de Broglie waves of the electrons are diffracted through the aperture and emerge as spherical waves. The intensity of the waves arriving at the photographic plate is a measure of what takes place at the plate.

According to Einstein, two interpretations of this experiment are possible. In the first, the Schrödinger wave function of the electrons does not represent individual electrons but only an electron "cloud," or ensemble; and $|\psi(r)|^2$ represents the statistical probability of finding at least one electron of the cloud at point r on the photographic plate— not an unsatisfactory theory to Einstein. But in the second interpretation, quantum mechanics *is* considered a complete description of each individual particle, and $|\psi(r)|^2$ represents the probability that a given electron will strike the plate at point r. The experiment proceeds until the electron is actually recorded at a point on the plate. At that very instant, the probability in the second case that the electron will strike any other point on the plate is reduced to zero. But this requires the propagation of an instantaneous signal to every other point on the plate — a direct violation of the relativity theory, since signals can travel only as fast as the speed of light. The fact that the electron actually does hit at a definite point and that a probability interpretation violates relativity theory strongly suggested to Einstein that the quantum-mechanical description of individual particles is incomplete. It must be amended so as to enable an exact prediction of where the electron actually does strike the plate.

No contemporary record survives of Bohr's response to Einstein during the 1927 Solvay meeting, but in Bohr's later recollection of the discussion he offered a cogent rejoinder.[19] As he recalled it in 1949, Bohr argued in 1927 that Einstein's analysis entails the assumption that the screen and plate are infinitely heavy and thus possess exact, well-defined positions in space and time. According to the uncertainty relations, since there is no uncertainty in the position and time when the electron emerges from the slit, the energy and momentum of the scattered electron must be completely indeterminate, as indicated by the perfectly spherical electron wave emitted from the slit. If, however, the screen is slightly movable, rendering its position and time slightly uncertain during the diffraction, the uncertainty in the momentum and energy transferred to the electron will be reduced by a corresponding amount and the prediction of where the diffracted electron will hit the plate is enhanced. Within the limits of the uncertainty relations, quantum mechanics is as complete a description of individual events as we can ever hope to obtain. We are simply limited by nature in what we can know about individual events on the atomic scale. The argument failed to convince Einstein. Quantum theory "may be a correct theory of statistical laws," Einstein wrote to Sommerfeld within weeks of the Solvay meeting, "but it is an inadequate conception of individual elementary processes."[20]

Einstein's thought experiment illustrated yet another essential feature of Bohr's Copenhagen interpretation: the blurring of the traditional distinctions between observer and observed, subject and object, physicist and nature. In Einstein's "reactionary" position, as Pauli called it,[21] physical reality exists independently of the physicist, and the observed phenomena arise from independent natural processes obeying precise, simple, and elegant mathematical laws — the "secret of the Old One." To Bohr, on the other hand (to borrow Bohr's favorite phrase), every observation made by a physicist destroys the independence of the observed phenomenon, forces the subjective choice of a descriptive scheme on the experiment, and introduces uncertainties in measurements according to the uncertainty relations. In Bohr's view, Einstein's aperture experiment for the diffraction of a single electron is not an experiment involving the single object under study—the electron—but an experiment in which the measuring instruments—the screen and plate—are unavoidably joined with the investigated object. Moreover, while the electron obeys quantum laws, the laboratory instruments that detect the observed data are subject to classical Newtonian physics. Because of this blurring of the distinction between object and instrument and the mixing of classical and quantum elements in one experiment, the uncertainty principle comes into play: predictions about one part of an experiment can be made more exact by allowing it to interact with another part, rendering that part more inexact. The physicist must choose how he or she wishes to balance the uncertainties and complementarities in any given experiment. To Einstein such seeming subjectivism and intellectual resignation were anathema.

The demonstrated self-consistency—if not the sufficiency—of the Copenhagen interpretation reinforced the Copenhagen penchant for sweeping statements. Convinced by the last day of the 1927 Solvay meeting that his opponents had been refuted, if not silenced, Heisenberg wrote home: "I am satisfied in every respect with the scientific result. Bohr's and my views have been generally accepted; at least serious objections are no longer being made, not even by Einstein and Schrödinger."[22] Nothing, they now believed, could stop the spread of their conveniently closed and unified quantum mechanics.

Having triumphed within their own profession, the Copenhagenites now moved outside it, attempting to take their doctrine to other fields. Yet when they announced the doctrine and even attempted to apply it outside their own discipline, they made claims for it that went far beyond its actual meaning and significance within the confines of quantum physics.[23] Bohr, for instance, used his contribution to the 1929 Planck issue of *Die Naturwissenschaften* to inform philosophers of the new doctrine. He then proceeded during the early 1930s to apply complementarity to practically everthing: biology, law, ethics, religion, and life itself. Some quantum physicists began exploiting the Copenhagen doctrine for vitalist, neoromantic, and even mystical philosophies. Their fundamentally antiscientific and antirationalist attempts continue to this day.

Yet while the philosophical implications of the Copenhagen interpretation were enormous, most philosophers later acknowledged complete surprise at what the physicists had wrought and were slow at first to respond. Like his mentor, Bohr, the unphilosophical Heisenberg lost no time in waking them up, and he did so in the most direct way possible: he began lecturing to assembled audiences of philosophers on how the new doctrine now applied to *their* field.

Heisenberg's philosophical works require caution regarding what they reveal about their author. For physicist Heisenberg, philosophical issues were never of such moment as to merit the devotion he lavished on the issues of quantum formalism. Systematic philosophical positions always were of lesser importance to him. Some readers have not fully appreciated that Heisenberg's systematic philosophical pronouncements were always tailored for public consumption and were thus informed and motivated to a great extent by his personal aims in addressing each particular audience. Only rarely did any of Heisenberg's philosophical writings not derive from one of his public addresses. Significantly, these writings began to appear only after the formulation of the Copenhagen interpretation and after his appointment to the Leipzig teaching chair.[24] The new and revolutionary physics provided a convenient basis from which to pursue broader horizons opened by his new position within his profession.

The earliest of Heisenberg's prescriptive addresses, a lecture to Leipzig philosophers entitled "Epistemological problems in modern physics," served as a practice run for the most influential and widely read of his early philosophical pronouncements—a paper entitled "Causal law and quantum mechanics." Heisenberg delivered the paper in 1930 to an early gathering of the influential Vienna Circle of "exact epistemologists" at the Königsberg meeting of the GDNA. It was published in the circle's widely read journal, *Erkenntnis*. This paper was followed by similar talks delivered at the University of Vienna in 1930 and 1935, the Saxon Academy of Sciences in 1933, the universities of Munich and Göttingen in 1933 and 1934, the 1934 meeting of the GDNA held in Hanover, and to readers of the *Berliner Tageblatt* in 1931.[25]

As theoretical physics came under increasing suspicion and disrepute with Hitler's ascension to power in Germany in 1933, Heisenberg's lectures became noticeably defensive of theoretical physics. Nevertheless, Heisenberg's philosophical positions had been set as early as his first unpublished lecture to his Leipzig colleagues. His uncertainty principle was a challenge to the notion of causality in atomic processes, and causality, or the lack of it, became his major public theme. His claims, like Bohr's for complementarity, went far beyond their narrow quantum-mechanical significance.[26] Heisenberg and Bohr were not alone. Max Born had earlier called quantum mechanics a deterministic theory of probabilities. Quantum mechanics yields only probabilities — not exact predictions — for the outcomes of atomic events, but the probabilities themselves evolve in a precise, determined fashion in accordance with the Schrödinger equation.[27] Thus, acausal laws in the microscopic world did not imply for Born indeterminism in the macroscopic world. But by the time of Bohr's Lake Como address, Born's remarks during the discussion contained no hint of his earlier restraint. "It seems to me important to emphasize that the new quantum mechanics gives up determinism, which has dominated natural research until now."[28] Lay audiences throughout Germany, apparently critical of physics for its supposed subservience to so-called mechanistic materialism, would have welcomed such a statement.

Heisenberg, too, had renounced microscopic, or atomic, causality in the closing paragraph of his uncertainty paper. By 1928 he was ready to take on nearly all of Kantian epistemology, an essential underpinning of classical physics. He told his first philosophical audience that the relativity and quantum theories challenged Kant's claims that space, time, and now causality are so-called synthetic, a priori judgments and hence are suitable bases for establishing a rational, objective science. The challenge called, he declared, for the "very difficult task of rolling out the Kantian basic problem of epistemology once again and, so to speak, starting all over again. . . . But this is your task, not that of the scientist, who should only deliver material with which you can work further."[29] Physicists, if publicly criticized for their deterministic mechanism, could at least render a service by pointing the way for others and at most could establish their preeminence by stimulating other professions with nonmechanistic notions that could not be easily rejected.

Heisenberg delivered the promised material to the assembled epistemologists in Königsberg and Vienna in 1930. Many in both audiences, such as Philipp Frank, Hans Reichenbach, and John von Neumann, were associated with the famous Vienna Circle, the philosophical school of logical positivism that has struggled ever since to construct a logically coherent model of an empirically grounded theory. Relying more explicitly on Bohr's analysis of quantum experiments, Heisenberg shifted the philosophers' task from a reevaluation of Kant's entire epistemology to a reevaluation of the concept of causality — the assignment of a specific physical cause to each individual phenomenon. After examining the notion of causality itself, Heisenberg argued that while classical physics remained in full accord with Kantianism, the blurring of the subject-object distinction in quantum physics rendered the objectification of the world of perception impossible, thus removing a cornerstone of Kantian epistemology and the basis for Kantian a priori causality. The impossibility of an objective world of perception was for Heisenberg the "source of indeterminacy,"[30] and it rendered the traditional definition of causality not invalid but simply "empty of content."[31] "I hope," he told the exact epistemologists, "to have made clear . . . to you that the situation created by atomic physics really does make a renewed discussion of the concept of causality necessary."[32] A year later, he told newspaper readers: "Now it is the task of philosophy to come to terms with this new situation."[33]

Numerous epistemologists heeded Heisenberg's call for a rethinking of causality. Not surprisingly, their methods drew heavily on the Copenhagen interpretation. One noted spoke of the Vienna Circle, Moritz Schlick, even sought extensive direction from Heisenberg in his analysis of quantum causality.[34] As philosophers finally took note of the implications of the Copenhagen doctrine, Heisenberg, his purpose achieved and his priorities set (he would pursue physics rather than philosophy), began to portray his own field as having little more to add to the discussion. "The researches of the physicist that have affected the field of philosophy came to a certain conclusion two years ago," he told readers of the Berliner Tageblatt in 1931.[35] Bohr's many discourses on complementarity through 1929, which manifested Bohr's own strenuous efforts to refine his originally vague definition of the principle, now served as the standard source.[36]

Heisenberg's more philosophical views on the Copenhagen doctrine evolved primarily out of conversations with his Leipzig student and close personal friend, Carl Friedrich von Weizsäcker, and with Grete Hermann, a former philosophy student of the Lenard Nelson school in Göttingen, who came to Leipzig in 1932 to study quantum acausality at its source. After Hermann's arrival, the conversations grew increasingly intense, while Heisenberg's own participation equally diminished.[37] As Weizsäcker recalls and as Heisenberg's own correspondence indicates, Werner did not concern himself with such matters beyond what he would need for his public addresses.[38]

With both physicists and philosophers now scrutinizing his claims, Heisenberg, while removing himself from discussion, decided to protect his position by checking the weakest point of his own argument—the gamma-ray microscope. By 1930, quantum mechanics had advanced to include relativistic quantum fields, which altered the details of the argument. Heisenberg assigned his student, Weizsäcker, the task of thoroughly analyzing the gamma-ray microscope experiment — an analysis that he himself had twice flubbed in the past — using the Copenhagen doctrine and the most recent formalism of quantum field theory. Weizsäcker's analysis, completed in 1931, yielded no reason to question the irrefutability and universal sufficiency of the Copenhagen doctrine.[39]

With their confidence reinforced and the acceptance of their views spreading rapidly, Heisenberg, Bohr, and their followers had little reason to fear any challenge. But complacency turned into shock and confusion in the wake of a powerful and subtle attempt to refute their doctrine in 1935 in the pages of the leading American physics journal, *Physical Review*. As Léon Rosenfeld, Bohr's collaborator, put it: "This onslaught came down on us as a bolt from the blue. Its effect on Bohr was remarkable."[40]

The jolting paper bore the title, "Can quantum-mechanical description of physical reality be considered complete?" Its authors were none other than their old adversary, Einstein, then at the Institute for Advanced Study in Princeton, and two collaborators: Boris Podolsky, a Russian physicist on a fellowship to the Institute who apparently actually wrote the paper, and Nathan Rosen, a recent MIT graduate who was now a research fellow with Einstein.[41] Despite the sudden shock, the now famous Einstein-Podolsky-Rosen (EPR) argument threatened less to undermine Copenhagen orthodoxy among its true believers than to hinder efforts to maintain a loyal following among the less devoted. "A certain danger . . . exists," Pauli wrote to Heisenberg, "of a confusion of public opinion — namely in America."[42] By 1935, the physics community in the United States was the largest in the world and it was producing first-rate work. American physicists constituted an audience that the Copenhagenites did not want to lose.

Published in May 1935, the EPR paper, consisting of a mere four pages, is to this day the subject of considerable scientific, philosophical, and historical study and debate.[43] Eight years after the Como and Solvay meetings of 1927, the paper and the responses to it revealed the persistence of deep-seated differences over fundamentals that the Copenhagenites, despite their pronouncements, could neither resolve nor silence. The new arguments of 1935 echo the earlier Bohr-Einstein debates, and in a sense the EPR paper represents the distillation of those debates. But again there was no universal resolution or agreement.

The authors of the EPR argument answer its title question with a definite no, but their argument devolves not from formalism or description but from certain preconceptions and prior set criteria fundamental to their position. The paper proceeds, as did Einstein's Solvay arguments, from a so-called a priori realist ontology: the assumption of a knowable physical reality independent of the physicist. The opening sentence reads: "Any serious consideration of a physical theory must take into account the distinction between the objective reality, which is independent of any theory, and the physical concepts with which the theory operates." Acceptance of this distinction underlay the validity of their postulated criterion for completeness two paragraphs later: "Every element of the physical reality must have a counterpart in the physical theory." This criterion required, however, a definition of an element of reality, hence the EPR reality criterion: "If, without in any way disturbing a system, we can predict with certainty (i.e., with probability equal to unity) the value of a physical quantity, then there exists an element of physical reality corresponding to this physical quantity." Bohr, on the contrary, maintained that such a disturbance cannot be avoided and that it limits our knowledge of reality. With their definitions and criteria, Einstein, Podolsky, and Rosen claimed that the disturbance can be avoided and that because quantum mechanics does not take this into account it is an incomplete theory.[44]

Their argument entailed, as usual, a thought experiment: two independent particles A and B interact for a finite time, then separate without further interaction. There is no quantum-mechanical reason that one cannot measure the momentum and position of A at two different times without disturbing B. If this is the case, then one should be able to predict with absolute certainty the simultaneous momentum and position of B at any time after the interaction. Definite values of the momentum and position of B at a given instant must therefore be elements of reality. Since the simultaneous measurement of the exact momentum and position are excluded in principle from quantum mechanics by the uncertainty principle, these elements have no counterpart in the theory. The theory is therefore incomplete.

The EPR argument provoked immediate public and scientific reaction. The *New York Times*, already attuned to public fascination with Einstein, carried a report under the headline "Einstein attacks quantum theory." But in that article, Princeton physicist E. U. Condon pointed out the weak link in the EPR chain of reasoning: "Of course, a great deal of the argument hinges on just what meaning is to be attached to the word 'reality' in connection with physics."[45] Pauli was even less generous: "If a student in the early semesters had made such objections to me, I would have regarded him as very intelligent and hopeful."[46]

Perhaps regarding Americans as somewhat less advanced than his imagined pupil, Pauli urged Heisenberg to respond immediately in the pages of the *Physical Review*. Heisenberg did draft a response (in German), entitled "Is a deterministic completion of quantum mechanics possible?" He sent copies to Pauli and to Einstein himself.[47] But he deferred to Bohr, who had already submitted a response to the *Physical Review*, and withheld the paper.[48] When Heisenberg learned several months later, in August 1935, of attempts by Schrödinger and Laue to build on the EPR argument, he revived his manuscript, but Bohr found fault with its logical coherence, which killed it for good.[49] Bohr apparently did not like Heisenberg's use of Grete Hermann's recent argument as to why additional, undetected (so-called hidden) variables cannot yield a deterministic theory.[50] Heisenberg dropped her argument when he incorporated the manuscript into one of his philosophical addresses at the University of Vienna at the end of the year.[51]

In their responses to EPR, both Bohr and Heisenberg emphasized actual laboratory measurements of atomic processes over the EPR preference for notions about the real world — epistemology and the actual operations of the researcher superseded ontology, the supposed nature of reality.[52] Yet Bohr and Heisenberg deployed their epistemologies in quite different ways. Bohr emphasized that all measurements of atomic phenomena must eventually be expressed in terms of classical laboratory apparatus. He argued that measurements of the position and momentum of the particles in the EPR experiment naturally entail an essential complementarity that is intimately associated with the disturbance introduced by the measurement. Treating both objects A and B as particles, if the position of the particles is measured exactly at a given time, using (for instance) a screen with two holes in it, an "uncontrollable" (probably meaning "unpredictable") momentum is imparted to the apparatus in the very act of measuring. This destroys any use of the momentum conservation laws to predict the exact momentum of the second particle at any future time. The same occurs in the conjugate procedure of measuring momentum first and then attempting to predict position. Again, the measurement process itself makes the classical apparatus part of the experiment, thus disturbing the system and destroying the precise simultaneous appearance of conjugate properties, such as momentum and position. And, wrote Bohr (for the benefit of those too reactionary to accept the new state of affairs), "This new feature of natural philosophy means a radical revision of our attitude as regards physical reality."[53]

Heisenberg made the same point differently in his unpublished manuscript and published lecture. Any experiment entails two types of laws: those applying to the laboratory measuring apparatus, which, as Bohr tirelessly argued, are thoroughly classical in nature, and those pertaining to the atomic phenomena to be studied, which are thoroughly quantum mechanical in nature. Each set of laws holds in its own domain and is precise and nonstatistical. Statistics enters only with the researcher's measuring experiment. Since every experiment attempts to bridge the gap between the classical laboratory domain and the quantum world of the atom, the experimental apparatus introduces a "cut" or interface between the two domains. Because physical laws are precise on either side of the cut, the statistical characteristics of quantum mechanics enter with and along the cut; that is, on the performance of the measurement. A deterministic completion of quantum mechanics must occur along the cut. But if new determining variables are introduced along the cut and the cut is subsequently moved, "then . . . a contradiction between the lawlike consequences of the new properties and the [precise] relationships of quantum theory will be unavoidable."[54] A deterministic completion of quantum mechanics is not possible, after all.

Despite — or because of — the profound arguments on both sides of the EPR debate, neither side ever managed to convert the other. For Heisenberg, Bohr, and their closest colleagues, EPR was more a nuisance than a serious objection — a threat only to maintaining the commitment of their followers. The tone of Heisenberg's unpublished manuscript is not defensive but condescending toward his critics and skeptics. The essentially statistical character of quantum mechanics had already been fully explored by Bohr, Heisenberg declared, and "the essential content of the following train of thought is already available in the first papers on the fundamental interpretation of quantum mechanics."[55] Everything had been discussed and settled years earlier.

Yet everything was far from settled for those outside the circle of Copenhagen devotees. Einstein and his followers, to the end of their lives, insisted upon various versions of the EPR argument. In his response to the contributors to a volume in honor of his seventieth birthday in 1949, Einstein reiterated the argument nearly word for word in asserting the incompleteness of quantum mechanics. He expressed his general views as follows: "Above all . . . the reader should be convinced that I fully recognize the very important progress that the statistical quantum theory has brought to theoretical physics. . . . This theory and the (testable) relations, which are contained in it, are, within the natural limits of the indeterminacy relation, *complete*. . . . What does not satisfy me in that theory, from the standpoint of principle, is its attitude towards that which appears to me to be the programmatic aim of all physics: the complete description of any (individual) real situation (as it supposedly exists irrespective of any act of observation or substantiation)."[56]

Heisenberg was equally adamant to the end of his life. His most extensive rebuttal of the attitudes espoused by the critics of the Copenhagen doctrine appeared in a series of lectures on the intellectual history of physics delivered at the University of St. Andrews, Scotland, during the winter of 1955–1956. These lectures were later published as *Physics and Philosophy: The Revolution in Modern Science*. The Copenhagen interpretation found its natural place in his sweeping account of physical notions since the Greeks. These notions entailed an ever increasing renunciation of classical realism and of the separation of observer and observed. Every attempt, including the EPR paper, to return to a classical realist ontology contradicted this historical trend and, a now Aristotelian Heisenberg asserted, represented nothing more than a desire to return to an outmoded ontology of materialism. "It cannot be our task to formulate wishes as to how the atomic phenomena should actually be," he declared, "our task can only be to understand them as they are."[57] The implications of the Copenhagen interpretation were true and irrevocable.

Pauli's worries notwithstanding, the physics community, especially the American community, never wavered in its preference for Copenhagen. Although Einstein spent the rest of his life in the United States, his objections never persuaded many of his American colleagues. The majority had already pledged their allegiance to Copenhagen by 1935. This came about not only because the Copenhagen interpretation always seemed to work when needed (regardless of philosophical disputes), while its opponents offered no viable alternative, but also because Americans were already receptive to the proselytizing influence of European quantum mechanicians during the late 1920s and early 1930s. American physics had come of age by the 1920s, and American physicists were eager to participate in new fields of research.[58]

Like Americans, audiences the world over eagerly welcomed the new European atomic physics, especially when received directly from its practitioners. Backed by well-endowed philanthropic foundations, Americans mounted aggressive and successful efforts to import the new physics and to establish it at American universities. Contrary to popular belief, American physics, though heavily oriented toward classical and practical fields, never lagged far behind European physics. Nor did the arrival of European refugees in the 1930s mark a major turning point in American physics, as has been supposed. Although the United States still lagged behind Europe in such exotic contemporary fields as relativity and quantum physics, the country was already building momentum in those areas by the early 1930s and actually led in such classical fields as acoustics.[59]

A fertile institutional setting, already in place by the end of World War I, provided a crucial element for the flourishing of imported European specialties. According to one study, the democratic, practical, and collaborative American university physics department proved more conducive to the further development of quantum physics than did the hierarchical German professorial institute, once American physicists had mastered the new discipline.[60] They achieved their mastery of it through a consciously conceived twofold strategy: dispatching bright students to Europe to learn the latest science at the source, while inviting guest lecturers to teach it to those who remained at home. Karl T. Compton, one of many Americans sent to Max Born's institute, reported: "In the winter of 1926 I found more than 20 Americans in Göttingen, at this fount of quantum wisdom."[61]

Americans continued to pursue European fashions after the creation of quantum mechanics and especially after most efforts to lure leading quantum theorists to American professorships had failed. Nearly every European atomic physicist of any renown lectured in the United States during the late twenties and early thirties. One report notes that by late 1929 American universities had invited at least eight of the leading European contributors — Heisenberg, Born, Dirac, Debye, Hund, Sommerfeld, Bragg, and Brillouin — to lecture to their physicists on quantum mechanics over periods ranging from six weeks to a full semester.[62] The benefits were mutual and lasting. By rapidly assimilating quantum mechanics, American physicists could, by the early thirties, work independently of or in equal partnership with their foreign colleagues. The annual Ann Arbor summer school sessions, sponsored by the University of Michigan and funded by foundations, helped to maintain the momentum until the late thirties. Heisenberg and other prominent Europeans were invited, often repeatedly, to deliver series of summer lectures to American physicists and their students on their very latest work. The arrival of refugee scientists from Europe during the 1930s only stepped up the already accelerated pace of American science, helping to ensure its lead by the time the world entered its second global war.

The benefits to visiting Europeans were of equal significance (even aside from their handsome salaries).[63] The contacts they established with foreign colleagues were essential to the further development of their field. These colleagues also provided important assistance during the massive relocation of German and other European scientists after 1932.[64] Lecture tours required the quantum mechanicians to develop (many for the first time) coherent and structured presentations of the new science for both learned and not-so-learned audiences. Textbooks by the masters proliferated, another means of spreading the new doctrine (and incidentally the author's fame). Sommerfeld's lectures in Pasadena served as the basis for a volume on wave mechanics, supplementary to *Atombau*, that enabled him to organize his thoughts on the subject.[65]

Heisenberg had a similar experience. As a condition for accepting the appointment to Leipzig in 1927, Heisenberg had negotiated an eight-month leave of absence in order to accept a series of lecture invitations from around the world. With bookings at MIT and the University of Chicago, then in Japan and India, Heisenberg boarded a ship at Bremerhavn in early March 1929 for his first transatlantic trip. Werner eagerly anticipated the journey, which would ultimately take him around the world, and he looked forward to meeting fellow visitors Sommerfeld, Dirac, and Hund "over there among the wild Americans."[66] But less than a day out to sea, the ship was trapped for a few days by an early spring fog and a solid sheet of ice. Adrift in the unending fog prison, Professor Heisenberg began to entertain second thoughts. He wrote to Bohr of how much he would rather spend his holiday in usual fashion with his youth group at their Bavarian ski hut "instead of traveling all the way to America — but one has to try everything."[67]

The trip must have been exhausting, even for a 27-year-old. After delivering the MIT and Chicago lectures, Heisenberg traveled to Washington for a few days to lead a session on atomic structure and spectral lines at the American Physical Society meeting. From there he went west to climb in the Rocky Mountains, writing excitedly of the unanticipated beauty of the mountain landscape, which reminded him of home, then through Colorado and Arizona to the Grand Canyon, on to Pasadena for a week of lectures and touring in the Sierras, then back to Chicago by June. From Chicago he wrote to Bohr that in this "unsettled life" he traveled 1000 kilometers per week — and in those days one went only by train.[68]

In Chicago (where he stayed with German-American physicist Carl Eckart), when not lecturing and working, Heisenberg sailed and swam in Lake Michigan, went on a fishing expedition to the northern Wisconsin lakes with Barton Hoag, a Chicago experimentalist, stopped in Madison to see Dirac, and engaged in playing tennis and music with other physicists. On July 20, he described for Pauli the rest of his exhausting schedule: he would meet Dirac in Yellowstone Park in mid-August and leave for Hawaii and Japan, where they would stay and lecture until mid-September. While Dirac headed from Japan to Siberia and Moscow, Heisenberg would travel through China to India to lecture, tour, and visit the Himalayas, finally arriving back in Leipzig in time to begin the winter semester in November 1929. "Then I hope to be able to really do physics once again."[69]

Aside from mountaintops, the series of ten lectures that Heisenberg delivered to the University of Chicago in March and April 1929 were the high point of his tour. The lectures were the basis of his first book, *The Physical Principles of the Quantum Theory*, which was probably the most influential and widely read early account of the Copenhagen doctrine and the inspirations behind it. Heisenberg's account is remarkable both for its clarity and for its overreliance on Bohr. While he frequently cites various versions of Bohr's complementarity paper, Heisenberg does not mention his own name anywhere in the text. Even though he portrayed the uncertainty relations as a consequence and clarification of quantum phenomena, citations of his works, particularly his uncertainty principle paper, are conspicuously absent from the bibliography. Bohr alone is cited, even for uncertainty. One reads in the (awkwardly translated) foreword: "The purpose of this book seems to me to be fulfilled if it contributes somewhat to the diffusion of that 'Kopenhagener Geist der Quantentheorie' . . . which has directed the entire development of modern atomic physics."[70] For Heisenberg, Bohr obviously embodied that "Geist."

Other itinerant lecturers outside Bohr's inner circle were less bullish on Bohr and complementarity. While Born and Jordan fondly dedicated their new textbook to Bohr, Sommerfeld neglected to include complementarity in his supplement on wave mechanics to his own textbook, *Atombau*. Pauli opened his 1933 handbook account of quantum mechanics with the uncertainty principle and complementarity as the formal basis for constructing the new physics, but Fermi founded his influential Ann Arbor lectures on the uncertainty relations alone.[71] Numerous textbook writers have since agreed with Fermi.[72]

The uncertainty relations lay closer to the derivation and to the actual applications of the quantum mechanics formalism than did the epistemology of complementarity. This circumstance, combined with the easier comprehension of uncertainty, facilitated the ready acceptance of uncertainty and other useful exotica of quantum physics by the more practical-minded Americans as the foundation of the new physics. At the same time, they tended to leave the more refined aspects of complementarity and similar rarefied issues of the Copenhagen doctrine to the so-called philosophers.[73] The EPR debate of the mid-1930s and beyond came too late to convert the already entrenched Americans, or to undermine widespread allegiance to the doctrine that has dominated quantum mechanics to this day.

With a full professorship, a successful and accepted physics, and worldwide recognition as a leader of the quantum revolution, Heisenberg arrived back in Leipzig from his world tour in the fall of 1929 having fulfilled most of his life's ambitions. In an exquisite recollection of their journey together to Japan that year, Dirac recounted how, while strolling one day with Heisenberg in the vicinity of a beautifully constructed Japanese pagoda, he watched in astonished amazement as young Heisenberg, without uttering a word, gingerly climbed to the top of the pagoda and, in triumphant defiance of certain injury and even death, balanced himself precariously on one foot on the very pinnacle of the building as a fierce wind swept wildly about him.[74] As he balanced in the wind beneath the sky, the daring and ambitious physicist could be satisfied in knowing that he had also attained at last the pinnacle of his life and work.

CHAPTER 14

Fresh Fruits

Within days of the October 1927 Solvay Congress, Heisenberg arrived in Leipzig as head of the Institute for Theoretical Physics. Heisenberg's research and his new institute propelled the rapid dissemination and acceptance of the Copenhagen interpretation of quantum mechanics. Leipzig physics epitomized the new quantum physics centers that would produce a new generation of physicists nourished entirely on the Copenhagen spirit.

Physicists everywhere looked to the enormous possibilities of applying and expanding the new science to numerous areas of physics and to other sciences. For many scientists, especially those in the United States, the obvious utility of quantum mechanics spoke louder than any lecture on epistemological principles. Heisenberg and the Leipzig institute took an early lead in demonstrating its uses. "The gate to that entirely new field — the quantum mechanics of the atomic shell — stood wide open," Heisenberg recalled, "and fresh fruits seemed ready for plucking." After nearly a decade of anomalies and crises, the five years following the 1927 Solvay Congress seemed to many physicists, including Heisenberg, "so wonderful that we often spoke of them as the golden age of atomic physics." [1] Even so, difficulties encountered in expanding the new physics to high-energy research and upheavals among the German public and students multiplied in that same period. The euphoria of those wonderful years often turned into deep despair.

Heisenberg's institute was a subdivision of the university's Institute for Physics, which was headed by Peter Debye. Heisenberg and Debye both arrived in 1927 and immediately transformed the institute and its physics. During the last years of the ill and aged Wiener and Des Coudres, once-powerful Leipzig physics had atrophied to near nonexistence. Although the grand old men had been at the forefront during the first decade of the century, neither had fully assimilated Einsteinian physics. Des Coudres concentrated on classical and technical subjects; Wiener, until the last, on a kinetic ether theory and its experimental confirmation.[2]

Gregor Wentzel, appointed to a theory position in 1926, summed up the sorry state of Leipzig's pre-Heisenbergian physics in a letter to Sommerfeld in the same year: "Meanwhile the old routine continues on in even worse form than before. Not much more can be made of the older student generation. What I have experienced in knowledge of physics on state final examinations is indescribable. Only one candidate has registered for the doctorate, and he appears to be hopelessly untalented. He is, by the way, a businessman, which is often the case here."[3]

By 1932, Leipzig physics had revived to such an extent that a civil servant in the Saxon Culture Ministry exclaimed: "In fact, the extraordinary representation of physics in Leipzig (Debye, Heisenberg, Hund) is one of the greatest internationally active posts in the university."[4]

Debye inherited from Wiener his dual position as professor of experimental physics and director of the physics institute. The turn-of-the-century institute building was located just outside the city center at Linéstrasse 5—midway between a cemetery and a mental institution. The horseshoe-shaped building housed lecture rooms, laboratories, offices, instrument collections, and staff apartments. Heisenberg's Institute for Theoretical Physics occupied the northern wing of the two-story building.[5]

Debye also inherited Wiener's aging loyal assistants, Möbius, Kämpf, and Sängewald, along with the three engineering sections that Wiener had established with tenured faculty—applied mechanics and thermodynamics, radio physics, and applied electrical science. Although the electrical section enjoyed some repute as the creator of the first successful television receiver, Debye ignored the technical sections and placed Wiener's assistants in charge of the basic beginners' laboratories.

As part of its deal with Debye, the Culture Ministry, located in Dresden, hired Debye's Zurich assistant, Heinrich Sack; recommended a Notgemeinschaft grant for his second assistant, H. Falkenhagen;[6] provided 3000 marks to fund an annual conference—the world-famous Leipzig Lectures held each spring—and empowered Debye to negotiate with Heisenberg to bring him to Leipzig. When Heisenberg accepted Dresden's appointment to Leipzig, Debye presented him with a former Zurich pupil, Felix Bloch.

Heisenberg started practically from scratch, with one student, no assistant, and no place to stay. Des Coudres, a lifelong bachelor, had occupied a small, sparsely furnished service apartment under the roof of the institute, which now fell to Heisenberg. During his last years, Des Coudres had hired a caretaker to look after him, Frau Gretschmer, who lived in an assistant's apartment in the institute. In early November, Heisenberg arrived to find Frau Gretschmer in Des Coudres's apartment, Wentzel in the assistant's apartment, and the institute reeking of dust. He ordered the doors and windows thrown open before going off to find bed linen.[7] By December Heisenberg had moved from a hotel to the institute and had bartered a Columbia University offer for free rent. He had hired a full-time cleaning woman to spruce up the institute, make his meals, and tidy his apartment.[8] Werner also had his apartment refurbished, installed a piano in his rooms, and ordered a Ping-Pong table for the basement. But, he regretted, the refurbishing had left his toilet across the hall from his apartment.

As his first professorial acts, Heisenberg tried to steal an assistant from Sommerfeld and began lecturing without formal announcement. Apparently he took over Wentzel's course on the theory of electricity, for which, surprisingly, 40 students showed up the first day. "That's indeed a good beginning," he wrote to his proud parents.[9] But his letter to Sommerfeld was less well received. "Your letter reeks of your bad conscience from beginning to end. You want to steal assistants, and of course only the best ones!"[10] He did without an assistant until the summer.

That summer of 1928, the first in which Heisenberg's name appeared in the course catalogue, a record-setting 150 students enrolled in his lectures on classical mechanics, 80 students attended his advanced course on atomic physics, and his advanced research seminar with Wentzel grew to 12 participants, thanks in part to 3 students whom Sommerfeld had urged to go.[11] Sommerfeld also furnished Guido Beck, who received a four-year appointment as Heisenberg's first assistant. Bloch, Heisenberg's second assistant, graduated that semester as his first doctoral student. Like most neophyte professors, he was nearly overwhelmed by the sudden strain of academic duties, prompting a complaint to his parents: "I have doctoral papers to correct, galley proofs of my own papers are here, and the lectures require a lot of preparation. Sometimes everything is almost too much for me; also I have unfortunately very little time for music."[12]

With the sudden revival of Leipzig physics, the number of physics students exploded. During the winter semesters of 1927–1928 and 1929–1930, their numbers shot up 166 percent (from 41 to 109).

Within one semester alone, from winter 1927–1928 to summer 1928, the number of physics students nearly doubled, while the total number of Leipzig students increased by only 28 percent.[13] Physics classrooms and laboratories overflowed, causing an irate student delegation to demand more laboratory instructors. Debye managed to obtain Rockefeller Foundation funding for a new laboratory, only to have the Culture Ministry refuse to fund new assistants.[14]

The growth of Leipzig physics far exceeded the tremendous general increase (71 percent) of physics students in Germany.[15] New developments in science had captured the imagination of German young people, and between 1927 and 1933 an extraordinary number of future notables flocked from abroad to work and study with Heisenberg, among them Edward Teller, Laslo Tisza, Lev Landau, George Placzek, I. I. Rabi, John Slater, and W. V. Houston. The flood continued nearly unabated until the start of World War II.[16]

Heisenberg built his institute on the Munich and Göttingen models. He transferred the experimental laboratory to Debye (but kept a technician for lecture demonstrations) and elevated the director of the section for mathematical physics — Wentzel, followed in 1929 by Friedrich Hund — to the rank of full professor (but with authority and salary below his own). Heisenberg gave the basic three-hour theory lecture each semester on classical mechanics, thermodynamics, electrodynamics, and optics. He also offered a weekly one-hour special seminar on an advanced topic of current interest, a research seminar with Wentzel or Hund on the structure of matter, and a weekly physics colloquium with Debye in which local and invited speakers lectured on their latest research.

In the early 1930s, Heisenberg began formal lectures on quantum theory and atomic physics, while Hund offered courses on basic theory or on his own research areas, primarily the properties of solid bodies. A lively exchange of assistants and postdoctoral students with Pauli's institute in Zurich and Bohr's in Copenhagen established a new European network of quantum research centers.[17] After his habilitation in 1932 and a stint as Pauli's assistant, Bloch joined the Leipzig professors with courses on general relativity, the quantum theory of magnetism, and the absorption of high-speed particles in matter. Bloch, of Jewish descent, returned to Zurich in 1933 and eventually emigrated to Stanford University.

Although Friedrich Hund, the assistant director of the theory institute, was five years older than Heisenberg, the two men had been good friends since the Göttingen days when they had worked and wandered together as Born's students and assistants. They had used the familiar "du" long before it was common among colleagues. Until the birth of his second child, Hund's family lived in the small assistant's apartment above Heisenberg's. But Hund was never happy with his role as Heisenberg's shadow and with often being assigned the second-rate students. He was also unamused by a standing joke in Leipzig: the course catalogue always listed the heads of the research seminar as "Heisenberg mit Hund" (Heisenberg with dog). Furthermore, Hund never got over being ordered by Heisenberg to return to Leipzig from a visit to Chicago in 1929, before he had taken up his post, so that Heisenberg could continue his travels.[18] But after Hitler's rise to power in 1933, Hund became one of Heisenberg's closest political confidants and supporters.

Hund's even more strained relationship with Debye reflected the wide divergence between theory and experiment in Leipzig. One American visitor to the physics institute wrote to a colleague back home in early 1933: "I see Bloch occasionally, but being an experimental physicist, I have very little to talk with him about. The categories of theoretical and experimental physicists are more sharply defined here in Leipzig than anywhere I have ever seen before."[19] Perhaps comparing Hund unfairly with Heisenberg, Debye always regarded Hund as a second-rate theorist. Hund, in turn, regarded Debye as "clever but lazy." Heisenberg reported that he often saw the cigar-chomping Debye in his garden, watering the roses during duty hours. "Debye had a certain tendency to take things easy," he once noted.[20] Hund stayed on in Leipzig nevertheless and eventually inherited the directorship of the institute after Heisenberg followed Debye to Berlin in 1942.

Because of their youth and their perpetually young life styles, Heisenberg and Hund shared an easy rapport with their students. Hund recalls that Heisenberg often ran their joint seminars like a youth-movement Thing (criticism session), to which Hund, a former Wandervogel, had little objection. For his advanced special seminars, Heisenberg often bought pastries at the corner bakery and prepared tea on his chemical stove.[21] Until the formation of political student groups after 1933, the two professors often hiked in the hills with their students on weekends (though never with Debye and his pupils), and assistants often accompanied Heisenberg to his Bavarian ski hut. During the summer months, Heisenberg took daily horseback riding lessons at 7:00 A.M. and frequently attended large swimming parties, once bragging to Bohr that he could still perform difficult dives.[22] Competition never lagged: student challengers found in Werner a ready opponent on the university tennis courts, and the basement Ping-Pong table became a special focus of institute fun. Following his trip to the Far East and his hours of practice with Asian experts aboard ship, Werner was the undefeated champion.

As his twenty-seventh birthday approached in 1928, the youthful professor occupied his thoughts during an important faculty meeting with a letter to his mother on what he wanted for his birthday: a new pillowcase or a bread basket. He thanked her profusely for sending both. "I am so incredibly happy when such a token of my home penetrates my all too physics-oriented Leipzig existence."[23] The familiar longing for his Munich home and the companionship of his youth-group comrades continued unabated, and, as usual, he threw himself into his work in order to repress such feelings.[24]

In spite of his earlier plaint to his parents, music occupied much of Werner's time. He practiced for hours alone in the evening in his apartment. Music was again a way of escaping the loneliness of Leipzig and a way of relaxing from the strain of work, of reestablishing his equilibrium. Even with many friends and colleagues in Leipzig, none had shared the intense experiences of his youth and few could accompany him to the mountaintops of reflection and physical intuition to which he drove himself. Those who could were not in Leipzig. An intensive correspondence continued unabated with Bohr and Pauli.

Still, not even music could cure him of his relentless drive. Werner could not let up, even while he was relaxing. Bloch once recalled listening one evening from his own room as Werner practiced over and over again just a few bars from a Schumann concerto. After finally getting it right, Heisenberg came down to Bloch's room for a chat before retiring. Bloch recognized that behind Werner's often adolescent behavior was an extraordinary seriousness, an intense concentration of purpose that went far beyond mere adolescent fun or the suppression of unwanted emotions. "We all knew the dreamy expression on his face," Bloch remembered, "even in his complete attention to other matters and in his fullest enjoyment of jokes or play, which indicated that in the inner recesses of the brain he continued his all-important thoughts on physics."[25]

Located in central Germany, south of Berlin between Halle and Dresden (the capital of Saxony), Leipzig enjoyed special status as a trading center for goods, culture, and people from the east. Leipzig's book trade was the largest in Europe, its library organization dominated the German library system, and the annual Leipzig Trade Fair served (and still does) as an international industrial showplace. One eighteenth-century visitor wrote: "The Leipzig inhabitants are to be regarded as a small cultural republic. Everyone produces his little original contribution. Wealth, knowledge, talents, possessions of all kinds lend the location its fullness."[26] Goethe chose Auerbachs Keller in Leipzig as the place where the obsessively learned Dr. Faust and a group of students met under the influence of the evil Mephistopheles.

Culture thrived in Leipzig, as it did elsewhere in Germany, during the late Weimar period. The university, one of the oldest in the world (founded in 1409), still housed the traditional faculties of medicine, law, theology (Lutheran), and philosophy (including science). The original legal and economic independence of the university within the free city of Leipzig had long since been replaced by the authority of the Saxon Ministry of Culture, but culture and learning enjoyed unhindered reign. Weizsäcker judged the cultural offerings the best in Germany, although Berlin offered more of everything.[27] Music was especially available; there were free concerts by the renowned Bach choir every Friday evening in the Thomaskirche and weekly Mendelssohn concerts by the equally renowned orchestra in the Gewandhaus auditorium. Heisenberg often wrote home of his visits to both. After one letter in which he told them of dancing until 1:00 A.M. following a Gewandhaus concert, his parents chided him for leading a life of pleasure in Leipzig. They had obviously missed the point. Work and career were his main preoccupations, he insisted. "Many things are much more serious to me than you imagine."[28]

Musical pleasures were indeed more serious than mere relaxation and escape: music also provided Heisenberg a direct entrée into the elite social circles of Leipzig. During that period, Leipzig social life was dominated by a plutocracy, as Hund called it, composed of book publishers, university professors, and judges and attorneys at the Reichsgericht (Reich Court).[29] Culture united them all. "Social life blossomed very nicely during those years in Leipzig, although in simpler style than before the war," the Leipzig philosophy professor Hans Driesch recalled. "We circulated with colleagues form all of the faculties, but in addition also with many families on the Reichsgericht."[30]

Politicians circulated in such elevated circles only on an individual basis. Political maneuvering and intrigue remained beneath the supposed dignity of cultured members of the social elite. The main exception was Dr. Carl Goerdeler, the monarchically inclined conservative mayor of Leipzig (and a later participant in the 1944 assassination plot against Hitler), who frequented the same circles as did Heisenberg.[31]

Music served those circles as a common denominator, and Heisenberg attended many musical evenings in the homes of Leipzig plutocrats.

One of his most frequent musical hosts was the distinguished Otto Mittelstaedt, an attorney at the Reichsgericht whose grandson, Peter Mittelstaedt, later studied physics under Heisenberg. Heisenberg and other physicists regularly visited the Mittelstaedt home and that of their in-laws, the Brückings, an old publishing family. Although it is difficult to imagine a genteel Werner in semiformal evening attire, chatting about the latest gossip over tea and cookies while waiting to join the evening's chamber ensemble, these circles of cultural acquaintances served him well — not only by ensuring his membership in the Leipzig social elite, but also by providing personal support and professional influence in the years after 1932. It was at one such musical evening that Heisenberg later met his future wife.

In the early 1930s, Professor Heisenberg also circulated in two other influential professional and cultural groups, which had overlapping memberships. In June 1930, the mathematical-physical section of the Saxon Academy of Sciences in Leipzig elected Heisenberg a full member.[32] At first, such academies were intended to stimulate scientific discussion, to promote science and learning, and to counsel ruling authorities, but with the revival of universities and the establishment of professional journals, their activities mostly resembled an academic men's club. Six months after his election, Heisenberg thanked his supporters by delivering a paper on energy fluctuations in a quantized electromagnetic field.[33]

Heisenberg and a few other members of the Saxon Academy also belonged to a Professorenkränzchen, or professors' club, called the "Coronella." This informal group of eight to ten younger male members of the philosophical faculty met regularly for evenings of music and academic chatter. Both types of cultural clubs grew in popularity as the self-styled Kulturträger (literally, culture bearers) of society felt themselves increasingly challenged and politically isolated in the last years of the Weimar era. Heisenberg never attained an administrative position in the Leipzig philosophical faculty, but several of his close colleagues did. These connections gave him access to a wide circle of influential people, access that would prove crucial after Hitler's rise to power.

Of course, Heisenberg's most immediate circle of acquaintances consisted of his professional colleagues, assistants, and students. By 1932, this circle encompassed mainly Hund; Bloch; B. L. van der Waerden, professor of mathematics; Karl-Friedrich Bonhoeffer, professor of biochemistry and brother of theologian Dietrich Bonhoeffer; and C. F. von Weizsäcker, who completed his dissertation in 1933. Werner was not especially close to Debye or his staff.

Heisenberg had met young Weizsäcker in Copenhagen when, as Bohr's assistant in the winter of 1926–1927, Heisenberg had been invited to a musical soirée at the home of the German ambassador to Denmark, Ernst von Weizsäcker, Carl Friedrich's father. When Heisenberg was appointed to Leipzig, Carl Friedrich, still a gymnasium pupil, decided to join him. But when he graduated in 1929, Heisenberg was abroad. Weizsäcker enrolled for a year at the University of Berlin to await his future teacher's return. Although ten years younger than his teacher, Weizsäcker became one of Werner's closest friends and companions until Weizsäcker transferred to Berlin in 1936. He often traveled with Heisenberg on weekend outings or joined him with other physicists and youth comrades at the youth group's Bavarian ski hut. By all accounts, their discussions centered on such elevated topics as physics and philosophy. When once asked if they ever discussed politics or economics during the early thirties, Weizsäcker replied: "Not much, actually no." Hund, however, responded to the same question: "Yes, daily."[34] If Weizsäcker, a student and assistant, was Werner's intellectual colleague and companion, Hund, a full professor, was his academic and political confidant.

As the political situation fell into disarray during the last five years of the Weimar republic, so did post-Solvay efforts to expand the successful quantum mechanics beyond its completed state. Although quantum mechanics provided the foundation for comprehending atomic events, and the Copenhagen interpretation provided the epistemological connection between the atom and the everyday world, an important element of the connection was still missing. Electromagnetic radiation, or light quanta, emitted or absorbed as atomic electrons jumped between stationary states, provided the main source of information on the internal workings of atoms. But what was the precise connection between the electron and the light quantum? Since the light quantum existed neither before it was emitted nor after it was absorbed, how could we understand what seemed to be the very creation and annihilation of electromagnetic energy by charged material particles? The answers to such questions lay beyond the capabilities of quantum mechanics in 1927.

Nor did any connection yet exist with the other great advance in physics during the twentieth century—the theory of relativity. Such a connection seemed warranted by the high speeds of electrons in atoms, by Einstein's reformulation of the basic equations of electrodynamics in the special theory of relativity, and by the demands of the relativity theory itself on every law of nature.

Since electromagnetic radiation is a manifestation of electric and magnetic fields, the search began almost immediately for what is now called a relativistic quantum field theory. This extremely technical and mathematical branch of quantum physics, the foundations of which were laid by Heisenberg, Dirac, Pauli, Jordan, and their colleagues during the late 1920s and early 1930s, continues to this day with much the same program and approach.[35] The effort to understand the essential nature of matter, fields, and their interactions, and the search for the proper way to wed the lessons of quantum mechanics with the fundamentals of relativity theory, have also spawned some of the most exotic and expensive research — both experimental and theoretical. The enormous high-energy accelerators, bubble chambers, and data analysis equipment that were built after World War II, employing vast teams of physicists, technicians, and students, are technological wonders in themselves. They were stimulated by the discovery of new types of nonelectromagnetic fields in the middle 1930s, the quantum behavior of which comes into play only when particles smash into each other at such extremely high energies that their internal workings become evident.

Large-scale experimental research was paralleled by the continuing search for a unified field theory — a quantum field theory encompassing all four types of fields known today, rather than just the electromagnetic field. Today, the implications of research into the smallest constituents of matter actually far exceed the minute realm of nuclei and elementary particles, directly influencing such specialties as materials engineering and our understanding of the very origin and fate of the universe. Werner Heisenberg was again a leading member of the small band of abstract theorists who established the program and laid the foundations of relativistic quantum field theory as it has been pursued ever since.

One of the primary difficulties of this effort was to find a formulation of quantum mechanics that would satisfy the demands of relativity theory. Einstein's special theory of relativity rested on two basic postulates: the constancy of the speed of light for all observers and the requirement that the laws of physics, and the equations expressing them, should look the same in all reference systems moving at uniform velocity, regardless of the relative velocities of those systems. In translating equations and laws between two such reference frames, the three dimensions of space and one dimension of time must follow a certain prescription — the so-called Lorentz transformation. Since time and space are relative concepts, they are not the same in all moving reference systems.

Even before Heisenberg left Copenhagen for Leipzig, he and his closest colleagues had already set themselves the tasks of reformulating the newly acquired equations of quantum mechanics to render them relativistic and of expanding the equations to include the emission and absorption of light. The tasks were made difficult by the appearance of two very different fundamental entities in nature to which the new equations should apply — charged matter and electromagnetic radiation (electric and magnetic fields propagating through space). These two entities, though so different in nature, were actually very similar and intimately related. Material electrons are subject to forces transmitted by fields, and electrons subjected to accelerating forces can generate electromagnetic fields that are detected as radiation. Furthermore, by virtue of the wave-particle duality, under certain conditions electromagnetic waves can appear as particles — light quanta, or photons — and electron particles can appear as matter waves.

The late nineteenth-century classical electrodynamics of Maxwell, Hertz, and Lorentz could account to a large extent for the interaction of nonrelativistic electrons and nonquantized fields, which gave rise to forces and radiation observed in a laboratory. Einstein's special theory of relativity, presented in 1905 in a paper entitled "On the electrodynamics of moving bodies," rendered the basic equations of classical electrodynamics relativistic, but they were not yet quantized — they represented continuous electromagnetic waves, not discontinuous, discrete light particles, or light quanta. By the same token, the Schrödinger equation, formulated in 1926 and assimilated into the interpretative framework of quantum mechanics, represented the statistical motions of material electrons, but it was not relativistic and it did not encompass the quantization of the matter-wave field into discrete electron particles.

Heisenberg and other quantum physicists began to search almost immediately for a relativistic quantum electrodynamics — a theory that quantized both matter and radiation, that encompassed the intimate relationship between the two entities, and that yielded equations that were fully relativistic. Pascual Jordan helped make the first step as early as the 1925 three-man paper on the matrix version of quantum mechanics. Pure electromagnetic radiation (without charged matter) in a box with perfectly reflecting walls appeared as a series of standing waves, or harmonics, each standing wave containing a different integral number of wavelengths. Jordan reinterpreted the number of wavelengths as the number of photons of a given energy in that particular wave — the waves could be broken up into light quanta. This led directly to the two results long regarded as the most characteristic of light quanta confined in a box: Planck's 1900 formula for the energy distribution of black-body radiation and Einstein's 1909 results on the fluctuations of the average energy and momentum of such radiation.

In early February 1927, as Heisenberg assembled the uncertainty principle in Copenhagen, P. A. M. Dirac, having returned to Cambridge from Copenhagen, built on Jordan's work and provided the basis of an approach to quantum field theory that has been followed ever since. In two papers on quantum electrodynamics submitted to the Royal Society of London,[36] Dirac treated matter and radiation, electrons and electromagnetic fields, as wholly separate entities. He then treated the connection between the two as a small perturbation of the individual energies of the electrons and fields alone. The interaction between matter and radiation could be treated (it was soon shown) by a process of successive approximations.

At this stage, Dirac considered only the first approximation and treated each of the two basic entities in a different way. The charged matter consisted of a single electron orbiting in an electrostatic field — a hydrogen atom. The electron remained nonrelativistic yet subject to quantum mechanics, thus orbiting in one of the quantum energy states represented by a Hamiltonian energy function, H_{atom}. The electrostatic field, on the other hand, was both relativistic and subject to quantization through the application of matrix mechanics to Jordan's harmonic wave components. Each component acted like a noncommuting quantum number, or q-number (as opposed to a classical, commuting c-number).[37] This yielded an energy function for the electrostatic field, H_{rad}, that consisted of numerous light quanta, but the light quanta could not be enumerated simply like marbles in a box. Since light quanta, unlike marbles, are totally identical and indistinguishable and carry no spin, Bose and Einstein had shown that they can be counted only in a certain peculiar fashion, known as Bose-Einstein statistics. Dirac derived this independently invented statistics directly from the quantum mechanics of light quanta.

In Dirac's theory, the interaction between the orbiting electron and the electrostatic field, or the force exerted by the field on the electron and the reaction force of the electron on the field, could be treated via a small perturbing energy function, H_{int}. In other words, Dirac had broken down this simple situation into three parts — the orbiting electron (or atom), the field, and the interaction — each represented by an energy, the sum of which equaled the total energy: $H_{tot} = H_{atom} + H_{rad} + H_{int}$.

Dirac showed that the effect of adding the small energy H_{int} to the other two energies could be interpreted as the emission or absorption of a single photon by the electron; that is, by the creation or annihilation of a photon as the electron moved through the field. Furthermore, he soon discovered that higher order approximations to the total interaction, involving the addition of energies $H_{int}H_{int}$, $H_{int}H_{int}H_{int}$, and so on, corresponded to the emission and absorption (creation and annihilation) of two, three, and so on, photons. A quantum field theory of electrodynamics could be built up by successive approximations, via a relativistic perturbation theory — a highly mathematical procedure that has been followed in principle ever since.

Wolfgang Pauli, still in Hamburg, was so enthusiastic about Dirac's work that he immediately wrote to Heisenberg of a program to build Dirac's theory into a complete quantum electrodynamics (QED) analogous to classical electrodynamics.[38] Throughout most of 1927, as the Copenhagen interpretation came to full fruition, Heisenberg and Pauli attempted — in whatever free time they had — to construct such a QED on the basis of the oscillating electric and magnetic field strengths, E and H, treated as Dirac's q-numbers but now subject to the uncertainty relations.[39] Since E and H are q-numbers, they should obey a commutation relation; that is, $EH - HE = F$, where F is some function, not equal to zero. Because the electric and magnetic fields, E and H, are different at different points of space and time, the commutation relation at the basis of their theory had to be rendered relativistically invariant.

A breakthrough occurred at the end of 1927. Pauli and Jordan, who would soon succeed Pauli in Hamburg, obtained the necessary commutation relations for E and H in empty space, away from all charges, and valid at every point in space-time; that is, for every inertial reference frame.[40] With these relations and the equations of quantum mechanics, Pauli and Jordan managed to obtain a relativistic quantum theory of the electromagnetic field, a theory that described the behavior of light quanta propagating through empty space or bouncing around inside a box with perfectly reflecting walls.

Early in 1928, as Heisenberg began lecturing in Leipzig, he and Pauli settled on a new program: generalizing the Jordan-Pauli theory such that relativistic charged matter could be included in the reflecting box or in the vicinity of the light quanta in the previously empty space.[41] Heisenberg recognized from the very beginning of Pauli's first QED program that they would have to be evenhanded in their treatment of matter and radiation — both matter waves and electromagnetic fields would have to be quantized, and the equations for both would have to be relativistic. The matter side of the task required the most work. Similar to the discrete light quanta of radiation theory, they had to obtain discrete particles of matter from the wave functions of quantum mechanics — what became known as second quantization (the first being quantization via Schrödinger's equation). But the Schrödinger equation alone would not do, since it was not relativistic. Heisenberg believed from the very start that the charge, the mass, and the special counting procedure, or statistics, of electrons would emerge from this second quantization, rather than entering as postulates of the theory.[42] A general relativistic quantum field theory would arise to encompass any sort of atomic problem involving matter, radiation, or both. This became by 1928 the aim of the second Heisenberg-Pauli program.

The ingenious Dirac again supplied a crucial impetus. On January 2, 1928, as Heisenberg prepared to deliver his formal inaugural address to the Leipzig faculty,[43] Dirac submitted to the Royal Society his best known achievement: the Dirac equation, a fundamental relativistic wave equation for electrons that replaced the nonrelativistic Schrödinger equation in situations (such as high speeds) where relativity is important.[44] The Dirac equation — actually four first-order differential equations for the four components of the relativistically invariant electron-wave function — also contained electron spin in a natural way. Since 1926, it had been thought that electrons rotate around their axes like toy tops but with a puzzling half quantum of angular momentum. Quanta, by definition, came only in integral units. Yet such electron spin was essential both to the Zeeman effect in spectroscopy and to the Pauli exclusion principle. It provided the necessary fourth quantum number for Pauli's rule that no two electrons could possess the same four numbers. Furthermore, by virtue of their supposed spin, electrons differed from spinless light quanta and were thus counted differently when grouped together. Electrons obeyed a statistics developed by Fermi and Dirac, which differed fundamentally from Bose-Einstein statistics. A wave function describing numerous light quanta displayed symmetry properties different from those of a wave function describing numerous spinning electrons.

While electron spin could be attached to Schrödinger's original wave equation only with great difficulty, spin was easily included in matrix mechanics after Pauli's invention of so-called spin matrices. Pauli's spin matrices appeared in Dirac's new relativistic electron equation as a natural consequence, not as an ad hoc appendage. At the same time, Jordan and Wigner showed in 1928 how Pauli's exclusion principle could be expressed as a so-called anticommutation relation ($AB + BA = F$) for the electron-wave functions. The Fermi-Dirac statistics, invented for spinning electrons, arose naturally from the Dirac equation, just as Bose-Einstein statistics for light quanta arose naturally from the Pauli-Jordan quantization of the electromagnetic field.[45] The natural derivation of such results, earlier invented and merely appended to quantum mechanics, greatly strengthened confidence among physicists that the breakthroughs represented by the equations of Pauli, Jordan, and Dirac were surely in the right direction. With the new Dirac wave equation describing relativistic, spinning electrons, the Jordan-Pauli theory for quantized electromagnetic fields, and Dirac's perturbation procedure for treating the interaction between the two, a full-fledged relativistic quantum field theory seemed ripe for plucking.

The fruit remained out of reach a good deal longer. Three problems seemed so fundamental that they shook the confidence of the usually optimistic Heisenberg even in the basic validity of quantum mechanics, and at a time when he and others were attempting to persuade their colleagues to accept their approach and their version of quantum mechanics as final and irrevocable. One of the three problems hounded this theory for two decades and is still a point of contention—the infinite self-energy of an electron (or any particle in its own field). A similar problem had arisen in classical theory. Its reappearance in quantum theory and in a more fundamental way suggested a more fundamental failing of the theory.

The problem arose from the puzzle of the electron's size—a puzzle that had also rendered electron spin problematic. The electron carries mass and charge, but it is usually considered to have no size—that is, to exist as a mathematical point, with zero radius. In classical electrostatics, an electron sitting by itself is surrounded by an electric field. The electric field contains energy, and a calculation of the amount of energy is related inversely to the radius of the electron. If the electron has no radius, then the energy, its so-called self-energy, becomes infinite—an impossibility. This suggested that a finite electric charge cannot be squeezed into an infinitesimal point. In quantum electrodynamics, this appeared as the interaction of the electron possessing mass with an infinite number of photons, which are massless.

In classical theory, one solution to the self-energy problem was to endow the electron with a size—treating it as a ball of matter, like a marble or billiard ball. The size could be chosen so that the self-energy turned out to be the energy contained in its rest mass according to relativity theory: $E = mc^2$, m being the rest mass of the electron and c

being the speed of light. But that required an explanation of why the electron did not explode under the repulsive forces of its own charge. Nor could a finite size alone, without a time variable, be relativistically invariant. Pauli recognized immediately that only a theory that did not treat matter and radiation as wholly different entities only barely interacting with each other would solve the problem — that is, a unified field theory, in which one wave function described everything, without perturbing interactions, was needed. He wondered aloud if the ever inventive Dirac could help find such a theory.[46] The search continues to this day.

A second difficulty for quantum electrodynamics seemed mainly mathematical. Since the variables that refer to matter and to fields are not conjugate to each other (unlike momentum and position), their commutation relations should vanish ($AB - BA = F = 0$, or $AB = BA$). They didn't vanish, a circumstance traced to the zero rest mass of the photon when treated like a particle. But the third difficulty seemed at the time the most devastating of all: Dirac's relativistic wave equation for the electron allowed the electrons to jump to stationary states of negative energy. Electrons could thus possess negative energy — which is not possible! The lowest possible energy of anything is zero. There seemed no way to exclude such states and no way to know what a jump to a negative energy could mean.[47]

Ironically, at the very height of the outward success of quantum mechanics, despair grew audible in Leipzig and Zurich — to which Pauli had recently moved as Debye's successor — especially after Dirac lectured on his equation during Debye's first annual Leipzig Lecture Week in June 1928.[48] Pauli again resigned from quantum physics to await "a fundamentally new idea."[49] He spent the rest of 1928 working on the old prequantum problem of the so-called H-theorem in statistical thermodynamics and writing a utopian novel — not necessarily in that order. Heisenberg likewise turned from "the more important problems" to applications of nonrelativistic quantum mechanics, "in order not to frustrate myself continually with Dirac."[50] Fruitful uses of nonrelativistic quantum mechanics suddenly blossomed amidst the despair in Leipzig and Zurich after Dirac's depressing lecture.

Inspired by Sommerfeld's recent semiclassical electron theory of metals, the theoreticians turned to the solid state.[51] Immediately following the formulation of quantum mechanics and the invention of electron spin, Pauli and Heisenberg had earlier considered the quantum properties of ions and electrons in metallic solids. A more captivating topic, the Copenhagen interpretation, had drawn them away, but frustration and

Bloch's arrival in Leipzig brought a sudden revival of Heisenberg's earlier ideas. In his 1928 doctoral thesis — the first under Heisenberg — Felix Bloch presented a quantum-mechanical electron theory of metals.[52] Most metals are conductors of electric current, which is a stream of electrons. One puzzle involved the alacrity with which the current is produced, even in very long wires. A current flows almost immediately after a wire is plugged into a voltage, even though the free electrons should be slowed by collisions with the metal ions in their long trek down the wire.

Bloch solved the problem by considering the electric potential formed in the wire by the metal ions. If the ions are assumed to be arranged in a crystal lattice, then the potential is periodic — a kind of wave of attraction all along the wire, a negative electron being attracted to each positive ion in a periodic fashion as it flows to the end of the wire. Solving Schrödinger's equation for a quantum electron moving in such a potential, Bloch explained (as Sommerfeld could not) how electrons are able to propagate long distances through metallic conductors in short time intervals. The chance of an electron colliding with an ion in a nearly perfect crystal lattice is very low, even at room temperature. As temperature decreases, the lattice becomes more rigid, structural imperfections diminish, and the electrons travel even more freely, thus vastly increasing the conductivity, as is observed. (Superconductivity, essentially zero resistance, occurs at the very lowest temperatures and requires a different quantum explanation.) Sommerfeld's theory, which treated the electrons as a nonquantized "gas," predicted just the opposite behavior — increasing resistance as temperature decreased.

In the spring of 1928, as Sommerfeld toured the world, he sent Rudolf Peierls, one of his students, to study with Heisenberg. Heisenberg presented Peierls with a problem: to consider the effect on Bloch's theory of an insight that had proven crucial to Heisenberg's earlier explanation of helium spectra (and that had enabled Heisenberg's Nobel prize-winning work). This insight was the notion of an exchange interaction, a force generated solely by the exchanging of positions of two totally indistinguishable quantum particles.[53] Because of the exclusion principle, no two electrons can be in the same quantum state. Since, unlike classical particles, quantum electrons are identical, they can be considered to exchange places continually, which gives rise to an additional attractive exchange force between the electrons. Heisenberg suggested that Peierls examine the effects of this force on Bloch's theory of metals. Heisenberg himself, avoiding quantum electrodynamics, examined the same effect as a possible explanation of one of the most puzzling phenomena exhibited by metals—the appearance of ferromagnetism.[54]

Ferromagnetic materials, such as iron, can be magnetized when they are placed in a magnetic field, and they remain magnetized even after the field is removed. One possible explanation of this effect derived from the spinning valence electrons of the iron atoms. Any rotating charge acts like a closed current loop, and a closed current loop behaves like a little magnet. A molecular magnetic field inside the iron—the so-called Weiss field (named for Zurich physicist Pierre Weiss)—maintains the alignment of all the spinning valence electrons in the same direction, and this renders the magnetization of a piece of iron. But aligned spins among electrons, all in identical states, are excluded by the Pauli exclusion principle. Heisenberg had solved the puzzle earlier by hypothesizing that his exchange force in a model of the metal molecules proposed by Heitler and London rendered the parallel alignment of all spins in a metal the most likely arrangement.[55] But he couldn't prove it until 1928. The published derivation, which he submitted to the *Zeitschrift für Physik* in May 1928, required the application of mathematical group theory to quantum mechanics, which had been recently achieved by Eugene P. Wigner, Hermann Weyl, and others. An exchange interaction involving myriads of identical spinning electrons could not be handled without the new methods. Once applied, the widely disputed, longtime puzzle of ferromagnetism readily succumbed to the completed quantum mechanics of the Copenhagen school.[56]

Further convinced a month later by Dirac's depressing Leipzig lecture that the search for a relativistic quantum mechanics of the electron is "the saddest chapter of modern physics," Heisenberg went out of his way to emphasize in two forums the success of quantum mechanics in comprehending ferromagnetism.[57] In 1929 he delivered an address on the subject to the German Metallurgical Society, which was published in their organ *Metallwirtschaft* (*Metal Economy*).[58] A year earlier, he contributed a summary of his theory to a volume honoring the sixtieth birthday of his teacher and mentor, Arnold Sommerfeld. Pauli offered for the same celebration not his utopian novel but his latest thoughts on the classical *H*-theorem.[59] While the practical uses of quantum mechanics were obvious, advance on the "more important problems" of quantum field theory still awaited a "fundamental new idea."

Heisenberg's ingenuity provided an idea at the end of 1928. He had hit upon a mathematical trick for dealing with the difficulty caused by the commutation relations for matter and radiation. He introduced into the equations a term proportional to a constant that circumvented the nonzero commutation, then he let the constant shrink to zero in the final result.

Pauli traveled from Zurich to Leipzig in January 1929 for a district meeting of the German Physical Society. By the end of the meeting, he and Heisenberg had settled on a joint paper to be completed by the time Heisenberg departed for the United States in March.[60] The paper, entitled "On the quantum dynamics of wave fields," constituted the first full-fledged relativistic quantum field theory. The paper and its sequel later that year were the only papers that Heisenberg and Pauli, who continued to work closely together for the rest of their lives, would ever jointly publish.[61]

The coauthors treated matter and fields, particles and quanta, both abstractly and equally. A quantum field theory is analogous to a mechanical system with an infinite number of degrees of freedom. The fields can be quantized by treating the variables as noncommuting q-numbers, which are subject to quantum mechanics. Matter also forms a field and must be described by a relativistic wave equation. Applied to electrodynamics—the physics of moving electrons and dynamic electromagnetic fields—the Jordan-Wigner relativistic commutation relations had to be used for the electromagnetic field, the Dirac equation—despite its difficulties—for the electrons. In their long paper, Heisenberg and Pauli developed the foundations and the entire mathematical apparatus of relativistic quantum field theory from first principles and an analogy with classical fields. Applying the apparatus to electrodynamics, the only force-bearing atomic field then known, they showed how all the expected and required properties of matter, fields, and their interactions, including their unresolved problems, resulted from this highly abstract theory.

In the sequel, published shortly thereafter, Heisenberg and Pauli showed how Heisenberg's trick could be replaced by a requirement regarding the definition of electric and magnetic potentials known as gauge invariance—the requirement that the theory should yield the same results even if the potentials are defined in a different but mathematically equivalent fashion. Their definition, known as the Coulomb gauge, remained the only gauge considered until the 1940s. Their formulation of field theory, though extremely abstract, served as a basis for relativistic electrodynamic research for over a decade and as an exemplar for field theories to this day.

Although Heisenberg's trick removed one of the main difficulties hindering a proper field theory, the other two difficulties—Dirac's negative energies and the electron's infinite self-energy—maddeningly resurfaced even in this most sophisticated of relativistic field formulations. Ironically, the man who would become Heisenberg's American Doppelgänger, J. Robert Oppenheimer, was at that moment on a Rockefeller stipend in Pauli's Zurich institute. Three years younger than Heisenberg, the American-born Oppenheimer had emerged, like Heisenberg, from the cultured upper-middle-class elite (although his was a Jewish-American elite), he had attended the best American universities, and he had recently studied with Max Born, receiving his doctorate from Göttingen in 1927.

Working for several months with Pauli before returning to professorships in California, Oppenheimer examined quantum perturbation theory. Dirac had explored the lowest (first-order) perturbation in quantum electrodynamics, which corresponded to the emission or absorption of a single photon by an electron. Oppenheimer tried the next step, involving two photons — the emission *and* absorption of a photon by an electron. Even using Dirac's relativistic equation for the electron in the Heisenberg-Dirac formulation of field theory, the self-energy of the electron diverged again to infinity. Infinite-energy electrons were, of course, an impossibility. Oppenheimer demonstrated that the divergence destroyed even such simple yet essential calculations as the spectrum of the hydrogen atom, a single electron orbiting a proton. Self-energy caused an infinite displacement of the quantum atomic states, rendering any application of the theory to atoms impossible.[62]

As indicated, the electron's charge squeezed into an infinitesimal mass point surrounded by an infinite, continuous radiation field caused the divergence of the electron's energy to infinity. Endowing the electron with a size, a finite radius, so that it looked like a ball of mass and charge, seemed to be the only way out. Yet there still seemed to be no way to fit a finite size into the theory. Bohr, while eagerly applying complementarity to almost every dilemma of life, began to concede the possibility of a limit to quantum mechanics. In his prestigious 1930 Faraday lecture to the British Association in London, Bohr suggested an analogy with the development of quantum mechanics. In penetrating to ever smaller sizes, down to the structure of atoms, and ever farther away from the familiar world of the laboratory, the usual classical concepts found their limit at the surface of the atom, where quantum concepts come into play. So too, he argued, quantum concepts may fail at distances on the order of the size of elementary particles, which are also about the size of the nucleus of most atoms, roughly 10^{-13} centimeter.[63]

Quantum mechanics, Bohr noted, had been constructed in close correspondence with classical theory, which admits only point particles.

Perhaps there is a minimum fundamental length, about the size of the electron or proton, below which quantum mechanics no longer holds. Bohr even began to speculate in private that a new quantum revolution might be necessary to handle the nucleus and elementary particles. He wrote Dirac several months after his lecture: "I . . . believe firmly that the solution of the present troubles will not be reached without a revision of our general physical ideas still deeper than that contemplated in the present quantum mechanics."[64]

While Pauli was correcting the galleys of their joint paper on quantum field theory, his warnings to Heisenberg, then touring the United States, over the self-energy problem grew shriller. But Heisenberg had apparently already imbibed his share of American pragmatism. Even after a long discussion at the California Institute of Technology with Oppenheimer, lately returned from Europe, Heisenberg wrote Pauli: "The catastrophal interaction of the electron with itself does not concern me very much, despite your warnings. You are of course correct that this interaction makes the theory temporarily unusable; but it is indeed that already because of the Dirac jumps [to negative energies]."[65]

Heisenberg's flippancy regarding self-energy may also have arisen from an inkling of what he might do about it. Within months of returning to Leipzig late in 1929, Heisenberg presented to Bohr another piece of extraordinary ingenuity. The contrast in Leipzig between failing field theories and successful lattice theories—those of Bloch, Peierls, and Heisenberg—may have prompted Heisenberg's newest flash of genius: why not turn the whole world into one big lattice? Space itself might be construed as a honeycomb of minuscule cubic cells of the size of an elementary particle. If such cells existed, they would constitute an absolutely minimum distance, below which the size of any elementary particle, or any size whatsoever, could not shrink. Such a minimum distance would make infinitesimal particles impossible—thus rendering self-energies finite—and, as Bohr declared in London, would supply a lower limit to all theorizing using current quantum concepts.

More precisely, Heisenberg suggested that the size of these hypothetical space cells could be fixed by the smallest indeterminacy of length that is allowed by the rest energy of the largest elementary particle, the proton. If the maximum uncertainty ΔE in the energy of a proton is just its rest energy Mc^2 (M being the rest mass of the proton, c the speed of light), and the minimum time associated with its size ΔL is $\Delta t = \Delta L / c$, then by the uncertainty relation $\Delta E \cdot \Delta t \geq h/2\pi$, the fundamental minimum length ΔL must be about $h/2\pi Mc$, h being Planck's constant.

In a series of letters preceding a visit to Copenhagen at Easter 1930, Heisenberg informed Bohr, whom he thought would sympathize with such a radical attempt to introduce a lower boundary to quantum mechanics, of his plan to construct a lattice world of cubic cells of size $(h/2\pi Mc)^3$ and see what happened. At least the self-energies, the source of the "most horrible things," would be rendered finite; at most a new quantum revolution would be induced.[66]

Reconnoitering his lattice world, Heisenberg relied on the lattice theories of his pupils, Peierls and Bloch. Most recently, Peierls, having reached a dead end with exchange forces in Bloch's theory, managed at Heisenberg's suggestion to explain the anomalous, or positive, Hall effect in metals by the notion of energy bands, groups of quantum energy states in which the metal electrons move.[67]

In his pre-Easter letters to Bohr, Heisenberg applied the relativistic Klein-Gordon wave equation — an equation somewhat analogous to Dirac's equation but for integral spin particles (instead of half-integral electrons) — to a one-dimensional lattice world. The brave new world yielded results quite similar to those of Peierls's band theory.[68] Heisenberg even obtained the correct electron mass at one extreme of the energy band, but expansion of the theory to three dimensions led to "thoroughly radical changes of our present quantum theoretical concepts."[69] At the very least, the theory required violations of the hallowed conservation laws of energy, mass, and charge. Bohr, now convinced of the need for radicalism, would tolerate the loss of energy and mass conservation, if necessary, but not of charge. Nor would he support such a crass proposal as a world composed of cubic cells. After much debate in Copenhagen, Heisenberg quietly dissolved his lattice world, and the infinite self-energy problem loomed larger than ever.

Self-energy was only one of several problems. In a paper on electron self-energy that summer of 1930, Heisenberg showed that, even if the electrons travel so fast that their infinite self-energies at rest can be neglected (because of relativistic effects), Dirac's negative energy states still destroyed any attempt to treat the most basic situation — one electron moving alone through empty space, surrounded by its own electric field. Because of the infinite number of negative states, the calculated energy of that field also exploded to infinity. From this he concluded: "It is not probable that we will arrive at a solution without considerable alterations of the quantum theory of wave fields."[70] Nuclear physics and the close approach of colliding high-speed particles, both of which already displayed puzzles, were now for Heisenberg, Bohr, and their colleagues beyond the reach of any quantum theory, relativistic or not, until alterations allowed the handling of smaller distances.

Renewed frustration echoed through Heisenberg's letters home: "My work is still not going as well as I wish; nature is apparently constructed in remarkable subtlety."[71] The untimely death of his father compounded Werner's slump. In early November 1930, Heisenberg's mother called to tell him that his father was suffering from typhoid fever, the disease Werner had contracted a decade earlier.[72] August had drunk some tainted water while attending an international philology conference in Greece. He died a short while later, on November 22, 1930, two weeks short of Werner's twenty-ninth birthday.

Despite his father's obvious support of Werner's career and their camaraderie over chess games whenever Werner was at home, their relationship had remained rather distant and strained since the close of the world war. August constantly fretted over his son's career and constantly worried how far the youth-movement business might take him. Werner, for his part, though reveling at times in rebellion, had known early what he wanted in his career, and by 1930 he had already achieved most of it. He did not always appreciate his father's meddling as he made his rapid climb to the top of his profession.

In his first letter from Leipzig to his mother after the funeral in Munich, Heisenberg seemed moved less by sorrow at the tragic loss of his father than by the reminder it brought of his own mortality and waning youth. A month before his father's death, the 28-year-old had attended a youth meeting from which he came away feeling "very old."[73] Looking back on his young life in his letter, he mused, "A saying occurs to me that ends thus — 'aber ging es leuchtend nieder, leuchtet's lange noch zurück' ['but if it sinks in shining splendor, still it shines a long way back']. I believe as long as we ourselves are in this world, we must be satisfied with feeling this shining-back. . . . I remember the time when I myself was at my liveliest, you know, about ten years ago; that was also the most beautiful time of my life, such that my happiness also transferred itself to others."[74]

Each year, on the anniversary of his father's death, near his own approaching birthday, Werner offered reflections to his mother. He would comment on the course of his own boyhood and life and his father's life. During the difficult thirties, Werner felt a heightened sense of mortality, nostalgia for his bygone youth, and a twinkling frustration with his urges "to build up something definitive that will last as long as I can at all work."[75]

Thwarted in 1930 by such impossibilities as infinite-energy electrons jumping to negative energy states, the "aging" professor turned his attention to his students and to his teaching program. He also appeared before public audiences, speaking to the epistemologists in Königsberg, to his colleagues at the Saxon Academy, and to British scientists assembled at the British Association in 1931. For research, he concentrated on further applications of quantum mechanics to "less important problems," those derived mostly from experiments performed in Debye's section of the Leipzig institute—theories of the magnetization curve, the incoherent scattering of X rays by nuclei, and a so-called formula collection for cosmic-ray phenomena induced by high-energy upper-atmospheric particles.[76]

"On the other hand," he told Bohr in 1931, "I have given up concerning myself with fundamental questions, which are too difficult for me."[77] In the Christmas issue of the *Berliner Tageblatt* that year, he wrote that progress on such fundamental questions as the quantum mechanics of small distances, such as in the nucleus, would have to await further probing of that minuscule ball of matter at the center of the atom: "Whether indeed the year 1932 will lead us to such knowledge is quite doubtful."[78]

He was wrong. Only six months later, Heisenberg stunned his colleagues with the first contemporary nuclear theory—a quantum mechanics of the nucleus that laid the foundations of nuclear physics as it is still practiced. Heisenberg's theory came, as he predicted, in the wake of new empirical data. In March 1932, British physicist James Chadwick announced the discovery of the neutron. Before that, only three so-called elementary (that is, not compound) particles were known: the electron, the proton, and the photon or light quantum. Since "beta rays" emitted by the nucleus had turned out to be electrons, most nuclear theories placed electrons in the nucleus. The mass of the nucleus arose solely from the protons, the charge from the cancellation of a portion of the positive charges of the protons by the negative electrons. But that nuclear geography led to many problems. The confinement of electron rest energy within the tiny nucleus violated the uncertainty principle. Moreover, Heisenberg and Pauli had shown in their paper that quantum electrodynamics could not account for beta decay—the emission of electrons from the nucleus with the transmutation of a neutron into a proton; nor did energy and momentum appear conserved in beta decay. Nuclear physics seemed a puzzle at every turn. As usual, the central actors responded to these difficulties differently. Pauli hypothesized a hitherto undetected neutral particle emitted during beta decay to account for conservation. Bohr, on the contrary, included energy and momentum conservation among the classical concepts to be discarded within nuclei. Heisenberg displayed his characteristic pragmatism.[79]

While Heisenberg doubted the existence of electrons in nuclei for reasons of uncertainty, he would support Bohr's discarding of conserva-

tion laws if necessity demanded it. He awaited the arrival of experimental data to suggest a new point of attack, to breach, as he called it, the
barriers to a relativistic physics of small spaces and particles. Chadwick's
neutron gave Heisenberg his long-sought chance, quickly enabling his
nuclear theory—a totally unexpected departure.

Heisenberg arrived at his new theory not only by exploiting his deft
sense of pragmatism but also by displaying his special brand of physics,
what is often known as style.[80] His legendary intuition permitted him to
leap to a bold solution without stumbling over intervening steps. Furthermore, he could easily tolerate contradictions with accepted principles to achieve his result—and he could even tolerate contradictions in
his own final results.

In contrast, Bohr failed to arrive at an acceptable nuclear theory that
year, even though he was aware of Chadwick's discovery. Characteristically, Bohr was tolerant of well-considered transgressions of fundamentals, but basic principles could not be discarded without due process—
as earlier, consistency and careful deliberation offered "the only hope
of advance." It also appeared that Bohr's thoughts were elsewhere. On a
trip to Copenhagen in April, Werner mentioned some of his new ideas
to Bohr, but Bohr was now concerned with quite a different matter, a
new theory of superconductivity.[81]

Just before the neutron discovery, Bohr and Heisenberg appeared to
be in quite different states of mind. In February 1932, Bohr was deeply
immersed in completing the manuscript of an overview of the sorry state
of nuclear physics. The manuscript was to be a more carefully constructed exposition of informal remarks he had made the previous
October to a conference on nuclear physics in Rome.[82] But Bohr was
still having trouble expressing his views on the difficult subject. Bloch
volunteered to meet Bohr in Salzburg in February to help him complete
the manuscript before going off together to Heisenberg's Bavarian ski
hut, where Heisenberg and Weizsäcker awaited their arrival for a skiing
vacation.[83]

Bohr's nuclear pessimism initially infected Heisenberg, who engaged
Bohr in an incessant inquiry into the difficulties of nuclear physics.
When Bohr finally returned to Copenhagen in March, he found Chadwick's letter announcing the neutron discovery.[84] A month later,
Werner was in Copenhagen, mentioning some vague new ideas on
nuclear physics. He then fell strangely silent for two months. The
Copenhagen visit apparently stimulated Heisenberg's thoughts in a new
direction. Gradually, he came to recognize the full potential of a new
point of attack centering on the neutron. On June 20, 1932, he broke
his silence with a letter to Bohr. It was accompanied by the first

installment of what would become a classic three-part paper, "On the constitution of atomic nuclei." [85]

"The basic idea," Heisenberg told Bohr, "is to shove all fundamental difficulties onto the neutron and to do quantum mechanics in the nucleus." [86] While Bohr awaited a revolutionary new physics to handle such small dimensions as in the nucleus, Heisenberg had been poised for an attack that would enable him to advance into the nucleus by side-stepping the usual problems, such as infinities. Heisenberg saw in the neutron not the route to a new physics but a vehicle for rendering the old nonrelativistic quantum mechanics applicable to the new realm, atomic nuclei. All puzzles, inconsistencies, and contradictions that stood in the way of such a theory could be "shoved onto" the neutron, where they could await their resolution once the future physics did arrive. Rather than attempting to imagine an all-encompassing theory — as Bohr might have done — Heisenberg exploited the neutron in terms of contemporary theories, offering it up as a scapegoat that would shoulder a multitude of sins.

Among its sins, in Heisenberg's theory, was the circumstance that the neutron was at once both a fundamental particle and a composite of two fundamental particles. In the opening paragraphs, Heisenberg treated the neutron first as an "independent fundamental constituent" of nuclei, then as an ill-defined composite of a proton and an electron. Such a composite of two particles to form one was incomprehensible in any theory, including quantum mechanics. Nevertheless, beta decay, which entailed the ejection of an electron from the nucleus with the transmutation of a neutron into a proton, seemed to suggest such a structure. Moreover, beta decay itself seemed to require, as Bohr suggested, that "the conservation laws for energy and momentum are probably no longer applicable." [87] In other words, Heisenberg's neutron was simultaneously indivisible (fundamental), compound, and a contradiction of both quantum mechanics and conservation laws! [88] As with the atomic core model a decade earlier, the ends richly justified the means — enabling no less than the beginnings of a nuclear theory.

Thanking Pauli for "valuable discussions" (presumably by letter), Heisenberg developed his theory of the binding of neutrons to protons and to other neutrons to form nuclei in close analogy with the chemical bonding of atoms to form molecules. Chemical bonding had been among the greatest successes of quantum mechanics, and Heisenberg had been a central contributor with his invention of the ever adaptable exchange force — a purely quantum-mechanical force arising for the exchange of identical electrons. If a neutron could be considered somehow a composite of a proton and an electron, then the neutron-proton force could be made analogous to the chemical bonding of the ionized molecule H_2^+. This molecule consisted of a single electron orbiting two protons. The electron can be considered to belong to each of the protons in turn, rendering the proton and electron together a neutral atom. In the nuclear case, the neutron is composed somehow of a proton and an electron. It binds with a proton by sharing its electron with the proton for a brief time allowed by the uncertainty principle. In a sense, the neutron and proton play a wild game of catch with the one available electron, the proton turning into a neutron when it catches the electron, the neutron turning into a proton when it releases the electron.

By the same token, the neutron-neutron nuclear force could be treated in analogy to the neutral H_2^+ molecule — two protons tossing two electrons back and forth to each other. However, unlike the molecular exchange force, which exchanged only a spinning electron, the nuclear exchange involved electric charge — a neutron playing catch with a proton changed back and forth into a proton, its charge changing from zero to a positive amount equal to that of a proton, then back to zero. To maintain the molecular analogy, Heisenberg treated electric charge like electron spin by inventing what was later called "isospin." Zero charge (neutron) was analogous to an electron spinning in one direction (the spin vector pointing up); positive charge (proton) was analogous to an electron spinning in the opposite direction (spin vector pointing down). (There was no negatively charged nucleon, nor did Heisenberg consider an exchange-type proton-proton force.)

Given isospin, place exchange, and contradictory neutrons, the entire apparatus of nonrelativistic quantum mechanics lay at Heisenberg's disposal. With these innovations, he easily accounted for deuteron binding (a neutron bound to a proton), the stability of helium nuclei (alpha particles, or two neutrons joined with two protons), and certain aspects of radioactive decay. In his second paper — submitted while lecturing for a second time in Ann Arbor during an unbearably hot summer — Heisenberg accounted for the stability of nuclear isotopes and derived the so-called Meitner-Hupfeld effect in the scattering of gamma rays by neutrons. Part three, completed in December 1932 after he returned to Leipzig, contained further studies of stability, including the use of a nonexchange force, also in analogy to H_2^+.[89]

Heisenberg had opened the door to the entire nucleus. The neutron provided the key and quantum mechanics the royal road. During the next few years, Wigner and Majorana would embellish Heisenberg's exchange force to include spin exchange, Pauli would propose his neutrino to account for energy and momentum conservation in beta decay, and Fermi and Yukawa would offer quantum theories of beta decay and nuclear forces based not on problematic neutrons but on two new types of nonelectromagnetic quantum field theories that accounted for the so-called weak and strong interactions. Beta decay was not the simple ejection of an electron that is somehow squeezed together with a proton to form a neutron. Rather, Fermi suggested in 1934, the proton, electron, and neutrino are created by the field in the process of decay, just as a photon and electron are created in electrodynamics by a single accelerating electron. By the same token, the force between protons and neutrons is not carried by an electron bouncing back and forth but by a new type of field, the quantum of which, Yukawa suggested in 1935, is a new heavy electron, or meson.

Heisenberg's proton-neutron model of the nucleus set in motion the field of contemporary nuclear structure studies, and it stimulated the contemporary branch of quantum field theory that focused on the two nuclear fields, in addition to electromagnetic interactions. (A quantum field theory of the fourth known field, gravity, still remains an even bigger problem.) Save for a brief paper on lightweight nuclei, summary reports on his theory and its applications to the seventh Solvay Congress, held in Brussels in October 1933, and to a celebration of Zeeman's fiftieth birthday in The Hague in 1935, Heisenberg did not write again on nuclear structure physics until the outbreak of World War II, seven years later.[90] He had achieved all that he wanted in nuclear structure. Although the nucleus continued to attract Heisenberg's students and visitors throughout the next decade, the discoveries of the positive electron (positron) and cosmic-ray showers, both occurring in 1932, returned Heisenberg's attention to the "more important problem" of finding a suitable fundamental quantum field theory, first of the electromagnetic field, then of the "weak" Fermi field, and, by 1939, of the "strong" Yukawa field—a program and an undertaking in which Bohr had little interest or faith.

No correspondence between Heisenberg and Pauli survives from 1932, when Heisenberg worked most closely with Bohr on nuclear structure, but their collaboration and correspondence rapidly escalated again thereafter, largely replacing those between Bohr and Heisenberg on "more important" matters. While Bohr continued to explore nuclear physics, developing the widely used liquid drop model of heavy nuclei, Heisenberg turned to high-energy physics and to the insights it provided into the fundamentals of field theory. The higher the energy with which two particles smash into each other, the closer their centers approach each other and the more information one can obtain about the behavior of matter and fields at distances even smaller than the sizes of elementary particles.

Besides the different direction in research that Heisenberg pursued after 1932 — research that brought him away from Bohr's line of research and closer to Pauli's interests — Bohr and Heisenberg also became intensely involved in the ominous events unfolding in Germany. Both men were soon drawn into activities far removed from nuclei and positrons. Less than six weeks after Heisenberg completed his third paper on nuclear physics, Adolf Hitler attained the position of German Chancellor. On the evening when Hitler came to power, Heisenberg happened to be visiting the Weizsäcker family's home in Berlin. As Heisenberg and his student, Carl Friedrich von Weizsäcker, looked out over the darkened Berlin streets that evening, wondering aloud to each other what the future held, a torchlight parade of Hitler's brown-shirted storm troopers marched below, row on row, in celebration of the "Führer's" ominous triumph.[91]

PART THREE

The Lonely Years

CHAPTER 15

A New Regime

At the end of January 1933, Germany's president, Field Marshal Paul von Hindenburg, appointed Adolf Hitler, then chairman of the influential National Socialist German Workers Party, German chancellor and head of a new cabinet in Berlin. Many Germans were relieved. A cabinet crisis was finally resolved, and with nationalist conservatives in control. The *Leipziger Neueste Nachrichten*, a conservative Leipzig newspaper not allied with the Nazis (as the National Socialists were disparagingly called), extolled the event: "The first day of the Hitler cabinet has closed in the brightest glitter. The day was dominated by a feeling of widespread joy at the unification on the [political] right. . . . It cannot be better expressed than in Hitler's own words at his first cabinet meeting: 'Faith and trust shall not be disappointed!'"[1]

From the moment Hitler gained control of the chancellery, he and his party held the "nation of poets and thinkers" in an ever-tightening grip. Within a day, the Reichstag was dissolved; within a month, the constitution was suspended. By the summer, thousands of Jews and political opponents had lost their jobs, and many were leaving the country. The first concentration camps—intended to concentrate opponents, criminals, and others in a common prison—were already in operation. Political efforts to halt the National Socialist takeover were thwarted by the imposition of one-party rule. A year later, by the end of August 1934, Hitler had created for himself the position of national Führer; Germany's first democracy had been stamped out by a Nazi dictatorship.

The frightening rapidity and seeming ease with which Hitler and his henchmen seized the German state resulted from a combination of unique demonical genius and the particular susceptibility of the populace to demagoguery. Although politically the National Socialists gained their greatest support from the unemployed and the economically threatened lower middle class, most observers agree that they could not have taken over so rapidly and completely after January 1933 had they not received the crucial support of the army and the initial acquiescence of the upper middle class—civil servants, bureaucrats, industrialists, professors.[2]

Through manipulation, propaganda, violence, and intimidation, the National Socialists outmaneuvered and overwhelmed great blocks of the all-too-willing German people. Uninspired by the impotence of the Weimar state in foreign affairs, disillusioned by heavy political infighting that caused a perpetual cabinet crisis in Berlin, and anxious for the future as Germany and the world sank into economic depression, many Germans—and academics most particularly—longed for the return of a powerful, unified Reich, free of so-called party chaos. Even more so than before, German academics shunned overt involvement in political intrigue as antithetical and detrimental to scholarly objectivity. Politics were beneath the dignity of an academic aristocrat. Thus, for many people, openly opposing the new regime meant descending into the dirty world of politics, while openly supporting or even tacitly acquiescing to one-party rule was seen as somehow apolitical and objective—as it had been in the days of the Kaiser. Heisenberg's earlier views that politics were a "money-business" and that the Weimar democracy seemed merely a transitory phase preceding a more fundamental system suggest that he too embraced, at least in part, such confused political notions.

Such notions are demonstrated by the numerous widely publicized written declarations by academics—many of whom were not party members—in support of Hitler and his evolving dictatorship.[3] Because of German reverence for professors, the public impact of these manifestos must have been significant; some were signed by as many as 68 professors at a time. Anti-Nazi academics could have issued their own equally powerful declarations—at least before Goebbels's Propaganda Ministry seized control of the public media late in 1933. Heisenberg's and Debye's names appear on an appeal, published late in 1932, for more financial support of science, which was signed by a total of 141 academics.[4] But not a single countermanifesto appeared from them or others in opposition to Nazi measures. Older academics, such as the influential Max Planck, were still smarting from the ludicrous manifestos of World War I, and they apparently discouraged any new appeal. Political manifestos smacked of party politics, while quiet diplomacy and judicious compromise seemed more certain of protecting what really mattered to them—German science and scientists.[5] But another reason for reticence was also at work—the use of violence.

During the early days of the Weimar era, some splinter parties advanced their programs by literally beating the opposition: they unleashed gangs of armed thugs organized into private armies. Hitler's party employed its own paramilitary organization, the brown-shirted Sturmabteilung (SA), or storm troopers. Although storm-trooper street violence waned in the last years of the republic because of official proscription and an effort to give the Nazi party an appearance of decency, violence suddenly intensified after the party came to power. The left, abandoned by voters and in disarray, resorted in some cities to a desperate urban guerrilla warfare that was quickly suppressed by superior numbers of Nazi troops and by the depletion of leftist ranks through constant arrest. In Leipzig, where the majority had voted consistently for the Socialist and Communist parties,[6] newspapers were filled with reports of street battles, followed by power displays by marching brigades of Nazi troops waving swastika banners. The establishment of the Dachau concentration camp just outside Munich with 43 leftist opponents received wide publicity.[7] (It was not yet the death camp it would become.)

With the suppression of overt political opposition, the suspension of civil rights after the mysterious burning of the Reichstag building on February 27, 1933—blamed, naturally, on a communist—and the appointment of Reich commissars and governors in each state to ensure "order and security," the new regime turned more openly in the spring of 1933 to the persecution and expulsion of Jews. To the political onslaught was now added a moral affront. Nazis employed their usual tactics of force and coercion to achieve their aims and stifle opposition. April 1933 opened with a "day of boycott" of Jewish stores. Then followed the enactment of the infamous anti-Semitic laws for the "restitution" of the civil service (April 7) and against "overcrowding" of German schools and universities (April 25).[8] The bloodletting had begun.

During the day of boycott, which actually took place over a four-day period, brown-shirted storm troopers harassed Jewish store owners and their customers and rampaged through laboratories, libraries, classrooms, and courtrooms, forcibly expelling Jews from their work. Not all Germans sympathized with the action.[9] Heisenberg's associate, the upright nationalist mayor of Leipzig, Dr. Carl Goerdeler, defended a Jewish shopkeeper from storm-trooper harassment.[10] Critics of the regime tended to keep silent about these and other dictatorial measures for fear of provoking more, but again there were exceptions. One particularly vociferous exception was Albert Einstein.

On a visiting professorship to the United States when Hitler came to power, Einstein made known his decision not to return to Germany and declared in an interview: "As long as I have any choice in the matter, I shall live only in a country where civil liberty, tolerance, and equality before the law prevail. . . . These conditions do not exist in Germany at the present time. Men, among them leading artists, who have made a particularly great contribution to the cause of international understanding are being persecuted there."[11]

Shortly thereafter, Einstein resigned his nonteaching professorship at the Prussian Academy of Sciences and called on all democratic nations to unite in opposing National Socialist Germany.[12] His appeal was in vain. Nazis responded by accusing Einstein of participating in a foreign horror crusade against Germany and raided his summer home on the pretext of seeking terrorist weapons. Master propagandist Joseph Goebbels declared that the boycott and harassment of Jews would continue until such anti-Nazi "propaganda" ceased.[13] Physicists Max Planck and Max von Laue, attempting at the time to ameliorate the situation through quiet diplomacy, informed their colleague that he was making matters difficult by violating the unspoken rule against what Laue regarded at the time as political involvement: "Here they are making nearly the entirety of German academics responsible when you do something political."[14]

Einstein of course did not keep silent. His response to Laue is refreshing for its candor: "I do not share your view that the scientist should observe silence in political matters, i.e., human affairs in the broader sense. . . . Does not such restraint signify a lack of responsibility? Where would we be had men like Giordano Bruno, Spinoza, Voltaire, and Humboldt thought and behaved in such a fashion? I do not regret one word of what I have said and am of the belief that my actions have served mankind."[15]

Despite Einstein's stature, the Prussian Academy announced on April 1, while Max Planck was away on his annual Sicilian holiday, that it had no reason to regret Einstein's resignation.[16] Refusing appeals to terminate his vacation, Planck returned at the end of April to an appalling situation — friends and colleagues removed from their jobs and preparing to leave the country. Worried about posterity's judgment of the academy's handling of Einstein, Planck inserted a positive statement on Einstein in the academy record when he became its secretary a month later. But by its action, the most prestigious of German academic institutions had indirectly endorsed the expulsion of Jews from academic positions.[17]

The law for the "restitution" of the professional civil service followed on the heels of the boycott. It confronted academics even more directly with the grim issues at stake — teachers and professors, as civil servants, were now personally affected. Paragraph 3 of the new law stated: "Civil servants who are not of Aryan ancestry are to be placed in retirement." The only exceptions, in deference to Field Marshal Hindenburg, were civil servants who had served at the front in World War I or had become civil servants before August 1, 1914. Paragraph 4, which allowed for no exceptions, stated: "Civil servants who by their previous political activity do not offer the guarantee that they will stand up at any time for the national state without reservation can be dismissed from service."[18]

The Reich commissar and cultural ministry in each state decided who should be dismissed and passed the decision along to the appropriate faculty for prosecution. At first there was some confusion regarding the enforcement of the new law and a faint hope that it would not be too harsh. German physicist Hans Kopfermann, asked by Niels Bohr to report on the situation in Germany, wrote in May 1933 that state bureaucrats were defining "non-Aryan" rather narrowly. Many dismissals were actually temporary leaves of absence until the authorities could decide each case individually. Although the party had unleashed its Nazi students in a campaign against the "non-German spirit," there was at first only little support for the violent crusade at most universities.[19] The ambiguous status of the affected faculty became clearer in the fall of 1933 — but it also worsened. The application of anti-Semitic laws grew ever more oppressive until the end of the Kampfphase (phase of struggle) in 1935.

Like the Leipzig mayor and other nationalist-oriented non-Jewish German academics, Heisenberg was at first appalled at the crudity of the new leaders and the "excesses" of their new regime, but he greatly sympathized with the long-term national revival promised by the National Socialists. "Much that is good is now also being tried," he wrote as late as October 1933, "and one should recognize good intentions."[20] He and others expected that the regime, like its immediate predecessors, would hardly last out the year. An urgent political response, had they with their "apolitical" attitudes even considered one, seemed to them unnecessary. Because of this, he and most other non-Jewish academics still in Germany did not make overt protestations like Einstein's. Nor did the obvious moral issue of anti-Semitism move Heisenberg and his colleagues to action. Although some no doubt sympathized with anti-Semitic policies, there is no direct indication that Heisenberg did so in the scientific sphere. Rather, as noted earlier, he and his colleagues tended to regard anti-Semitism as a mere political issue — and thus an issue to be avoided entirely. The benefits brought by official anti-Semitism did not encourage them to do otherwise. Nevertheless, the plights of their closest Jewish colleagues and the damage to the German physics profession demanded a response.

Many Jewish academics, reading the signs, had left Germany as soon as the law was enacted. Nobel prize-winning scientists Fritz Haber and James Franck resigned outright, refusing to take advantage of the "leniency" the law provided for front veterans. Max Born, barred from the classroom and placed on permanent leave, left for his summer home in northern Italy, intending never to return. In June 1933, he informed Einstein and Heisenberg that he did not want his children to live as second-class citizens in the country of their birth. Nor could he excuse the parting anti-Semitic accusations and epithets hurled at him in Göttingen.[21] Göttingen, one of the world's premiere science centers, the town that had nurtured Heisenberg to the top of his profession and had witnessed some of his greatest achievements, was soon stripped of many of its world leaders in science.[22] In this crisis, uncertain and confused, Heisenberg turned to Max Planck for counsel and advice.

The match seems unlikely. Planck was then 75 years of age, Heisenberg 31. Planck, though a founder of quantum theory, could hardly be called a radical thinker. He had opposed the Copenhagen interpretation so dear to Heisenberg. Like Einstein, he had called for the revival of determinism and objectivity in atomic physics. Heisenberg lived in a world that glorified youth and rejected what was seen as the ossified ideas of the older generation. He embraced the radical elements of Copenhagen physics and excelled in a profession that richly rewarded new and successful ideas. At the same time, Heisenberg's world hated hypocrisy and revered integrity and devotion to duty above all else.

Planck, it was universally agreed, represented the epitome of ethical fortitude and consistency and was considered a paragon untainted even by the usual human frailties. As Prussian Academy secretary and president of the prestigious Kaiser Wilhelm Society, Planck, a wiry-looking physicist kept spry by his long devotion to strenuous mountain wandering tours, had presided for years as unofficial dean of Heisenberg's profession. He had chaired committees that provided Heisenberg's postdoctoral grants and had recently ameliorated his criticism of Copenhagen physics. Heisenberg's support of Planck's quiet efforts to protect their profession in the face of the worsening political situation was evident to readers of Heisenberg's reviews of Planck's collected essays in this period.[23]

The April dismissals, accompanied by the uproar over James Franck's resignation in protest, prompted an unsettled Werner to seek out Planck even before Planck's return to Berlin from his Sicilian holiday.[24] Throughout the remainder of 1933 and into 1934, Heisenberg would meet and correspond often with Planck and occasionally with Planck's Berlin colleague Max von Laue, the courageous anti-Nazi president of the German Physical Society, as they worked together to defend and protect their profession against regime policies. Broadscale opposition to the regime in general was not a consideration.

Since the main professional concerns in the spring of 1933 were the dismissal of leading Jewish physicists from teaching positions and their emigration from Germany, Heisenberg and his costrategists concentrated on two goals: persuading the dismissed physicists to remain in Germany and working behind the scenes to rescind the dismissals. Evaluating such a strategy — and the situation in which the strategists found themselves when it ultimately failed — requires the advantages of historical hindsight. At that time, the Western world had no experience with a regime such as Hitler's and no conception that such a regime could lead its people inexorably into the evil nightmare that the Nazi dictatorship became. Nor did German academics have much experience with moral action or political commitment.

With the regime in control of the legal apparatus and the SA in control of the streets, perhaps the most effective opposition Heisenberg and other academics could have launched in those early months would have been the political mobilization of the middle and upper middle classes against the regime and its policies. But given the initial popularity of the regime, opposition would have been possible only if it had been mobilized well before 1933. That would have required a political sensitivity and commitment to democracy that did not exist. Individual professors who did attempt to spark a democratic opposition were regarded as examples of the futility of individual protest — precisely because they were alone. A year earlier, Gerhard Kessler, a non-Jewish professor of economics in Leipzig, had sought to reach students and the public with pamphlets and lectures against National Socialist doctrine. His efforts were rewarded with mass demonstrations against him and the violent disruption of his lectures by rampaging students.[25] He was dismissed from his post in April 1933, was arrested by the Gestapo in July, and emigrated soon thereafter to the United States under the threat of ending up in a concentration camp. The several other attempts in Leipzig to speak out against the Nazis were equally silenced.[26]

It could be argued that a mobilization of mass opposition could have been achieved in early 1933 had more non-Jewish academics followed the examples of Einstein, Franck, and Haber: that is, had there been a simultaneous resignation of professors in moral indignation at the dismissal of their colleagues and the treatment of Jews in general. With their international reputations and without fear of being victimized by anti-Semitism abroad, they could easily have found jobs elsewhere. The upright Otto Hahn, in fact, suggested just that to Max Planck. Such blatant persecution and antidemocratic measures certainly warranted the action. Planck, however, refused; his view of politics led him to believe that their protest would go unreported and that their jobs would be filled by unworthy riffraff—an even greater damage to their profession in their beloved homeland.[27] A noted Leipzig geophysicist made the same argument to an American visitor: many lower ranking academics had joined the Nazi party and were only waiting for an opportunity "to force dismissals so they can get jobs."[28]

Actions often bespeak self-image. Since the new regime had come to power and was pursuing its policies under the guise of absolute legality, and since professors did not recognize the utter contempt in which Hitler and his cronies held intellectuals, they must have seen themselves and their situation not much differently than they had before 1933. Heisenberg, Planck, and other elite German physics professors were conscious of their positions as the prestigious creators of quantum physics, heads of major scientific institutions, and leading representatives of German culture. They felt that their importance to their nation transcended their personal preferences. They felt a special responsibility to defend their profession—and by extension their culture—against the intrusive excesses of a regime whose overall goals seemed otherwise admirable, even when defense required such personal sacrifices as engaging in political intrigue. For them, resigning in protest did not square with that responsibility. Instead, during the first year of Hitler's reign, Planck, Laue, and now Heisenberg responded as German professors had done for decades, relying on their high social standing and personal diplomatic skills to influence the new leaders in the right direction.

One retrospectively bizarre expression of the German academic position was the upright Max Planck's audience with the new chancellor, Adolf Hitler, just two weeks before Heisenberg's first meeting with Planck. Planck, as president of the Kaiser Wilhelm Society, a network of government-sponsored research institutes, had returned to Berlin from his Sicilian holiday on April 28 to find seventy-fifth birthday greetings from Hitler. He also learned that many of his closest Jewish colleagues had lost their positions and that the society was being forced to impose the new laws on Fritz Haber's Kaiser Wilhelm Institute for Physical Chemistry. Haber having resigned in protest and having left the country, Planck used his thank-you to Hitler and his status as society president to approach the new chancellor, hoping to convince Hitler that his policies needed correction.

According to chancellery records, Planck paid his visit on May 16, 1933.[29] Hitler, still dressed in the respectable suit and tie of the German politician, received the balding and moustached physicist in the chancellor's office. Reports differ about what actually transpired. After the war, Planck recalled that he had argued that Jewish scientists could be good Germans too and that some should be spared for the sake of German science. Haber, for instance, had shown his dedication to Germany by introducing gas warfare during World War I. Hitler responded that he had nothing against Jews — it was the communists he was against — and then flew into such a rage that Planck could do nothing but leave. Einstein heard in 1934 that Hitler had threatened the old man with imprisonment in a concentration camp.[30] But Heisenberg, who visited Planck at his home in the elegant suburb of Berlin-Grunewald at the end of May, transmitted only a positive report.[31]

In a June 1933 letter to Max Born soon after Born had left Göttingen for northern Italy, Heisenberg told of a pledge that Planck had supposedly received from Hitler: "Planck has spoken — I think I can pass this on to you — with the head of the government and obtained the assurance that nothing will be undertaken beyond the new civil service law that will impede our science."[32]

Perhaps Planck had not told Heisenberg everything that had happened at the meeting; perhaps Heisenberg did not convey everything to Born. Whatever actually transpired at the Hitler meeting, Planck, his Berlin colleague Max von Laue, and now Heisenberg seemed far from discouraged. They began intensive efforts working through bureaucratic channels to prevent, delay, and cancel job dismissal orders.[33] Optimism and faith in reasoned diplomacy abounded. Kopfermann recounted the mood and strategy to Bohr: "Laue, for example, who is probably the most optimistic, is attempting to delay all decisions for as long as possible [to provide time to argue against dismissals]. In this way he hopes to be able to save a large portion of those threatened."[34]

Although appeals by scientists and academics to public opinion remained unthinkable — both too crass and too confrontational — private petitions to state bureaucrats multiplied. In June 1933, Planck and Heisenberg circulated a petition in support of the Göttingen mathematician Richard Courant, who, unlike Franck and Born, had decided to stay and fight his dismissal.[35] The same month, Heisenberg issued a personal appeal to Born to return to Göttingen to await improved conditions: "Since only a very few will be affected by the law—you and Franck certainly not, Courant probably not either—the political transformation could take place by itself without any sort of damage to Göttingen physics. . . . Certainly in the course of time the ugly will separate itself from the beautiful. . . . Therefore, I would like to ask you not to make any decisions yet, but to wait and see how our country looks in the fall."[36] The few could be sacrificed for Göttingen and the hope of a better future. Born sent a typed transcript of this portion of the letter to Ehrenfest as an example of the position of "our well-meaning German colleagues."[37]

Born found it difficult to respond to Heisenberg. He seemed genuinely touched by Planck's and Heisenberg's efforts on his behalf; he was willing, he said, to postpone until autumn his decision to leave Germany for good. He was aware that resettling in a foreign country would be difficult for him and his wife, given their age. But as one who had been unjustly placed on forced leave from his tenured position, he felt a responsibility to support others who found themselves in a similar situation. He felt especially allied with Franck, who had publicly resigned to protest the pernicious new law. Perhaps Heisenberg could also understand how hurt Born felt at being rejected by his own countrymen because of his religious heritage and how worried he was about his children's future: "What shall I do now? I will see if I can delay my decision."[38]

Heisenberg showed Born's letter to Planck, who expressed relief at Born's willingness to postpone his decision until autumn. "Perhaps then the situation will be more reasonable," Planck declared.[39] Encouraged by the effect that the Courant petition seemed to be having, Heisenberg initiated a petition in support of Born, which Arnold Eucken circulated in Göttingen.[40] Heisenberg also visited former Leipzig professors in the Prussian Culture Ministry in Berlin at least twice in support of his Göttingen colleagues.[41]

But Born could not wait after all. Pauli, in Zurich, headed one of the international refugee organizations that had sprung up to facilitate the emigration of German scientists to positions abroad. These organizations quickly developed into sophisticated conduits to freedom for persecuted scientists and science students. With generous foundation support, they maintained files on the plights of threatened individuals and provided stipends, at first primarily to British universities, to create three-year positions. At the end of the period, the committees helped the universities create more permanent arrangements. Soon, however, Britain and other European countries could not or would not absorb any more refugee scientists, many of whom moved (though also with difficulty) to the United States.[42]

Born had earlier worked in England, and through Pauli's emergency committee, in early July 1933 he accepted a three-year appointment in Cambridge. Born originally intended to remain on leave from his Göttingen post, still hoping to return one day.[43] But after his oldest children, left behind in Göttingen, experienced further indignities, he gave up that dream. On the eve of his departure for Cambridge, Born resigned. "One cannot serve a state that treats one as a second-class citizen and that treats children even worse," Born told Sommerfeld. Planck, who was vacationing at a nearby Italian villa, was, as might be expected, "very depressed."[44]

The autumn of 1933 did not bring the hoped-for return of reason — the amelioration of Nazi policies, if not the collapse of the Hitler government — but it did bring news of interest to Heisenberg: he was to receive the prestigious Max Planck Medal at the September meeting of the German Physical Society in Würzburg. Heisenberg chose not to attend the meeting and went instead to a concurrent Copenhagen physics conference, held annually at Bohr's institute.[45] Perhaps he wanted to reaffirm the Copenhagen connection for refugee physicists. During his travels to and from Copenhagen in September 1933, Heisenberg stopped in Berlin for daylong consultations with Planck and Laue on the state of their profession.[46] Laue, then head of the German Physical Society, had for months been gathering lists of dismissed physicists of all ranks and passing them to Bohr, who was a member of and fund-raiser for a refugee organization and was preparing to establish a Danish committee.[47] Heisenberg acted as an occasional conduit of information to Bohr.[48]

Planck and Laue were also fighting hard to ward off new attacks on their profession. A decree had been issued revoking the habilitation of non-Aryans. At the same time, the Nobel prize-winning Nazi physicist Johannes Stark attempted to gain control of the German Physical Society and the Notgemeinschaft and to join the powerful Prussian Academy as the new Führer of German physics. Laue publicly and courageously opposed Stark at every turn. Although he succeeded in thwarting many of Stark's ambitions, he soon found himself defending the very mention of Einstein's name and the teaching of relativity theory in Germany. His actions prompted praise from Einstein and fears for his safety.[49]

Planck, soon to display the courage of his convictions by holding a forbidden public funeral service for Fritz Haber, who died of heart disease in exile, submitted a memo to Heisenberg's former Leipzig colleagues at the Prussian Culture Ministry on behalf of non-Aryan habilitants in general and Lise Meitner in particular.[50] Although women scientists had been excluded from German laboratories until after World War I—they would distract the men from their more important work, it was thought—Meitner, an Austrian-Jewish physicist, had worked for decades under Otto Hahn's protection in the Kaiser Wilhelm Institute for Chemistry in Berlin. With Planck's support, she had become one of the first female physicists to habilitate in Germany and even received Einstein's adulation as "our Madame Curie." Now, Planck's renewed efforts on her behalf failed. Meitner lost her habilitation and was banned from her classroom, isolated in Hahn's semiprivate laboratories until she was forced to flee when Nazi Germany annexed Austria in 1938, just weeks before the most momentous discovery to come out of Hahn's laboratory—the discovery of nuclear fission. To formalize policy, a Reich statute, first promulgated in 1934, restricted habilitation—hence future professors—to Aryans, and then, in a 1938 revision, only to those who would prove their political reliability by attending an indoctrination camp for university lecturers.[51]

Besides working to combat these affronts, Heisenberg, Planck, and Laue had to confront a new challenge to their professional strategy of holding dismissed colleagues in Germany while attempting to have the dismissal rescinded: the possibility that some professors would break ranks and resign of their own accord, even without threats from above. Their fears were realized when non-Jews Hermann Weyl in Göttingen and Erwin Schrödinger in Berlin left their posts in early September, citing health and working conditions. Heisenberg, still preoccupied with his profession and believing that others should show similar dedication, was especially angry with Schrödinger, "since he was neither Jewish nor otherwise endangered."[52] A shaken Max Planck wrote to Max von Laue with a stiff upper lip: "I regard Schrödinger's resignation as a new deep wound to our Berlin physics, which we must endure with all of the energy available to us."[53]

On his return from Copenhagen in early October 1933, Heisenberg met with Planck and Laue to discuss the situation. Both of the older men despaired of their ability to halt the exodus of physicists, including

those who were apparently not even in danger.[54] They tried on Schrödinger their previous strategy, even though it had failed in every case—
buying time by turning the resignation into a temporary leave, then
making a strong personal appeal to the individual to return to his post
while maneuvering with bureaucrats behind the scenes. Heisenberg took
the lead. He followed up successful visits to the former Leipzig professors in the Prussian ministry with a long exchange with Schrödinger,
which he again shared with his mentor, Planck.[55] But again the strategy
failed. Schrödinger accepted a temporary position at Oxford, arranged
through a refugee organization, then held a position in Graz, Austria,
until the German army invaded in 1938, and ended up in Dublin,
Ireland, where, as with Einstein in Princeton, an institute for advanced
study was built around him.

Having failed in their efforts to keep their colleagues in Germany,
Planck, Laue, and Heisenberg turned to what seemed to them a positive
response under the circumstances—finding worthy replacements to fill
teaching chairs left empty by dismissed or departing scientists. Heisenberg again played a leading role. While Planck attempted to have Max
von Laue appointed to Einstein's former position at the Prussian Academy, Heisenberg made inquiries of non-Jewish foreign physicists, especially among the Dutch, concerning their interest in a German chair.[56]
Planck considered Heisenberg himself a candidate for a key position.
Heisenberg thought that Planck would have him replace his archrival,
Schrödinger, in Berlin. "Among the blind, the one-eyed man is king!" he
exclaimed.[57] But the decimation of Göttingen physics and mathematics
and the ultimate failure of the petition on behalf of Courant (who, after
a brief interlude in Cambridge, went to New York University) apparently convinced the dean of German physics that Heisenberg should fill
Born's now vacant position.[58] The optimism of the previous spring gave
way to grim determination. Recalling the height of Göttingen physics
only a decade earlier in a letter to Franck (then in Copenhagen) in early
1934, Heisenberg declared: "I fear that a long time will pass before such
a time of scientific enthusiasm will be possible once again in Germany.
But I want to hold out here. That I will do everything in my power for
our Göttingen, you may be sure."[59]

The advantages of historical hindsight are again apparent. Although
the exodus of leading scientists enhanced the influence of those who
remained behind, there is no reason to doubt at this point that Heisenberg's conscious intentions and those of his advisors were less than
upstanding. Planck's sense of duty and ingrained commitment to upright
behavior would not permit otherwise.[60] Heisenberg, Planck, and Laue,

after all, regarded themselves as representing German's highest cultural ideals. Rather, it is in their focus on profession instead of individuals and instead of broader issues, and in their failure to take a moral stand (if not a political one), that Heisenberg and his advisors went astray. Their new response to regime anti-Semitism—filling vacated positions —may appear reasonable when viewed in terms of preserving German physics. Yet there is no indication that they ever reflected on a broader implication of this tactic, that the preservation of decent science under the Nazi regime would support the arguments that National Socialism was not so bad after all and that it was not fundamentally incompatible with the ideals of scientific inquiry. Nor did they appear to consider that the preservation of decent physics in such places as Göttingen might play into the hands of those who said that the Jewish professors were not needed anyway. Most disturbing is the ethical implication that was ignored: to participate in finding a replacement for a man who had been unethically dismissed or who had resigned in protest could be seen as tacit acceptance of the grounds for dismissal and denial of the legitimacy of the protest. The Nazi regime confronted Germans with most difficult moral and political decisions for which even the most upright among them were thoroughly unprepared.

Heisenberg traveled to Göttingen in early 1934 in pursuit of the new policy. After delivering a lecture entitled "Atomic theory and knowledge of nature"—scheduled for a time when some younger faculty and students did not have storm-trooper drill—Heisenberg met in the home of the head of the Universitätsbund (University League) with local representatives of industry, science, and the Notgemeinschaft (Emergency Association for German Scholarship).[61] In view of the "crisis situation which has surprisingly developed in the mathematical-physical fields at the University of Göttingen," they decided to revive an old academic-industrial alliance set up at the turn of the century by Felix Klein for the support of mathematics and physics and, in addition, to offer Heisenberg an immediate appointment as Born's successor.[62]

Although Heisenberg's colleague Debye did not believe that he would leave Leipzig,[63] Heisenberg readily accepted the position. But his appointment at Göttingen first required approval by higher authorities—the local Nazi University Teachers League and the new Reich Education Ministry (REM). Both caused trouble. Nazi opponents of the plan had little use for modern theoretical physics. They were alerted when, at Heisenberg's suggestion, Sommerfeld's recent student Fritz Sauter applied for and received faculty approval to serve as Heisenberg's assistant in Göttingen.[64] After bureaucratic maneuvers and the recovery of (deliberately?) mislaid files, the REM finally approved Sauter's appointment. He began lecturing on theoretical physics in the summer of 1934, the first course on the subject given at Göttingen in nearly a year. "The only thing missing is you as institute chief," he wrote to Heisenberg.[65]

But such great opposition to the appointment of a leading theorist had arisen in the wake of Sauter's approval that the REM, preferring to avoid a confrontation, simply refused to act further. Robert Pohl, the only remaining Göttingen physics professor, reported that he, the dean, and even the rector had pressured the REM, but to no avail: "I told you, the difficulties lie in Berlin."[66] Even Heisenberg's invitations to prestigious professorships in the United States could not be turned to his favor.

A year later, a frustrated Heisenberg finally agreed to drop the Göttingen offer, which was still pending. But by then there was plenty of consolation and another opportunity to occupy a key position: Sommerfeld had officially selected Heisenberg as his successor in Munich. At the same time, ominous signs were appearing on the horizon. Heisenberg's calls to Göttingen and Munich so incensed Nazi physicists that they mounted an increasingly nasty, increasingly public campaign against theoretical physics that temporarily prevented Heisenberg's call to any German physics chair.[67] The Nazis' transition from surreptitious meddling with science education to public denunciation of theoretical physics marked the end of the Planck-Laue-Heisenberg replacement strategy and the beginning of a new phase of the Nazi challenge to academic physics.[68]

But at the end of 1933, the replacement strategy seemed a good one, and Planck was extraordinarily pleased at the dedication shown by his younger colleague.[69] During the plenary session of the German Physical Society meeting on November 3, Planck conferred on his younger colleague the highest award a German physicist could receive, the Max Planck Medal. The aged Planck had knighted a new member of the German scientific elite. Less than a week later—just a month short of his thirty-second birthday—Heisenberg received a telegram from the Royal Swedish Academy. He had been awarded the Nobel prize for physics.

CHAPTER 16

Herr Professor

People fortunate enough not to have lived under a dictatorship such as Hitler's would find it difficult to imagine life in Nazi Germany. Nazi control extended not only horizontally — over social, political, and legal institutions — but vertically, from professional and regional strata down to the daily lives of individual citizens. It was never static. Widespread sympathy with national revival and weak response to repressive Nazi measures encouraged tighter control, which then created an even deeper sense of helplessness. Because of this vicious circle, the social and political horizons of the individual German began to shrink from the global expanse of nation and profession to the private sphere of one's closest friends and associates. This privatization of public political experience entailed what Hannah Arendt has called the paradoxical "atomization" of the new "mass society" into isolated individuals, existing in a world of twisted reality, inverted ethics, and continual fear. For us of the post-Nazi and post-Stalin era, it may be easy to recognize where such a world will ultimately lead. But by those who were a part of this first modern Western encounter with what would culminate in the incarnate evil of the death camps, the potentialities of unchecked demagoguery and character defamation could not be imagined during the early years of that regime.

Heisenberg's experiences and reactions were typical of those of most educated Germans. Like most, he was at first sympathetic toward the regime's nationalistic aims, if not all its individual policies.[1] At the same time, his letters to his mother display a narrowing sphere as his professional efforts were thwarted and his personal life was subjected to scrutiny and control. In 1933 and 1934, he maneuvered to counter the effects of the dismissal policy — the "cleansing" of the universities — on the German physics profession, but by 1938 he referred to a political assault against the whole of modern physics and equally against himself, a leading representative, as "our small, private political problems."[2]

State intrusion into private life intensified with the dismissal policy, but again without cogent resistance. In June 1933, Heisenberg, like other public servants, succumbed to the indignity of submitting his parents' birth and marriage certificates to state authorities to determine ethnic origin. "I *must* turn them in," he lamented to his mother.[3] Travel abroad now required state approval; spies operated in lecture halls and laboratories; party members informed on friends and associates; attendance was required at official ceremonies and marches and occasionally at a political indoctrination camp. Heisenberg did not openly resist such requirements. He nonchalantly informed Mrs. Bohr of his required attendance at a weekend indoctrination camp in February 1935.[4] A month later, universal military service was introduced for all men to age 45, again without objection.[5] A year earlier, Werner had already planned to volunteer for an "army sports camp," in order "to acquaint [myself] a little more with this politics."[6]

Psychologically, the most devastating form of control was the solemn oath of personal allegiance to Hitler required of all civil servants and soldiers. The pledge, instituted by law in August 1934, was a final step in the imposition of dictatorial rule: control over the personal ethics of civilians and soldiers alike.[7] Such a matter was not taken lightly, even though many who swore the oath found they could not abide by it. According to university records, Heisenberg finally signed his civil-servant oath by January 1935.[8] Later, as an army reservist, he probably also swore the soldier's allegiance to lay down his life for Führer and fatherland.[9]

Like most other Germans in his situation, Heisenberg apparently rationalized his compliance with the more obvious measures of control by regarding them as the least compromise required to achieve his higher aims. Yet compliance, for whatever reason, did not hinder the even more devastating measures that inevitably followed, confusing as well as stunning most Germans. Heisenberg's future wife, then a new and unmarried gymnasium graduate, later wrote of the mood in 1933: "Most people were insecure and frightened and were not really sure what things were leading to, or what to believe."[10] Young people — many of whom were in the vanguard of the Nazi movement — looked to their elders for guidance in such matters and often found them equally bewildered. Many turned to those who did offer answers—National Socialist demagogues.

Heisenberg's unit of the youth movement, the Neupfadfinder, since 1926 members of the united Deutsche Freischaar, had dreamed for over a decade of an approaching third Reich, headed by a saviorlike Führer who would lead Germany out of Weimar decadence, greedy capitalism, and national disgrace. By the late 1920s Hitler was winning over a new generation of students and youth-movement members with similar phrases as the National Socialist Party experienced a steady upsurge in support. Heisenberg's generation of youth-movement Altmannen (alumni) warned against such seductions.

At a 1931 Easter camp of the Freischaar Jungmannschaft (young men's group), one of Heisenberg's comrades presented a fireside lecture, "The national movement and our political situation," in which he tried to make the distinction between the youth movement's apolitical ideals and Hitler's empty political phrases.[11] Receiving such apolitical encouragement and holding such ominously idealistic notions as Reich and Führer, it is little wonder that when independent youth groups were outlawed in 1934, nearly all younger Neupfadfinder members joined the Hitler Youth, while older members joined the SA (Sturmabteilung) or SS (Schutzstaffel). Most founding elders, the Altmannen, quietly retired. Altmann Werner met with his aging group one last time that summer before dissolving their formal assembly forever.[12]

The bewilderment of the older generation is evident in the diversity of their modes of personal response. Werner's future wife, Elisabeth, recalled that many "now felt called upon to add themselves to the equation by joining either the party or the SA."[13] Party records and postwar affidavits indicate that, like many middle-class professionals in those years, a number of individuals close to Heisenberg chose this route.[14] Some joined the party. Several key associates in Heisenberg's professors' club also became party members, as did Heisenberg's close colleague and friend, Pascual Jordan. Some family friends joined for fear of losing their jobs. One colleague of Heisenberg's father found himself a member of the SA when, in 1934, the SA absorbed the Stahlhelm, a paramilitary veterans' organization to which he belonged. He quit six months later but still had to face postwar denazification.

Despite much that has been written by and about Heisenberg during those difficult years, his private position remains difficult to reconstruct in detail. The written record, the principal stuff of history, loses its reliability when viewed against the backdrop of dictatorial politics. In contemporary accounts, critical statements may be absent because they could have been used against their author. Positive remarks may have been made to win the favor of a bureaucrat. Third-party reports must be read through the opportunist lens of the informant. Postwar accounts of events may be clouded by the terrible suffering brought on by the regime and the war. Actions may speak louder than words, but only when they are reported in their own time and viewed within their own context.

While captured documents show that Heisenberg never joined the Nazi party, he did hold high scientific positions throughout the Third Reich and performed required duties without complaint. He never openly broke with the regime and continued to pursue his science. Still, as indicated previously, on a personal level he disapproved of what he regarded as Nazi excesses and defended his profession as conditions changed. In retrospect, many of Heisenberg's responses are frustratingly weak, insensitive, even repugnant. Of course, there are limits to what any individual or group of individuals can do and can be expected to do under such circumstances. Not everyone is a hero or a martyr. Once recognizing those limits, however, why didn't Heisenberg at least offer a protest or eventually emigrate in disgust at the new regime? How did his own responses appear to him at that time?

As with many Germans of that period, whether directly persecuted or not, Heisenberg's lifelong attachment to his homeland could not be broken except under the most extreme circumstances. His entire life and career were bound to the culture and people of Germany. After the war, Heisenberg attempted several times to describe the situation as he and others perceived it. The most extensive of his statements, of which the others seem derivative, is an unpublished manuscript dated November 12, 1947.[15] It was written apparently in support of Ernst von Weizsäcker, accused and convicted at Nuremberg in the so-called foreign ministries case. The manuscript displays the expected surfeit of postwar apologetics and special pleading.[16] Its purpose was, after all, to exonerate the (directly and indirectly) accused. Moreover, much of what Heisenberg wrote applies to the war years, for which Weizsäcker had been brought to trial and during which home-front dictatorial control was at its zenith. Nevertheless, a sense of the world in which Heisenberg lived, a vestige of the difficulties he endured even before the war, seems to rise through the special pleading and overrationalizations.

An intensifying dilemma characterized the situation of those who, like Heisenberg, sympathized with the nationalistic aims of the regime but did not condone the repressive assaults against their profession and their friends. With most Germans supporting Hitler, those non-Jewish Germans who opposed Nazi measures were forced, wrote Heisenberg rather simplistically, to choose between two alternatives: passive or active opposition. For Heisenberg, passive opposition meant emigration, either literal emigration "to enjoy security from persecution in a foreign country" or "inner emigration," withdrawal "from all responsibility." As the wording suggests, passive opposition was, for Heisenberg, tantamount to desertion. After the war, he told one critic: "I have never been able to have the least sympathy for those people who withdraw from all responsibility and then tell you during an innocuous table conversation: 'Well, you see that Germany and Europe are going to pieces; I always told you so!'" [17]

In Heisenberg's reconstruction, active opposition offered only black and white alternatives. One side entailed direct opposition with armed resistance and public protest. This response, which Leipzig socialists and the coup d'état conspirators of 1944 had chosen, was for Heisenberg admirable, but at best it was hopeless and at worst irresponsible. You would only end on the gallows or in a concentration camp, he reasoned, and your sacrifice would go unreported and unknown. Consider the tragic example of the White Rose student group whose leaders were all executed during the war for the heinous crime of distributing antiwar leaflets at the University of Munich. Nevertheless, Heisenberg's argument exaggerates the extent of Nazi control over the news media and daily life during the early years of the regime. When pressed on this point, Heisenberg's Leipzig colleague, Friedrich Hund, simply shrugged in exasperation: "But what could a physicist do?" [18]

Heisenberg wrote that Ernst von Weizsäcker and others like him, including himself, followed another form of active opposition — the acquisition of a certain amount of influence that from the outside appeared to be collaboration. "It is important," he wrote, "to be clear that this was actually the only way really to change anything."

From the vantage point of history, Heisenberg's chosen mode of opposition — or rather, his chosen rationalization for staying at his post — manifests the dilemma and the debilitation experienced by the subjects of the Third Reich. While one could indeed help to ameliorate conditions somewhat, and could do so better the higher the office one attained, this type of opposition also implied full acceptance of the situation — the appearance of collaboration was of less consequence than internal intentions; what little one could change was more valuable than how much one could not. Such thoughts reveal the political ambivalence and misjudged perception of effectiveness in which Heisenberg and others indulged. Yet because of this position, many non-Jewish Germans later felt that their own courage and suffering were never fully appreciated by those who had never experienced a dictatorship. Theirs was not so much the fear of persecution and murder—to which there can be no comparison—but of discovery, the constant fear that the appearance of collaboration would be suddenly stripped away to reveal their inner opposition, an opposition that only their courage prevented from breaking out into open resistance. The failure of the regime to strip away the appearance preserved the larger illusion of opposition that made collaboration a way of life for so long.

Theirs was a life confronted daily with painful contradictions and compromises, all endured for the sake of appearance in order "really to change something." In his 1939 novel of the Jewish exile, *Exil*, Lion Feuchtwanger captured the feeling in the complaint of his fictional character Riemann, a highly regarded German musician, who visits his expatriate Jewish friend Sepp Trautwein in Paris. In one scene Riemann, after tying to explain why he will not leave Germany, conveys to Trautwein what he must endure for his decision: "Do you think that it's pleasant? . . . When you have to stay there day by day, and everything is much worse than you read in the newspapers, and you have to keep your mouth shut and on top of that play music, *mein Lieber,* that's no small thing. When, after you've played Beethoven, you have to play the Horst-Wessel-Lied, try that sometime, Sepp, and see how it is." To which Sepp replies: "But who put you up to it, *mein Lieber?*"[19]

Within days of the appointment of a Reich commissar for Saxony, the education ministry appointed a National Committee for the Renewal of the University of Leipzig. The committee consisted of Nazi students and faculty, charged with overseeing the transformation of the university into a "leaders' school of the German Volk."[20] Students formed the vanguard at Leipzig, as at most universities, in carrying it out.[21]

Simultaneously with the April 1933 anti-Semitic laws, the National Committee abolished all student organizations except for its own Nazi Students League, which already controlled the student government. On April 12, the national office of the Students League unleashed its infamous campaign against the "non-German spirit" at universities, which in Leipzig took the form of open intimidation of Jewish and democratic professors, one of whom suffered a fatal heart attack.[22]

An Italian visitor to Heisenberg's institute reported home his observations soon after arriving in Leipzig in 1933: "Leipzig, which has a social-democratic majority, accepted the revolution without difficulty. . . . Brown uniforms are not much in evidence, but swastikas can be seen everywhere. The Jewish persecutions delight most Aryans. The number of people who will find jobs in public and private establishments as a result of the liquidation of the Jews is considerable."[23] By the end of May, two American visitors to Leipzig reported the situation at the university in relatively favorable terms: "It is clear that conditions in Saxony are not so fanatical as in Prussia, and the professors dare to speak out much more. One sees comparatively few flags on the street." [24] Because of poor press, the Saxon authorities even ordered the student leader, SS-Sturmführer Friedrich, to rein in his student minions. [25]

During the first semesters under the new regime, the National Committee acted as liaison between the Saxon authorities and the university in the matter of dismissals. According to one count, in 1933 the university lost 21 professors to dismissal, resignation, forced retirement, and death. [26] The Physics Institute lost one professor with the unfortunate name of Marx, who headed the Radio Division, two assistants—Felix Bloch and Heinrich Sack—and one technician, Heisenberg's lecture assistant Herr Dornfeld. [27] The minutes of meetings of the Leipzig philosophical faculty, to which the Physics Institute belonged, yield little evidence of the behind-the-scenes diplomacy that resulted. An entry for the meeting of May 17, 1933, attended by Heisenberg, reads simply: "The dean [geophysicist Ludwig Weickmann] reports on the steps he has undertaken with the government as a result of the new professional civil-servants law. A discussion about it is not desired." [28] The dean's reports of dismissals and ministerial orders likewise appeared in the official record without comment. [29] Nazi professors and state and party officials naturally enjoyed full access to such records.

Other less public (though written) sources yield a somewhat fuller picture. News of agitation against Jews and foreigners in the Physics Institute reached Dresden by May. On May 30, 1933, a student member of the National Committee audaciously wrote the interior minister himself, demanding an investigation of Bloch, Sack, Marx, and mathematician B. L. van der Waerden on the basis of religious and political rumors about them. [30] Both Bloch and Sack, whom Debye had brought from Zurich, had gone home for the spring break, and both had already decided not to return to Leipzig. Sack, Debye's assistant, had asked for a year's leave to wait out further developments. Debye managed to obtain a fellowship for him in Brussels, but the interior minister recommended dismissal anyway. [31] A protest by the Swiss consul in Dresden and a threat of retaliation against Germans in Zurich were without avail. [32] Sack was replaced in October.

Heisenberg's assistant, Felix Bloch, had, like Sack, realized that little future remained for him in Germany. [33] Bloch had already applied for a Rockefeller stipend to work in the United States before he learned of the anti-Semitic laws. According to the rules of the Education Board, which administered the stipend, a fellow had to specify where he or she would work after the end of the grant. At Heisenberg's urging, Bloch indicated Leipzig. As the new law became public, Bloch wrote a worried letter to Bohr, then in the United States, asking Bohr to intervene with the Education Board to change his future location to Copenhagen.[34] Bohr agreed, prompting a thank-you from Werner "for . . . your efforts on behalf of our young physicists, whose well-being lies on all our hearts," and an apology "for all of that which is now happening in this country."[35] A year later Bloch, then in Rome, accepted an invitation from Stanford.[36]

Little remains to illuminate the extent of Heisenberg's role as conduit to Bohr and Pauli, both of whom were active in international refugee organizations. Most of the correspondence on such matters was destroyed when Pauli and Bohr themselves joined the exodus from Europe during the war. Records do remain regarding Heisenberg's reactions to the plights of three young physicists: Gerhard Herzberg, Herta Sponer, and Guido Beck. Because his wife was Jewish, Herzberg received notice in 1934 of dismissal from his assistantship in Darmstadt. The physicist-philosopher Michael Polanyi, then in Manchester, England, appealed to Heisenberg, who made inquiries in Darmstadt. Herzberg should not have been dismissed under the civil service laws, but because of his state-defined position, local bureaucrats could do as they pleased. Heisenberg could do little more for Herzberg in Germany and suggested that Polanyi might do better for him in England.[37] Herzberg and his wife soon settled in Canada.

Heisenberg was more directly successful in helping Sponer and Beck — but again only by looking abroad. Though not Jewish, Sponer, a member of Franck's Göttingen Institute, realized that her future as a woman physicist in Nazi Germany was bleak. "The German woman belongs in the kitchen!" as Kopfermann facetiously put it.[38] Sponer lost her job, and Heisenberg learned of her decision to emigrate during his attempts to salvage Göttingen physics. After several inquiries abroad, he managed to find her a position in Madrid.[39] She later moved to Chicago, where she and Franck were married.

As an Austrian Jew who was unemployed when Hitler came to power, Guido Beck was harder to place. Beck's assistantship with Heisenberg had already ended in early 1932 and could not be continued. With no job offer and no position in Leipzig, Beck's future was dark. Bohr had secured funds for him to go to Copenhagen until Easter 1934, but it was only a temporary remedy. Throughout most of 1933, Bohr and Heisenberg sought other funding for him. By the end of the year, they finally created a unique position for him at the German University in Prague with the combined support of Dutch, Danish, and Swedish foundations. Beck remained in Prague only a few years, traveled to the Soviet Union, then went to France, from which he escaped to Latin America during the German occupation.[40]

Erich Marx's dismissal, announced to the faculty on September 27, 1933, came in the midst of a power struggle in the Physics Institute and at a time when the university itself was undergoing internal battles.[41] More than a month before Marx's removal, and probably in anticipation of it, Karolus and Schiller, heads of the two other technical divisions in the institute, petitioned the Saxon ministry for independent institutes of their own.[42] With technical physics in the ascendant in the new state and Marx in trouble, the technical physicists made their move. Debye, apparently regarding defense of Marx as hopeless, countered with a replacement and a redefinition of his field from radio to radiation physics. With the help of Dean Weickmann, Fritz Kirchner, a nuclear physicist, was quickly appointed. The independence of the technical divisions was prevented, but not without the festering resentment of the division chiefs. A year later they aimed their hostility at Heisenberg.

While Heisenberg substituted as institute head during Debye's absence in Brussels in 1934–1935, Kirchner was invited to assume a full professorship with his own institute in Cologne.[43] The offer was too tempting to refuse. Heisenberg was now faced with thwarting the renewed pretensions of the technical physicists. He managed to hold them in check, but their close connections with Debye's longtime party-affiliated assistant may ultimately have worked against the institute heads. Heisenberg's Nazi opponents later enjoyed inside information on institute affairs.[44]

During the first full school year under the new regime, intensified university reorganization went far beyond the exclusion of unwanted faculty. Characteristically, a new university leader proclaimed: "The university in the new Germany will be *political,* an educational institution for political persons who place their knowledge and abilities in service to the nation."[45] Politics, not in the democratic sense of party politics but in the fascist sense of "service to the nation," meant subservience to the new representatives of the nation, the National Socialist German Workers Party. To realize this aim efficiently, Nazi student leaders organized their followers into close-knit work and study teams.[46] A "teachers corps" was formed for middle- and upper-level

Werner and Erwin backed up by their Osanbrück relatives, with Aunt Grete at far left.

Werner and Erwin in Würzburg.

Werner (standing) and his family at the end of World War I.

Niels Bohr and Max Planck (seated), in the 1920s. (*Courtesy NBA*)

Heisenberg in Göttingen, c. 1924.

Dirac and Heisenberg in Cambridge, 1930s.

Heisenberg with his landlady, Mrs. Maar, Maar's niece, and Pascual Jordan in Copenhagen, c. 1927. (*Courtesy NBA*)

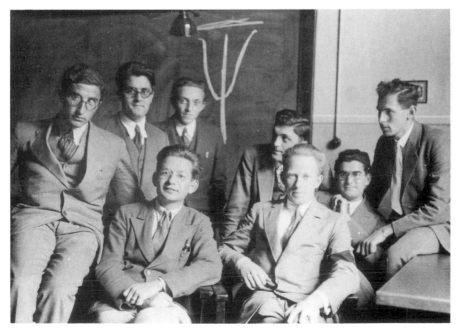

Heisenberg with his seminar students shortly after his father's death in 1930. From left: Placzek, Gentile, Peierls, Wick, Bloch, Heisenberg, Weisskopf, Sauter.

Solvay Congress of 1927 (*Courtesy AIP Niels Bohr Library*)

A. PICCARD   E. HENRIOT P. EHREN
P. DEBYE   M. KNUDSEN   W.L. BRAGG
I. LANGMUIR   M. PLANCK   Mme CUR

ERZEN Th. DE DONDER   E. SCHRÖDINGER E. VERSCHAFFELT   W. PAULI W. HEISENBERG R.H. FOWLER L. BRILLOUIN

MERS   P.A.M. DIRAC   A.H. COMPTON   L. de BROGLIE   M. BORN   N. BOHR

H.A. LORENTZ   A. EINSTEIN   P. LANGEVIN   Ch.E. GUYE   C.T.R. WILSON O.W. RICHARDSON

Absents : Sir W.H. BRAGG, H. DESLANDRES et E. VAN AUBEL

Heisenberg, seated to right, with the Leipzig professors' club "Coronella."

Heisenberg in the 1930s.

Nobelists and family members at the Stockholm train station, 1933. From right: Schrödinger, Heisenberg, Dirac, Dirac's mother, Schrödinger's wife, Heisenberg's mother.

Nobel Prize ceremony. King of Sweden with Schrödinger and Heisenberg.

Elisabeth and Werner Heisenberg in Göttingen, 1947.

Copenhagen Physics Conference, June 1936. First row (left to right): Pauli, Jordan, Heisenberg, Born, Meitner, Stern, Franck; second row: Weizsäcker and Hund behind Heisenberg, Otto Frisch (second from right); third row (left to right): Kopfermann, Euler, Fano; fourth and fifth rows, second from left: Peierls and Weisskopf; standing along wall: Bohr and Rosenfeld. (*Courtesy NBA*)

Heisenberg with two of his sons, late 1940s.

Heisenberg, Max von Laue, Otto Hahn, late 1940s.

Surveying the Urfeld scenery, late 1960s.

Heisenberg talking with Edward Teller at the Feldafing physics conference, summer 1965.

lecturers to control senior faculty through the approval of new habilitations and faculty appointments.[47] Partially unsuccessful attempts were made to introduce the Führer principle into a faculty that was for centuries a stronghold of free thought and independence.[48] The new Führer of the university, Arthur Golf, joined with the student SS-Sturmführer to proclaim an end to the "conflict" between the apolitical professors and their "political" students. Professors and students now joined hands as "comrades under Hitler."[49]

Throughout all this, Heisenberg maintained his academic position and offices but withdrew privately into his physics, his music, and his circle of friends and colleagues. In his memoirs, he recalled a challenging encounter early in the regime with a leader of the Leipzig Hitler Youth.[50] The session must have taken place before June 1933, when Heisenberg moved from the institute into a small house.[51] Heisenberg remembered meeting the student after he had finished practicing a Schumann concerto on the piano of his institute abode. He invited the student, who had been listening in a window well in the hallway, into his living room. The student, dressed in the starched brown uniform of the Hitler Youth, was puzzled. Why did Heisenberg, still a young man and a close associate of the youth movement, now hold himself so aloof from the younger generation, "like one of those conservative old professors who live completely in the past"? Heisenberg's sympathies seemed to lie with the youth leader, who claimed that "Germany needs political liberation from her state of inner corruption."[52] In his recounting of the meeting, Werner even appears to admire the youth's passion for bringing about change through direct action. But Heisenberg recalls drawing the line at mixing means and ends. "You must realize that I cannot help you when Germany is being ruined," he declared. "It's as simple as that."

But, of course, things were not quite that simple, for shortly afterward Werner turned to Planck for advice. Events also challenged his preferred aloofness from National Socialists and from those outside his immediate circle of friends, associates, and institute members. At the beginning of the 1933–1934 winter semester, Heisenberg learned that he would receive the prestigious Max Planck Medal, and on November 9 he received the exciting news that he had been awarded the 1932 Nobel prize for physics.[53] At the same time, Heisenberg refused to participate in a highly publicized national rally held on November 11 in Leipzig under the auspices of the National Socialist Teachers League. The rally, a widely publicized "demonstration of German scholarship," supported the nation's withdrawal from the League of Nations, to be decided ostensibly by a referendum and an election on November 12. Heisenberg informed the rally organizer, physicist Johannes Stark, that he would not attend.[54]

Numerous teachers and students, four university rectors, and six professors did attend, among them the noted philosopher Martin Heidegger (also a rector). A vindictive Stark informed Leipzig students of Heisenberg's refusal to join the "acknowledgment by professors to Adolf Hitler." Students, delighted by Heisenberg's prestigious prizes but angered by his failure openly to support the cause, were thrown into confusion. "How vehemently the debates swirled about you in those days," one lecture student recalled, "when at the beginning of the winter semester 1933 your refusal to participate in the election rally resulted in a small scandal in the institute! And how much support for you among the students finally outweighed everything else!"[55]

Elisabeth Heisenberg recounts an episode she heard only later about a befriended Nazi youth leader—perhaps the same young man Werner had met earlier—who warned Heisenberg that a band of students planned to disrupt his main lecture the day after the announcement of his Nobel prize. The student managed to defuse the demonstration by deftly leading the protesters out of the lecture hall. The remaining students broke into a standing ovation.[56] Only a few days later, on the foggy evening before the scheduled teachers' rally, nearly the entire Leipzig student body honored the new Nobel laureate with a torchlight procession—the admiring students marching row on row to his home in storm-trooper formation.[57] A week later, Heisenberg invited the district leader of the Nazi Students League to his home to defuse any lingering animosity. The young man officially exonerated the professor of any fundamental opposition to Führer and state.[58]

Following the November uproar and a mysterious trip to Zurich on behalf of refugee physicists, Heisenberg traveled by train to Stockholm with his mother to receive his Nobel prize from the hand of the reigning king of Sweden. Stopping in Copenhagen on the way to thank Bohr personally for collaboration that had led to the prize (and, probably, for recommending Heisenberg), Heisenberg and his mother arrived at the Stockholm train station where they were greeted by two other physicists, Dirac and Schrödinger, who were there to share the 1933 Nobel prize for physics.

The conferral of prestigious prizes and the attribution of discoveries to individuals are always crucial matters for scientists—recognition by colleagues and posterity are more important for scientists than for most other professionals. A special aura of prestige surrounds the Nobel prize, and the Nobel committee, which selected the recipients, was acutely aware of this. The physics prizes conferred in 1933 were especially noteworthy in this regard. Reflecting the domination of physics by experimental work and the committee's preferences for tangible results, most physics prizes had previously gone to experimentalists or to those whose work greatly influenced experimental research. For the first time, three theoretical physicists were chosen primarily for their contributions to theoretical physics. The official citations, however, indicated the peculiar manner in which the committee viewed the nature and relative contributions of these three men. Heisenberg was awarded the reserved 1932 Nobel prize "for the creation of quantum mechanics, the application of which has led, among other things, to the discovery of the allotropic forms of hydrogen" — at least a partly tangible result. Dirac and Schrödinger shared the 1933 prize "for the discovery of new and fruitful forms of atomic theory."[59]

Bad feelings abounded. Max Born's contributions to quantum mechanics had been slighted, as were those of Jordan, Pauli, Bohr, and others;[60] Schrödinger must have winced at the designation of Heisenberg as the creator of quantum mechanics, while he and Dirac seemed only half worthy of a full prize for "new and fruitful forms" of the theory. Although the awkward situation may have been made worse by the professional situation — Born and Schrödinger had both recently left Germany, despite objections from Heisenberg, and were now without permanent jobs — their level of recognition and the apparent overrecognition of Heisenberg seemed the largest difficulty. Anecdotal claims to the contrary, Heisenberg maintained a cordial relationship with both Born and Schrödinger to the end of their lives.[61] He expressed his own unease with the entire affair on several occasions in 1933. After writing an apologetic note to Born from Zurich in November, he wrote to Bohr on November 27, 1933: "Concerning the Nobel prize, I have a bad conscience regarding Schrödinger, Dirac, and Born. Schrödinger and Dirac both deserved an entire prize at least as much as I do, and I would have gladly shared with Born, since we have also worked together."[62] The Nobel committee eventually made up for its oversights, awarding Nobel prizes to Pauli in 1945 and Born in 1954.

The distress and excitement of Stockholm behind him, Heisenberg returned to his work and to his small circle in Leipzig. With Bloch now gone, Werner's recent doctoral graduate, Carl Friedrich von Weizsäcker, emerged at the center of his closest associates. During this period, Weizsäcker met with Heisenberg daily for long discussions and soon became his teacher's closest confidant. Their relationship was so close at times that Heisenberg would often feel abandoned when Carl Friedrich was away.

The son of a long line of aristocratic German diplomats, Carl Friedrich von Weizsäcker, who maintained his noble title, Freiherr von Weizsäcker (free lord from Weizsäcker), could always turn a diplomatic phrase when needed.[63] While his brother, Richard von Weizsäcker, studied law (he eventually rose to the presidency of postwar Germany), Carl Friedrich intended to study philosophy. He added the field of physics to his interests after meeting Heisenberg in Copenhagen during the twenties. Although eleven years younger than Werner, Weizsäcker, with his high forehead, narrow chin, and aristocratic demeanor, was more involved than Werner with philosophical matters. They discussed philosophical issues endlessly in Leipzig and on their frequent hiking and skiing expeditions, often with former youth-movement members, in the Bavarian hills. As the political situation deteriorated after the National Socialists' rise to power, Werner came to rely on Carl Friedrich and his intellectual understanding of worldly matters. "Only the friendship with Carl Friedrich, who struggles in his own serious way with the world around us, leaves open to me a small entry into that otherwise foreign territory," he wrote to his mother in October 1934.[64] Hitler had consolidated his hold on higher office just two months earlier.

Weizsäcker remained in Leipzig as Heisenberg's friend and assistant until Weizsäcker habilitated and transferred to Berlin in 1936. During that period, Heisenberg's retreat into his personal circle hastened in the wake of the Leipzig events of 1935. Armed with the anti-Semitic Nuremberg laws and no longer restrained by a now deceased Hindenburg —the president who had appointed Hitler and who demanded special exemptions for Jewish front veterans—the regime introduced its second civil service purge to prepare for its first four-year plan—economic and military mobilization in anticipation of World War II. The new purge provoked a crisis in the Leipzig philosophical faculty. On May 1, 1935, at the request of the Saxon Education Ministry, Mutschmann, the Reich potentate for Saxony, ordered the immediate dismissals of Professor Wach (theology) and Assistant Levi (mathematics) and the forced retirements of Professor Landsberger and Professor Wiegert.[65] Being only part Jewish or World War I front veterans, the four had been spared earlier.

Heisenberg reacted with shock and dismay. Not only did he believe that the worst was already over, but he was also well acquainted with the victims, Wach being a member of the professors' club. The order now made it obvious that there was no protection for Jews (or anyone else, for that matter) in public service, whatever their accomplishments or status. Heisenberg also realized that the university would never be free from bureaucratic meddling.

Heisenberg reviewed the situation, which would prove to be a turning point in his relationship to the state, with his closest colleagues — Weizsäcker, Hund, B. L. van der Waerden, and Karl Friedrich Bonhoeffer (brother of theologian Dietrich Bonhoeffer). He then met with another member of his professors' club, Dean Helmut Berve, to formulate a response. Berve, history professor and party member, had served as dean of the entire philosophical faculty since late 1933. They all agreed to express their disapproval at the next faculty meeting, knowing full well that a report of their protest would eventually reach Dresden, the capital of Saxony. A more direct confrontation with the authorities would probably not succeed, they reasoned, and might result in even more dismissals, including their own.

The somber professors assembled on May 8 in their faculty's oak-paneled branch library for the carefully planned and orchestrated meeting. Dean Berve opened the session with a formal announcement of the dismissals.[66] Under the steady gaze of the portraits of their predecessors lining the library walls, Heisenberg, Hund, van der Waerden, Schweitzer, and Reichert — all closely associated with Berve and the club — voiced their prepared objections. Reichert inquired about the grounds for the dismissals; Rector Felix Krueger responded that his own inquiries in Berlin had been fruitless. Berve complained that the faculty had been kept completely in the dark. Van der Waerden, who wondered if there had been other grounds for the dismissals, was followed by Heisenberg. "Herr Heisenberg doubts that the measures now being taken are consistent with the intention of the law, according to which front veterans also belong to the Volk community, and he holds it a duty to his friends to help those affected in any way he can." Hund noted that nonveterans should be ashamed at the treatment of the veterans. Finally, Berve interjected — for the record — that he was permitting the discussion only to report to Dresden the faculty's mood.

The meeting adjourned to await Dresden's response. It was not long in coming. Van der Waerden recalls that within a day, even before Berve submitted his report, the education ministry had been informed. As might be expected, the protests had little effect. The dismissals and retirements remained in force. Several days later, van der Waerden and probably Heisenberg received formal reprimands via the rector from Mutschmann himself.[67] On May 12, Heisenberg wrote his mother that the rector was pressuring him to enter the Reichswehr as a reserve officer in order to demonstrate his loyalty to the Reich, despite his voiced objection to this application of regime policy. His hope of becoming dean of the science section of the philosophical faculty seemed doomed.[68]

Berve was also punished for his performance. Having allowed his faculty to express its displeasure, the Reich commissar himself ordered his replacement as dean at the end of the semester. Berve recalled after the war that the local head of the Nazi University Teachers League subjected him to considerable abuse for merely meeting regularly with his professors' club friend Wach in a Leipzig restaurant. On the day on which the club made its last outing with Wach before his departure from Leipzig, Berve received the order from Dresden to step down as dean.[69]

Perhaps because Heisenberg and the others were so shocked by the dismissals — which seemed so arbitrary, yet so symptomatic of life under Hitler — this episode emerged as a crucial moment in their prewar lives. Heisenberg, Hund, and van der Waerden remembered these events as having occurred two years earlier, when the dismissal policy was first introduced. Apparently it was not until 1935 that they fully understood how arbitrary the Nazis could be and how powerless they were to object by traditional means. Van der Waerden recalls that Heisenberg had expected at that time an even worse punishment than a reprimand — dismissal.[70] But in view of his Nobel prize, international fame, and Max von Laue's example of survival in office despite outspoken opposition to Nazi measures, he was unlikely to suffer that fate.

Years later, Heisenberg recalled that with the faculty protest failure, he and his colleagues at last contemplated resigning from the university.[71] Since they had been unable to turn back the dismissal policy or to halt the dictatorial transformation of the university during the preceding two years, resignation had become by then an appropriate — probably the only appropriate — political and moral alternative. Even as lower ranking incompetents waited in the wings, the Nobel laureate's actions would send a signal far beyond the Leipzig faculty. A fundamental decision was now necessary. Torn and uncertain, Heisenberg turned again to the aged Max Planck for direction.

The 77-year-old Planck received the 33-year-old physicist in the dark, old-fashioned living room of his home in the upper-class Berlin suburb of Grunewald. All that seemed to be missing, Werner wrote, was an antique oil lamp on the table in the center of the room.[72] The tired and somber Planck seemed somehow years older since their last meeting.

Despite their differences in age, background, and opinion regarding quantum mechanics, the two physicists had grown closer in outlook during the last few years under the mounting repression of academic physics. As in 1933, Planck saw in Heisenberg a hope for the future of German physics; Heisenberg saw in Planck a man to respect, a symbol of the esteemed tradition, achievement, and integrity of German science. During the first phase of dismissals, Heisenberg had traveled often to Berlin to seek political advice from Planck and Laue, both of whom were well known for their rectitude, but this time Planck himself seemed bewildered.

Resignation in protest was never an option for Planck. In 1933, he had talked his Berlin colleague Otto Hahn out of such a move in favor of fighting from within. At that time, he still hoped to influence government policy through personal diplomacy. But in 1935, as he and Werner sank into the overstuffed chairs in the gathering darkness of Planck's dimly lit living room, Planck's only counsel was one of political acquiescence. Planck saw events unfolding as an avalanche, out of control, even out of the government's control. For Planck, further political or diplomatic maneuvering was now futile. Resigning a position in protest seemed to Planck equally hopeless. Heisenberg's resignation would surely go unreported and unknown.

Instead, according to Heisenberg, Planck argued, "it is to the future that all of us must now look." They should withdraw from the present climate, but not from their posts. Planck believed they were now charged with a new responsibility — to preserve enclaves of decent German culture and science as "crystallization points" for a better world.[73] It was a return to the well-practiced politics of Weimar, to the very position that Heisenberg himself had settled on long ago amidst the throes of revolution and coup d'état: the dissociation of good German culture from the ephemeral world of bad political rulers, the separation of nation from state, profession from administration, intentions from appearances. In the end, such distinctions played right into the hands of their enemies.

Planck's advice was the kind that Heisenberg understood and welcomed, if indeed he had not mixed his own views with his memory of Planck's advice. It gave Heisenberg new impetus to fulfill his rationalized role of active collaboration — refusal to desert his high post to accept one of the many tempting offers he was receiving from abroad. A captain cannot abandon his ship, no matter how violent the storm — especially not as the storm worsened. Planck may have prepared Werner with similar advice in 1933. In his June 1933 letter to Born, just after having visited Planck in Berlin, Heisenberg wrote of his desire to work amidst "a circle of people who understand that we live for them and for the science entrusted to us."[74] By late 1935, Heisenberg's horizons had shrunk to the circle of his science, students, friends, and colleagues.

In the fall of 1935, as a new semester started and new troubles mushroomed, Werner wrote his mother of his new task, as he perceived it: "But I must be satisfied to oversee in the small field of science the values that must become important for the future. That is in this general chaos the only clear thing that is left for me to do. The world out there is really ugly, but the work is beautiful."[75] Heisenberg had made his fundamental decision. He would remain in his homeland and at his post no matter how difficult this became.

CHAPTER 17

Political Science

Heisenberg's retreat at the end of 1935 into the circle of his students, his physics, and his music reaffirmed his previous position. He was a physicist, and a good one. The beauty of physics, like the beauty of classical music, offered an escape from the "ugliness" and "general chaos" of the "world out there," and his contributions to this field would last for years. The nature and intensity of his correspondence throughout the 1930s, especially with Wolfgang Pauli, indicate a fervent devotion to these ideals. Only occasionally do hints of personal and political conflicts filter through their abstract discussions of technical minutiae. Physics and physicists, the scientists told themselves, were meant to exist in a world of their own.

Yet events "out there" continually intruded. Science possesses both a form and a content—personnel, workplaces, funding, and communications media, as well as theories, experimental data, and accepted criteria and procedures of advance. All were under siege in those years. Preservation of the form, or professional dimension, of science—a duty that no German physicist could refuse—placed Heisenberg and his colleagues on a collision course with Nazi demagogues and in a position that required enormous compromises. Before settling on what he called his "island of existence" in Leipzig, Heisenberg was forced to contend with these real-world conflicts and compromises—but not before his science received a new impetus.

Late in the first year of Hitler's regime, soon after the news of the Schrödinger and Weyl resignations, the discoveries of the positron and of cosmic-ray showers turned Heisenberg's physics in a new direction,

toward high-energy interactions in relativisitic quantum electrodynamics. Both discoveries arose from research on cosmic rays, the extremely high-energy particles and photons entering the atmosphere from
outer space. Using a newly designed cloud chamber as a detector,
Patrick Blackett and Giuseppe Occhialini, working in Cambridge,
England, discovered that when a cosmic-ray particle hit the top of a thin
lead plate in a cloud chamber, a "shower" of particles was ejected from
the bottom of the plate.[1] Shortly before, Carl Anderson, working in
California, had announced the discovery of a new subatomic particle.
He had found a new "positive electron," or positron—a particle having
the same mass as the electron but an opposite charge—in studies of
cloud-chamber photographs of cosmic-ray events.[2]

Dirac's relativistic quantum theory of the electron, published in 1928,
suddenly warranted new attention.[3] Because this theory, though well
reasoned, incorporated the paradoxical existence of negative energy
states, it had hitherto attracted much interest but few adherents. Electrons could theoretically jump to states of negative energy—whatever
such states might be. Dirac had proposed a way out. He had suggested
that the negative states are actually so filled with negative-energy electrons that positive-energy electrons could only rarely jump to a negative
state. An occasional unfilled state would appear as a positive particle,
which Dirac identified in 1930 as a proton—the only known positive
elementary particle. But protons did not possess the correct mass or
other properties for that purpose—hence his proposal a year later of a
new kind of particle, the "antielectron."[4]

Anderson's new positron looked identical to Dirac's antielectron,
inspiring renewed interest in Dirac's theory. A positron could be interpreted as a hole, or bubble, in an infinite sea of negative-energy electrons. An electron falling into a hole and disappearing in a flash of
photons could be observed in the laboratory as pair annihilation—the
destruction of an electron-positron pair, with the appearance of
two photons carrying their energy. The inverse would appear as pair
creation—the disappearance of a photon with the appearance of a pair
of particles. Because of the hypothesized negative sea, a bare charge
alone in empty space could no longer be considered alone at all; it was
surrounded by an infinity of negative-energy electrons.

This vastly complicated matters for theorists. Previously, two major
hindrances blocked the road to a satisfactory relativistic quantum field
theory: the appearance of Dirac's negative energy states, and the infinite
self-energy of the electron—the divergent interaction of a point electric
charge with its own electromagnetic field. The positron accounted for

the appearance of negative energies, but it added a new infinity to be contended with — the infinity of interactions between a single positive-energy electron and the infinite sea of negative-energy electrons.

Cosmic-ray showers — the numerous positive and negative particles produced when a high-speed cosmic ray hits a thin lead plate — presented an equal challenge to theorists. Dirac's Cambridge colleagues, Blackett and Occhialini, suggested that the creation of electron-positron pairs might be the origin of their observed showers. Upon hitting a lead plate, a high-energy electron would be deflected and absorbed by radiating a photon (a process dubbed *Bremsstrahlung*, "brake radiation"), which would in turn produce pairs that might be observed as a shower or be absorbed by ionization.[5] Theoreticians working with the Cambridge experimentalists produced a series of calculations on Bremsstrahlung and pair creation for application to the stopping of high-speed particles in matter. Heisenberg, however, was preoccupied at this time with his research on nuclei, with the preparation of his report on the subject for the Solvay Congress of October 1933, and with his efforts to respond to the Nazi-ordered dismissals of his colleagues. He did not learn firsthand of the new experimental results and of Dirac's latest theorizing until the Copenhagen physics meeting in September 1933.[6]

The theoretical results on particle stopping in lead plates, published soon after the Copenhagen meeting by Wendell Furry and his mentor, J. Robert Oppenheimer, in the United States and by refugees Hans Bethe and Walter Heitler in Britain were only partially encouraging.[7] Bremsstrahlung and pair creation could account for the stopping of particles in matter, but only for energies well below those observed for showers. As it had earlier, the theory diverged to infinity at high energies or small distances of approach. Collision cross sections increased without bound as energy increased, in direct conflict with the observed penetration of high-speed electrons through large blocks of lead — long path distances through matter implied instead low cross sections for collisions.[8] This contradiction, together with prior expectations and arguments, led most researchers to perceive a breakdown of the theory at an upper energy limit. They established this limit, as before, by the wavelength of a matter wave equal to the size of a classical electron — or, in energy terms, by the corresponding minimum uncertainty in energy equal to the rest energy of an electron. This yielded the condition that the theory broke down for all energies greater than $137\ m_0 c^2$, or about 10^8 eV, where m_0 is the rest mass of the electron, c is the speed of light, and the factor 137 arises from the inverse of the so-called fine-structure constant, $2\pi e^2 / hc$.[9]

Returning to Leipzig from Copenhagen, with a Berlin stopover to meet with Planck and Laue, Heisenberg was both optimistic about reversing Schrödinger's resignation and excited about the positron. "What do you say about the positive electrons?" he asked Sommerfeld. "It appears that once again Dirac was much more right than we had previously thought. . . . Hopefully you will find enough rest from politics during the winter semester to enjoy the positive electrons."[10]

During his three-week stay in Copenhagen, aside from discussing the situation in Europe, Heisenberg had discussed Dirac's hole theory with Dirac, Bohr, and Pauli and had come to the conclusion that since the theory diverged as usual, it would not help in accounting for nuclear forces and beta decay, the ejection of electrons from nuclei as neutrons turned into protons.[11] An entirely new theory was still needed to handle minute distances comparable to the size of a neutron. Nevertheless, wrote Heisenberg, "for a series of questions: Meitner-Hupfeld effect, Bremsstrahlung, etc., the theory appears to work completely. Even Bohr was very enthusiastic."[12] Dirac, now a full professor in Cambridge, provided a practical reason for excitement during the Solvay Congress in Brussels at the end of October. In his report to the Congress, attended by all principals — Heisenberg, Pauli, Dirac, and Bohr — Dirac introduced a so-called density matrix formalism for hole theory that made possible the calculation of laboratory observables, such as electric fields.[13] But it contained the same infinities as before and even added a new one: the infinite polarization of the negative-energy sea. Dirac artificially diked the sea by cutting off the infinity at $E = -10^8$ eV or by subtracting it as the limit of the difference of two infinities.

Judging from the volume of letters exchanged between Leipzig and Zurich during the first half of 1934, after Heisenberg's return from Stockholm with his Nobel prize, Heisenberg, Pauli, and their associates seized on Dirac's density matrices with a passion that made up for lost time.[14] Although Pauli did not hide his disgust at Dirac's "limit acrobatics and 'subtraction physics,'"[15] he and his assistant, Victor Weisskopf, joined Heisenberg in an intensive effort to develop Dirac's abstract, complicated formalism into a complete electrodynamic positron theory. At the same time, Heisenberg pursued with equal vigor his response to the political assault on theoretical physics. The two endeavors, though in separate spheres, were for him united by their common though ultimately unattainable aim: to preserve a first-rate, politically untainted physics in the new Germany.

While the abstractions of positron theory occupied Heisenberg's scientific thoughts in early 1934, during their negotiations in Göttingen regarding the filling of now vacant teaching chairs Heisenberg and Dean Brandi sought new ways to preserve theoretical physics in its former stronghold and throughout Germany. One way, they decided, would be to convince engineers and industrialists of the intrinsic value of theoretical physics and of the advantages of hiring young theoreticians. With the regime emphasizing research for only immediate, practical ends — nearly the antithesis of Heisenberg's work at the time — theoretical physics required bolstering.

Heisenberg arranged in Göttingen to present his case for theory personally to the chairman of the powerful Association of German Mining Engineers and later in a lecture called "Science and technical progress" at the annual convention of the association in Düsseldorf, which appeared in the association's periodical, *Stahl und Eisen (Steel and Iron)*. His efforts paid off. Diplomacy and a personal appearance convinced the influential chairman and many of his followers "that the existence of a good theoretical physics in Germany is of decisive significance for the collaboration of technology and science."[16]

Pauli was less convinced of the positron physics that Heisenberg produced in this same period. By June, Pauli's proclaimed disgust with the Leipzig-Zurich hole theory formalism led him to resign from the positron program, and Heisenberg, encouraged by a ten-day visit in England with Dirac in April, published their "extraordinarily ugly" hole theory alone.[17] While Weisskopf demonstrated that the electron self-energy diverged to infinity much more slowly (logarithmically) than expected, Heisenberg could show that the subtraction physics formalism was at least compatible with the usual conservation laws and invariance requirements.[18] His paper arrived in the editor's office of the *Zeitschrift für Physik* just three weeks after he delivered his defense of theoretical physics to the engineers.

Pauli had resigned from the positron program in part because of "far-ranging negative results from the physical point of view" — infinite negative-energy seas and self-energies.[19] Two days after Pauli's resignation, Heisenberg informed him that at least one positive physical phenomenon did result: the scattering of photons on photons, a finite fourth-order effect characteristic of hole theory.[20] Debye suggested that this effect might be the origin of the sun's corona. While Pauli and Weisskopf turned to the formulation of an "anti-Dirac theory," a quantum field theory of bosons — that is, a field theory of particles with integral amounts of spin, rather than half-integral spin (fermions) — Heisenberg turned the scattering problem over to his students, Hans Euler and Bernhard Kockel, for calculation.[21] Although he kept abreast of their work, Heisenberg was soon embroiled once again in other matters: "There are so many other urgent and remarkable things here in Germany with which one must concern oneself for the time being."[22]

Among those other things were an even more direct and public defense of theoretical physics and a personal response to Johannes Stark, who still held a grudge against Heisenberg for his refusal to rally for Hitler the previous November.[23] Within days of Hindenburg's death in early August 1934, Hitler, strengthened by a bloody purge of the storm troopers in late June, announced the uniting of the offices of chancellor and president in himself, giving himself, as Führer, control over the state and the political apparatus of the Reich. Hitler's officials scheduled a plebiscite on the plan for August 19, and the Nazi party launched a major propaganda campaign to ensure a favorable vote — which could hardly have been otherwise. Stark sent telegrams to his fellow German Nobel laureates, inviting them to join in a public declaration of support for Hitler. Among physicists, Heisenberg, Laue, Planck, and the experimentalist Walther Nernst all refused, using once again their old argument that science and politics should not mix.[24]

Stark responded angrily that public support for Hitler was not a political act at all, but merely an "avowal of the German Volk to its Führer." At the same time, Stark claimed that the laureates were already taking a political position by refusing to honor Hitler while speaking publicly in favor of Einstein.[25] It was the kind of perversion of science and politics that so characterized Stark's thinking — and so encouraged scientists all the more to maintain the fiction of a separation of science from politics. Laue told the man not to write him any more personal letters.

A month after the Stark episode and Hitler's reported overwhelming victory as Führer, Heisenberg joined other leading scientists in presenting again the case for modern theory, including Einstein's theories, to the broader scientific public. The theories should be defended, even if the theorists were not. This time, the prestigious Society of German Scientists and Physicians (GDNA), meeting in Hannover in September 1934, served as a ready forum. Heisenberg delivered the keynote address to the plenary session of the society, entitled "Transformations in the foundations of natural science in recent time." In this survey of recent developments in physics, Heisenberg informed his audience that the newest theories, particularly the relativity and quantum theories, were not the products of purely speculative, "revolutionary ideas" foreign to science — as Stark would have it — but were rather the necessary results of years of careful research based on cogent empirical methods and cherished elements of classical physics: relativity and quantum theory "have been forced upon research by Nature in the attempt to carry the program of classical physics consistently to its end."[26]

Indeed, Heisenberg declared, the latest transformations in science had occurred step by step in concert with experimental studies. For instance, a revision in classical notions of space and time had become necessary only after the null results of Albert A. Michelson's ether-drift experiments; Philipp Lenard's investigations on the photoelectric effect had been adequately interpreted only through Albert Einstein's light-quantum theory. By referring to Jewish physicists by name and in the same sentence with the name of Nazi physicist Lenard, Heisenberg was defending the very mention of Jewish physicists and their work—if not the Jewish physicists themselves—in German lecture halls. The professional audience in this particular lecture hall did not seem to object. In fact, two other speakers seconded his remarks—experimental physicist Walther Gerlach and the famous surgeon Ferdinand von Sauerbruch, then president of the GDNA. As a show of further support, Heisenberg's lecture, with names, was published in near record time in the pages of the widely read *Naturwissenschaften* and avidly reprinted elsewhere. Heisenberg seemed pleased with the result. The lecture "went decently," he reported to his mother, "and I was very satisfied with the behavior of the public; if it helps at all is another question."[27]

Einstein's name and his theories also appeared in Heisenberg's Leipzig course lectures that winter.[28] But by then the mention of Einstein or the explicit teaching of his theories were more dangerous acts than Heisenberg may have realized. After the Hannover lecture, one band of aged Nazis in Hamburg threatened to provoke the disruption of his classes unless he publicly retracted the lecture.[29] A Nazi functionary complained directly to party ideologue Alfred Rosenberg, suggesting that "the concentration camp is doubtless the suitable place for Herr Heisenberg." Rosenberg agreed but regretted that because of Heisenberg's international stature, he could only issue a reprimand.[30]

After the inevitable tension of the Hannover meeting, Heisenberg headed for a brief respite in Bohr's Copenhagen institute, returning to Leipzig in October 1934 to find his pupils well into their study of the scattering of light on light. The phenomenon arose in hole theory from the virtual excitation and decay of electron-positron pairs during the collision of two electromagnetic light quanta, or photons. The effect could be handled by higher terms in quantum perturbation theory. The higher order terms giving rise to the effect appeared as small, nonlinear correction terms to Maxwell's electromagnetic field equations. Since Maxwell's equations were thus no longer linear in the field strengths and their derivatives, a light quantum colliding with another light quantum did not simply result in the addition of the fields in each light quantum. Instead, scattering occurred, and within the limits of the uncertainty principle, electron-positron pairs were created for an instant before annihilating each other.[31]

Again the demand of relativistic invariance — the constancy of the equations under a transformation to another reference frame moving at constant velocity relative to the first frame — had to be met. Heisenberg and his Leipzig assistants even believed that, once they found a suitable relativistically invariant modification of the old Maxwell equations to handle the new quantum scattering of light on light, they would have important clues to a complete modification of Maxwellian electrodynamics, a hint of the proper formulation of a relativisitic quantum electrodynamics. The different characteristics of charged matter and uncharged, massless electromagnetic fields still caused divergence of the self-energy of the electron — the calculated energy of an electron interacting with its own electromagnetic field. The close relationship between high-energy photons and electron-positron pairs — between matter and fields — in the context of the Dirac-Heisenberg-Pauli hole theory promised new insights into the puzzles of matter and fields and into ways to render the theory of their interaction acceptable — tasks that Heisenberg called "the central problem of relativistic quantum field theory."[32]

By 1935 hole theory and its mathematical puzzles were for Heisenberg no longer ends in themselves but means to achieve an entirely new theory. The severe limitations of hole theory at high energies and short distances were evident and unavoidable. The situation seemed to Heisenberg analogous to an earlier one: "With regard to quantum electrodynamics, we are still at the stage at which we were in 1922 with regard to quantum mechanics. We know that everything is wrong. But in order to find the direction in which we should depart from what prevails, we must know the consequences of the prevailing formalism better than we do."[33]

For his dissertation, Euler, with help from Kockel and Heisenberg, pursued the Maxwell modification in the case of photon-photon scattering. He submitted the results in November 1935 to the glowing approbation of his mentor, Heisenberg. The two then published the results in a joint paper a month later.[34] Euler and Kockel had earlier argued that the best way to proceed was to start from the covariant Lagrangian form of the classical Maxwell equations, $L = \frac{1}{2}(E^2 - B^2)$, where E and B are the electric and magnetic field strengths, and to find appropriate additional terms for L to encompass the nonlinearities. The problem was complicated not only by the extremely complex mathematics (up to sixth order in the fine-structure constant, $2\pi e^2/hc$), but also by a physical limitation on this approach: the Lagrangian was valid only for fields that were nearly constant over a Compton wavelength, $\lambda = h/mc$, m being the mass of the electron. The Leipzig physicists had again encountered a lower boundary to their physics and at about the same distance — on the order of the size of the electron.

Heisenberg reacted to this result in early 1935 as he had before: by searching the lower distance limit of the theory for clues. He first attempted to build a cutoff of the equations into the formalism by introducing a nonlocalizing operator that smeared the electron over its volume. Bohr and Pauli both rejected it.[35] A more acceptable alternative seemed to result from the efforts of Born and Infeld. Born, now in Cambridge, England, and his assistant, Polish physicist Leopold Infeld, had been attempting to find a unified theory of matter and fields — a theory that would encompass both entities and thus avoid infinities. They searched for a theory by working from the opposite direction: introducing a spherical electron into the classical Maxwell equations, then attempting to quantize the severely complicated result.[36]

Euler and Heisenberg discovered that an analogous procedure could work in their case as well. The nonlinear terms in the Lagrangian could be expressed using a "critical field strength" equal to $1/137$ (the value of $2\pi e^2/hc$) times the field at the surface of an electron ball. For smaller field strengths, the nonlinear term in the Lagrangian yielded events up to fourth order, such as the scattering of photons on photons. For higher field strengths, the expression could yield, they thought, "an important clue for the further development of the theory."[37] Returning to Copenhagen from a visit to Leipzig concerning the plights of refugee physicists, Bohr was at least not pessimistic about Heisenberg's physics. Heisenberg, still in touch with his former Göttingen teacher, wrote to Born that the Born-Infeld theory of spherical electrons "represents a correct correspondence analogy to the theory still to be sought."[38]

But the ever critical Pauli had become more skeptical than ever. He too believed that quantum physics was in a muddle similar to its state just before the invention of quantum mechanics. Each physicist reacted in typical fashion. While Heisenberg, though aware of the problems, struggled with the mathematical formalism in search of clues, Pauli demanded new physical insights, not new mathematics. In November 1935, he nearly repeated his famous remark to Kronig of ten years earlier: "It is very bad at the moment for theoretical physics, and I would grasp at any straw that offers itself to me, who is drowning in the Heisenberg-Dirac hole formalism."[39]

Heisenberg recognized that they had reached the limits of hole mathematics in the new Lagrangian. After listing the still unresolved difficulties and after comparing their equations with those of Born and Infeld, Heisenberg and Euler concluded that further advance required more knowledge of events occurring at tiny distances shorter than the electron radius. An analysis of experimental data on cosmic-ray showers—which still seemed to indicate a breakdown of positron theory and which involved energies far above that required to penetrate electrons—"is probably indispensable."[40] Leipzig physics had now come full circle.

Yet while circling through the formalism, Heisenberg and Euler persisted in their refusal to acknowledge, at least publicly, any nonscientific influence on their work. The names of Jewish scientists continued to appear without apology in their German publications (publications that most Nazis, save perhaps Stark, would not read anyway). But the "ugly" outside world intruded nonetheless into their Leipzig "island of existence." As early as 1934, the authorities had forced Kockel to resign as Heisenberg's deputy assistant because he had once belonged to a socialist student group.[41] Although he was permitted to stay in Leipzig, his dissertation and his career were delayed by the official disapprobation.

Hans Euler experienced similar indignities. After Euler received his doctorate in 1936, Heisenberg wanted him to replace Weizsäcker as his assistant. Weizsäcker had recently habilitated and could thus no longer serve as assistant. Also, as Heisenberg wrote to Debye, now director of the new Kaiser Wilhelm Institute for Physics in Berlin, Euler's talents corresponded more closely to Heisenberg's current interests, high-energy field theory, than did Weizsäcker's. A falling-out with Weizsäcker and his family may have stimulated Heisenberg's push for Euler, as well.[42]

Heisenberg also wanted to protect his new protégé with the assistant's position. Euler privately sympathized with aspects of Soviet communism, but his public position was one of noncooperation with Nazi authorities. He attracted regime attention when he steadfastly refused to join any party organizations or to attend an indoctrination camp for prospective teachers. When Weizsäcker, with Debye's help, moved to Berlin as assistant to Lise Meitner, the Saxon Culture Ministry used Euler's nonparticipation to deny him Weizsäcker's post.[43]

After the ministry's refusal, Heisenberg managed to have Euler named to the temporary post of "extraordinary assistant," while working to have the position upgraded. Lengthy negotiations a year later between

the ministry and the newly named prorector of the university, Heisenberg's professors' club colleague Helmut Berve, finally resulted in
Euler's promotion to full assistant.[44] But a plan to habilitate him in 1938
caused further problems. After demanding repeated statements from
Euler on his political and religious affiliations, the authorities again
insisted that he attend a teachers' camp per regulation, and Euler still
refused per political scruples. Heisenberg finally advised submission to
what seemed to him a mere formality — before he reported for his own
required military training that summer.[45] The sensitive (and sensible)
young man followed his teacher's advice, only to return to Leipzig
thoroughly unsettled by the experience. The weekend of paramilitary
drill and frightening lectures by Nazi demagogues on National Socialist
science and education ideology must have been unbearable. It was the
first of several events portending Euler's tragic fate in World War II.

The treatment accorded Heisenberg's disciples is indicative of the
kinds of pressures that the Nazi regime exerted on young academics, the
future teachers, scholars, and scientists of the new Reich. It also points
up one of the party's main education goals during the middle years of its
reign: the control of teaching personnel and their curricula.

Having lost the battle over dismissals by the end of 1935, physicists
now turned their attention to the major remaining issue, other than the
actual content of scientific research: the control of the physics profession through the control of academic appointments, curricula, publications, and professional organizations. By protecting the teaching of
science — who taught what to whom from what ideological standpoint
— they reasoned, they could protect the future of German science.

Professors were not alone. A number of state and party authorities
and individual pretenders to authority also vied for control of the
physics profession. The main institutional contender was the Reich
Education Ministry (REM) under Bernhard Rust. Hitler established this
cabinet-level ministry, an expansion of the Prussian Culture Ministry, in
May 1934 with the initial consent of those with similar dictatorial
interests in the area of education: Rudolf Hess (Hitler's deputy as party
boss), Alfred Rosenberg (party ideologue), and Heinrich Himmler (SS).
The REM functioned ostensibly as the highest authority for the naming
of university rectors, faculty, and student leaders and for the promulgation of degree and curriculum regulations. But in actuality, under the
weak and vacillating Minister Rust, it also served as a battleground for
rival factions and their representatives.[46]

As usual, ideology served the political aims of its protagonists. As the
pretender to authority in all matters academic, the REM attempted to
kill the protective notion of the apolitical professor — the renewal of the

German Volk demanded submission of the individual to the aims of the Volk. Professors must be leaders and molders of young people, Rust declared in a public address, and would therefore be chosen for their ability to guide young people to their future roles in the Reich.[47] In practice, however, ability counted as much with Rust as ideology.

Nevertheless, Rust's homage to Nazi ideology placed academics in a difficult position, which their enemies could easily exploit. The academics' traditional response to ideological demagoguery — that the apolitical and objective nature of science and scientists necessitated neutrality in research and faculty selection criteria — could be considered not merely annoyingly uncooperative but overtly antithetical to Nazi aims. For Nazi ideologues, to be acceptable, science, along with all other forms of learning, was to be perverted into a racially defined "völkisch," "deutsche" (German), or "Aryan" science; in every field, those seeking personal or political gain propounded their own Aryan doctrines.[48] One of the most pernicious of these doctrines was the "Aryan" or "deutsche" physics propagated by the Nobel prize-winning experimentalists Philipp Lenard and Johannes Stark.

These wretched individuals and their ideology have been discussed and analyzed at some length.[49] They typified to an extreme the predicament faced by many late nineteenth-century physicists transported into the strange world of twentieth-century relativity theory and quantum mechanics. Both had enjoyed honor and recognition as experimental physicists at a time when experimental physics was preeminent and the German empire was in order. Stark had even been one of the precious few early supporters of Einstein's light-quantum hypothesis, had performed experiments confirming the hypothesis, and had helped Einstein publish an early review article on light quanta. But as the monarchy collapsed in socialist revolution and the new field of theoretical physics suddenly gained preeminence, Stark and especially Lenard reacted with bewilderment and anger. That their field was no longer dominant was made painfully clear to them by the world fame accorded Einstein for a new and strange relativity theory that Lenard, at least, despite his Nobel prize, probably never fully comprehended. Both Einstein and Heisenberg later surmised that Stark's chagrin exacerberated existing pathological elements in his personality, rendering him a paranoid anti-Semite and, like his friend and cohort Lenard, an early follower of Adolf Hitler.[50]

Stark's paranoia was confirmed for him when, after he succeeded Willy Wien in Würzburg in 1920, he repeatedly failed to secure another suitable teaching chair anywhere in Germany, despite his Nobel prize.

The difficulty was his own doing. In 1922, he had left his Würzburg chair and used his Nobel prize money to buy a nearby porcelain factory, where he set up a private research laboratory—all in violation of the Nobel Foundation rules. While railing against general relativity and the Bohr atom from his porcelain works, he accused theoreticians of sabotaging his efforts to return to an academic appointment. For his part, Stark's cohort, the black-bearded Lenard, refused in the same year, 1922, to follow a government decree to close his Heidelberg laboratory and lower the institute flag in mourning after the assassination of the liberal Jewish cabinet minister Walther Rathenau. Leftist students decided to persuade the crusty old physicist by dragging him from his laboratory and dunking him in the nearby Neckar River—an act that fired his paranoid hatred for democrats and Jews even more.

Stark made his final pre-Reich attempt at an academic post in 1928, this time in Munich, where he hoped again to succeed the now lately deceased Willy Wien. But a faculty committee headed by Sommerfeld refused to place him on the list of candidates. Nobel laureate Stark seemed to Sommerfeld unqualified for the job: he had renounced his teaching position, violated Nobel Foundation rules, and produced little of scientific value from his private lab. Nor could Sommerfeld imagine working with a man whose scientific and political views were so far from his own.[51] The Munich faculty settled on the conservative Nobel prize-winning experimentalist Walther Gerlach, who had worked with Otto Stern on a famous 1922 atomic-beam experiment that lent overwhelming support to the notion of half-integer quanta of angular momentum in atoms. Stark regarded Sommerfeld from then on as his enemy.

By the late 1920s, Stark and Lenard were turning out an increasing number of speeches and pamphlets on their anti-Semitic dogma of "deutsche" physics.[52] This dogma, though never formally articulated, had at its center the Nazi belief that human creations, such as science, are the sum of individuals' contributions, which are in turn expressions of each individual's race and ethnic origin. It was the antithesis of the Marxian concept of science, according to which science is the nonindividualistic product of the economic conditions prevailing in a given society.[53] In counterpoint to then current Soviet notions of capitalistic versus communistic science, Lenard and Stark propounded "völkisch" notions of a "deutsche" versus a "Jewish" physics. In the foreword to his 1936 textbook on classical mechanics, entitled *Deutsche Physik,* Lenard proclaimed the racist foundation of physics: "'Deutsche physics?' one will ask.—I could have also said Aryan physics or physics of the Nordic-natured man, physics of the searchers of reality, of the seekers after truth, physics of those who have founded natural research. — 'Science is and remains international!' someone will object. He is however fundamentally in error. In reality science, like everything else that mankind produces, is conditioned by race, by blood.''[54] For Lenard and Stark, "Jewish" physics found its ultimate expression in the so-called formalistic relativity and quantum theories, which, they claimed, were not only intellectually obscure but in fact contradicted nature — that is, their naive realist conception of it. "Deutsche" physics, on the other hand, was based on the truth of Newtonian mechanics. This distinction, however, became explicit in their writings only as conditions changed after 1933.

As Nazi old warriors — both literally and figuratively — Stark, then 59, and Lenard, then 71, suddenly found themselves close to the seats of power in 1933. "At last the time has come when we can bring our conception of science and of scientists into play," Stark wrote to Lenard, and Lenard, though in retirement, boldly suggested to Hitler that in all university personnel matters Hitler should "obtain my counsel before making a decision."[55] Hitler did not need the advice.

Like regime policies as a whole, Lenard's and Stark's ideological writings through 1935 reflected less concern for academic dogma than for self-serving political manipulation — the freeing of physics from "Jewish-Marxist domination" through the appointment of themselves to leading positions. In a collection of articles published in 1934, a copy of which Heisenberg kept for evidence in his private files, Stark railed against the "Jewification of German science" during the Weimar years. This had supposedly resulted in "the holding down or exclusion of individual, German-conscious scientists" — presumably referring to himself — and in the propagation of "great dogmatic theories . . . Einstein's relativity theories, Heisenberg's matrix theory, and Schrödinger's wave mechanics."[56]

While Lenard remained in official retirement and served as unofficial patron and advisor to Nazi physicists, Stark was well rewarded for his polemics. In 1933, an acquaintance of Stark, Hitler's new interior minister Wilhelm Frick, appointed Stark head of the prestigious Physical-Technical Reich Institute (PTR) in Berlin (the German bureau of standards) — even though Stark's colleagues had opposed and still opposed the appointment. At the Würzburg meeting of the German Physical Society that fall, Stark unveiled his fantastic plans for the reorganization of all scientific research and publications under the general direction of the president of the PTR and submitted his candidacy for president of the society and for membership in the Prussian Academy as

Einstein's successor.[57] With significant segments of the physics community supporting the new regime and the rest optimistically preoccupied with quiet diplomacy, the physics profession as a whole failed to mount broad-ranging, public opposition to a physicist backed by the regime. As president of the German Physical Society, Max von Laue's single-handed but only partially successful opposition to Stark's dictatorial pretensions drew Stark's ire and earned Laue the praise and gratitude of physicists everywhere. But a year later, neither Laue nor awakened scientists from numerous fields could prevent Stark's appointment as president of the Notgemeinschaft (Emergency Association of German Scholarship), the highly successful funding agency of the Weimar era.

Stark's appointment ultimately led to his undoing. He soon found himself in the middle of the competition among competing regime power blocs represented in the Reich Education Ministry. As described by others from varying perspectives, hostilities began when Stark's personal plans to administer German science clashed with the government's decision to establish the REM for the same purpose.[58] One author argues that Theodor Vahlen, then head of REM office W-I for university affairs (W stands for Wissenschaft, scholarship), had appointed Stark to head the Notgemeinschaft in order to co-opt the vociferous physicist.[59] But plans by Erich Schumann and Rudolf Mentzel in REM office W-II (army research) to reorganize scientific research along military lines using the Notgemeinschaft—newly named the Deutsche Forschungsgemeinschaft (German Research Association)—met with violent opposition from their intended subordinate, Stark. Hardly envisioning himself a mere pawn of the German army, Stark summarily threw the two men out of his office. This alienated not only REM office W-II but also the army and the SS, of which Schumann and Mentzel, respectively, were members. In his efforts to become physics Führer, Stark now realized that he would need the active support and protection of a power center within the regime. Stark turned to Hitler's caretaker of party ideology, Alfred Rosenberg, whom Stark had earlier befriended and whom he now named a patron of the Research Association.

Late in 1935, as Heisenberg's attention turned again to cosmic-ray showers and high-energy physics, Stark's attention turned to the control of REM appointment policy and to Heisenberg in particular. The Munich faculty had begun to seek a successor for Stark's old nemesis, Arnold Sommerfeld, and Sommerfeld made no secret of his top choice for the position—Werner Heisenberg.

CHAPTER 18

Sommerfeld's Successor

If Planck was considered the dean of German theoretical physics, then Sommerfeld was the professor. Ever since his 1906 appointment as Boltzmann's successor in Munich, Arnold Sommerfeld had produced a steady stream of world-class theoretical atomic physicists. His textbook, *Atombau und Spektrallinien,* had educated an entire generation of physicists, and even after Hitler's rise to power his institute remained a center for theoretical research. But two years into the Third Reich, his tenure in Munich neared its end. On January 21, 1935, Reich Minister Rust promulgated a new law, under Hitler's signature, mandating retirement for university teachers over the age of 65 at the end of the semester.[1] Sommerfeld was 66.

Professor Sommerfeld must have learned privately of the new law before its official enactment: in early January he repeated the choice he had made for his successor back in 1928 — Werner Heisenberg. Heisenberg, barely half Sommerfeld's age, again could think of no better place to settle than his home town, where his friends, his mother, the blue skies and Alpine hills of Bavaria, and the Sommerfeld tradition all beckoned his return. Leipzig had been only a temporary way station until he could realize his dream of returning to the place of his formative youth. Once again, he gratefully accepted the honor as Sommerfeld's chosen heir. "If fate should place me in this position," he wrote his former mentor, "I will make every effort to hold up the tradition of the 'Sommerfeld school.'"[2] Anticipating difficulties with the party, he outlined for Sommerfeld his political background in case the information should be needed—as indeed it was. But the initial hurdles to Heisenberg's appointment were bureaucratic, not political.

Heisenberg was still Göttingen's choice as intended successor to Max Born. In addition, Debye was in line to succeed Nernst in Berlin and had just accepted the directorship of the soon-to-be-built Kaiser Wilhelm Institute for Physics, Hund was under consideration for a call to Königsberg, and Otto Scherzer, Sommerfeld's assistant, was likely to move to Darmstadt.[3] Despite Stark's ravings against theoretical physics, both the Saxon and Bavarian ministries worried that such a reshuffle would leave Munich or Leipzig without a theoretical physicist. (Their concerns, however, did not deter the new round of dismissals that spring.)

Cause for optimism abounded, nonetheless. With Rust's approval, the Bavarian Culture Ministry accepted an urgent request from the Munich faculty to allow Sommerfeld to substitute for himself until a successor could be found.[4] The law permitted such a bureaucratic circumvention. More important, the local Nazi lecturer Führer—appropriately named Dr. Führer—who had assigned himself the task of evaluating and approving all personnel matters offered no objection.[5] After Debye assured Sommerfeld that he would not take Heisenberg with him to Berlin and that Heisenberg truly wanted to go to Munich,[6] the Munich faculty submitted its list of candidates to the Bavarian ministry in early summer 1935. According to one source, the list consisted of a single name— Heisenberg. "We would like to help arrange for Heisenberg to work in the place of his choosing," it proclaimed.[7]

Neither the Munich rector nor the Bavarian ministry objected to the prospect of employing one of the most famous physicists remaining in Germany. Heisenberg's candidacy was duly forwarded to the man who now controlled all university appointments in the new Reich—the chemist Franz Bachér, mathematician Theodor Vahlen's deputy chief in office W-I of the Berlin Reich Education Ministry (REM).

Anyone who remained in Nazi Germany after the first wave of dismissals was well aware that compromises were required. This was especially so for professors, who were civil servants—that is, employees of the National Socialist state. Those who chose to remain in their teaching positions and those who attempted to advance their careers by obtaining an appointment to another post also chose to accept obvious compromises. The Hitler salute was required before all public lectures, official correspondence had to be signed "Heil Hitler," and participation in faculty marches, outings, and indoctrination camps could not be avoided. One of the more significant compromises involved professional association with influential government bureaucrats, most of whom had attained their positions solely through their Nazi connections.

Since by late September no decision had been made regarding Heisenberg's call to Munich, Debye visited Werner Studentkowski, the Dresden bureaucrat in charge of Saxon university affairs, on Heisenberg's behalf. Studentkowski had earlier demonstrated his "concern" for objective scholarship by founding the Nazi Students League in Leipzig. He had served as Germany's first professor of Nazi indoctrination, also in Leipzig.[8] Debye found in Studentkowski a surprisingly ready supporter of his former colleagues, the famous scientists under his administration. After discussing Heisenberg's transfer to Munich at length, Debye and Studentkowski decided to visit Bachér in Berlin. Both discovered in their separate visits that, with encouragement from Göttingen, Bachér had finally settled on Heisenberg as Born's successor in Göttingen. Because of this, Bachér had sent Munich's request for Heisenberg back to Munich with a demand for the usual list of three or more names from which to choose.[9] Debye and Studentkowski could only object to the bureaucrat "that Heisenberg himself does not want to go [to Göttingen] and prefers Munich. . . . Another acceptable candidate will have to be found for Göttingen."[10]

On November 4, 1935, just as Heisenberg and Euler were completing their last contribution to positron theory, the nonlinear Lagrangian discussed in Chapter 17, the Munich faculty duly submitted to the Bavarian Culture Ministry (and ultimately to Bachér) the requested list of candidates for Sommerfeld's chair. At the top was Heisenberg and, "far behind Heisenberg," a list of practically all of Sommerfeld's available former pupils and colleagues: Hund, Wentzel, Kronig, Stückelberg, Fues, Sauter, Unsöld, and Jordan.[11]

Faced with renewed pressure from Debye and Studentkowski and the additional support of the dean of the Munich philosophy faculty, Leopold Kölbl, in which Sommerfeld's institute was located, Bachér and the REM were now inclined to approve Munich's choice. Chemist Dr. Rudolf Mentzel, the SS officer in charge of REM office W-II (office for military research), had already found another candidate for Göttingen — Richard Becker, a hapless theorist at the highly regarded Technische Hochschule (TH) Berlin (Berlin Technical College), who had run afoul of the local army ballistics professor. Upon complaints from the ballistics professor, Mentzel dispatched Becker to Göttingen without appeal. This move left the TH Berlin with the largest number of physics students but without a lecturer in theoretical physics — and it clearly demonstrated to German physicists the willful attitude of their masters.[12]

But it also left an undeterred Heisenberg free at last to move to his beloved Munich.

The decision was postponed again. Old warriors Stark and Lenard, wielding their sharpest battle axe — "deutsche" physics — suddenly unleashed a public campaign to thwart Heisenberg's appointment and to strengthen their own influence on REM policies. As Munich again settled on Heisenberg and as Heisenberg and Euler sent their latest positron theory to press in December 1935, Werner was forced again to deal with "ugly" things.

The "deutsche" physics campaign has been described in varying detail and perspective.[13] It was closely related to several significant changes in the Third Reich that took place at about the same time. The first was a major shift from internal policies to police-state tactics, as exemplified in June 1936 by the shift of domestic power from Hitler's political interior minister, Wilhelm Frick, to his new Reich chief of police, Reichsführer-SS Heinrich Himmler. The second change concerned the beginning of overt preparations for war in the first four-year plan, which was aimed at achieving economic and military independence and self-sufficiency by 1939. Concurrent with these events, and following the second wave of dismissals in 1935, the Nazis escalated their suppression of any remaining opposition among non-Jewish Germans by practicing a new form of mental tyranny: accusing critics and opponents of simply thinking or acting "Jewish," thus threatening them with the same persecutions as actual Jews faced. Himmler's deputy, Reinhard Heydrich, spoke in 1935 of a "spiritualization" of the Nazi struggle for domestic control now that many of the Jews and overt political opponents had been ostracized — of a Kampf der Geister (struggle of the spirits) for the hearts and minds of those who remained.[14] Stark and Lenard, though not allied with the Himmler-Heydrich SS power bloc within the regime, eagerly joined the battle.[15]

Stark and Lenard opened their ideological assault in December 1935 at the dedication of the new Philipp Lenard Institute for Physics in Heidelberg. In his laudation for the now retired Lenard, Stark — with his high-pitched voice and meanly contoured face — raved not only at Einstein and other Jewish physicists, who, in his view, had produced the "Jewish formalism" of the relativity and quantum theories, but at "their non-Jewish pupils and imitators" who were teaching and using such physics.[16] Lenard's long struggle against Einstein and "Jewish" physics had, said Stark, only partially succeeded: "Now Einstein has disappeared from Germany; but unfortunately his German friends and supporters continue to act in his spirit." Planck was still head of the Kaiser

Wilhelm Society; Laue was still a physics referee in the Berlin Academy; "and the theoretical formalist Heisenberg, spirit of Einstein's spirit, is now even to be rewarded with a call to a chair." Such a situation could not be tolerated in Nazi Germany. "May Lenard's struggle against Einsteinism be a warning," Stark bellowed. "And it is to be hoped that the responsible referees in the Culture Ministry allow themselves to be guided by Lenard in the filling of physics teaching chairs, including those for theoretical physics."

These were ominous threats and serious accusations and demands. Of course, Stark and Lenard had raved earlier against "Jewish" physics and had even included Heisenberg's name in their ravings, but now the situation was much more serious. With the "spiritualization" of the struggle, the SS in ascendance, and war on the distant horizon, the political orientation of non-Jews was of paramount concern to the regime, and to call a non-Jew the spirit of Einstein's spirit was to call him an enemy of the Reich. Perhaps a parallel can be made with the demogogical blacklisting of liberals accused of being communists during the McCarthy era in the United States. To extract as much political gain as they could from their attack, Stark and Lenard turned to party ideologue Alfred Rosenberg, their ostensible patron in Hitler's cabinet.

Among the many ruthless competitors for power in Hitler's Third Reich, Rosenberg was no match for Himmler, Hess, Göring, or Goebbels. His primary claim to the title of party ideologue was his rambling anti-Semitic diatribe, *The Saga of the Twentieth Century,* which, like Hitler's *Mein Kampf,* few had ever bothered to read. Although Hitler continued to keep him in his cabinet, by 1936 his star was already in decline as others rose in power and stature, but he still held one point of light in his crown—he was editor-in-chief of the official Nazi party newspaper, the *Völkischer Beobachter (Völkisch Observer),* or *VB.* Like official party organs in other totalitarian regimes—*Pravda* in the Soviet Union or *Neues Deutschland* in former East Germany—the *VB* exerted an influence beyond its estimated circulation in 1936 of 500,000. All party officials were required to subscribe to it; rank-and-file party members showed their support by buying the paper; ordinary citizens read the newspaper to learn the regime's official line.[17] If an ideological subgroup could place an article in the *VB,* it would lend official approval to their position and influence regime bureaucrats to act on their behalf.

Soon after the Heidelberg meeting, Lenard complained to Rosenberg of the lack of due attention accorded "deutsche" physics in the *VB.* Rosenberg, willing to support the Nazi physicists only under pressure, responded by inviting Lenard to name a science editor to the paper, which Lenard apparently never did. Instead, at the end of January 1936 an article appeared in the *VB*, entitled "Deutsche Physik und jüdische Physik" ("German physics and Jewish physics"). In February, a transcript of Stark's Heidelberg speech calling Heisenberg the spirit of Einstein's spirit appeared in Rosenberg's monthly magazine for the party faithful, the *Nationalsozialistische Monatshefte (National Socialist Monthlies)*.[18]

Since the aim of Stark's and Lenard's political onslaught was control over university physics teaching, hence over the future of physics in Germany, an otherwise unknown Nazi physics student at the TH Berlin, Willi Menzel (not to be confused with Dr. Rudolf Mentzel), wrote the *VB* plea for the teaching of "deutsche" physics at German universities.[19] The young man was obviously well coached. Quoting at length from Stark's unpublished Heidelberg address and from the unpublished foreword to Lenard's forthcoming *Deutsche Physik,* Menzel tried, as did his mentors, to distinguish a supposedly overly speculative and mathematically abstruse (formalistic) "Jewish" physics from a solidly researched and experimentally grounded "German" physics. "Jewish" physics — the relativity and quantum theories — was not just wrong, it was bad — and it was bad because it was Jewish. For Heisenberg, Planck, Laue, and others who had worked at preserving decent physics at German universities — even if they did not preserve the teachers to teach it — Menzel's closing cry in an official organ of the Nazi party must have been unsettling: "We young people today want to continue the struggle for a German physics, and we will succeed in making its name just as esteemed as German technology and scholarship have been for years." "Jewish" physics was to be suppressed at German universities.

While Stark's Heidelberg diatribe sank to personal attacks on individuals, Menzel's assault focused on the teaching curriculum. He mentioned Heisenberg only once, in passing — as the founder of the formalistic matrix theory — and did not refer at all to Heisenberg's Munich appointment. Menzel's mentors probably reasoned that while a student could plead for curriculum, it would not do for him to complain about a professor, no matter what that professor's position might be. Although the article was a clear setup, a stunned Heisenberg was unsure of its import and unsure whether or how he should respond.[20] At a meeting in February with Munich Dean Kölbl, which lasted until 1:30 A.M., Kölbl confirmed Werner's suspicions: that the article "was expressly meant as an attack against me." Because of bureaucratic opposition raised by the article, his appointment at Munich would have to be postponed for the time being, he told his mother, "but I can wait; this complete idiocy cannot last forever."[21]

While waiting for the irritating idiocy to pass, Heisenberg was not idle. At about the same time as the Kölbl meeting, a befriended party district leader in Leipzig advised Heisenberg to protest the Menzel and Stark articles by visiting the Saxon Culture Ministry and by demanding an audience with the Reich Minister, which he did.[22] Rosenberg—or more likely an editorial staff member—agreed to print a response from Heisenberg in the party newspaper. The article appeared, together with a comment by Stark, in the February 28, 1936, issue of the *Völkischer Beobachter* and was so widely read that the *New York Times* took note.[23] On the same day, Dean Rudorf of the Leipzig science faculty directed a comprehensive letter to Studentkowski, urging him to active support of the physicists' cause.[24]

A now more worried Heisenberg wrote his mother of his apprehension about his future: "I have to have a lot of luck if anything more is to be made of my life."[25] He had begun to personalize the attack even more than was warranted. Of course, the language and the intent of the articles implied a great personal danger for Heisenberg, but they were ultimately aimed not at him but at the physics profession as a whole, of which Heisenberg was being held up as a representative. Nor did his future success in life really depend on whether or not he succeeded Sommerfeld. He began to associate his own fortunes with the fortunes of his profession.

Heisenberg took the lead in responding to Menzel in the pages of the *Völkischer Beobachter*. As he had in his Hannover address to the GDNA in 1934, Heisenberg offered party newspaper readers a carefully worded explanation of the nature and value of contemporary theoretical physics and of its importance to the education of young Germans. This time, however, he did not dare mention any Jewish physicists. The goal of physics, he wrote, is not only to observe nature but to understand it. Mathematics is often the most suitable way to formulate natural laws, but theoretical systems of concepts are more essential, and these often must be adapted as science progresses.

Relativity and quantum theory served Heisenberg as examples. Further research on these theories in particular, "from which perhaps the strongest influences on the structure of our entire spiritual life will arise, is one of the most important tasks for German youth," Heisenberg proclaimed in the Nazi newspaper. "Evidently courage is not quite dead in the universities," the *New York Times* editorialized.[26] But Stark's response, preceded by an introduction by Rosenberg himself, left no doubt that party orthodoxy stood squarely behind Stark's reiterated demand: "The type of physics that Heisenberg defends ought no longer, as it has until now, exert a decisive influence on filling physics teaching chairs."[27]

Two days after the *VB* exchange, Studentkowski met with Dr. Mentzel, the REM official, to discuss Leipzig physics. A memo that Studentkowski wrote to himself after the meeting was to the point: "Professor Heisenberg remains in Leipzig. Professor Hund likewise remains in Leipzig."[28] Despite his penchant for Nazi demagoguery, Studentkowski apparently had no use for "deutsche" physics and had cultivated friendly relations with leading Leipzig scientists who, because of their reputations, could perhaps work to his advantage in Saxon affairs. For whatever reason the Nazi bureaucrat supported the scientists, previous studies of this episode have overlooked the crucial role he played in the affair at this stage.

In a long reply on March 24, 1936, to Leipzig Dean Rudorf's February 28 request for assistance, Studentkowski told of successful meetings with various bureaucrats involved in the affair.[29] First, he wrote, he had managed to find "cover" for Heisenberg in the person of Dr. Lahr, head of the Saxon State Chancellery, who had direct access to the Reich plenipotentiary for Saxony. Heisenberg would not be molested as long as he remained in Saxony. Second, Studentkowski had made inroads with Alfred Rosenberg, who, as readers of the *VB* already knew, had backed Stark before declaring the entire controversy closed.[30]

Most importantly, Studentkowski had Mentzel's ear. Mentzel's earlier conflict with Stark and the Stark-Lenard bid for control of physics appointments would, one might think, have made Mentzel eager to support the physicists. But Mentzel appeared more cautious to Studentkowski. "He does not at all approve unreservedly of Stark's attacks, but he also has certain reservations about the present methods of theoretical physics, or better, about overreliance on them." Stark's and Lenard's connections with Hess's office in Munich through the Munich-based Nazi leagues for students and lecturers, in the person of Dr. Führer, probably gave Mentzel pause. Mentzel and the REM were then seeking improved relations with Hess, Hitler's party deputy.[31] Perhaps, too, Mentzel, acting the impartial bureaucrat, sought to play both sides against each other to his own advantage. For whatever ulterior motive, he willingly granted an audience to Heisenberg — following proper bureaucratic procedures, of course — and asked Studentkowski to relay the message. Studentkowski informed Rudorf: "Mentzel requests Heisenberg through me to visit him sometime, and to make an appointment as soon as possible." Rudorf passed the letter to Heisenberg with his best wishes for an early meeting with the SS officer.[32]

Heisenberg's meeting with Mentzel, probably in early April 1936, was a turning point. According to a form letter sent by Heisenberg, Max Wien, and Hans Geiger to the entire German physics professorate shortly after the meeting, Mentzel, ostensibly speaking for Rust, had requested during the meeting a memorandum signed by "most" German university physicists describing their understanding of the relationship between experimental and theoretical physics.[33] This memorandum — in effect, a petition for theoretical physics — would provide Rust "a suitable means for his instruction, which will then enable him to alleviate the unpleasantly tense situation that has recently arisen." In other words, Mentzel (and possibly Rust) was willing to move against Stark and Lenard, but only if he had the support, in writing, of the overwhelming majority of German physics professors. Heisenberg, Geiger, and Wien wrote a carefully worded memo and attached it to their letter, requesting its return to Heisenberg's private address — with or without signature — by May 19, 1936.[34] Not surprisingly, Lenard and Stark were livid.[35]

Geiger and Wien made ideal coauthors. Both were well-known, politically conservative experimentalists, at once sympathetic toward modern theory and acceptable to REM officials.[36] Wien's cousin, Willy, had been close to Stark, and Geiger, the famous coinventor of the Geiger-Müller counter, was being considered by Mentzel as a replacement either for Debye in Leipzig or for Nobelist Gustav Hertz in Berlin, who, being part Jewish, had resigned in the face of indignities.[37] Wien had already privately expressed some concern for theoretical physics and physics in general in the wake of the first round of dismissals. Encouraged by Debye, he had responded two years earlier by helping to formulate a memo on the declining state of German physics due to regime policies.[38] If that memo was ever actually submitted, it was almost certainly ignored. One SS Security Service (SD) officer assigned to the REM later recalled an office cabinet stuffed with similar memos.

Whether sent or not, the 1934 Wien memo served as the basis for the Heisenberg-Wien-Geiger petition of 1936, which, unlike its predecessor, enjoyed a huge success. Its cardinal point was utilitarian: Germany faced a critical situation regarding physics teaching and personnel that it could ill afford, especially as the four-year plan came into full swing. "The great demand for physicists in technology and the military is met by a lack of suitable candidates. Empty teaching chairs are filled only with great difficulties, and the number of physics students in the beginning semesters is much too small." Moreover, read the memorandum, theory and experiment were both essential for scientific progress and for future technological gains. Public attacks on theory were only frightening students away and damaging Germany's reputation abroad. Attacks against theory must cease in order to stimulate science and to maintain a useful physics profession in Germany.

These were arguments that few could refuse, and few did. Seventy-five physicists — nearly all German physics professors — signed the petition. Included were theorists and experimentalists, pure and applied physicists, party members and nonmembers.[40] It was an outpouring of support for physics and especially for Heisenberg and theoretical physics, support that had been building since 1933. After three years of hand-wringing, at last, it seemed, something could be done — even if that something were to point up the utility of physics to regime policies.

Rust received the momentous memorandum by October 1936 and handed it to his politically adept state secretary for evaluation. The secretary, thoroughly ignorant of physics, could only respond with a criticism of his weakling boss.[41] In the professors' conflict, he wrote in a private memo, one side — Heisenberg's — had sought help from the REM and had even referred to Rust by name in its cover letter and by title in its petition. If this were a purely scientific controversy, Rust should stay out of it; if it were political, which it was, then he, as culture minister, had even less reason "to mix in these things."

Rust decided to leave matters to Mentzel, who by that time had already bested Stark on another front. The cabinet power shifts of 1936 had greatly reduced the influence of both Frick and Rosenberg, leaving Stark and Lenard without powerful supporters. But Stark ultimately did himself in through his foolish mismanagement of Research Association funds. The rabidly empirical physicist had heavily sponsored a speculative project based on an old Teutonic myth in an effort to extract gold from south German heaths. The scandal gave the REM an edge in forcing Stark's resignation in November, and his replacement as head of the Research Association was none other than the indomitable Dr. Mentzel.[42] A visit from Munich Dean Kölbl to the Berlin REM in the same month brought welcome news, relayed to Heisenberg by Kölbl's colleague Sommerfeld: "Kölbl reported to me from the Berlin ministry that you have won in your controversy with St[ark] and L[enard] both 'scientifically' and 'morally.'"[43] Heisenberg would replace his mentor at last.

Closely following the twists and turns of Heisenberg's political fortunes during 1936 was a sudden shift in the prospects for his physics. At the same time as he hit back at the professional assault with a successful memorandum of support, Heisenberg firmly believed that he had found an explanation for the appearance of high-energy cosmic-ray showers that would once again revolutionize quantum physics. The two achievements reinforced his optimism in both areas. Since laboratory accelerators had not yet reached high energies — that is, above about 10^8 electron volts (eV) — during the 1930s high-energy physics concentrated on cosmic-ray events. By 1936, every experiment on the absorption of cosmic rays in matter had seemed to indicate — mistakenly, as it turned out — a breakdown of quantum electrodynamics (QED) near a theoretically expected upper energy limit of about 137 MeV.[44] High-energy cosmic rays penetrated much farther through matter than they should have according to QED. This suggested that the particles simply stop radiating away their energy, thus slowing down, at that energy limit. Internally, QED contained no such limit and required particles to radiate at all energies, without any supposed breakdown. QED would thus not do for shower theories, since showers involved energies far above the supposed upper limit for validity of the theory. Experimental evidence seemed so strong that physicists greeted with frank disbelief the discovery by Weizsäcker, Williams, and Landau in 1934 that no good theoretical reason existed to expect any breakdown.[45] Some incredulous physicists — notably Oppenheimer and Nordheim in the United States — even began introducing cutoffs into QED to match the theory to the data.[46]

Heisenberg took a different tack, pushing existing theory to its limits to try to perceive what lay beyond. He caught his first glimpse of a new frontier in May 1936. Exactly one week after the stated May 19 deadline for receipt of the Heisenberg-Wien-Geiger petition at Heisenberg's home address, Heisenberg reported to Pauli from his institute that an alternative to QED yielded new insights.[47] Italian physicist Enrico Fermi, then head of the internationally acclaimed nuclear research group in Rome, had recently formulated an alternative field theory to handle nuclear beta decay — the ejection of an electron and a neutrino from a radioactive nucleus as a neutron turns into a proton. (The emission of a positron, as a proton decayed into a neutron, and the distinction between neutrinos and antineutrinos were later discoveries.) When taken out of the nucleus and applied to the high-energy collision of a proton with a nucleus, the new field, Heisenberg discovered, produced an explosion of particles as soon as the particles approached closer than a minimum length, about the size of an electron. "It thus appears to me," he wrote, "that one can understand the existence of cosmic-ray showers immediately from the Fermi β theory."

Nuclear beta decay involved an electron, which was also a primary player in quantum electrodynamics and the basic source of electric and magnetic fields. Although QED could account for the creation of an electron out of electromagnetic energy, it could not account for the decay of neutrons nor for the nonelectromagnetic nuclear force holding neutrons and protons together in a nucleus. These were not electromagnetic processes. Inspired by the discussion of Heisenberg's neutron-proton nuclear theory, which Heisenberg had presented to Fermi and others at the 1933 Solvay Congress, Fermi had returned to Rome, where, in December 1933, he developed a theory of the neutron-proton force that relied on an entirely different field.[48] In form, however, it differed only slightly from Heisenberg's nuclear theory, according to which the neutron-proton force arises from the virtual exchange of an electron ("virtual" because energy conservation is violated for an instant). Together, an electron and a proton constituted a neutron.

Unfortunately, Heisenberg's exchange force violated energy and momentum conservation laws, unless one included the hypothetical massless "Pauli neutrinos" in the theory (the word *neutrino* is Italian for "little neutron"). Fermi showed that both Heisenberg's force and conservation laws could be maintained if the neutron and proton exchange not just an electron but an electron and a neutrino.

Fermi constructed his theory in close analogy to QED, but it involved an entirely different type of field. Much as an electron and a positron are attracted to each other in QED by tossing back and forth a photon, the quantum of the electromagnetic field, so a neutron and a proton are attracted to each other in a nucleus by "playing catch" with an electron and a neutrino, the quanta of a new "Fermi field." The free emission of such a field yielded the observed beta decay, just as the emission of light quanta could be observed as spectroscopic radiation.

In Fermi's formulation of field theory, an analogy is made with Hamilton's old formulation of classical mechanics, in which an energy function, the Hamiltonian, carries all the properties of the system. The analogy had been used earlier in Schrödinger's formulation of the wave equation and in Dirac's quantum perturbation theory. In the perturbation formulation of field theory, separate Hamiltonian functions (actually densities) are obtained for the matter and the field, while the interaction between the two — which gives forces — is treated as a small perturbation of the separate agreement. The perturbation is represented by a Hamiltonian density, H_{int}, containing wave functions representing the interacting matter and field quanta.

Fermi's Hamiltonian density for the interaction, H_{int}, may be written in modern relativistic form:

$$H_{int} = f(\overline{\psi}_P \gamma_\mu \psi_N)(\overline{\psi}_e \gamma_\mu \psi_\nu)$$

where ψ_P, ψ_N, ψ_e, and ψ_ν are the proton, neutron, electron, and neutrino wave functions, respectively; γ_μ is the 4×4 relativistic gamma matrix; f is a constant; and a bar over a wave function indicates the complex conjugate adjoint. (This version of the theory did not contain antineutrinos). The size of the coupling constant, f, represents the size of the interaction. Fermi made his constant analogous to its counterpart in QED, the so-called fine-structure constant, $\alpha = 2\pi e^2/hc$. In Fermi's theory, $f = 2\pi g/hc$, where g is a constant, but unlike the dimensionless α, f has the dimension cm^2.

Fermi's field was widely studied in Leipzig and elsewhere as an explanation both for nuclear forces and for the observed properties of beta decay.[49] Heisenberg, who read Fermi's first paper in Italian, was at first very excited by it. But his enthusiasm waned when he discovered in 1934 that the resulting nuclear force was much too weak to account for the size and range of the neutron-proton force in nuclei or even for the observed distribution of beta-decay energies. Indeed, during a four-part lecture series on nuclear physics, delivered in the spring of 1934 to Dirac's institute in Cambridge, England, Heisenberg found the force to be as much as 10^{10} too small.[50] Various ad hoc proposals were put forward to fix the Fermi force, and these were intensively studied in Leipzig and Zurich thereafter. All proved unsuccessful. The Fermi force — which was later relegated to only the weak interaction within protons and neutrons — remained an unsatisfactory account of the nuclear force and, at the time, even of beta decay.[51]

Even worse, because of the size of the matter-field coupling constant, f, the self-energies of the neutron and proton — the calculated energy of a proton or neutron sitting alone, surrounded by its Fermi field — diverged even more rapidly to infinity than did QED; a discouraging situation, to be sure, but Werner thrived on challenges. Having exhausted positron theory and having turned to cosmic rays for clues on how to handle distances shorter than the size of an electron, Heisenberg reported to Pauli in May 1936, as the physicists' memorandum circulated throughout Germany, a new and momentous discovery that could potentially revolutionize physics. In a perturbation expansion of Fermi's

H_{int} in analogy to QED, the contribution to the higher order terms diverged to infinity as they did in QED but in a fundamentally different way that accounted for one phenomenon that QED could not — the appearance of cosmic-ray showers.

A cosmic-ray shower arises when a high-speed particle hits a piece of matter, producing in the course of the collision numerous new particles. The creation of material particles had to be understood in field theory in a very fundamental way, as the transfer of energy from the collision to the field quanta excited by the collision, which are then detected as real (as opposed to virtual) particles in a cloud chamber or Geiger counter. In QED, the creation of a certain number of particles (electrons and positrons) is represented by the corresponding order of interaction — the octave of the harmonic vibration mode. For n number of particles, the cross section — essentially the probability — for exciting the nth-order interaction is proportional to $(\alpha)^n = (1/137)^n$. As n, the number of particles, increases, the probability rapidly decreases, since $1/137$ is a small fraction. Large bursts of particles, such as those that had been detected by Geiger counters, or even small showers containing a handful of particles, would thus be extremely rare.

Not so for the Fermi field. Here, the analogous cross section for the creation of n particles depends on a parameter $(\lambda_c/\lambda)^n$, where λ is the wavelength of the incoming proton and λ_c is a critical minimum wavelength, equal to $\sqrt{f}$. As the energy of the incoming proton increases, its wavelength decreases. For low-energy collisions, less than about 10^8 electron volts, $\lambda > \lambda_c$, and from the parameter higher interactions involving the creation of any new particles are rare. But as soon as $\lambda < \lambda_c$, the parameter increases — the interaction cross sections do not decrease but on the contrary grow ever larger, the larger the value of n. An instantaneous explosion of particles occurs within the minute area of collision, resulting in a shower of particles emanating from a single point. The number of particles produced when a proton hits a nucleus is governed only by the conservation of energy.[52]

Heisenberg had hit upon a new and astounding property of Fermi's beta-decay theory: when applied to cosmic rays, it could account for the observed creation of particles in a way that the reigning quantum electrodynamics could not. Despite his astonishment, Pauli responded with typical skepticism. He did not dispute Heisenberg's new use of Fermi's theory, but he doubted that it offered anything new. At very small distances, Fermi's theory went to infinity even faster than did QED. This feature actually unleashed the particle explosion, but to Pauli it indicated the need for a new revolution in physics to handle high energies, not the usefulness of alternative field theories.[53] Heisenberg actually agreed, but he believed that the Fermi formalism gave an important clue as to where to look for this revolution. He quickly submitted a manuscript on the subject in early June 1936 and took a copy with him to the annual Copenhagen physics conference later that month.[54]

As he wrote in the paper and argued in Copenhagen, the clue to the future lay in the critical wavelength λ_c. Since it controlled the creation of showers of new particles, and in a way that he had expected from his positron work with Euler, Heisenberg saw in it a fundamental characteristic of the future theory. This theory would contain a new fundamental constant, a critical length $l_0 = \lambda_c$, which would serve as the long-sought link between matter and fields and thereby remove the mathematical divergences that occurred in attaching fields to matter. The task for theorists now was to find a consistent way to introduce this new critical minimum length into physics and thus achieve the new theory. The result would be revolutionary. Indeed, he wrote (avoiding mention of Einstein's name), "It must probably be connected with a fundamental alteration of the formalism [the mathematics], just as, for instance, the introduction of the constant c gave rise to a modification of prerelativistic physics."[55]

But the need for revolution — and the direction in which to look for it — was not easily accepted outside the Leipzig-Zurich axis. During the Copenhagen meeting, attended by numerous refugee physicists, Walter Heitler, then in Bristol, stunned the audience with his announcement of Anderson's latest results on cosmic-ray absorption.[56] Despite theoretical arguments to the contrary, physicists were still convinced by cosmic-ray absorption data that QED was invalid for energies at which showers of new particles begin to appear. In May, as Heisenberg was busily submitting his Mentzel memo and uncovering the revolutionary shower properties of the Fermi field, Anderson was privately informing Bethe and Heitler in England and Oppenheimer at Berkeley of his new data on the energy loss of electrons. Unlike all his previous experiments, Anderson's new cloud-chamber arrangement was no longer biased against large energy losses. His data suddenly agreed with, rather than contradicted, the theoretically predicted absorption of high-speed particles using quantum electrodynamics! "As you can see," he wrote to Heitler, who passed it on to the Copenhagen meeting, "these data do not show any breakdown of the theoretical formulae for energies ≈ 300 MeV. The experiments are not accurate and the numbers measured are small, but certainly no large disagreement appears."[57] If QED did not break down, there was no need for the revolution that Heisenberg was espousing or even for the introduction of any other field theory to account for high-energy experiments.

Anderson's new experiments also clearly distinguished for the first time the two very different components of the primary cosmic rays at sea level: the "soft," easily absorbable, electron-photon component of the incoming rays; and the "hard," long-ranged component. While the soft component now showed precise agreement with QED to energies far beyond the supposed breakdown of the theory, the hard component still offered problems and was set aside for the time being. It penetrated such large layers of lead blocks that Anderson surmised that either QED eventually did break down at some very high energy or these particles were of an entirely unknown composition.

Anderson's new data on the soft electron-photon component of cosmic rays, which he published with his assistant, Seth Neddermeyer, arrived at the *Physical Review* just one day before Heisenberg submitted his shower paper to the *Zeitschrift für Physik*.[58] These two papers initiated two conflicting approaches to showers and their field-theoretic implications that were rivals into the war years and beyond. While Heisenberg's explosions attracted seekers of a new and even more radical quantum revolution — mainly the leaders of the first revolution, Heisenberg, Pauli, and their remaining Central European collaborators — the Anderson-Neddermeyer paper confirmed the Bethe-Heitler theory of absorption for the soft component and catalyzed the calculation of "cascade" showers using QED. For the supporters of this rival account of high-energy particle creation — mainly Oppenheimer, Heitler, and their collaborators in England and the United States — no new revolution was necessary at all: QED would do just fine. Heisenberg and Oppenheimer, so similar and yet so different, would never reach agreement on this or other matters during the coming years.

According to the cascade theory — presented independently by Oppenheimer and his student, J. F. Carlson, and by Heitler and Dirac's former Cambridge student, Indian physicist Homi Bhabha — a cosmic-ray shower is formed not by an instantaneous explosion but by a succession of well-understood, individual Bremsstrahlung (braking radiation) and pair-creation events as an incoming high-energy electron, positron, or photon bounces between atoms after hitting a thin piece of matter.[59]

Yet until the Oppenheimer and Heitler groups published in early 1937 their detailed calculations of the formation of cascade showers, Heisenberg's theory of "explosion" showers — the formation of a shower in a single event using Fermi's alternative electron-neutrino field — remained the most plausible account of showers, and his physics remained the most likely replacement for QED. After crudely manipulating the Fermi interaction to obtain the correct size of the nuclear force and the correct critical length for showers, Heisenberg bolstered his claims for the new physics with empirical predictions that were astonishingly consistent with available data. According to Heisenberg's derivations, the critical energy for the creation of showers of particles would be about 100 MeV, and the cross section would be about 10^{-26} cm^2, both of which agreed well with observations. Moreover, a shower would occur in a single act (an explosion), not in a series of individual events (a cascade). This seemed to be confirmed by the sudden bursts of ionization occurring in Geiger counters and by Blackett's photographs showing showers emerging from a single point in the cloud chamber. The showers themselves, Heisenberg predicted, would contain mostly electrons and slow Pauli neutrinos—the quanta of the Fermi field. The electrons produced would constitute Anderson's observed soft component within the showers themselves. The created neutrinos would constitute the hard component, which, according to Heisenberg's theory, could penetrate thick absorbers to produce further showers by induced beta decay.[60] Heisenberg's Fermi explosion-shower theory seemed to cover every observation.

The empirical consequences, Heisenberg's firsthand experience in struggling with the problems of quantum electrodynamics and Dirac's positron theory, and especially the new vistas opened by the Fermi field convinced Heisenberg both of the need for a new physics and that he really had a grasp of it. Quantum electrodynamics had numerous problems from the very start and could not, in early 1936, account for the phenomenon of showers. Therefore, Heisenberg wrote to Pauli, "I truly believe that the connection between the Fermi theory and shower formation is a very central point in the theory of matter."[61] In the politically charged atmosphere of 1936, Heisenberg's work had an additional significance. It was an indication that, as Gerlach responded to one foreign critic, "the scientific life here is absolutely not dead at all but is actually growing."[62]

The new theory could also serve as an object lesson on the productivity of the much-maligned theoretical physics. Between mountain outings and summer military camp, Werner wrote an article informing the educated German public of the potentially significant consequences of his new theory of showers. These, he emphasized, "are capable of experimental verification."[63] Far from generating purely abstract mathematical formalisms, Heisenberg claimed that he had discovered nothing less than a new universal constant whose "introduction requires a reformulation of the entire theory, as was the case for the constants h and c." As any reader who had experienced an introduction to modern physics knew, these two constants were the basis of quantum theory and relativity theory, respectively. Further theoretical analyses of empirical data on cosmic rays — not mathematical manipulation — "promise the most important contributions to the fundamental physical questions."

The political and cultural implications of the new physics may have encouraged Heisenberg's more grandiose claims for it. However, there is no evidence that any of his German colleagues saw it that way, even while expressing their overwhelming support for his petition in defense of their profession. Even petition coauthor and cosmic-ray investigator Hans Geiger remained unconvinced of Heisenberg's explosions. Geiger learned of explosions and their possible rival, cascades, not from Heisenberg but at a Zurich nuclear physics conference in July 1936, attended by Sommerfeld, Bhabha, Pauli, Schrödinger, and others.[64] He preferred cascades, both in theory and in experiment. Although the possibility of explosion showers remained open, Geiger explained in a public lecture as late as 1940, other cloud-chamber photographs taken by British physicist Blackett appeared to show a buildup of showers — despite the occurrence of bursts in Geiger's own counters![65] If the vivacity of German theoretical physics was to be demonstrated by Heisenberg's new revolution, Geiger would not support the cause. Physics and politics had to be held separate at all costs.

While German scientists were less than enthusiastic about Heisenberg's new theory, Pauli, a foreigner, once again became Heisenberg's chief collaborator and critic. Their collaboration was not at all empirical — as Heisenberg's public may have expected from his report — but it was thoroughly abstract and mathematical. As usual, the two sought to push their formalism to the breaking point and beyond. As a trick for doing so, Pauli resurrected Heisenberg's 1930 lattice world, discussed in Chapter 14 — whereby Heisenberg had sliced up space into cubes formed by the size of an electron as a minimum length. Pauli brought it with him to the Copenhagen meeting in June 1936.[66] Using Pauli's trick, Heisenberg and Pauli analyzed the relation of the fundamental length to showers and infinities in nonlinear, Fermi-like field theories, hoping to find a new theory that contained the length in a natural way. Pauli was skeptical that any consistent theory could arise in this way, but the possibility that an empirical proof for his still hypothetical neutrinos might appear from Fermi showers did not escape Pauli's notice.

Heisenberg did not join Pauli's effort until after completing his required eight weeks of military training in the summer of 1936.[67] Heisenberg was a corporal in a mountain infantry unit in Bavaria near the Austrian border. The rigorous training not only interrupted his work, but also forced the cancellation of a planned trip to the United States to attend the annual Ann Arbor summer school in physics and the elaborate tricentennial celebrations at Harvard University. Werner's abrupt cancellation of his American trip to attend the military training camp made the *New York Times*.[68] The political situation in physics was too volatile to permit his absence from Germany, he explained to his former colleague Samuel A. Goudsmit, then in Ann Arbor.[69] Apparently adherence to German military requirements came before foreign conferences — he could have found an excuse to dodge the military duty, had he so desired. The old Neupfadfinder seemed to welcome the physical challenges of military mountaineering and, he wrote his mother, the escape from responsibility: "I am physically healthy and I very much enjoy the duty itself. It is nice not to have to think for a change, but only to obey. . . . The duty agrees with me in every respect."[70]

Returning to work in Leipzig in October 1936, Heisenberg received a long letter from Pauli proving that the Fermi formalism led nowhere — with or without a hypothetical lattice world.[71] Both physicists reacted in familiar fashion. Pauli declared that manipulations of divergent field theories were similar to quantizing mechanical models in the early 1920s. Since both contained unobservables, he suggested that Heisenberg declare the eigenvalues of all field Hamiltonians to be fundamentally unobservable, then see what happens.[72] Applied to atomic models, it was in just this way that Heisenberg had formulated the breakthrough to quantum mechanics in 1925.

As in the twenties, Heisenberg wanted first to exhaust all lattice-world models, and he presented Pauli with various constructions for them to study and reject together. He revealed his plan to Max Born in November: "The showers are still occupying Pauli and me very much. I am very anxious to know how the work there will proceed. Pauli continually tries to prove that wave quantization always diverges, even *with* a universal length; I privately believe that Pauli is right but temporarily maintain the opposite, and in this way we get to know the mathematical properties of a nonlinear quantum field theory, which are highly interesting."[73] Six days later, Sommerfeld informed Heisenberg of his victory over Stark. Physics and politics, though running on separate tracks, ran along parallel lines for Heisenberg.

19

An Unending Loneliness

Heisenberg's two major successes in 1936—his response to the Nazi dogma of "deutsche" physics and his construction of a potentially revolutionary physics of high-energy cosmic rays—were, in the end, short-lived. By early 1937, the alternative cascade theory of cosmic-ray showers, based solely on the accepted physics of positrons and quantum electrodynamics, had accounted for nearly all of the data on cosmic-ray showers, leaving little need or expectation for any new physics. By that summer, a new and even more vicious attack by the proponents of "deutsche" physics made it impossible for Heisenberg to take his Munich chair and nearly impossible for him to remain in Germany at all. Even Werner's native optimism seemed at times outmatched by the course of events. His optimism received a "quantum" boost during the first half of 1937, when he met and married his wife, the future mother of his seven children. But even with marriage and a growing family, Heisenberg recalled the period before the outbreak of World War II as one of "unending loneliness."[1]

Heisenberg's frequent travels abroad—to Denmark, England, the United States—revealed to him Germany's marked isolation. As nearly every domestic challenge to Nazi policies failed, most Germans who found themselves in similar situations felt discouraged politically, without, wrote Heisenberg, "the slightest hope of a change from within." While he and others rode the storm to its end, the regime continued to engender distrust of all but one's closest friends, vastly increasing the "isolation of the individual," as Heisenberg called it. But it was an isolation to which Heisenberg's personality made him particularly susceptible.[2]

Throughout his life, Heisenberg's confidence in matters scientific contrasted sharply with his lack of assurance in personal affairs. This became manifest whenever circumstances separated him from his younger male companions, even as he approached middle age. His diffidence grew especially evident during the middle 1930s as his youth began to wane and after his only known premarital romance — a relationship with Carl Friedrich von Weizsäcker's charming younger sister, Adelheid — was brought to an abrupt end by her parents. Although Werner is remembered to this day in Leipzig as having dated a variety of eligible local socialites, little is known of his romance with Adelheid or of the reasons for its breakup. No doubt the family objected to the age difference between the professor, then 34, and the young woman, who was still in her teens and only recently graduated (1934) from a literary secondary school in Bern, Switzerland. Her father, Ernst von Weizsäcker, probably also preferred an aristocrat over an academic — even as impressive an academic as Heisenberg — as a suitor for his daughter.

During his semester break in March 1936, Heisenberg joined Carl Friedrich and the Weizsäcker family for a mountain-climbing excursion near their home in Bern, where the elder Weizsäcker was at that time Germany's ambassador to Switzerland. Werner later wrote his mother that he was so distraught over his renewed encounter with Adelheid that he could not even stop in Munich on the return trip to Leipzig.[3] Her father's resistance was the main obstacle, he wrote, but the teenaged Adelheid herself also seemed uncertain, not knowing what she wanted.[4] Two years later, she married an ennobled Army captain, Botho-Ernst zu Eulenberg, who was then on active duty with the Prussian infantry. He was later killed in action on the Russian front, leaving Adelheid alone in war-torn East Prussia with two small children.[5]

Heisenberg returned from Bern to the "horribly desolate" Leipzig in a depressed state. There, an uncertain future and a physics still swamped by cosmic-ray showers awaited him. His Munich appointment was, as of March, postponed by the newspaper onslaught, and the revolutionary prospects afforded by Fermi showers were not yet in sight. Even worse, his closest companion, Adelheid's brother Carl Friedrich, definitely intended to end his awkward situation as Heisenberg's assistant by habilitating as soon as possible and moving to Berlin.[6] Werner's mother again worried that his life was limited to too few people, to which he responded as before: "But it just doesn't work out any other way. And in physics I have also not done it any other way."[7]

Scientific and political success later that year raised Heisenberg's spirits briefly, but by November 1936, as the days shortened into winter and a new semester began, depression cast a pall over his victory in the public battle with "deutsche" physics. Weizsäcker's absence and the probable severing of relations with the entire Weizsäcker family left Werner lonely indeed. In his annual reflection on his life on the anniversary of his father's death, Heisenberg rededicated himself to his work, "for the sake of which I appear to have come into this world." But, hard as he tried to make work substitute for personal relationships, as he had so often in the past, he recognized that now that could be only a temporary solution. As he wrote his mother, "The single life is bearable to me only through my work in science, but for the long term it would be very bad if I had to make do without a very young person next to me."[8] Three months later, Werner was engaged to a woman 13 years younger than he. The pattern was already set by his close relationships with his younger companions and most recently by his attraction to Adelheid: he needed a much younger person to keep him in touch with the outside world and with his own youth; for marriage, only a much younger woman would suit the professor's tastes, "even though," he admitted, "I am not very young any more and no longer as lively as I want to be."[9]

While immersing himself in the antidote of physics, Werner remained connected to cultured social circles in Leipzig through music. One of those circles had revolved for several years around the home of Otto Mittelstädt, a Leipzig publisher. Part of Mittelstädt's house was occupied by his in-laws, the Bückings, also a publishing family. Toward the end of January 1937, Werner participated in one of his frequent evenings of chamber music, this one held in the Bücking home.[10] The piece to be played was Beethoven's Trio in G Major, which Werner had known from his youth. He was accompanied on the cello by Bücking and on the violin by his former professors' club colleague, Erwin Jacobi, who, though dismissed from his teaching post because he was Jewish, held on in Leipzig in private legal practice.[11]

Among the small number of invited guests for the concert that evening was a young woman, a tall and slender book dealer with a warm smile and a pretty figure who had recently arrived in Leipzig. She had become acquainted with the publishers through her occupation. This was the first time she had attended one of the Bücking affairs, and, as the trio began the performance, her bright eyes met those of the pianist as his practiced fingers caressed the keys. Heisenberg and the young lady, Elisabeth Schumacher, soon struck up a conversation that did not go unnoticed by their hostess, Mrs. Bücking. At the end of the evening, as the guests prepared to leave, she politely asked, "Mr. Heisenberg, would you please accompany Miss Schumacher home?"[12] Within a week the professor and the well-educated book dealer were planning a trip together to Heisenberg's Bavarian ski hut (with Hans Euler along as chaperon). Two weeks later, on February 11, they were engaged, and less than three months after that, on April 29, 1937, they were married in Berlin.[13] Heisenberg was then 35 years old, his bride 22.

Elisabeth was the youngest of five children born to Hermann and Edith Schumacher. Elisabeth's father was a Bonn professor of political economy, best known for his annexationist proposals during World War I.[14] Elisabeth's brother, Fritz, was also an economist. He emigrated to England in 1939 where, in contrast to his father, he was well known for his theories of small-scale enterprise.[15] A sister, Edith, married the widely read German journalist and war correspondent Erich Kuby.[16] In 1917, as the world war ended, Professor Schumacher had assumed the chair for global political economy at the University of Berlin and moved his growing family from Bonn to the elegant upper-class suburb of Berlin-Steglitz. The family occupied a spacious turn-of-the-century mansion in Arno-Holtz-Strasse near the botanical garden on the Fichte-berg, where the professor enjoyed his daily walk.

Elisabeth's father, like Heisenberg, had married a much younger woman relatively late in life. His granddaughter's description suggests that he was typical of his generation: "Single-mindedly wrapped up in his own life: dogmatic, authoritarian and dedicated to the pursuit of his career." Life in the Schumacher home bore a striking similarity to the environment in which Werner grew up: "Disciplined and regular, dic-tated by the professor's needs."[17] No wonder that, upon graduating from a Berlin gymnasium in 1933, Elisabeth sought her freedom at the other end of the Reich, in Freiburg, Baden, in southwest Germany. She studied German literature at the University of Freiburg, graduating in 1936, and while in Freiburg became romantically involved with the physicist Wolfgang Finkelnburg, who would later play a role in the response to "deutsche" physics. When Finkelnburg asked Elisabeth for her hand in marriage, she respectfully declined and moved to Leipzig, then the center of German publishing, where she entered the book trade.[18] Although women were hardly accepted in business at that time and especially under that regime, exceptions were made if one knew the right people, as she apparently did.

Meeting another physicist so soon after moving to Leipzig naturally awakened mixed feelings in Elisabeth. Finkelnburg had spoken highly of a Heisenberg whom he knew, but she did not appreciate until later that he was referring to *this* Heisenberg.[19] As children of oppressive upper-middle-class professional homes and rebounding at the time from failed romances, they both felt from their first meeting at the Bückings' that they were right for each other in many ways. "That evening decisively changed our lives," she later wrote. "We both felt that we had encountered 'our fate.'"[20] As Werner portrayed it to his mother, a conversation that had started out superficially soon revealed "a close agreement of opinions between her and me on the essential things. . . . This mutual understanding . . . soon went so far that it appeared to me natural to ask Elisabeth if she wanted to remain with me."[21] This time, Elisabeth accepted without hesitation.

Of course, more than just fate or a meeting of the minds must have been at work to precipitate such a sudden and lasting attraction for each other. A rebound romance is usually a prescription for disaster. Aside from the flowering of any strong feelings of love, one of the strongest attractions seems to have been mere circumstance — each, and Werner in particular, felt terribly alone and isolated in the Nazi Germany of 1937. Werner had just celebrated his thirty-fifth birthday in December, and without a younger companion to serve as a social stimulus and point of reference, the world around him seemed even more hostile and unbearable in those early months of 1937 than it had at any time previously during the Third Reich. In addition, a premarital blood test later revealed that Werner had been suffering from an acute case of anemia, which may have contributed greatly to his mental condition.

The lonely, lost, and emotionally floundering state in which Werner existed when he first met Elisabeth was made even worse by another depressing event that occurred just before their meeting: his compulsory collection for Winter Aid on the cold, grey streets of Leipzig. The sense of stability and acceptance that Elisabeth suddenly brought to his life must have meant everything to the emotionally desperate Werner.

The Winter Aid Society, a well-regarded charitable organization that provided the urban poor with food and blankets, had been subordinated to the Nazi party after 1933. Government officials required the socially elite professors, who were state employees, to demonstrate their concern for their lower "Volk comrades" by periodically joining student and other party organizations in soliciting donations from passers-by on city street corners. This enforced public begging was just one more sign of the contempt in which Nazi bureaucrats held the previously esteemed professors. Heisenberg had participated at least once before without overt complaint,[22] but this time, already shaken by the

Weizsäcker rejection and without the support of Carl Friedrich or his youth comrades, the Nobel laureate could hardly bear the humiliation of rattling a tin can on the dismal streets of wintry Leipzig.

Nazi propagandist Joseph Goebbels had ordered a special effort that year to collect additional funds in celebration of the fourth anniversary of the Reich on January 30.[23] As he shivered on his street corner, Heisenberg was overwhelmed by a feeling of "the utter senselessness and futility of what I was doing and of what was happening all around me."[24] He later recalled sinking into and out of a disturbing mental state that, from his description, bordered on psychosis. "The houses in these narrow streets," he wrote long after the bombing raids of World War II, "seemed very far away and almost unreal, as if they had already been destroyed and only their pictures remained behind; people seemed transparent, their bodies having, so to speak, abandoned the material world so that only their spirits remained behind."[25] Two months later, when Elisabeth went to Berlin for a few days to make arrangements for the wedding, Heisenberg barely survived until her return. "When I am by myself," he wrote her, "I now easily fall into a very strange state, which belongs neither to the past nor to the future and neither to you nor to physics, and with which nothing can be done."[26]

Just before leaving for the ski hut in early March, Werner accompanied Elisabeth to Berlin to meet her family. Although Professor Schumacher was acquainted with his university colleagues Max Planck and Max von Laue, he had never heard of Heisenberg or of his political troubles. When informed of his daughter's engagement, the economist's response was to ask her intended: "How do you expect to feed my daughter?"[27] Werner's income was deemed sufficient, and after her domineering father conferred his approval on the plan, the happy couple headed by car for the Bavarian hut — whereupon their chaperon discreetly absented himself — and a visit with the bride's future mother-in-law in Munich.[28]

As soon as Elisabeth and Werner returned to Leipzig in mid-March 1937, plans for the wedding went forward quickly. Heisenberg enlisted his old youth comrade Wolfgang Rüdel, now a Lutheran pastor, to perform the ceremony in Berlin, but youth-movement violinist Rolf von Leyden, whom Heisenberg wanted to play, had to be written off in favor of one of Elisabeth's musical friends.[29] As the details were being settled, Heisenberg received a telephone call from the Reich Education Ministry (REM) in Berlin with a surprise Easter egg: the REM was ready to appoint Heisenberg to Sommerfeld's Munich position, if he still wanted it.[30]

Stark's setback and the start of the four-year plan had resulted in a reorganization at the REM that brought Rudolf Mentzel and the SS into stronger positions vis-à-vis the party organizations on whose support Stark and Lenard had relied. In January 1937, Reich Minister Rust united the two offices for university affairs under SS officer Otto Wacker, the former education minister of Baden, who turned over matters of "deutsche" physics to his deputy — Dr. Mentzel. SS officer Mentzel had already replaced Johannes Stark as head of the German Research Association, a foundation, and was preparing to set up a Reich Research Council to garner funds from the budget for the four-year plan when he decided to appoint Heisenberg to Munich. With the 1937 summer semester approaching (it started in May) and the dogmatists of "deutsche" physics in retreat, Mentzel's deputy, Dr. Wilhelm Dames, called in March to offer Heisenberg the Munich post immediately.[31] By the time opponents of the plan could voice their displeasure, Dames argued, they would be confronted with a fait accompli.

Heisenberg refused. With wedding plans in full swing and little time to prepare for Munich lecturing, Heisenberg insisted on an August 1 appointment and the REM grudgingly acceded. Although the delay would afford Werner's enemies an entire semester to mount an opposition, the optimistic physicist saw little cause to worry. Nearly the entire physics profession had rallied to Heisenberg's side against his opponents. Future plans quickly came into focus. With lecture and travel engagements set for October, the move to Munich would have to occur before then. Heisenberg and his bride decided that, after a brief honeymoon, they would stay in Leipzig until almost the end of the summer semester in July, then move into a beautiful little house that Heisenberg had purchased in the Isar River valley just outside Munich. As soon as their furniture arrived, Elisabeth would arrange their new home alone, while Werner made another demonstration of his loyalty in his annual eight weeks of military training. In the meantime, the tenants in Heisenberg's Leipzig house were given notice, and, as 38 wedding invitations went into the mail at the end of March, Euler helped his boss prepare the Leipzig living quarters and his bedroom for the new bride.[32] The two-story house, located at Bozenerweg 14 in the suburban section of Leipzig, was damaged by Allied bombs during World War II and was eventually destroyed altogether. Elisabeth's favorite part of the house was its well-kept backyard garden, which came into full bloom just as the couple exchanged their wedding vows in April.

With his marriage approaching and an appointment at last to Sommerfeld's chair in his beloved Munich, Heisenberg's hopes soared. Although he worried openly to Bohr about the mixing of physics and marriage, he had Bohr's own successful balance of the two as a model. Bohr's example, combined with his own new power as administrator of a major institute, gave him more courage, he wrote, than Bohr could ever know.[33] Werner was certain that, once he settled at last in Munich, he would be able to achieve much of the same success as physicist, administrator, and husband that Bohr, his idol and father figure, had achieved in his career — no matter that the Nazi Reich hardly compared with Bohr's Denmark. The new Munich post, he wrote in his typically flat prose, "is nice because I can now have the feeling of building up something permanent that will last as long as I am able to work at all."[34] And to Pauli he wrote three days before the wedding: "I also have the feeling that a security will come into my life that can only be encouraging for all types of work. It seems to me as if I could now in a certain sense begin life all over again from the very beginning."[35]

Werner and Elisabeth exchanged vows at 3:00 P.M. on Thursday, April 29, in the small St. Annenkirche in the neighboring suburb of Berlin-Dahlem. Their wedding preceded an elegant reception and dinner in the nearby Schöneberger Ratskeller. After dining on turbot filet, roast lamb, and Rhoner Hofberg Auslese 1934, the newlyweds headed by auto for southern Germany and Austria for an Alpine honeymoon.[36] Exactly nine months later, the new Mrs. Heisenberg gave birth to fraternal twins, Wolfgang and Maria, the former named for Pauli, who duly congratulated the new father on his "pair creation."[37]

Motherly Elisabeth — she would bear five more children over the next twelve years — though herself often insecure and much dependent on Werner's drive and sense of professional purpose to buoy her during those difficult years, apparently provided the stability and sense of belonging that Heisenberg so desperately needed then and would need even more during the terrible months and years ahead. Their marriage had been as much one of personal necessity brought on by circumstances as it was one of love or passion. Heisenberg did not easily express his feelings or readily share his personal and professional problems with her, especially if he felt they would only upset her and the family. Their relationship never really filled the void of loneliness and alienation that often opened deep within him, nor did Werner ever really allow his wife and family to displace the central position in his life that his career and duties had long occupied. His work and duties as a leading German scientist frequently took him away from home for long periods — first during the war, then during his intensive efforts to rebuild German science after the war. It was not until the 1960s, as the children left home and Werner slowed his pace of work, that he and Elisabeth could at last freely travel, vacation, and simply be together undisturbed. But in many ways they could never make up for lost time. As age and illness took their toll on Werner, Elisabeth quietly regretted to herself that, after all those years together, they had never really gotten to know each other in a fundamental way.[38]

Having stabilized his personal life, Heisenberg immediately turned to building "something permanent" in his physics. But during those depressing winter days in early 1937, his work was already under attack abroad and in retreat at home. Carl Anderson's new data on the absorption of high-energy cosmic-ray electrons in matter confirmed the application of the Bethe-Heitler theory to large energies far beyond the energy at which the theory was supposed to break down. It was a theory that attributed the stopping of charged particles in matter solely to electromagnetic processes. News of the discovery that quantum electrodynamics did not break down after all had reached Leipzig and Zurich in January 1937, just ahead of the Bhabha-Heitler and Carlson-Oppenheimer papers presenting a complete theory of cascade showers—a buildup of particles and photons by a series of elementary electromagnetic interactions, during which the incoming particle dissipates nearly all of its initial energy.[39]

Immediately after Anderson's data were first reported in June 1936, Oppenheimer and Carlson in the United States had turned to cosmic-ray showers concurrently with but independently of Bhabha and Heitler in England. Anderson's data, though still preliminary, confirmed the application of electrodynamic pair creation and Bremsstrahlung (braking radiation) to the formation of showers by the soft, easily absorbable rays—electrons and photons—in cosmic radiation. A shower would occur in a sequence of elementary processes. For instance, a high-speed electron approaching a lead nucleus would be deflected by electromagnetic repulsion, releasing some of its energy as Bremsstrahlung. The photon thus emitted would approach another lead nucleus and induce the creation of an electron-positron pair. Each of these particles would then release Bremsstrahlung, which would induce the creation of more electron-positron pairs, until finally the energy of the incoming particle was distributed among a shower of particles and photons emerging from the back of the lead plate.

On the other hand, the hard, penetrating component of incoming cosmic rays still defied any explanation or identification. Both cascade theories were constructed at first only for application to the obedient soft component. Anderson and his Cal Tech assistant, Seth Neddermeyer, reasoned "that either the theory of absorption breaks down for energies greater than about 1000 MeV, or else these high energy particles are not electrons."[40] Since the newly rehabilitated radiation theory accounted more easily for other absorption properties, Anderson and Neddermeyer preferred the second alternative and began an immediate search for new particles.

Both cascade theories, while offering a complete account of the soft component, made room for Heisenberg's alternative proposal — explosion showers derived from an application of Fermi's nuclear field theory of beta decay to the hard component of cosmic radiation. Since explosion showers occurred in a single event, while cascade showers involved a series of events, Bhabha and Heitler suggested that explosions might be necessary for the soft component in two instances. First, large bursts — showers of myriads of particles in Geiger counters — required enormous energies in excess of 10^4 MeV, which might still entail a possible breakdown of the normal quantum theory of showers at very high energies. Second, a large cascade shower containing 100 particles required at least 1 centimeter of lead to form. Larger showers might appear in thinner plates more frequently than statistical fluctuations allowed. If either case were true, they wrote, "we should have to fall back on a completely new theory of showers, which introduced a new interaction between the particles themselves and also, possibly, light. An elegant theory of this type has been put forward by Heisenberg."[41]

Carlson and Oppenheimer conceded little. Heisenberg's theory, they declared, "is without cogent experimental foundation; and we believe that in fact it is an abusive extension of the formalism of the electron neutrino field."[42] The Fermi nuclear field was still problematic, and the suggestion that, when taken out of the nucleus and applied to cosmic rays, its divergence to infinity could somehow render characteristics of a future theory seemed to Carlson and Oppenheimer implausible and methodologically unsound. They preferred instead to study cosmic-ray events through "the more usual procedure of avoiding divergences by the formal device of reducing the coupling between heavy and light particles for high relative energies."[43]

To prove his point and to meet the challenge of Heisenberg's almost simultaneous alternative, Oppenheimer suggested to Lothar Nordheim and his wife, Gertrude, both recently arrived émigré physicists, a test of Fermi-field explosion showers for the hard component, construed to be protons. Since protons were components of nuclei, a high-speed cos-mic-ray proton hitting a nucleus in a lead plate should stimulate the nuclear Fermi force. While Oppenheimer and Carlson completed their paper in the fall of 1936, the Nordheims altered the Fermi force to reduce the interaction of the field with the proton in order to give the correct range and magnitude of the nuclear force.[44] They could then estimate the probability for the explosion of an incoming proton into electrons and neutrinos in the vicinity of the nucleus.

The outcome was as expected: Heisenberg's theory "no longer af-fords any explanation of showers."[45] The probability of proton decay in the field of the nucleus turned out to be vanishingly small. Obviously Fermi showers depended on how one tinkered with the Fermi force. Heisenberg, who had done his own tinkering, remarked that any cutoff introduced into the theory to reduce interactions should not cut off any "qualitatively new phenomena."[46]

Heisenberg and Pauli took immediate note of the new developments but remained undeterred from "our main problem, the quantization of waves"—as Heisenberg put it.[47] Of course, Heisenberg and Pauli wor-ried as much as Oppenheimer about accounting for cosmic-ray data. After all, their theory and its development ultimately rested on the complicated data of cosmic radiation. But their letters to each other in this period suggest that "the practical questions of cosmic radiation"—the actual unraveling of shower phenomena—were of a lower priority than their—especially Heisenberg's—primary goal, the search for "a future theory of elementary particles."[48] For them, cosmic-ray data were a means to an end, not an end itself. The irony of these prefer-ences at that time is the reversal of the expected influence of the social criticism of science. While "deutsche" physics demanded a retreat from abstract theories to more empirical or semiempirical physics, the empiri-cally inclined quantum theorists were to be found not primarily in Germany but in the other camp—in the United States and Britain—even though many were in fact German émigrés. Although theorists remaining in Germany (such as Heisenberg) did offer a more public appreciation of empirical and practical physics, they did not swerve noticeably in their private work from the internal trajectory of their theorizing.[49]

As soon as Heisenberg saw Anderson's new data and the Bhabha-Heitler cascade-shower manuscript, he readily conceded nearly all cos-mic-ray showers to cascades, extended the Bethe-Heitler theory to all energies, and admitted to Bhabha, then in Zurich, that perhaps applying the Fermi theory of nuclear forces to the subject of cosmic-ray showers was a "wide extrapolation from beta decay."[50] But he held out for at least one "genuine shower formation" per 1000 electron cascades and for the detection of neutrinos as the shower-producing hard component.[51] Both propositions were extraordinarily difficult to test.

By early 1937, Heisenberg's explosion-shower revolution was in fragments. Pauli claimed to have proven that every shower theory was divergent, even with a fundamental length.[52] Heisenberg iterated the Hamiltonian as a series of observable terms, but he quickly became bogged down in mathematics.[53] At about the time Werner was collecting coins for Winter Aid, Pauli's coworkers, Fierz and Kemmer, thought they had proved that neither Fermi's theory nor any relativistic variation of it could reproduce the size and charge independence of nuclear forces, the asymmetry of the beta emission curve, or finite energy values.

Heisenberg was undeterred. On April 26, he wrote to Pauli not only of feeling new security in his personal life as an almost married man but of an indomitable optimism in his work. Although all their attempts so far to obtain a satisfactory quantum field theory using the fundamental critical length had failed, at least they yielded "a feeling for the formal possibilities which lurk in the wave quantization, and I am still rather optimistic." And, he continued, "I am now also much more optimistic concerning the discovery of Fermi processes in cosmic radiation."[54] Three days later the optimist married his fiancée in Berlin.

Pauli resigned from the elated groom's physics by return mail. Werner's suggested fundamental length — supported by the seeming appearance of explosion showers in cosmic rays, which were induced in turn by Fermi's beta-decay field — did not yield any progress in "*the* fundamental problem of present-day physics" — wave quantization. Homi Bhabha's return visit to Zurich that spring only reinforced Pauli's "theoretical doubts" about the fundamental length. "I think 'we must be prepared' [an echo of Bohr] to find that the universal length of the β-decay theory will prove to be incorrect, and therefore the corresponding multiple processes [explosions] will not play any significant role at all in cosmic rays."[55] Heisenberg's entire proposal, based on the disputed existence of explosion showers, might have to be given up entirely.

Learning of Pauli' retreat upon his return from his honeymoon, Heisenberg rejected Pauli's doubts that explosions would ever be found as "a terribly defeatist attitude. . . . On the contrary, I am more convinced than ever that there the key to the wave quantization is really to be sought."[56] In a long letter to Bohr in early July 1937, ten days before moving to Munich, Heisenberg reviewed the experimental evidence for explosion showers, including a new discovery by Neddermeyer and Anderson: the heavy electron in the hard, penetrating component of cosmic rays.[57] This new elementary particle — the first to be discovered since the neutron — possessed the charge of an electron but carried nearly double its mass. Since, according to quantum field theory, every elementary particle is related to a specific type of field, in the years ahead the heavy electron became the subject of much research and controversy regarding its proper place in field theory. To Heisenberg, it provided a welcome support of his "program for the future" — even if the new particle did not seem to fit in anywhere in his nuclear physics of cosmic rays. New particles would appear as stationary states in a unified theory of matter and fields. But the further pursuit of this program, he admitted in one of his occasional letters to Kramers, would have to await more supporting data.[58]

Heisenberg wrote his letter to Bohr while he and Elisabeth planned the move to Munich. Immediately after the move, Heisenberg had intended to leave for military training, but circumstances forced a change of plans. Premarital blood tests had turned up Werner's anemia, and the now pregnant Elisabeth was suffering from both an injured knee and morning sickness.[59] The doctor forbade military duty and ordered a mountain cure for both patients.[60] The couple decided to stay in Munich for two weeks in July, then to spend most of August in beautiful Engadin in southern Switzerland before settling into their new home.

In early June, as the peach trees blossomed and heavy electrons filtered through Anderson's lead blocks, Leopold Kölbl, now rector of the University of Munich, informed Heisenberg of an ominous development. Stark had learned of Heisenberg's appointment and had told Kölbl during a meeting in his office that he planned to take the matter to "higher authorities." The threat did not puncture Werner's optimism. Stark could no longer be taken seriously, he wrote. "And in the long run, I will surely win the contest with Herr Stark, if I don't do something wrong."[61]

Five weeks later, on Thursday, July 15, the Heisenbergs arrived in Munich by automobile, as planned. They unloaded the car at the old Heisenberg home on Hohenzollernstrasse, where the boyhood battleship still stood docked on a dresser in the bedroom. From his mother's phone, Werner called Kölbl to announce his arrival. The rector's only response was to ask if Heisenberg had seen the latest issue of the SS newspaper *Das Schwarze Korps* (*The Black Corps*). "There's a long article on you in it. Buy it and read it," he said. "Then we can talk." [62] Werner found a copy of the July 15 issue at a local newsstand and opened it to find a full-page broadside attack on science and himself entitled "'Weisse Juden' in der Wissenschaft"—"'White Jews' in science"—signed at the bottom by Stark. Another long and lonely skirmish had begun.

20

Himmler's Henchmen

The victory of racial anti-Semitism is to be considered only a partial war. . . . For it is not the racial Jew in himself who is a threat to us, but rather the spirit that he spreads. And if the carrier of this spirit is not a Jew but a German, then he should be considered doubly worthy of being combatted as the racial Jew, who cannot hide the origin of his spirit. Common slang has coined a phrase for such bacteria carriers, the 'white Jew.'"[1]

So began — replete with grammatical errors and non sequiturs — in the July 15, 1937, issue of the SS weekly *Das Schwarze Korps*, one of the most vicious and repulsive attacks on science in general and on Heisenberg in particular to appear during the Third Reich. It precipitated an even more violent struggle between academics and Nazi demagogues for control of physics ideology and appointments, which came to an end only in the depths of World War II. Heisenberg and others attacked in this way ultimately prevailed, yet the episode left him even more oriented toward the position he would take in the war. As with all political intrigues, the battle over physics had both public and private consequences.

Heisenberg's impending appointment to the University of Munich as Sommerfeld's successor rekindled Johannes Stark's efforts to prevent the appointment and to gain control over physics teaching chairs. He received new encouragement during the Heidelberg celebration of Philipp Lenard's seventy-fifth birthday on June 7. With Alfred Rosenberg out of the picture and the SS in the ascendant, Heisenberg's opponents turned to Himmler and his organization for support in imposing their will on the Reich Education Ministry. Younger SS officers in Lenard's and Stark's circles facilitated the effort; two in particular played key roles. Dr. Hermann Beuthe, physicist and government councilor, was Stark's right-hand man in the Physical-Technical Institute in Berlin. Dr. Ludwig Wesch, an SS-Obersturmführer who had received his physics doctorate and habilitation under Lenard, was now the old man's assistant.[2] Beuthe and Wesch were also members of the SD (Sicherheitsdienst), or security service, a branch of the SS under the direction of Reinhard Heydrich, later notorious as the "hangman of Lidice."

Hitler had originally established the SS (Schutzstaffel) as his elite personal bodyguard, selected from the ranks of his brown-shirted storm troopers, the SA (Sturmabteilung). But after Hitler's bloody purge of the SA in 1934—apparently in order to gain the backing of the army, which the SA rivaled in military matters—the black-shirted SS began to take over a wider range of police and storm-trooper functions. Within the SS, the SD served as a secret police. In June 1936, Hitler reorganized police activity under Bavarian police chief Heinrich Himmler, whom he elevated to the cabinet-level position of Reichsführer-SS in Berlin. Under Himmler, the Gestapo (Geheime Staatspolizei) took over secret-police work, while the SD turned to intelligence gathering and reporting on social and cultural affairs, such as science.[3]

Journalist Gunter d'Alquen, an SS officer and SD man, had worked on the editorial board of the *Völkischer Beobachter* under Alfred Rosenberg until the reorganization of the SS brought him a new opportunity: editorship of *Das Schwarze Korps*.[4] His black-shirted comrades Beuthe and Wesch probably convinced d'Alquen to publish an article submitted by the proponents of "deutsche" physics, entitled " 'Weisse Juden' in der Wissenschaft"—" 'White Jews' in science."

The page-long diatribe is divided into three sections: the Nazi definition of "white Jews" excerpted at the beginning of this chapter; a personal attack on Heisenberg, called "The dictatorship of the grey theory"; and a political attack on the university professorate, " 'Science' has failed politically." The last section was signed "Stark" and was phrased to give the impression that Stark had been invited by the newspaper to express his views on the matter.[5] The second part, the one on Heisenberg, although unsigned, was written nevertheless in Stark's unmistakable venomous style. He was probably assisted by his subordinate, Beuthe: the latter told Wesch that he had helped with the article and that he had gathered most of the information for the Heisenberg section.[6]

After vilifying "the Jews Einstein, Haber, and their like-minded comrades, Sommerfeld and Planck," and accusing them again of manipulating physics appointments to exclude "Germans," the article proceeded to mount a full-scale assault on "that white Jew" and "representative of the Einsteinian 'spirit' in the new Germany" Werner Heisenberg. Stark's perceived archenemy, Sommerfeld, had chosen Heisenberg as his successor, which made Heisenberg an heir apparent to the "white Jewish" establishment and thus a lightning rod for Stark's ruthless efforts to gain influence, at last, over government appointments to German physics professorships. By exploiting the dominant hate ideology of the day, character assassination of a key individual could be used to further the political ambitions of those who had otherwise failed to achieve the influence they craved. The denunciations of the McCarthy era in the United States perhaps offer again a distant parallel.

To demonstrate "how secure the 'white Jews' feel in their positions," Stark delivered a litany of Heisenberg's offensive actions, as researched by Beuthe (whose name in German means "booty"). The list was presented to make Heisenberg appear a covert enemy of the state. He had "smuggled" his article defending the teaching of relativity theory into a party newspaper; he had circulated a petition among physicists in order to influence wrongly a state agency (the REM) and to silence his legitimate critics; he had refused to join his fellow Nobel prize winners in the 1934 declaration of support for Hitler's presidency; his appointment to the Leipzig chair in 1927 was unearned, since he was clearly too young to have accomplished anything of value — a circumstance that "proved" that he gotten the chair only because he was backed by the "white Jewish" establishment; he had dismissed a "German" assistant in his institute in favor of the Jewish physicists Bloch and Beck; his institute continued to harbor an inordinate number of Jews and foreigners to the exclusion of "Germans"; and so on.

For those who failed to catch the inference of this catalogue of sins, a large-print subhead calling Heisenberg the "Ossietzky" of physics made the point. Carl von Ossietzky, a courageous pacifist opponent of the Nazis and winner of the 1936 Nobel peace prize, was at that moment imprisoned as a traitor in the Dachau concentration camp. He would die of torture and malnutrition within a year.[7] Heisenberg, "white Jew" and "Ossietzky" of physics, was only one example among "many others," the article declared. "They are all representatives of Judaism in German spiritual life who must all be eliminated just as the Jews themselves" must be dispatched.

Blasted full force by the violent threat, German physicists, especially those likely to be counted among the "many others," immediately rallied to their colleague's defense and indirectly to the defense of their profession. This time, there were no faculty protest meetings or even a circulated petition — such acts were too likely then to lead to a fate like Ossietzky's. Instead, Heisenberg's colleagues submitted official letters of complaint to local bureaucrats for transmission up the chain of command to ministerial authorities. They also supported him with their official recognition. In the months immediately following the article's publication, the Göttingen Academy of Sciences elected Heisenberg a corresponding member. The Saxon Academy, populated by most of the non-Jewish members of the former professors' club, elected him deputy secretary of the math and physics class at a time when the REM was attempting to coordinate academies behind the National Socialist state.[8]

Local deans and their state overseers in nearly every instance forwarded the letters of protest to state ministries and beyond, which indicates the broad sympathy Heisenberg and his colleagues enjoyed in this matter. Many of these local administrators carried the title Pg., party comrade. Stark may have gained the backing of elements within the powerful SS, but he obviously did not enjoy the backing of the REM or the party faithful among academics, over whom Stark was ultimately attempting to gain control. They realized that more was at stake than the fate of one individual.

Sommerfeld wrote a long letter to Munich rector Kölbl in which he complained that he and Heisenberg had been slandered. He demanded an end to Stark's ravings "in the interest of the reputation of German science" — a plea that would appeal to German nationalists.[9] Pg. Professor Kölbl forwarded Sommerfeld's letter to the Bavarian Culture Ministry, adding a remark of his own: "It is outrageous that an active professor [Heisenberg], a civil servant of the National Socialist state, should be attacked in this way in a newspaper."[10]

Heisenberg's Leipzig colleague Friedrich Hund, who also considered himself one of the "many others," fired off letters both to Leipzig rector Koebe and to REM chief Rust. He requested that Rust take steps to ensure that Stark "can no longer ruin the honor of our science in this way."[11] Hund also informed Debye of this protest. Debye, now in Berlin, had already taken a copy of the SS article to a meeting of the Kaiser Wilhelm Society senate where, he wrote Hund, "it was condemned by everyone with whom I spoke."[12] Debye subsequently added his own letter of complaint to the growing pile.[13]

The most significant of the letters to arrive at the REM came from the victim himself. Heisenberg demanded that Minister Rust — still vacillating between the SS and the professors — take a stand against the SS regarding Heisenberg's personal honor, a request that, ironically, Nazis would understand.

It is difficult to comprehend the painful and debilitating situation in which Heisenberg now found himself. The great, apolitical Professor Heisenberg was accused of being, like the hated Jews, an enemy of the state and a subversive bent on undermining regime policies. How should we take Heisenberg's official protestations against such accusations and his demand for restored honor — that is, renewed recognition by the regime and a dropping of the charge that he was a traitorous "white Jew"? Surely it would have been a true badge of honor to be guilty of such "crimes." His official complaints notwithstanding, from what we know of Heisenberg and his views, he did not fully support the regime, but he had come to accept it and the failure of his efforts to do anything about it. He was not attempting to dissociate himself from Jews for reasons of anti-Semitism; nor did he believe that German Jews were really non-Germans or that they had been enemies of the state before the state had forced them into that role. But other than expressing displeasure through bureaucratic channels and protecting those whom he could in his institute, these were matters of so-called private, active opposition — the rationalization that allowed him to accept the situation and to maintain his prominent scientific position in Hitler's Third Reich. By its very nature, the compromise of so-called active opposition required the public suppression of any private moral or political scruples regarding the regime.

Regardless of Heisenberg's private views, by 1937 the time was long past when one could openly express any opposing position in letters to Nazi bureaucrats (or even in private letters that could possibly fall into the wrong hands) — that is, if one wanted to remain in Germany and continue to work as a civil servant of the National Socialist state. Moreover, in this situation, as Planck had counseled long ago and Heisenberg had already accepted, they should focus only on the future — batten down the hatches and wait for what they regarded as the inevitable catastrophe to blow over.

Now, in 1937, the storm was howling right over Heisenberg's head. He was being accused of holding antiregime views and of allying himself with pro-Jewish subversives of like mind. If his enemies could ever make these accusations stick, they would not only render his continued work and teaching impossible, but even worse, he would be placed in grave personal danger, politically isolated and branded a traitor.

On the other hand, if he could successfully refute these accusations and force his enemies into retreat, he apparently reasoned, he could use his victory as protection against further insult and as a basis for improving the lot of German theoretical physics, if, indeed, anything could really counter the enormous damage—scientific, political, moral— already done to it by the regime. If he did succeed, whatever private objections he harbored would be buried even deeper in his private self, where they would be even safer from discovery.

The political assault had focused on an individual; the individual now demanded a new compromise—exoneration and rehabilitation of his honor in return for not vacating his public office and voiding all the compromises it required. Privately, however, that raised the specter of emigration, for life and work for Werner would be impossible as long as he bore the humiliation and abuse of the Nazi regime—and that would be difficult indeed. In a revealing personal letter to Sommerfeld, Heisenberg wrote of the dilemma in which Stark, as well as his attachment to Germany, had placed him, a position that seems so reminiscent of the dilemma of which Max Born had attempted to enlighten Werner nearly four years earlier. Wrote Werner: "Now I actually see no other possibility than to ask for my dismissal if the defense of my honor is refused here. However, I would like to ask you for your advice in advance. You know that it would be very painful for me to leave Germany; I do not want to do it unless it must be absolutely so. However, I also have no desire to live here as a second-class person."[14]

On the day the SS article appeared, Werner arrived in Munich with his pregnant wife, Elisabeth, on their way to the Alps for a rest. Within two days of their arrival, he composed and sent an official letter, along with a copy of the SS article, to Dean Helmut Berve, his Leipzig friend and colleague. In the letter, to be forwarded through channels to the education ministry, Heisenberg demanded that the REM make a fundamental decision: "Either the ministry regards the standpoint of *Das Schwarze Korps* as correct, in which case I request my dismissal. Or, on the other hand, the ministry disapproves of such attacks, in which case I believe that I have the same right to protection that, say, the Wehrmacht itself would render its youngest lieutenant in such a case."[15] Pg. Berve immediately forwarded the request to Leipzig rector Koebe, along with his own request for a stand from REM minister Rust.[16]

The Saxon Education Ministry duly forwarded the letter and similar ones to Berlin, but Saxon bureaucrat Studentkowski—previously so active in Heisenberg's defense—decided to stay neutral. For one thing, the SS was now involved, and he was not an SS member. More important, Munich's difficulty in replacing Sommerfeld and Leipzig's difficulty in replacing Debye with a suitable experimentalist meant that there was probably little chance of finding a comparable replacement for Heisenberg in Leipzig.[17] The search for Debye's successor had lasted for more than a year before Gerhard Hoffmann, the cosmic-ray experimentalist who was last on everyone's list, was finally asked to take the post as of April 1, 1937.[18] Studentkowski did not want a repeat of such difficulties in Heisenberg's case.

With or without the numerous letters of complaint—which, of course, any Nazi agency could easily ignore, since it did not depend on professors for support—the REM remained favorable toward Heisenberg despite the presence of many SS officers in its ranks. Stark could not be allowed to dominate REM policy, even if he could claim SS support. Accordingly, SS-Stürmführer Dr. Otto Wacker, now head of REM office W (for Wissenschaft, scholarship), immediately responded to Heisenberg's letter with a request for supporting evidence against the charges. Heisenberg forwarded a detailed rebuttal of each accusation.[19] Wacker duly opened an investigation, but it soon became clear that neither Wacker nor his REM boss, Rust, would act until a decision had been reached in another investigation.[20]

This time, Heisenberg, too, had taken the matter to "higher authority." Exactly one week before his first letter to Wacker in July, Heisenberg had written directly to the Reichsführer-SS, Heinrich Himmler, requesting in nearly the same words as his letter to the REM a similar fundamental decision: either approval of Stark's attack, in which case Heisenberg would resign, or disapproval, in which case he demanded the restitution of his honor and protection against further attacks. Not wishing to alarm his pregnant wife with the enormous gamble he was taking—of which he himself may not have been fully aware—Werner did not inform her of it until much later, at which point she was overwhelmed with shock and anger.[21]

If Heisenberg had sent his letter to Himmler through normal channels, as he did the REM letter, it probably would never have arrived. At his mother's suggestion, Heisenberg chose a safer route. As indicated earlier, Heisenberg's grandfather, Nickolaus Wecklein, former rector of the Maximilians-Gymnasium in Munich, had belonged to a hiking club of like-minded Bavarian gymnasium rectors. One of the members of this group was Himmler's father, Joseph Gebhard Himmler, assistant rector in Landshut, who died in late 1936. Heisenberg's mother had become acquainted through her father with Mrs. Himmler, who now lived in Munich. Mrs. Heisenberg offered to take Werner's letter to Mrs. Himmler to deliver to her son, Heinrich.

According to Heisenberg's much later account of the meeting of the two mothers, which must have occurred in late July or early August 1937, Mrs. Himmler politely received the visitor in the living room of her small, respectably furnished apartment. A crucifix was nestled prominently in one corner of the room with freshly cut flowers reverently arranged in front of it. Mrs. Himmler was at first rather skeptical of Mrs. Heisenberg's request, not wanting to interfere in her son's affairs. Mrs. Heisenberg ultimately gained her confidence by saying, " 'Oh, you know, Mrs. Himmler, we mothers know nothing about politics — neither your son's nor mine. But we know that we have to care for our boys. That is why I have come to you.' And she understood that." [22]

Mrs. Himmler probably gave Heisenberg's letter to her son during his visit to Munich in August.[23] But Himmler did not respond until November 4, 1937, after learning the result of a preliminary internal investigation.[24] No discussion of the matter appears in Himmler's surviving correspondence with his mother.[25] In his letter of November 4, Himmler simply asked for a defense against the charges. Heisenberg immediately provided a point-by-point rebuttal, almost identical to the one he had provided to Wacker.[26] First, Heisenberg explained to Himmler, the "German" in his institute was replaced by Bloch and Beck because of his complete lack of interest or ability in modern physics. Second, Heisenberg had refused to sign Stark's declaration in support of Hitler because he had doubts about Stark himself — which were shared, incidentally, by SS functionaries — and because of his long-held belief that scientists should, as they had done in the past, remain apolitical.[27] Third, he had participated in the preparation and circulation of the Wien-Geiger-Heisenberg petition at the explicit request of REM official Mentzel, an SS officer.

The best way to settle the matter, Heisenberg suggested to Himmler in closing, was to arrange a face-to-face confrontation between himself and Stark, whom he had never met. Heisenberg was almost always successful in personal encounters, and, he naively believed, personal diplomacy still offered a good chance of inducing Stark to withdraw his charges and retreat from further interference in professional matters. The meeting never took place, but because of the severe nature of the charges, his mother's urging, and the favorable result of the preliminary investigation, Himmler decided in November 1937 to consider the case in depth. That decision and Heisenberg's written defense set in motion an intensive SS investigation that lasted more than eight months and profoundly affected Heisenberg's personal and political life.

Heisenberg also had the welfare of his new family to consider, but for him the paramount issue was whether or not Himmler would bestow his stamp of approval, thus facilitating his continued work and unhindered life in the otherwise stifling Nazi environment. Heisenberg, the eternal optimist, now turned suddenly cautious. In the summer of 1937 he entered into secret negotiations with a visitor from Columbia University in New York City.[28] "When I thought about New York," he later wrote his mother, "it was less the thought of the intrigues of Herr St[ark] that was decisive, but more the prospect of living for many more years in an environment that makes work—for the sake of which I now exist— almost impossible."[29]

Not wanting to prejudice the two ongoing investigations into his activities, Heisenberg did not disclose the negotiations to his faculty until required by law to do so. His plan was to visit Columbia in the second half of 1938—either temporarily or permanently, depending on the situation at that time.[30] Yet despite this flirtation, the eternal optimist never truly believed that he would have to resign—nor would he unless Stark made matters much worse by making work and teaching absolutely impossible.[31] Heisenberg was still the ministry's official choice for Munich, and by the end of October he learned that the REM investigation had come to a conclusion in his favor. Elisabeth wrote Werner's mother of the good news; she believed they would be in Munich by early 1938.[32] But, unknown to her, the SS investigation had only just begun, and Rust did not dare act until Himmler had approved of Heisenberg.[33] Heisenberg would have to wait a while longer.

Himmler handed the Heisenberg case over to the SD "cultural division" of his personal staff.[34] At the same time, Nazi physicists maneuvered to prevent Heisenberg's appointment to Munich even if the Reichsführer did stamp his approval. Munich, the "capital of the movement," was headquarters to academic organizations established by Hitler's party deputy, Rudolf Hess, in order to exert his own competing influence on academic affairs. Among these Hess organizations were the Nazi Students League; its headquarters, the Reich Student Leadership; and the Nazi University Teachers League.

SS physicist Ludwig Wesch may have played a role in stirring up opposition to Heisenberg among the Munich groups. As a Lenard pupil and cofounder and occasional head of the Nazi Students League in Heidelberg, he maintained close ties to the Munich Student Leadership and to its subcommittee for natural science.[35] Nazi science students required little encouragement to join the fray. Their journals, *Deutsche Mathematik* (*German Mathematics*), founded in 1936, and *Zeitschrift für die gesamte Naturwissenschaft* (*Journal for the Entirety of Science*), founded in 1933, produced a rising torrent of anti-Semitic Nazi science propaganda. An attempt by party member Pascual Jordan, one of the founders of quantum physics, to temper the radicals' fury against modern physics with his monograph *Physik des 20. Jahrhunderts* (*Physics of the twentieth century*) only precipitated even longer assaults on theoretical physics in the pages of both journals.[36]

Heisenberg and his publications were eventually targeted specifically, especially after Hugo Dingler, an aged Austrian Nazi physicist, decided to join the student cause.[37] The Reich Student Leadership maintained an SD office for the evaluation of candidates for university appointments.[38] In the matter of Sommerfeld's successor, the students were in agreement with their comrades in the Teachers League, who had already gained influence by opposing Sommerfeld's choice — Heisenberg. Interconnections among all these groups facilitated concerted action.[39]

One historian has concluded that Wilhelm Führer, then head of the Munich branch of the University Teachers League, was the driving force behind the opposition to Heisenberg's appointment, which in turn drove the rising fortunes of the Teachers League.[40] Führer was joined by Bruno Thüring, Teachers League representative to the Munich philosophical faculty, and Fritz Kubach, historian of science and Reich student leader for natural science, all of whom, like Hess, were headquartered in Munich.[41] But Führer's actions were themselves instigated from even higher up, by Hess himself, and in concert with the publication of Stark's article in the journal of Hess's competition, Himmler's SS. Just one day before the article on "white Jews" appeared in *Das Schwarze Korps,* Führer received an order to reject Heisenberg as a candidate for Sommerfeld's chair.[42]

In the fall of 1937, after the SS onslaught had delayed Heisenberg's appointment once again, the Munich Teachers League began proposing its own candidates to succeed Sommerfeld, all of whom Sommerfeld and colleagues rejected as unqualified. Most had little training in theoretical physics, and none could be considered suitable as a successor to the great Sommerfeld. The danger that one of these individuals rather than Heisenberg would nevertheless occupy Sommerfeld's chair vastly increased after a conference of university rectors in December 1937. At that meeting Wacker, apparently seeking closer ties with Hess's organizations, agreed to consider "political reliability" a specific criterion for faculty appointments. He further agreed that Hess, with the willing assistance of his Teachers League, should politically evaluate candidates for professorships.[43] The implication was clear. Heisenberg would never succeed Sommerfeld unless the Nazi students and teachers could be convinced of his "reliability," and that was impossible without SS "exoneration."

Heisenberg himself chose not to discuss the investigation in any of his memoirs — he did not even discuss it with his wife at the time. Moreover, the SS records of the investigation were apparently lost in the war — SS functionaries burned as many documents as they could get their hands on in the last days of the Third Reich. Other sources indicate that Himmler's investigators apparently focused on two areas: Heisenberg's ideological standpoint in scientific matters and his personal and political orientations. The SS — hardly an objective agency bent on exonerating those falsely accused of traitorous actions — employed its already infamous methods to discover the "truth." Heisenberg had to endure long and exhausting interrogations; spies were planted in his classroom and throughout the institute; the Gestapo bugged his home. The SS also used another tactic it had perfected: bringing an even more serious charge against the victim, who would then be all too eager to "confess" to the lesser, original charge in order to escape the greater danger.

The more serious charge brought in this case indicates that Heisenberg was indeed in grave personal danger. Hints regarding the charge are found only in letters surrounding the investigation. The accusation is spelled out in one such letter in November 1937 regarding the article in *Das Schwarze Korps*: "Not everything, however, is in the article; for example Heisenberg is not clean with respect to §175; he indeed married quickly but only to cover this up."[44] The reference is to section 175 of the old Weimar criminal code, in effect to this day, making male homosexuality a crime.[45] If convicted of this crime in 1937, the offender landed immediately in a concentration camp.

The imputation requires careful handling. First, it is an accusation made by a dedicated SS functionary engaged in a campaign of character assassination. Such a source is hardly reliable or objective. Second, the SS often used the charge of homosexuality to extract confessions to lesser crimes. Third, it is true that Heisenberg did prefer the company of younger men, and one or two in particular. The investigation, however, which no doubt involved interrogations of some of these younger companions, apparently yielded no evidence of homosexuality; if it had, that evidence would certainly have been used against him. Moreover, if the SS functionary did think he had such evidence, it may have concerned the Wyneken affair in the Bavarian Neupfadfinder, which had caused a considerable scandal in the early 1920s. Apparently it had not involved Heisenberg's group. Heisenberg did marry rather precipitously for various reasons, but there is no indication that concealing homosexuality was one of them.

The agony that such accusations must have caused Heisenberg is evident in his annual assessment of his life. In a long letter to his mother in November 1937, he expressed his feelings more openly than usual: "I wish for the coming year a clearing away finally of these horrible things, for, as unwillingly as I admit it, such a struggle poisons one's entire thoughts, and the hate for these fundamentally sick individuals who torment one eats into one's soul."[46] A week later, as he looked forward to his first Christmas with his new wife, he had entirely withdrawn into quiet family pleasures: "In this we realize once again how important living together with decent people is."[47] Nevertheless, the accusations and investigations of 1937–1938 had a lasting effect on Heisenberg. Even toward the end of his life, long after the regime had passed, the imagined sound of Nazi jackboots and visions of black-shirted Gestapo functionaries marching up the stairs to his bedroom still occasionally awakened him, perspiring, from a fitful sleep.[48]

Heisenberg's correspondence in that period indicates several difficult trips to Berlin to further his case. At least one of these was for an official interrogation in the notorious basement chambers of the SS headquarters at Prinz-Albert-Strasse 8. A cynical sign reading "Breathe deeply and calmly" hung on the bare cement wall as a constant reminder to the victim of his or her predicament.[49] Of the three known SS investigators assigned to Heisenberg, one worked with the Sipo (Sittenpolizei), or morals police, and all three had some training in physics. Heisenberg had even participated in the final examinations of one of them for his Leipzig doctorate in physics![50] Convinced by Heisenberg himself, his diplomatic Berlin supporters, and their own conscientious investigation, all three turned into strong and valuable supporters thereafter.

Perhaps the most influential of the three was Mathias Jules, born in 1911 and just completing his doctorate in mathematics and theoretical physics under Max von Laue at the University of Berlin.[51] After receiving his doctorate in June 1938, he was appointed assistant in Laue's institute. He later held the same post under Heisenberg, who succeeded Laue as head of the institute in 1943. As a member and leader of the Natural Science Group in the Nazi Students League of the university, Jules organized the first indoctrination camp for mathematics students in 1939.[52] In the same year he was appointed an honorary member of SD Lead Unit Berlin in the new Reich Main Security Office (RSHA) under Reichsführer-SS Heinrich Himmler.[53] In early 1938, a member of Himmler's personal staff asked Jules to prepare a comprehensive report on Heisenberg and theoretical physics.

Jules performed his duty with enthusiastic thoroughness. "Above all," he later wrote, "on the one hand I had to assimilate as wide a technical foundation as possible, and on the other hand I had to know well the technical-personal connections."[54] Conversations with Heisenberg and Heisenberg's closest colleagues pointed him toward a favorable conclusion. In gratitude, one of the more well-known colleagues helped Jules to reestablish himself after the war and the denazification trials by writing a physics textbook with him.[55] Jules later took a teaching post in Hannover.

Thanks in part to Jules, the SS investigation ended positively for Heisenberg. Although the final report of the investigation has not been found, most of the bureaucratic memos on Heisenberg's views thereafter contain the same evaluations in many of the same words, which suggests they had the same source, the SS report. The earliest of these memos, a long affidavit on Heisenberg's views sent in 1939 from Himmler's office to the REM, gave Heisenberg a clean bill of political and scientific health. He was found to be neither the raving follower of "grey theories" nor the traitorous, Jewish-inspired enemy of the Reich that Stark had painted, but rather a harmless apolitical academic who reportedly even expressed favorable views regarding the Nazi regime.

However, it is impossible to determine to what extent the views attributed to Heisenberg in this report were really his own, to what extent they were formulated by Jules or someone else, and to what extent they were extracted under Gestapo duress. According to the 1939 report: "For Heisenberg, theoretical physics is merely the working hypothesis with which the experimenter inquires of nature in suitable experiments. The evaluation occurs first and foremost through the experiment. The theory that is confirmed by experiment is thus the clear description of the observations made in nature using the exact means of mathematics." And: "Heisenberg's personal character is decent. Heisenberg is typical of the apolitical academic. . . . Over the course of several years, Heisenberg has allowed himself to be convinced more and more of National Socialism through its successes and is today positive toward it. He is however of the view that active political activity is not suitable for a university teacher, save for the occasional participation in indoctrination camps and the like."[56] Whether completely accurate or not, this remained the Nazi regime's official assessment of Werner Heisenberg until the end.

By the spring of 1938, as the SS report neared completion, two events suggested Heisenberg's imminent exoneration: state approval of his earlier election as class secretary of the Saxon Academy of Sciences, a state institution, and permission granted by the REM for a two-week lecture tour to England in March 1938 — just a week after Germany's Anschluss (annexation) of Austria.[57] Neither activity would have been approved had the SS investigation gone against him.

Yet upon returning to Leipzig in early April, Heisenberg had still received no official word from the SS. As he would often do in the future, Heisenberg returned to Jules, who was the probable source of this disconcerting news: "The decision lies with the Reichsführer-SS who in your case, in my opinion, does not want to do anything more."[58] Himmler, preoccupied with the Austrian Anschluss, was letting the "professors' controversy" wither away for lack of attention. Heisenberg again spoke of resignation.[59] Nevertheless, two months later and still without an answer from Berlin, he prepared again to undergo active military training.[60]

Himmler finally acted in July 1938, just over a year after the onslaught in the pages of *Das Schwarze Korps*. But before he did, further influence was required from another quarter. Earlier that year, when Heisenberg and his colleagues had learned that the SS investigation had concluded favorably but that Himmler was ignoring the case, Heisenberg and colleagues took recourse in a tactic they would often use in the future: working diplomatically through respected applied physicists. Personal diplomacy was, as always, the German professor's strongest suit.

As practical men contributing to Germany's economic and military buildup, which was now in full swing, applied physicists could hardly be considered overly abstract or formalistic. A key player in this strategy, who had known and respected Heisenberg since his early days in Göttingen, was Ludwig Prandtl, the Göttingen professor of applied mechanics. As a specialist in hydrodynamics and aircraft aerodynamics, Prandtl was familiar with both modern theoretical physics and applied practical research.[61] He was also well practiced in negotiation — thus the perfect man to approach Himmler regarding Heisenberg. Prandtl did so at the March 1 banquet celebrating Hermann Göring's new German Academy for Aeronautical Research, located in a new and spacious airplane hangar on the outskirts of Göttingen.

Conveniently seated next to the Reichsführer at the banquet, the aerodynamics professor discreetly raised the topic of unjustified attacks against theoretical physics and "especially the personal distress of Herr Heisenberg."[62] In a carefully worded five-page memorandum to Himmler on July 12 (after the Austrian Anschluss), Prandtl again defended Heisenberg and theoretical physics and suggested several elements of a resolution to the case.[63] First, Prandtl recommended adopting a compromise that Himmler himself had casually mentioned over dinner: that in teaching Einstein's physics Heisenberg should be asked to take great care to separate the man from his work. Second, Himmler should write a letter personally disavowing the smear of Heisenberg's character so that Heisenberg's effectiveness as a teacher could be restored. Third, for the benefit of Nazi physics students, Heisenberg should be allowed to publish an article on theoretical physics in the Nazi student journal *Zeitschrift für die gesamte Naturwissenschaft*.

Prandtl's diplomacy finally prodded the Reichsführer into action. On July 21, 1938, Himmler sent an official letter to Heisenberg's private address in Leipzig informing the professor of his personal decision: "I do not approve of the attack of *Das Schwarze Korps* in its article, and I have proscribed any further attack against you." He then invited Heisenberg to join him for a "man-to-man" discussion of the matter in Berlin in November or December and ended with a postscript admonishing the physicist in the future to separate the personal and political characteristics of the researcher from his research. He did not need to mention specific names.[64]

On the same day, Himmler also sent a memo to Heydrich, his SD chief, enclosing a copy of Prandtl's letter. In his memo, Himmler ordered Heydrich to encourage the student science leader, Kubach, to allow Heisenberg to publish in the *Zeitschrift* and to refrain from further assaults on Heisenberg. Nazi physicists would have to find a way to advance their cause other than character assassination of Sommerfeld's heir. For, Himmler declared, "I believe that Heisenberg is decent, and we could not afford to lose or to silence this man, who is relatively young and can educate a new generation."[65]

Himmler's letter reached an elated Heisenberg two days later, on July 23, in the Bavarian mountain village of Fischen. Werner had settled his family there while he prepared for military duty in nearby Sonthofen on the Austrian border. Heisenberg immediately thanked Himmler for his letter, "which freed me from my great concern,"[66] and readily agreed to the new compromise that Prandtl had struck with Himmler, a compromise that Prandtl had probably prearranged with Heisenberg and one that would endure through the period ahead. Heisenberg would take pains in the future to separate the scientist from his science. The useful results of work by Jewish researchers would be allowed in German classrooms — and could be readily exploited to German technological advantage — but any public mention of Jewish scientists themselves would not.

Most compromises are entered into as a means to an end, the acceptance of an undesirable situation for the sake of a greater benefit. In this case, the perceived benefit to Heisenberg was the restoration of his so-called honor, enabling him to remain in Germany and to continue his work and teaching in contemporary theoretical physics.

Given the poisoned atmosphere of Nazi Germany in 1938, the enormous new compromise to which Heisenberg and his colleagues agreed — the expunging of Jewish names — was probably the best they could expect. How difficult was it to accept? Not very. After all, the battle for retaining Jewish physicists at universities had been fought and lost over three years earlier, by the summer of 1935, and that unconscionable loss was soon accepted as work returned as close to normal as it could. By accepting Planck's counsel and remaining in Germany at that point, Heisenberg had already entered into a pact with the devil — a pact that the SS onslaught threatened to nullify. When Heisenberg demanded exoneration from Himmler, he had signaled his readiness to continue in the pact — he would accept the overall situation as long as he were left alone to work, to teach, and to act as before on behalf of his science. Officially agreeing not to mention the names of prominent Jewish physicists in lectures and papers involved some regrets but little additional anguish, especially if their work could now be used and taught to future generations without hindrance. In reality, many of their names, including Einstein's, were freely used within private circles of trusted colleagues and students and even occasionally in publications. But in the official public arena, the moral element was again sacrificed for the sake of professional advantage.

In retrospect, was the bargain worth it? Hardly. Did Heisenberg see it that way? Not exactly. To him, remaining in Germany was apparently worth almost any price, as long as he could continue to work and teach. Like many of those who were forced into emigration, his entire life and upbringing had instilled in him an unbreakable attachment to Germany that he could not easily deny, even temporarily. Only if the Nazis made life and work unbearable for him, only if they humiliated him into less than second-class status or worse — imprisonment — would be contemplate a move across the German border. (Family and children were, of course, also a consideration, but they were never threatened the way he had been.) Heisenberg obviously felt an enormous attachment to Germany and an enormous drive to continue his work in Germany and for the future of German physics.

In the end, these two perspectives came together. Heisenberg came to regard his personal survival in Nazi Germany as tantamount to the survival of decent physics, and the continued survival of some elements of decent physics provided the grounds for Heisenberg's personal continuation of the struggle. Heisenberg's exoneration meant to him exoneration of theoretical physics itself. And, by seeing himself in such grandiose terms, he more easily succumbed to further compromises and ingratiation with the regime.

This became clearer in the years ahead, during the struggle for two pieces of tangible evidence of exoneration that Heisenberg demanded as early as his first letter of response to Himmler in July 1938. First, he wrote, he looked forward to the proposed meeting with Himmler, so that "any sort of form could be found in which it could also be made publicly clear that the attacks on my honor were unjustified." Heisenberg demanded publication of an article in the Nazi science student journal as a suitable public concession of unjustified attacks. Second, he wanted SS headquarters to send copies of Himmler's letter to the REM and to the rectors of the universities of Leipzig and Munich.[67] Accepting the call to Munich would serve as his ultimate exoneration and evidence of the victory of decent physics over Stark and his Nazi minions. Neither demand was easily fulfilled, and the longer they were delayed, the more importance they seemed to take on. Not until the depths of World War II did Heisenberg receive the evidence that he regarded as proof of his exoneration.

Like Faust, the personal price Heisenberg was willing to pay for his bargain with the devil could also have meant his life. In July 1938, nine days after Himmler cleared his name, Heisenberg was marching in German uniform. Six weeks later, he nearly marched off to war.[68] After successfully annexing Austria, Hitler had set his sights on the Sudetenland, a Germanic territory in Czechoslovakia that was now nearly surrounded by the new German Reich. As tensions mounted in September, active military units, including Werner's, were placed on full alert. Heisenberg's superiors extended his tour of duty into October and distributed weapons to the troops. Heisenberg was sure of an impending invasion and certain that, because of the alliance arrayed against German expansionism, this attack would touch off a second world war. If war had indeed broken out then, it would have found him on the front lines, risking his life for a nation that had just subjected him to a year of inquisition and humiliation and that really cared little for him or his science.[69]

Fortunately for Werner, war was narrowly averted at the end of September by the infamous Munich treaty entered into by Hitler and Czechoslovakia's allies, Italy, France, and England. British Prime Minister Chamberlain returned to London proclaiming "peace in our time." It was the last appeasement Hitler would receive. A much relieved and now unburdened Heisenberg was back at his desk in Leipzig by mid-October.

But Czechoslovakia was dismembered, and Germany had emerged as the most powerful state in Europe. The European Allies had reached a tentative and uneasy compromise with the Nazi regime, a compromise that Hitler would exploit and betray within a year. Maintaining his own uncomfortable truce with the regime, a meditative Heisenberg wrote his mother from his army barracks as the guns were carefully returned to their racks, "It is strange to think how the fate of every individual and the deaths of many hundred thousand can hang on the decision of one man."[70] Within a year, this experience would become very real indeed.

CHAPTER 21

German Physics

Himmler's letter allayed Heisenberg's concern and removed the SS threat, but it did not stop "deutsche" physics. Heisenberg's personal victory could not translate into a professional victory for science as a whole. In the end it could not even win him the prize he originally sought, the incident that had precipitated the whole affair — appointment as Sommerfeld's successor. Himmler's SS was only one of three competing state bureaucracies involved in academic appointments, and in this sphere it was the least influential. Of the other two agencies, Hess's party academic organizations — the student and teachers leagues — insisted on making independent recommendations to the Reich Education Ministry (REM), which made the final decisions.

Matters came to a head after the resolution of the Sudeten crisis in the fall of 1938. Sommerfeld, nearly 70, announced his intention to cease teaching after the winter semester. He hoped by this action to confront the REM with the necessity of appointing Heisenberg without delay. The REM asked the Munich faculty once again for its recommendations. By October, the faculty had two lists from which to choose. One, submitted by Sommerfeld and Gerlach, named Heisenberg, Weizsäcker, and Richard Becker, in that order; the other, submitted by Bruno Thüring, the Teachers League's new university representative, named three applied physicists to the theoretical physics chair — and mediocre ones, at that.[1]

The dean, Friedrich von Faber, a backer of the Teachers League, chose the second list and forwarded it to Dr. Wilhelm Dames, the REM officer in charge of mathematics and physics appointments. Since the university rector, a state appointee, was noncommittal, Faber attempted in December 1938 to prod him into support with a long official letter supporting the Teachers League's choices. In order to counter the ossification of theoretical physics and to enable a return to the natural and the pictorial, he wrote, "under no circumstances shall men be named whose works lie in the train of thought of Einstein's relativity theory or that have been written in the spirit of pure formalism."[2]

Heisenberg's was a lost cause even before the dean's letter. SS officer Dr. Dames had been appointed to an honorary position in Stark's Berlin institute, and like his boss Otto Wacker, he had joined what Heisenberg called "the opposing party."[3] By early November 1938, Dames had already made his decision. On personally inquiring in the REM's Berlin offices about the Munich chair, Heisenberg learned that Dames would choose Sommerfeld's successor from Thüring's list. However, to avoid alienating Heisenberg and his supporters entirely, the diplomatic Dames promised Werner an appointment to a newly vacated chair in Vienna, in recently annexed Austria. Heisenberg's first choice remained Munich.[4] And so the skirmish continued, with the Sommerfeld faction, SS, REM, and party all jockeying for position and determined not to lose political ground.

Hess wrote twice to Rust to inform him that, because of "the previous political behavior of Professor Dr. Heisenberg," the candidate was unsuitable for appointment to any chair, least of all Vienna's.[5] When on the basis of these letters the REM wanted to deny Heisenberg any new appointment, Himmler promptly wrote Rust to the opposite effect: Heisenberg was especially suited for the Vienna chair, because he could be influenced in the proper direction by what Himmler saw as the strong Nazi character of the university after the annexation of Austria. He even enclosed the SS affidavit on Heisenberg's views to support his point.[6] Himmler continually avoided the promised "man-to-man" meeting with Heisenberg—probably in order not to appear unduly influenced by the physicist. But he did meet separately in Munich with Hess and with Walter Schultze, the head of Hess's Teachers League, to try to sway them in Heisenberg's favor—either for Vienna or for Munich.[7] Meanwhile, Sommerfeld and Gerlach were tirelessly arguing Werner's case in Munich faculty meetings—but to no avail.[8] In the end, Dames chose the least qualified of Thüring's three candidates as Sommerfeld's successor. Wilhelm Müller, the victor, was best known for his textbook on engineering mechanics and for his recent essays on "deutsche" physics.[9]

Heisenberg's contacts in Berlin informed him privately of the tangled web of political interests that still militated against his appointment to Sommerfeld's chair. As Heisenberg reported it to Sommerfeld,[10] Hess and Schultze, as party leaders, could not back down under pressure from Himmler because the party had already committed itself publicly against Heisenberg. Himmler himself did not want to force the party to accept Heisenberg because Heisenberg was not a party member, was at best ambivalent toward the regime, and if placed in the "capital city of the movement" might embarrass Himmler before the party faithful. Dames, though supportive of the "deutsche" physics faction when it worked to his advantage, found it prudent to counter its rising influence on REM appointments by choosing the weakest candidate from among the party's choices. Finally, the appointment of Nazi scientist Tomaschek to the faculty of the Munich Technical College in early 1939 strengthened the district Teachers League and made Munich university officials unwilling to risk further battles with them. When provoked, the Teachers League could call down the wrath of party, state, and students on its opponents.

Müller succeeded the great Arnold Sommerfeld in the fall of 1939, just as war broke out. An enraged Walther Gerlach declared that theoretical physics was now dead in Munich.[11] The Vienna chair was now also out of the question for Werner — Hugo Dingler, converted in his old age to "deutsche" physics, had joined the student cause and blocked Heisenberg's appointment.[12] As consolation, Himmler promised him a recommendation to another highly placed chair. Such an appointment, together with the promised publication of his views in the Nazi science student journal, still constituted for Heisenberg tangible proof that he was officially rehabilitated in the new Reich. "But it appears," Heisenberg wrote Sommerfeld, "that the last word on this has not at all been spoken."[13]

Heisenberg's SS victory also had little ultimate impact on the decline of his institute or on the deterioration of theoretical physics under the Reich. Wilhelm Führer, who had been recently elevated to an REM position, soon joined Dames in promulgating new university physics curricula and examination criteria that would specifically and radically reduce concern with theoretical physics in favor of "practical" training.[14] The number of theoretical physics students and doctorates granted continued to decline sharply in Leipzig and throughout the Reich. For instance, of the eighteen doctorates in theoretical physics that Heisenberg produced during his tenure in Leipzig (1927–1942), sixteen graduated during the Third Reich. But only one of them, Erich Bagge, received a doctorate between the SS attack and the start of World War II and only three achieved doctorates during the war, of whom two were foreigners.[15] REM chief Rust himself noted a similar nationwide decline and ordered Wacker—who in turn ordered Dames —to undertake a study of the possible elimination of some theoretical physics chairs. Dames, though delighted at the prospect of abridged theoretical physics, refused nonetheless to recommend reduction in his own area of administration.[16]

Throughout the losing battle for Sommerfeld's chair and the painful and humiliating investigation to which Heisenberg was subjected following the ugly SS assault, Germany was steadily sinking ever deeper into the noisome pit of Nazi barbarism. The SS and the Gestapo increased their police-state activities and expanded their use of the concentration camp to include not just detention but forced labor for all "undesirables."[17] Slave labor and genocide were not far behind. Himmler's exoneration of Heisenberg in the summer of 1938 occurred simultaneously with, in the words of one historian, a massive "tidal wave of terror" against Jews: intensified expropriation of property, arrests, expulsions, and street violence.[18] Just one month before Himmler wrote his letter to Heisenberg, he had ordered an expansion of the forced emigration of Jews, and his henchman Heydrich had issued quotas on the arrests of both Jews and "antisocials."

The descent continued into the autumn in step with Hitler's plans to provoke a war over the Sudetenland. Frustrated on that front by Allied appeasement, the regime received an unexpected opportunity to vent some of Germany's escalating war fever. On November 7, 1938, just days before the fifteenth anniversary of Hitler's beer-hall putsch, a distraught Jewish teenager whose Polish parents had been recently expelled from the Reich shot a third-rank German diplomat in Paris. The diplomat's death two days later served as a pretext for unleashing nationwide violence against German Jews.

Goebbels suggested that "spontaneous" protest demonstrations against Jews be held throughout the Reich. Hitler apparently concurred. Unbeknownst to Heisenberg, Himmler, while pressing for Heisenberg's call to Munich, was arranging for SS support of Goebbels's plan.[19] With war frustratingly averted and average citizens lusting for action, local party and political officials, spurred on by the network of state and party organizations, had little trouble inciting mobs in every city and town across the Reich into a bestial frenzy of violence against Jews and Jewish property during the night of November 9–10, 1938. David H.

Buffum, the American consul in Leipzig, described the violence as "a barrage of Nazi ferocity as had had no equal hitherto in Germany, or very likely anywhere else in the world since savagery, if ever."[20]

That terrible night came to be known as Reichkristallnacht, or the night of broken glass, for the tons of shop-window glass that littered every German street the following morning. Finally, the Nazi regime had revealed in unmistakable terms to the world and to its own citizenry its true nature. It left Heisenberg, like other nonparticipants whom Buffum had observed, "benumbed over what had happened and aghast over the unprecedented fury of Nazi acts." In his report to Washington, Buffum described how in Leipzig Jewish homes and apartments were invaded and demolished by ravening mobs who threw everything and everybody out onto the streets — often from upper-story windows. Men, women, and children were paraded through the streets and parks in humiliation. Hundreds of Jewish shop windows were smashed; the three Leipzig synagogues were burned to the ground; and, following Heydrich's order, police rounded up the first of several thousand Jewish men who were transported over the next two weeks to concentration camps. There they would "prove" their usefulness to the Reich by "contributing" their forced labor to the four-year plan.[21]

A visibly shaken Werner wrote to his mother on November 12 that he and Elisabeth were "still completely in shock from the last nights."[22] The beautiful tree-lined streets of Leipzig had been turned into heaps of trash; the large Bamberger and Hertz department stores in the center of town were smoldering ruins. Worst of all, a friend had told them of a horrible scene on the morning of November 10 as entire Jewish families were dragged, screaming, to the train station, shoved onto passenger trains, and expelled from Germany. The same day Heisenberg wrote of that scene, Hermann Göring, architect of the four-year plan, signed an order systematically excluding Jews from the German economy and from schools and universities. Jews could not enter retail stores, sell goods and services anywhere in Germany, or hold any managerial or executive position.

The grim weeks and months following the frenzy of Kristallnacht took their toll on Elisabeth, now pregnant with the Heisenbergs' third child.[23] To recuperate from the turmoil, the family retreated during the spring semester break in April 1939 to the peaceful village of Badenweiler in the beautiful Black Forest region south of Freiburg. From there Heisenberg continued to campaign by mail for his Munich appointment — Müller's official call to the chair was still a year away.[24] Tucked away in their Black Forest village, Werner and Elisabeth, realizing that the turmoil would only get worse, decided to search for a permanent country retreat for the difficult times ahead. Crossing over the nearby Rhine into southern France one day, they sat quietly on a hilltop bench looking back across the Rhine toward the verdant hills of their beloved Germany. After beholding the sight for some time, Heisenberg quietly whispered to his wife, "How can I ever leave?"[25]

A friend of the family had already told them via Werner's mother of a cottage for sale in the Upper Bavarian village of Urfeld, south of Munich. The wood-framed cottage with its three bedrooms, kitchen, and large veranda had belonged to the well-known late impressionist painter Lovis Corinth. It stood at the foot of the Herzogstand mountain range —the veranda overlooked beautiful Lake Walchen, and beyond, the towering, snow-capped Isarwinkel Mountains. The Heisenbergs had not yet seen the cottage, but Werner wrote his mother and the family friend to say that they were definitely interested.[26] The house and its price— 26,000 marks, furnishings included—sounded ideal, and Werner's mother began negotiations in Munich. Their Isar Valley house, which they had rented out, was now up for sale. Both sales took longer than expected, extending into the summer of 1939, when Heisenberg was scheduled to visit Columbia and several other American universities. The deal on the cottage was finally closed just before he set sail for the United States that summer. The family, he wrote his mother from aboard ship, would move into the cottage for the remainder of the summer immediately after Heisenberg returned to Germany.[27]

The transatlantic trip and its impact were also outgrowths of the SS newspaper assault on Heisenberg and theoretical physics two summers earlier. The onslaught had drawn the attention of the entire physics community to Heisenberg's science and to his personal views of theoretical physics and its practitioners. While Heisenberg submitted himself to SS and REM scrutiny, he did not remain passive. Although many of his close colleagues had rallied around both Heisenberg and their profession, many others had not. To encourage his supporters and to convince the wavering, Heisenberg hit the lecture circuit, speaking mainly to other physicists and those in closely allied fields. "It is probably good if I don't refuse such invitations," he wrote to his mother in November 1937.[28]

The dogmatists of "deutsche" physics had repeatedly called for the return of theoretical physics from a supposed overreliance on abstract formalism to a close reliance on experimental data. In both nuclear and cosmic-ray physics, there existed a close relationship between theory and experiment, and the two fields contained problems and results of both intellectual and practical interest. Although Heisenberg himself concentrated on the more theoretical problems of both fields, between the summer of 1937 and the spring of 1938 he tirelessly repeated two lectures emphasizing the theoretical-experimental and the intellectual and practical aspects of nuclei and cosmic rays. One lecture, entitled "The present tasks of theoretical physics," he presented to audiences in Münster, Freiburg, Stuttgart, and Ludwigshafen.[29] The other, a more technical talk entitled "The transit of very energetic particles through the atomic nucleus," he gave with variations to physicists in Frankfurt, Dresden, and Bologna; to readers of *Die Naturwissenschaften;* and on his election as deputy class secretary of the Saxon Academy.[30]

In January 1938, Heisenberg reviewed the theoretical and empirical evidence in support of his still controversial contention that a universal length must appear in nuclear and cosmic-ray physics and must be connected with the possible existence of cosmic-ray explosion showers. His only moderately technical article appeared in the pages of the *Annalen der Physik,* a physics research journal usually reserved for original technical reports of new results. In his opening paragraph, Heisenberg apologized for not meeting those standards with the excuse that the issue was dedicated to Max Planck, who, readers well knew, had only six months earlier been vilified along with Heisenberg as a "white Jew." He closed his essay to Planck by assuring his readers that "in theoretical physics it can always only be a question of mathematical connection between observable quantities. . . . But probably a considerable expansion of the available empirical material would be the necessary precondition for the carrying out of such a program [of the universal length.]"[31]

Heisenberg lectured at a nuclear physics conference in Bologna just days after he spoke at a colloquium on probability theory in Geneva.[32] Hess's organizations and the SS diligently opposed such traveling abroad, including travel to the United States, but the REM Congress Office, in consultation with Saxon officials, made the final decisions on foreign travel. The REM was especially favorable toward Heisenberg's foreign tours after receiving reports on his good behavior abroad — the SD and German citizen spies had kept a close watch on the traveling physicist.[33]

In the spring of 1938, the REM again granted Heisenberg permission to travel abroad, this time to England, and this time over no objections. However, the REM resolutely refused to allow him to attend a theoretical physics conference in Warsaw. Hitler had set his sights on Poland for his next campaign and did not want any German academics there to stir up trouble. At that time, the SS investigation was close to its positive conclusion. After having spent a year in painful SS and REM investigation and humiliation, Heisenberg would not jeopardize the impending exoneration over a conference. The REM intrusion would be another compromise that he would have to accept. Heisenberg regretfully complied with REM demands and canceled his travel plans to Warsaw. In his absence, Léon Rosenfeld read a French translation of Heisenberg's prepared paper to the conference, but the regime did not even permit the paper to be published in the conference proceedings.[34] According to one participant, the Warsaw conferees, who knew nothing of Heisenberg's predicament, formed a low opinion of their German colleague who, it seemed, had all too easily capitulated to the unreasonable demand of his government to dissociate himself from their conference.[35]

Fortunately the regime had not prevented Heisenberg's important trip to England several months earlier. There he achieved a revitalization of his physics of the fundamental length shortly before Himmler provided the exoneration that enabled him to continue work in Germany with less hindrance. The revitalization occurred just as difficulties rendered untenable the application of Fermi's beta-decay theory to proton-neutron forces within the nucleus and just as the elusive evidence for Heisenberg's explosion showers seemed actually nonexistent. Heisenberg's Planck paper in this same period was actually a defense of the fundamental length, despite the slim evidence.

Werner's breakthrough involved the shift to a new force field, away from his earlier use of Fermi's field to an alternative field proposed in 1935 by Japanese theorist Hideki Yukawa.[36] Fermi had devised his electron-neutrino field to account for the decay of neutrons and protons. The Fermi field, when applied to the attractive force between neutrons and protons in the nucleus, carries a charge between the two particles as required, but the masses of the exchanged electron and neutrino are too small to account for the large size and short range of the nuclear force. In most cases, the more massive the exchanged particle, the larger the force and the shorter the range. As a remedy, Yukawa invented a new field represented by a new, heavier charged particle. A neutron and a proton would attract each other by rapidly emitting and absorbing not electrons and neutrinos but "virtual" Yukawa particles.[37] Because each nucleon has a spin of $1/2$, the new particles — variously called yukons, dynatons, mesotrons, and mesons — had to carry integral spin (bosons). The only available quantized boson field for charged particles was the Pauli-Weisskopf scalar field for zero-spin particles.

Yukawa adopted this field and, in analogy with QED, he assumed that the density of the interaction Hamiltonian could be written

$$H_{int} = (g/\sqrt{hc/2\pi})\,\overline{\psi}\psi\phi$$

where ϕ is the scalar meson field, $\overline{\psi}$ and ψ are the nucleon wave functions, and $g/\sqrt{hc/2\pi}$ is a matter-field coupling constant analogous to $2\pi e^2/hc$ in QED (the symbol f was already taken by Fermi).

In order not to neglect beta decay, Yukawa suggested that his particle, once freed from nucleons (if the energy is great enough), could spontaneously decay into an electron and a neutrino. Not wishing to neglect cosmic rays, Yukawa supposed that "the massive quanta [mesons] may also have some bearing on the shower produced by cosmic rays."[38] The 1937 discovery of the heavy electron in the cosmic-ray hard component, possessing the same charge (e) and about the same mass ($200\ m_e$) as Yukawa's massive quanta, suddenly turned attention in the West to Yukawa's ideas. But Yukawa's simple, spinless, charged mesons would not do—much more had been discovered since 1935 about the spin dependence and the charge independence of the nuclear force.[39]

Yukawa's ideas nevertheless inspired a flurry of theoretical activity on nuclear forces—most of it in England, most of it by German émigrés, and most of it during the months just preceding Heisenberg's visit there in March 1938. Recently arrived Nichols Kemmer, Pauli's former assistant, served as a prime catalyst. Shortly after seeing Yukawa's paper in 1937, Kemmer derived all of the mathematical possibilities for boson fields and compared their predictions with data on the deuteron, a bound neutron and proton. He convinced his colleagues Fröhlich, Heitler, and Bhabha that only the spin-one vector mesons, mesons represented by a vector wave function, agreed with deuteron data.[40] Kemmer and colleagues managed to derive practically the entire range of available nuclear data before turning to a phenomenon involving nuclear forces arising outside the nucleus: cosmic rays. "We think, therefore," they wrote, "that it might be a reasonable policy to try to link up nuclear properties (forces and magnetic moments) with the cosmic-ray phenomenon of the hard component rather than with the beta decay."[41]

Of the two constituents, or components, of cosmic radiation, the easily absorbable soft component had been identified as electrons and photons obeying quantum electrodynamics—the quantum mechanics of charges and the electromagnetic field. The penetrating hard component had been identified as a new particle, a heavy electron. Because of its greater mass, it can penetrate greater distances of matter than can electrons and photons. Because the heavy electron and the Yukawa meson seemed so similar (and because there were no alternatives), most physicists, like Kemmer and colleagues, proceeded on the assumption that the Yukawa meson and the hard-component heavy electron were, in fact, identical. Unfortunately, they were wrong, but physicists did not discern the existence of *two* mesons in cosmic rays until after the coming war. The penetrating heavy electron is actually the mu-meson (muon), associated after all with the Fermi field. After penetrating long distances through matter, it decays into an electron and a neutrino. It is the decay product of the pi-meson (pion) of the Yukawa field, which is created high in the atmosphere and is not seen at sea level. This distinction, not realized at the time, would have resolved numerous puzzles throughout this period.

Regardless of which meson was actually observed, theoreticians immediately encountered an old obstacle in the nuclear physics of cosmic rays: the divergence of the self-energy and of interactions at high energies or small distances of approach. In fact, outside nuclei, higher order interactions—the multiplication of virtual particles—took on increasing importance, as they had in Fermi theory. And as in Fermi theory, for distances or wavelengths smaller than a certain critical minimum length —or collision energies larger than a certain value, the rest energy of the meson, 100 MeV—the number of newly created particles multiplied without bound (save for available energy). All this had a familiar ring. "According to the views developed here," Kemmer told his readers, "Heisenberg showers should be expected to occur."[42]

A month later, the inventor of the theory of Heisenberg showers arrived in England. He immediately launched into nonstop discussions with local theorists; with his host, cosmic-ray experimentalist Blackett; and with the hordes of physicists from all over England who came to Cambridge and Manchester to hear him speak. "I hardly ever get to bed before 12:30," he wrote his wife, "get up around 8, and talk the entire day through with physicists almost without a break. It is important to me now to lose myself entirely in physics."[43]

Heisenberg learned that the vector-meson field could produce multiple, or "explosion," processes that seemed to account for all the main features of cosmic-ray data for the hard component. According to the scheme developed in England, incident gamma rays (γ) from outside the atmosphere hit an atmospheric nucleon—a proton or neutron—

producing in a multiple (explosion) process the plentiful hard component in the upper atmosphere. According to this scheme,

$$\gamma + P \longrightarrow P + nY^+ + nY^-$$

the Ys being positively and negatively charged mesons and n being the number of mesons of each type, as determined by the available energy. Most of the mesons would reach sea level to produce explosion showers of more mesons. For example,

$$Y^+ + N \longrightarrow P + nY^+ + nY^-$$

N being a neutron. After penetrating large blocks of matter, the mesons would then decay into cascading electrons and undetectable neutrinos or disappear in matter by ionization.[44]

Returning to Leipzig, Heisenberg immediately revamped his old physics with new arguments, replacing the critical length of Fermi's theory with the critical length of meson theory as a new fundamental constant of nature. The new length, $l_0 = h/mc$, where m is the meson mass, now acted even more explicitly than in 1936 as a fundamental constant of nature, as a defining feature of both the current quantum theory and the future refined theory. The fundamental length marked the lower boundary, the "limits of applicability of the present quantum theory."[45] All the present quantum-mechanical methods and field theories, Heisenberg argued, ceased to apply as soon as collision energies and momenta crossed the boundary established by the new length; that is, as soon as colliding particles touched closer than the Compton wavelength, h/mc, of the Yukawa meson. Above that energy boundary, or below that distance of approach, quantum mechanics ceased to be valid and meson explosion showers burst forth as the first phenomena associated with the future quantum theory for that region.

Heisenberg had often discussed his views regarding a fundamental length with the sympathetic Bohr. In May of 1938, Werner sent a copy of his manuscript, "The limits of applicability of the present quantum theory," to Copenhagen, intending it to be something of a new uncertainty principle.[46] While Heisenberg prepared his manuscript for publication, Hans Euler, his assistant, attempted to distill meson explosion showers from available data on large bursts of particles behind thick and thin absorbers. This work would earn him his habilitation later that year.[47]

That summer, as Himmler drafted his letter of exoneration and as Heisenberg contemplated the possibility of marching off to war during the Sudeten crisis, Euler and Heisenberg began a thorough analysis of all data available on cosmic rays, including cascade and explosion showers of all types, to support their physics.[48] In a scene reminiscent of his reading of Plato during the suppression of the Munich soviet republic, Heisenberg, sequestered in the Sonthofen army barracks that fall with machine gun at the ready, submitted a highly technical spinoff analysis of the penetrating cosmic-ray component for an issue of the *Annalen der Physik* devoted to another "white Jew" — Arnold Sommerfeld.[49] Heisenberg apologized to Pauli for the circumstance that, for political reasons, only "Aryans" were allowed to celebrate Sommerfeld.[50] Pauli helped organize an alternative celebration in the *Physical Review* and a boycott of the *Annalen*.[51]

The linchpin of the Euler-Heisenberg argument was an apparent agreement between their theoretical value for the half-life of the meson, 10^{-6} seconds, and the result of their analysis of the decay rates of heavy electrons in air and matter.[52] With such a short half-life, mesons could not have come from outer space — they had to be produced in the upper atmosphere, surely by the multiple or explosion process. By the same token, they could not survive the long journey to sea level in that time unless their half-lives were extended. The only way that could happen was through the celebrated time dilation of relativity theory. Time slows down for fast-moving particles.

In the winter of 1938–1939, while Himmler pressed to have Heisenberg appointed to the Munich chair, Heisenberg took his new results on another tour of the lecture circuit. He spoke at physics conferences in Hamburg and Leipzig and to the Bavarian section of the German Physical Society in the "capital city of the movement." Carefully avoiding Einstein's name, he told each audience of his new success as a theoretician in accounting for empirical data and of the meson half-life result. "This is an immediate consequence of the relativity principle — and, at the same time, a striking confirmation of it."[53]

Unfortunately, this confirmation and Heisenberg's plans for a future theory of the fundamental length did not go unchallenged. Other analyses of available data, undertaken mainly abroad, seemed to confirm instead a discrepancy between the theoretical and experimental half-lives for the cosmic-ray meson.[54] Multiple processes and explosion showers still eluded direct detection at sea level, the penetrating component was observed once again to penetrate much further than it should, even with Yukawa forces,[55] and the discovery of correlated extended bursts of cascades led to the suspicion that Euler's explosion bursts were really a multitude of simultaneous cascades set off by penetrating rays, rather than single-event explosions. British, French, and American teams confirmed the suspicion.[56] As a result, one American team told the June 1939 Chicago cosmic-ray symposium (which Heisenberg attended), "We shall adopt the more interesting and extreme position, and deny the existence of explosions until we are forced to recognize them."[57]

Most physicists who denied the existence of explosion showers believed at the same time that quantum electrodynamics, which rested on the reigning quantum mechanics, might encounter a boundary of validity at some very high energy, but this boundary would have nothing to do with Heisenberg's critical fundamental length associated with the rest mass of particles or with the corresponding relatively low critical energy.[58] Quantum mechanics should be fully applicable to the empirical mysteries of both nuclear and cosmic-ray physics. Practically alone in his views by 1939, Heisenberg believed in the existence of just such a length, establishing a boundary, and the need for a new revolution in physics, similar to the one that had led to quantum mechanics, to handle nuclei and cosmic rays. Although some empirical data were open to interpretation either way, most physicists, such as those associated with the Oppenheimer group in the United States and the Heitler-Bhabha-Kemmer group in Britain, perceived little support for Heisenberg's contentions. In 1939 and 1940, both groups also advanced a theoretical challenge to Heisenberg's fundamental length, the earlier of which prompted Heisenberg's most mature prewar views and the starting point for his postwar work.

In early 1939, Homi Bhabha, now working in Cambridge with P. A. M. Dirac, argued that divergences in field theories had nothing to do with explosions or with Heisenberg's supposed limits of quantum theory. They reflected only procedural problems: "These limitations are probably due to the fact that this is not the quantization of the correct classical equations for point particles, and are not due to the existence of a fundamental length r_0." Bhabha claimed that the correct classical equations were those that Dirac had derived earlier—when searching for "some new physical idea"—by invoking his trick of subtracting infinities.[59]

Bhabha considered a simplified meson field, an unquantized, neutral, spinless field that differed from electrodynamics only in the rest mass of the meson. When he subtracted infinities, multiple processes and the critical length (arising from the rest mass) simply disappeared—and he attained a direct correspondence with Dirac's equations. This proved,

Bhabha declared, that Heisenberg's limit, "if true, cannot be based on explosions as derivable from the theory of the meson," nor could explosions, "if they exist . . . be connected with the rest mass of the meson."[60]

The only boundary to Bhabha's classical theory was a familiar one: the onset of radiation reaction caused by the recoil of the nucleon acting on the scattered meson field, which required a quantum theory. But the energies involved were far above Heisenberg's boundary, thus opening "the whole field of nuclear phenomena and an important part of cosmic-ray phenomena" to Bhabha's theory.[61] Moreover, because only a weak coupling between mesons and nucleons occurred, low orders of perturbation theory were sufficient. This reduced the interaction of mesons with matter—and the likelihood of explosion showers —even more. Because of this, Bhabha's mesons were less easily absorbed than are electrons and photons. The lower absorption yielded almost exact agreement with data on the higher than expected penetration of the penetrating cosmic-ray component.

Oppenheimer and his colleagues also employed a weak-coupling argument to refute Heisenberg's critical length—in this case for meson-electromagnetic interactions. As they had in 1936, they followed the usual procedure of avoiding divergences by reducing the coupling, the attachment of the field to the particles. In a paper published in 1940 in the *Physical Review,* the leading American physics journal, they too found that they could account for collision events far beyond Heisenberg's boundary: "The evidence indicates that it is the largeness of the coupling, and not the occurrence of lengths smaller than the critical $h/mc \approx 2 \times 10^{-13}$ cm, that limits the applicability of the quantum mechanics."[62]

Heisenberg reiterated just the opposite in both the United States and Germany. He could not suffer Bhabha's "Cambridge subtraction methods," he told Pauli, and, as a correction of Bhabha's theory, he introduced a new critical length in May 1939 derived from the opposite direction—reliance on a strong-coupling theory containing the divergent terms that Bhabha had obliterated.[63]

In a paper entitled "On the theory of explosionlike showers in cosmic radiation II," Heisenberg also used an unquantized classical neutral theory, but with spin.[64] But, unlike Bhabha, again he obtained higher order interactions, or explosions, controlled by a length that contained both the rest mass and the spin coupling:

$$l = g\sqrt{4\pi/hc} \; (h/mc)$$

where g is the coupling constant for spin. The spinning mesons attached to a bare nucleon imparted a special field to the nucleon that caused the strong coupling with an incoming cosmic-ray meson. After working out the equations, Heisenberg found that, as a meson smashed into a nucleon, strong coupling caused an enormous recoil of the nucleon that strongly dampened the radiated meson field. If the nucleons possessed a minimum size, chosen to make their self-energy convergent, the radiation reaction also converged. This reduced the meson-nucleon interaction, thus yielding, in the opposite way to Bhabha's, almost exact replication of the observed higher penetration of penetrating particles. At the same time, when the particles collide, the nonlinear field interaction arising from the strong coupling at distances below the critical length acted like a turbulent liquid confined in a small container, while the radiated meson field, like a liquid sprayed into space, condensed into an explosion of droplets, or particles. Divergence and an associated critical length had again provided Heisenberg "an access to the region in which the present quantum mechanics fails."[65]

Heisenberg brought his revolutionary new shower theory to the United States in the summer of 1939. REM officials now offered little objection to travel by the recently exonerated physicist, but his hosts strongly objected to both his scientific and his nonscientific views.[66] During his monthlong travels to New York, Chicago, Ann Arbor, and Indiana, American physicists and their émigré colleagues had a last look at the great physicist who had chosen to stay in Germany and who would in coming years assume a prominent position in the rival German nuclear research effort during World War II. Their impressions of him at that time helped to shape their attitudes toward him both during and after the war. There was the question of his reargued science, which differed so greatly in import from their own. But the main controversy surrounding Heisenberg was why he wanted to remain in Germany. The Reichkristallnacht of the previous November, his ill treatment at the hands of the vicious Nazi regime, and the impending world war seemed more than sufficient reason for emigration. Mystification at his continuing to live in Germany may even have affected his colleagues' perception of his science. In their view, if Heisenberg had been in the United States during the previous year, he would surely have seen that American experimentalists and theorists had proved his physics wrong. American science was, after all, on an ascendant trajectory, while German theoretical physics seemed in obvious decline.

These unarticulated views may have informed the response to Heisenberg's newly developed arguments for the universal length at a cosmic-ray symposium held in Chicago at the very university where, three years later, Enrico Fermi would produce the world's first sustained nuclear chain reaction. Most of the papers presented during Heisenberg's session of the symposium opposed the identification of the meson with the heavy electron and contained arguments against the proposal of a universal length. According to Heisenberg's recollection of the meeting, the animated discussion following his session soon degenerated into a shouting match between himself and J. Robert Oppenheimer, the doyen of West Coast physics and the future head of the Manhattan Project.[67]

The reasons for Heisenberg's refusal to leave the Reich were much more complex than his American hosts or the Warsaw conferees before them probably realized. In both instances sentiment toward Heisenberg tended to be negative, and from the vantage of retrospect it is a sentiment with which one may well agree. Of course, Heisenberg did have ample opportunity to emigrate: after the SS newspaper assault, two major American universities had offered him tailor-made positions. Heisenberg had refused both; as incredible as it seemed, he did not want to leave Germany. Heisenberg's secret negotiations with physics professor George Pegram of Columbia University had led to a lucrative offer by the summer of 1938. But Werner had by then received Himmler's exoneration, and he informed Sommerfeld from a village near his military barracks: "I have written to Columbia University that I want to remain in Germany, and that I would like to come over there sometime, but only for a brief period."[68]

Apparently A. H. Compton, head of the American Physical Society, had gotten wind of Columbia's inquiries — probably from Pegram, who was also an official of the society.[69] During Heisenberg's American travels in 1929, Compton's brother, a Princeton professor, had offered him a position there, but for patriotic reasons he preferred to return to Germany. "Because of the recent developments we think that he may now be interested in an offer," Compton informed his son, who was in Germany shortly after the 1937 SS assault. The son was instructed to repeat the invitation, this time for the University of Chicago, and Compton prepared to meet any demands Heisenberg might make.[70] Heisenberg again refused — he still would not leave Germany. When Heisenberg finally arrived for his monthlong visit in 1939, his colleagues in the United States were more than curious about the possible reasons, other than devotion to Nazism, a rational person could have for not wanting to leave Germany, faced as he was with the terrible social and professional conditions of 1939 and with an impending war.

Those reasons were still patriotic and, it seems, still predicated on the absurd hope that Hitler and his henchmen either would somehow be replaced or would become more reasonable in due course. Before that happened, however, if war were inevitable, then the German nation had to be protected from defeat, even if that meant giving this current regime the victory. And with Germany's industrial potential and her massive war preparations, victory seemed to most Germans a likely outcome. Recollections abound of Heisenberg's statements during his American tour regarding the approaching war and his decision to stay. Heisenberg himself recalled expressing the conviction that Germany would lose the war and that he would be needed there to pick up the pieces.[71] However, British physicists Mott and Peierls (an émigré) write that "in the recollection of his colleagues, he appeared to foresee a German victory. Was this a failure in communication, or did the views appear to him, or to the others, in a different colour in retrospect? One knows the fallibility of human memory."[72] His colleagues sound more accurate.

Beyond patriotic attachments to national revival, adversity had already bred tenacity. Heisenberg's colleagues could not know that absolutely no rational argument, however cogent, could outweigh the psychological grounds for his decision to return to Germany. Heisenberg had already survived six and a half years of the Nazi regime. He had decided long before, even as the situation worsened, that he would not leave his homeland unless the authorities made his work and teaching absolutely impossible. Moreover, his perceived duty to his students and to the future of his profession in Germany coincided with and justified this decision. Nor, apparently, did he seriously contemplate moving his wife and family to safety abroad. Just before he boarded the ship for the United States, he had closed the deal on the mountain cottage, and while he debated with his colleagues in the United States, his family anxiously awaited his return to Germany so that they could move to their mountain retreat ahead of the gathering political storm.

By first threatening his work and his teaching and then officially protecting both, the SS affair in the end actually reinforced Heisenberg's decision to stay in Germany and to accept his lot under the regime. With the SS affair settled, promises of protection and support from Himmler, a newly acquired mountain refuge, and his potentially revolutionary science on an upswing, Heisenberg boarded the ship for New York thoroughly content with himself and with his decision to return to his troubled country. "I have the feeling," he wrote his mother, "that now everything is in place [in my life], as far as this depends on me. Of course many difficulties could still come from the outside. But I will deal with them much more easily than with the inner difficulties."[73]

After lecturing in Chicago to overflow audiences and spending a tiring two weeks of endless seminars and discussions with numerous faculty and students at Purdue University in Indiana, Heisenberg accompanied former Göttingen students and meson theorists Lothar Nordheim and his wife to Ann Arbor for a few sticky days during a sweltering mid-July. There he stayed in the home of Samuel Goudsmit, an organizer of the annual Ann Arbor summer school in physics.[74] Talk among the many students and faculty attending the summer school soon turned inevitably to emigration. Heisenberg recalled one such conversation in his memoir, *Physics and Beyond*.[75] To Fermi's arguments for emigration, Heisenberg responded that he had long since gathered about him a circle of young people whom he saw as the hope of the future, and "if I abandoned them now, I would feel like a traitor. . . . [Besides] I don't think I have much choice in the matter. I firmly believe that one must be consistent. . . . People must learn to prevent catastrophes, not to run away from them. Perhaps we ought even to insist that everyone brave what storms there are in his own country."

Other participants remember the same conversation somewhat differently. Goudsmit, then on the Michigan faculty, recalled: "Enrico Fermi and I asked him the question many others had asked: 'Why don't you come here?' He answered: 'No, I cannot, because Germany needs me.' He believed that the Hitler excesses, of which he strongly disapproved, would soon blow over. He felt that he would be needed to repair the damage made by the regime."[76] Max Dresden, another participant (then a Michigan student), recalls that Heisenberg seemed at first to waver in answering the question put by Fermi and Goudsmit. At that point, Mrs. Fermi, who had been driven from Italy by German-inspired anti-Semitic laws, said that anyone must be crazy to stay in Germany, whereupon Heisenberg launched into his vehement objection.[77] The patriot obviously would not budge.

Heisenberg returned to a hot New York for the last week in July, lecturing at Columbia and visiting his Uncle Karl and Aunt Helen in the suburbs. During his stopover, Columbia professor George Pegram, the fatherly experimentalist who had done everything he could to get Heisenberg to join the Columbia faculty, tried one last time to convince him to stay, and Heisenberg tried once again "to get him to see my point of view."[78] For one final time, the answer was no. Heisenberg set sail in early August, leaving a thoroughly puzzled Pegram on the dock as the nearly empty luxury liner *Europa* steamed toward the German Reich. One month later, the German Reich was at war.

PART FOUR

A World at War

CHAPTER 22

A Fine Line

On September 1, 1939, war broke out; on the day after, towards evening, our son Heinrich was killed." On this unemotional note Ernst von Weizsäcker, then state secretary in the German Foreign Office in Berlin, began his memoirs of World War II.[1] Lieutenant Heinrich von Weizsäcker, platoon leader in the Ninth Infantry Regiment, fell the evening of September 2 on the Tucheler Heath near Danzig. His brother Richard, in the same regiment, watched over him until morning, when he was brought back to Stuttgart for a funeral presided over by his other brother, Heisenberg's colleague and confidant, Carl Friedrich. "And there he now lies, under the wooden cross from the Tucheler Heath" — one dead among the millions to follow during the next six years, the carnage sparing neither innocent nor educated nor witting belligerent.

Two days after Hitler unleashed his army into neutral Poland, England and France declared war on Germany. For the second time in less than a generation, Europe — and soon the world — was at war, the savagery and brutality of which the world could barely have imagined in those early days when Heinrich fell.

The Weizsäckers were not the only family among Heisenberg's cultured acquaintances with fathers and sons at, or on their way to, the front. The musical Mr. Bücking was on the front lines that day as the German army smashed its way through Poland toward the Vistula. By December he was back in Leipzig with an Iron Cross and orders to join the forces at the Western frontier, whence Hitler would soon unleash his second Blitzkrieg.[2]

Many of Heisenberg's physics colleagues also reported for duty that September. On September 16, seven aging experimentalists turned up as ordered at the Army Ordnance Office in Berlin, toting their military arm bags packed with underwear and toiletries for the front.[3] They were more than a little relieved to learn that they were not headed for the front. Kurt Diebner, army research expert for nuclear physics and explosives, and his assistant, physicist Erich Bagge, had ordered them instead to a meeting on the potential applications of a recent German discovery—nuclear fission. At the meeting, the scientists explored the technical aspects of exploiting fission, both controlled and uncontrolled, but it was soon clear that much more research was required. Dr. Bagge suggested to the assembled experimentalists that his Leipzig mentor, Professor Heisenberg, be included in their newly formed "uranium club" in order to provide a theoretical foundation for their work.[4]

Bagge's mentor, meanwhile, had been waiting impatiently for his marching orders since early September.[5] The slightly built physicist looked much younger than his 37 years, and when he put on the uniform of his reserve infantry brigade, the innocence betokened by his friendly face, blond hair, and disarming smile contrasted with everything that grey uniform and its insignia represented. Ever since he had nearly marched into battle during the Sudeten crisis a year earlier, Heisenberg had been convinced that war was inevitable and that he would be in it; like his acquaintances, he was ready for the fight. Orders to the front must be on their way, he felt. Orders finally did reach him later that month—but not to the front. Bagge arrived in Leipzig on September 25 to inform the physicist that he was to attend the second uranium club meeting in Berlin the next day and that Bagge had arranged for Heisenberg's mobilization, not for the infantry but for research under the auspices of Army Ordnance.[6] Heisenberg traveled to Berlin that night and soon joined the German club. Unlike their counterparts in the previous world war, Heisenberg and other German scientists would fight their war on the research front.

On September 26, Heisenberg, Otto Hahn, Carl Friedrich von Weizsäcker, and several other nuclear scientists reported as ordered for the uranium club meeting held in the research office of Army Ordnance on Hardenburgstrasse, directly across from the Berlin Technical College. Once again the scientists reviewed the practical means of exploiting fission, and the theorists provided what they knew about the fission process, but they all agreed that continued research, both theoretical and experimental, was required before the possibilities could be definitely determined. After discussing the types of questions to be answered, the scientists dispersed to their various institutes. With funding provided by Army Ordnance, they began to implement a research program developed and coordinated by Diebner and Bagge. The German nuclear fission project had begun.

Both German and Allied interest in nuclear energy, controlled and otherwise, had quickly mushroomed after the 1938 discovery of fission —the splitting apart of a heavy nucleus with the release of enormous amounts of energy—by Otto Hahn and Fritz Strassmann in Berlin.[7] After the annexation of Austria, Hahn's longtime colleague, Austrian Lise Meitner, had fled to Sweden just before the big discovery. Meitner and her nephew, Otto Frisch, who had also fled the Reich, soon showed how fission could occur on the basis of Niels Bohr's liquid drop model of heavy nuclei. Upon absorbing a neutron, a heavy drop would become unstable and split in two, releasing energy and particles. Frisch told Bohr in Copenhagen of the discovery, and Bohr brought the news to America in January 1939. While Bohr and John Wheeler worked out a complete theory of nuclear fission in Princeton, a Paris nuclear research team under Frédéric Joliot (son-in-law of Madame Curie) confirmed in April that on average more neutrons were released per fission than were absorbed. An energy-producing chain reaction could occur.

Physicists on both sides of the coming war alerted their governments to the prospect of a new weapon made possible by the discovery of fission. In March 1939, several months before Heisenberg arrived for his visit in the United States, Heisenberg's discussion partners, George Pegram and Enrico Fermi at Columbia University, contacted the U.S. Navy about the remote possibility "that uranium might be used as an explosive."[8] But like most scientists, Fermi was skeptical that an uncontrolled chain reaction could actually be achieved. The navy shelved the idea until, later that fall, after the outbreak of war, Einstein's famous letter to Roosevelt, written in August at the urging of Szilard and a now enlightened Fermi and Pegram, reached Roosevelt's hands.[9] Several months later, Frisch and Rudolf Peierls, both German refugees in England, alerted the British government to the possibility of nuclear fission. By mid-1940, as the German army rolled across Europe, two Allied nuclear fission projects were already under way—one in the United States, the other in Britain. They and the French team later merged to form the Manhattan Project—the Allied effort to construct an atomic bomb.

Several German scientists likewise alerted their superiors to nuclear developments. Two Göttingen professors informed the Reich Education Ministry, which turned their letter over to Abraham Esau, the incongruously named head of the physics section in the Reich Research Council. Esau organized a study committee. Meanwhile, two other professors in Hamburg, Paul Harteck and Wilhelm Groth, informed Erich Schumann, head of the weapons research office in Army Ordnance (and a descendant of the composer), of the possibility of an enormous new explosive. A skeptical Schumann handed the matter over to his explosives expert, Dr. Diebner, who had earlier worked on nuclear physics. Diebner enlisted Bagge to the cause.

Since Diebner and Bagge were the least skeptical about the potential of fission, by the outbreak of war in September 1939 Germany was the only nation with a military project on nuclear fission. The Reich also controlled the world's largest supply of uranium ore, having seized the rich Joachimsthal mines in occupied Czechoslovakia. As the nation that discovered nuclear fission and because of the open publication of results in Allied nations until as late as June 1940, Germany was also privy to all the necessary basic research. Moreover, Siegfried Flügge, one of Heisenberg's former pupils, broadcast Germany's interest in nuclear energy with a widely read article, "Can the energy content of nuclei be made technically useful?"[10] No wonder that in years to come Allied scientists were convinced of Germany's head start in the race for an atomic bomb.

Judging from his rate of production during the first few months of the war, Heisenberg was indeed working with enormous energy on the theoretical possibilities for exploiting fission. The prospect of performing such research surely came as no surprise to him. During his visit to the United States in the summer of 1939, Heisenberg had had long discussions with Fermi and Pegram, both of whom were by then well aware of the theoretical possibility of an explosive and, perhaps further induced by their encounter with Heisenberg, would soon approach Einstein concerning the letter to Roosevelt. Heisenberg recalled in his postwar *Physics and Beyond* that during their conversation in Ann Arbor Fermi raised the prospect that after the outbreak of war scientists in all nations would "be expected by their respective governments to devote all their energies to building the new weapons."[11]

Heisenberg recalled conceding the truth of Fermi's point: "You are only too right in what you say about our participation and responsibility." But he offered that no matter how feverishly governments and scientists strove to achieve an atomic bomb, "for the present I believe that the war will be over long before the first atom bomb is built."[12] In other words, as Heisenberg recalled his position in the summer of 1939, he and other German scientists would readily work on a nuclear project established by their government, but they would not achieve a bomb—

not because they would refuse at this point to produce one on moral or political grounds but because the technical difficulties would be so great that the war would end before they could overcome them. These circumstances and his remembered reactions to them would change over the coming years and decades. As with his research and activities before the war, each step of Heisenberg's participation in the German nuclear project must be seen in its own context and in the light of the inherent tensions of an individual attempting (however misguided this was) to make the best of life under a vicious, antiscientific dictatorship now at war.

Heisenberg's approach to the first phase of the nuclear project is obvious — he immediately immersed himself. Within three months of receiving orders to report for the Berlin meeting, he produced the first of two parts of a secret, comprehensive theoretical report to Army Ordnance entitled "The possibility of the technical acquisition of energy from uranium fission."[13] Using the scant available data on nuclear properties and the basic Bohr-Wheeler fission theory, available openly in the *Physical Review,* Heisenberg surveyed every aspect of the practical exploitation of fission in a uranium "machine." The conclusion of his detailed report confirmed the possibility that a controlled fission reactor was technically feasible and that one of the uranium isotopes, when obtained in sufficiently enriched form, would constitute a tremendous explosive. Heisenberg's report immediately made him the leading German expert on nuclear fission, and it served as a basic guide for the German project throughout the war.

For his fellow researchers, Heisenberg's report also confirmed and reiterated many essentials. As Flügge had already noted in his article, natural uranium consists of two main isotopes, uranium 238 (U-238) and the very much rarer uranium 235 (U-235) (less than 1 percent of natural uranium). In a flash of genius, Bohr had realized that each isotope responds quite differently to bombardment by neutrons.[14] The rarer U-235 is easily fissionable by low-energy thermal neutrons; the more abundant U-238 fissions only with difficulty and only for very high-energy neutrons. It also absorbs neutrons at certain resonant energies, producing the unstable isotope U-239. Since fissioning nuclei emit more neutrons at all energies, a chain reaction is best achieved in a lump of natural uranium by slowing down the faster neutrons to thermal energies with a moderator — a substance that slows the fast neutrons without absorbing too many of them. At the lower energies, the neutrons will not be absorbed by U-238 but will probably fission one of the rare U-235 nuclei, producing more neutrons and more fissions. The chance of fission improves if the U-235 content of natural uranium can be enriched by separating out some of the more plentiful U-238.

Using preliminary nuclear data, Heisenberg theoretically examined the use of several types of moderators with different amounts of natural uranium in two basic arrangements: spherical and cylindrical configurations of alternating layers of uranium oxide and moderator—a so-called pile. Size and shape were crucial—the optimal arrangement would prevent the circumstance that too many neutrons could escape the pile without fissioning another nucleus. Pure carbon and "heavy water" (deuterium oxide, or water in which each hydrogen atom has an extra neutron) seemed to Heisenberg the best moderators to slow the neutrons so that they would cause the U-235 nuclei to fission, and the use of both moderators in a cylinder (or cube) filled with alternating layers of uranium oxide, heavy water, and carbon seemed at first the best configuration. But such a device required enormous amounts of each substance. Assuming a pile of about 1 cubic meter in volume, Heisenberg predicted that a chain-reacting layer configuration could be achieved with 600 liters of heavy water, 1000 kilograms of pure carbon, and 2000 to 3000 kilograms of pure uranium oxide.[15]

Heisenberg also predicted that, because of the absorption of neutrons by U-238, the reaction would reach equilibrium by itself at a temperature high enough to generate large amounts of electricity, if the reactor were used to heat steam to drive an electric dynamo. He did not realize that if more material were used than the minimum amount necessary, equilibrium would require a much higher absorption of neutrons through the presence of a control substance—otherwise, the chain reaction would increase without stopping, creating a very messy meltdown. The earliest German piles never contained such controls; neither did they go critical.[16]

By enriching the U-235 content of natural uranium, a smaller, mobile reactor could be built at a higher temperature—which, Heisenberg later suggested, could be used to drive German tanks and submarines; if enough U-235 were separated entirely from a block of natural uranium and compressed into a ball, fission would be nearly instantaneous: an explosion would occur. Isotope enrichment was the only way to obtain a mobile machine, Heisenberg told the German army in his report, and isotope separation was "the only method for producing explosives, the explosive power of which exceeds that of the strongest available explosives by several powers of ten."[17]

In the second part of his secret report, submitted at the end of February 1940, Heisenberg seemed less optimistic about the practical realization of the possibilities opened by nuclear fission.[18] He did not again mention an explosive and cautioned about the engineering. First, the enrichment and separation of rare isotopes such as U-235 were beyond Germany's (or any nation's) technical capabilities at that time. Since an element's isotopes (which differ from each other only in the number of neutrons in the nucleus) are chemically identical, highly sophisticated techniques are required to separate and identify each isotope. In the coming years, the regime's racist policies would blind the scientists to a crucial alternative. Since the late twenties, Nobel laureate Gustav Hertz had been perfecting the gaseous diffusion method of isotope separation. But in 1935, because his famous uncle, Heinrich Hertz, the discoverer of electromagnetic waves, was of Jewish descent, he was forced out of his position as head of the physics department at the Berlin Technical College. Although Hertz remained in private industry in Berlin until the end of the war, the Germans never developed his isotope separation method. It was one of the successful methods used by the Allies in the Manhattan Project and later, thanks to Hertz, by the Russians.[19]

Furthermore, although Germany possessed large quantities of uranium ore, it still lacked techniques to process it on an industrial scale into usable uranium oxide and eventually into metal plates, cubes, and powder. Nor did Germany possess the heavy water required for a self-sustaining critical reactor. The moderator problem was made worse by Heisenberg's new conclusion, in which he was encouraged by Heidelberg's imprecise data and the preference of others for heavy water. He had determined that the more plentiful element carbon, even in purified graphite form, probably would not do: the estimated cross section — the effective size of a carbon atom as seen by a fast neutron, which would slow the neutron by collision — was much too small. "It has therefore become doubtful," Heisenberg declared in his second report, "whether the uranium machine could be built with pure carbon." Calculations by Weizsäcker's Berlin assistants promptly supported this conclusion.[20] Bothe and his Heidelberg assistant mistakenly confirmed the inappropriateness of carbon a year later, and the Germans did not reconsider carbon as a moderator until late in 1944.[21] By then they were far behind the Allies. Fermi's Chicago pile first went critical in December 1942 with a graphite moderator to slow the neutrons and cadmium control rods to slow the reaction.

One of Weizsäcker's assistants predicted that the two possible configurations — horizontal layers or concentric spherical shells — would both go critical with only uranium and heavy water. Since it was smaller, a "spherical machine" seemed preferable, he wrote, requiring only about 720 kilograms of uranium oxide and 400 liters of heavy water, packed in seven alternating layers to a radius of about 1 meter. Heisenberg's coworkers in Leipzig immediately began constructing shells for a spherical machine.

Another piece of the puzzle fell into place with the discovery of alternatives to U-235 as the fissionable isotope. Otto Hahn's Berlin team had already discovered that U-239, derived from U-238 by the absorption of a neutron, decays in 23 minutes to the new element 93 (uranium being element 92), which the Hahn team named Eka Re (now called neptunium). In a secret report, Weizsäcker suggested to Army Ordnance that Eka Re should be fissionable by thermal neutrons; since it was easily separated by chemical means from uranium, it would enable the construction of a very small machine or a very explosive bomb.[22]

Weizsäcker was on the right track. Before the Allies banned the publication of fission research results, an American research team in the June 15, 1940, issue of *Physical Review* reported that Eka Re is itself unstable, decaying in 2.3 days into the long-lived element 94, now called plutonium.[23] Everyone soon realized that this new element would be equally suitable as an explosive and was easily obtainable from the transformation of U-238 in a working natural uranium machine. In theory, at least, a working reactor would produce not just energy but the material for an atomic bomb.

It is difficult today to comprehend the motives and rationale that allowed Heisenberg and his colleagues to place their great abilities so easily at the service of the German army at war. What would compel Heisenberg to report immediately to the research section of Army Ordnance the workings of a new energy-producing device and to confirm the possibility of a new explosive in technical reports marked "Geheim" (secret) in his own hand?

Critics and supporters of Heisenberg are sharply divided. Two former members of British wartime nuclear research, Sir Rudolf Peierls and Sir Nevill Mott, offer the following explanation: "It is reasonable to assume that [Heisenberg] wanted Germany to win the war. He disapproved of many facets of the Nazi regime, but he was a patriot. . . . Most citizens of most countries at war participate in the war effort when called upon, and the few who do not require exceptional courage and exceptional strength of conviction."[24] A Heisenberg supporter offers just the opposite conclusion: he did display exceptional courage by gaining a leading scientific position on the project, thereby taking a large share of responsibility for the direction of research and for keeping other, less scrupulous scientists from constructing a weapon that would indeed have enabled Hitler to win the war.[25]

A closer look at both of these views enables a fuller appreciation of Heisenberg's actual position. The supporter's account is one that Heisenberg and his colleagues offered after the war, and it assumes a level of control that Heisenberg never really possessed. Although he did develop many of the theoretical aspects of the project and did direct the Leipzig and Berlin branches of the undertaking, the overall project was in the hands of Army Ordnance until 1942 and under the Reich Research Council and other agencies thereafter. In addition, before the fall of 1941 every practical application of fission was still only a vague theoretical possibility. Even if Nazi scientists, acting on the slim hope that they could soon achieve a bomb, seized control of the project, it is unlikely that they could have pushed the project any faster than could the capable scientists (some of whom were party members) who were performing the research at that time.

The issues of patriotism, participation in the war on Germany's side, and the desire to defend the German nation — if not the Hitler regime — against defeat were already settled for Heisenberg long before the outbreak of war — as they obviously were for most of his cultured acquaintances. The interpretation offered by Mott and Peierls coincides most closely with this view, but there were other inducements, in addition to helping Germany win the war.

As with their Allied counterparts, scientific curiosity and a more utilitarian motive were also at work with Heisenberg and his colleagues. The outbreak of war and the interest of German Army Ordnance in nuclear fission suddenly offered Heisenberg and the atomic scientists a unique opportunity to prove their worth at last to their rulers. At the same time, they would have the protection — from political meddling and economic cutbacks — offered by a significant government project run by the victorious German army. After seven years of a depressing, losing struggle against regime intrusion, and now faced with a weakened profession, the continued ravings of "deutsche" physicists, and the further blockage of Heisenberg's professional ambitions, these were valuable gains.

Heisenberg did take the lead, not in keeping others away from the project but in seizing the opportunity and the initiative with his comprehensive two-part report to Army Ordnance on applied nuclear fission theory, produced in amazingly short order. In the process, he apparently convinced himself again that, amidst the technological and material conditions prevailing in early 1940, he could advance toward a useful energy-producing machine during the course of the war—useful to himself, to his profession, and to Germany—while disregarding the possibility of an explosive, which lay in the undefined future. He could ensure continued recognition of himself and of nuclear research by tantalizing regime officials with the prospect of a bomb, without concerning himself at this point with the prospect of actually building one. Even isotope separation should be supported, he argued, although, should it succeed, it would enable the extraction of a readily explosive isotope.

These are the reasons, it seems, for the incongruous circumstance that Heisenberg referred explicitly to isotope separation and to an enormous nuclear explosive in the conclusion to his first report to German Army Ordnance, while the paper itself actually concerned only the application of nuclear fission theory to the construction of a reactor. Patriotism, utility, curiosity, and support of the German war effort united to produce the extraordinary effort that Heisenberg invested in nuclear fission research during the early months and years of the war.

Physicist Peter Debye painted a similar picture of the motives and attitudes of the German atomic scientists in a long conversation in Berlin with Warren Weaver, an official of the Rockefeller Foundation. Within days of the outbreak of war, Army Ordnance had invoked military prerogatives to wrest control of fission research from Abraham Esau and the REM. Seeking to centralize research in Berlin, Schumann and Diebner took over Debye's Kaiser Wilhelm Institute for Physics. Debye, a Dutch national, refused either to resign or to accept German citizenship. He was placed on leave from government service at full pay for the duration of the war and eventually emigrated to the United States. Debye immediately informed the Rockefeller Foundation (which had built his institute in 1936) of the turn of affairs.[26] Weaver headed for Berlin to see for himself.

According to Weaver's log of his meeting with Debye in February 1940, Debye and the uranium club were well aware that the army hoped to achieve an "irresistible offensive weapon" from nuclear research. The researchers themselves conveyed their own, very different aim to Debye: "With D[ebye] they consider it altogether improbable that they will be able to accomplish any of the purposes the Army has in mind; but, in the meantime, they will have a splendid opportunity to carry on some fundamental research in nuclear physics. On the whole D[ebye] is inclined to consider the situation a good joke on the German Army."[27]

Heisenberg recalled after the war: "I, like several of my colleagues, was *told* to work on the technical exploitation of atomic energy"—an unlikely story; and elsewhere: "The official slogan of the government was 'We must make use of physics for warfare.' We turned it around for our slogan: 'We must make use of warfare for ˌ ˌysics!'"[28] At first Heisenberg probably did doubt the practical realization of theories that he readily provided the authorities—as indicated by his reaction when he realized in 1941 that his theoretical predictions would come true. And for the next three years the only fundamental research recorded by the curious theorist did consist of applied nuclear reactor theory. Heisenberg did invoke the rationale of using warfare for physics, but he did so while pursuing just the opposite: making physics useful for warfare in order to render it and himself acceptable to the rulers of the Reich. Whether or not they thought they could or would build the weapon the army wanted, Heisenberg and his compatriots saw themselves as walking a very difficult fine line. We, in retrospect, see them diligently performing basic research that they would have performed at this stage whether or not any such line existed. Who, we may ask, was fooling whom?

Until more pile components became available, most of the uranium club work focused on confirming the details of Heisenberg's theoretical predictions and obtaining precise measurements of various properties of the materials. This information would be used to improve theoretical estimates and to guide reactor construction. Three technical problems also required solving: the scientists had to develop suitable methods of isotope enrichment and separation; they had to obtain large quantities of heavy water and uranium oxide; and they had to discover the right geometry and size for a self-sustaining critical pile.[29]

Of the nine task-oriented research groups scattered among German laboratories and coordinated by Diebner, Heisenberg worked closely with two, one at his own physics institute at the University of Leipzig and the other at the Kaiser Wilhelm Institute in Berlin. By October 1940, Heisenberg was dividing his weeks equally between Leipzig and Berlin, then about two and a half hours apart by train.[30] After removing Debye, Schumann had appointed Diebner, the project administrator, to be provisional head of the institute, located in the quiet, tree-lined western suburb of Berlin-Dahlem. But unfortunately for Diebner, Debye's institute staff, most of whom were close to Heisenberg, were still in place. They included Weizsäcker and his assistants, as well as experimentalists Wirtz, Fischer, Bopp, and Debye's faithful technician, Gretschmer. The staff considered the energetic Diebner, a man with strong party connections but a weak grasp of nuclear theory, unworthy to administer their work. Weizsäcker and Wirtz increasingly involved

Heisenberg as an outside advisor in institute affairs, hoping that one day the famed physicist would supplant Diebner as permanent head of the institute.[31]

Heisenberg, for his part, used his Berlin connection to gain an influential role in all of the theory and most of the main reactor experiments in Germany, which were performed by the Berlin and Leipzig groups. In order to house the Berlin B series of radioactive cylindrical reactor models, the Berlin team constructed an outbuilding — named the Virus House to keep away the curious — on the grounds of the neighboring Kaiser Wilhelm Institute for Biology. At the same time, Heisenberg and his Leipzig team prepared the L series of spherical reactor models, which lasted through 1942.

Under Heisenberg's supervision, Robert Döpel, Kirchner's successor as Leipzig professor of radiation physics, directed the experimental work at the Leipzig institute. He was assisted by his talented wife Klara, a trained lawyer, and by the institute's able technician, Wilhelm Paschen, who actually built the various contraptions. Karl Friedrich Bonhoeffer, brother of courageous anti-Nazi Dietrich Bonhoeffer and professor of physical chemistry, collaborated in the resolution of heavy-water problems. But Gerhard Hoffmann, Debye's successor as Leipzig professor of experimental physics, had little to do with Leipzig pile research, even though he was a charter member of the club. Heisenberg's disdain for Hoffmann's former pupil, Diebner, who administered the entire research effort, did not endear the theoretician to him, nor would the often difficult Döpel countenance interference from the crusty old experimentalist.

The uranium club soon suffered shortages. In early 1940, Heisenberg put in a request to Diebner for up to a metric ton (1000 kilograms) of pure uranium oxide, and experimentalist Paul Harteck asked for up to 300 kilograms for his pile research in Hamburg. But because it had not been deemed important until then, only 150 kilograms of industrially pure uranium oxide existed in all of Germany. Diebner assured the impatient Heisenberg that by the end of June 1940 he would have his metric ton. The Berlin Auer Company was working at top speed on the uranium ore from the seized mines at Joachimsthal.[32] Later that year it would tap the stores of captured Belgian ore mined from the Belgian Congo.

Heavy water also posed a problem. In a letter to Heisenberg in January 1940, Harteck asked if anything was being done to procure heavy water.[33] Again Heisenberg had the jump on everyone else. Meeting with Diebner just after the New Year, Heisenberg recommended the construction of an industrial heavy-water plant as soon as he had experimentally confirmed his theoretical prediction that heavy water would make a suitable moderator. Döpel and Heisenberg supplied the confirmation in August, but by then German army conquests had again provided an outside source.[34]

In April 1940, Germany, attempting to outflank Britain, marched into Denmark and Norway. On May 3, German troops captured the Norwegian town of Vemork, near the world's only heavy-water production plant, operated by Norsk Hydro-Elektrisk. Rather than build their own plant, the Germans would simply use Norway's. With improvements in electrolytic production demanded by the Germans, the plant was producing 300 liters of heavy water a month by the time British and Norwegian commandos put it temporarily out of service in 1943.[35] By then, Germany had extracted probably just enough moderator to make a reactor.

Hitler's conquests also provided German scientists with a cyclotron. Thanks to the Rockefeller Foundation, a cyclotron had just been completed in Niels Bohr's Copenhagen institute, and a second was nearing completion in Joliot's Paris laboratory. A cyclotron would enable the researchers to measure the necessary nuclear constants and to create fissionable elements such as plutonium, otherwise obtainable only from a reactor.

When Paris fell to the German army, Schumann and Diebner immediately headed for Joliot's lab. By April 1941, Bagge and Gentner were working alongside Joliot—who remained in Paris out of devotion to the survival of French science—to complete the cyclotron. Heisenberg urged Bagge to make the most of the opportunity by performing some fundamental nonfission research with the esteemed Joliot.[36] With the successful exploits of the German army, everything seemed to be in place. German scientists now possessed all the uranium, heavy water, and cyclotrons they would need—a circumstance that seemed to bother no one, even though only the invasion and exploitation of neighboring countries had made it possible.[37]

The astonishing successes of Hitler's Blitzkrieg also left Leipzig relatively undisturbed until the war came home in 1943. Of course, reminders of the war were everywhere: frequent nighttime air-raid alarms (but few actual raids at first) and annoying shortages of food and fuel. Elisabeth Heisenberg, now coping with five children, had to find necessities by herself, while Werner commuted weekly to Berlin. Yet the Heisenbergs managed to live a more or less unperturbed life. There were the births, baptisms, and usual illnesses of their children, musical evenings with their friends the Jacobis (who somehow survived in Leipzig to the end), Christmas celebrations, and the annual blooming of Elisabeth's flower garden.[38]

During the first two summers, from school's end until late fall, Elisabeth, the children, and a nursemaid moved to the Urfeld retreat in the Bavarian Alps. Werner lived alone in the Leipzig house or with his in-laws, the Schumacher family, in Berlin-Steglitz, not far from the Dahlem institute. While working intensely on fission research in two cities, Heisenberg seemed satisfied with his personal life. "If one can keep life in one's small circle in order," he told his mother, "one must be content."[39]

The university also remained in relative order after Helmut Berve, Heisenberg's friend from the professors' club and the earlier demoted dean, became university rector in 1940. In his inaugural address, Berve quoted Goethe to the student Führer: "Bilde Künstler, rede nicht" — "Educate artists, hold your tongue."[40] Academic matters, not politics, predominated under Berve. Heisenberg pursued his nuclear research, his lecture duties, and his supervision of two doctoral candidates and was generally left alone.[41] Party member Berve even blocked the promotion of one associate professor who attempted to advance on the strength of his party connections rather than the quality of his academic credentials. The injured man complained bitterly to the office of party ideologue Alfred Rosenberg in 1942: "The rector, dean, and a large portion of the faculty are humanistically oriented and try to do everything to label revolutionary efforts as unobjective."[42]

But there could be no doubt that the regime was very much in control of German life, academics, and nuclear research—and there were constant reminders of its power. On entering the university, students were divided into close-knit indoctrination squads under a Nazi student leader, who oversaw nearly every detail of their personal lives and studies. Party members among advanced students and faculty were required to wear party badges everywhere, even in the laboratory. On two bitterly cold days in January 1942, the Gestapo publicly rounded up some of the remaining Jews in Leipzig—men, women, and children— stripped them of their coats, and drove them 18 kilometers in an open truck to a small town for brief internment on the first leg of what would be their final journey—to (as we now know) the eastern death camps. Heisenberg's acquaintance Carl Goerdeler, the former mayor of Leipzig who would take part in the failed attempt in 1944 to assassinate Hitler, recorded the events after watching helplessly from his home.[43]

The regime's influence on institute affairs—and Heisenberg's attempts to counter it—are illustrated by the case of Edwin Gora. Gora, a Polish student of German descent, was studying theoretical physics in Warsaw when Germany invaded Poland. Warned of the imminent arrest of intellectuals, he returned to his hometown in southern Poland and wrote to Heisenberg, the leading German theorist, of his predicament. Heisenberg invited the young man to Leipzig and helped him to enroll at the university and to obtain a job as a tram conductor. But in 1941 the Gestapo ordered Heisenberg to bar Gora from the institute—they had received a less than favorable report from Edwin's hometown regarding his attitude toward the Reich. Heisenberg complied without resistance, but with encouragement from his wife, who felt a motherly compassion for the young man, Heisenberg quietly took Gora under his wing, gave him private lessons in their home, and eventually enabled him to pass his doctoral examination under Hund in 1942.[44] Gora worked until the end of the war with Walther Gerlach in Munich, before moving on to the United States.[45]

Despite his assistance to Gora, Heisenberg, who worked so hard to preserve the insularity of his institute, was in the end often powerless to control the impact of events on his young colleagues. Especially calamitous were the fates of his brilliant assistant, Hans Euler, and Euler's close friend, Bernt Olof Grönblom, a promising Finnish physics student. As noted earlier, Euler, a Soviet sympathizer, had survived in Germany under Heisenberg's protection. Like many "fellow travelers" of that era, Euler was badly shaken by Stalin's 1939 pact with Hitler and by the German-Soviet partitioning of Poland later that year. When Stalin invaded Finland a year later, Grönblom left Leipzig to defend his homeland, leaving his close friend Euler a changed man. Stalin's actions left the sensitive Euler completely unsettled about himself and his political orientation in a world now dominated by two ruthless dictators. Heisenberg recalled inviting Euler to join the uranium club, fearful that Euler would be drafted despite his fragile health. To Heisenberg's surprise, the unbalanced Euler had already signed up for the Luftwaffe.[46] After flight training, he served as a meteorologist in a reconnaissance plane, flying missions over Crete, Egypt, and England in 1940 and 1941. Heisenberg's efforts to persuade his assistant to return to work, with or without joining the club, were of no avail.

Euler's letters to Heisenberg in this period are heartbreaking for their carefree tone. In his last note on June 16, 1941, Euler wrote: "We often recall to each other the ocean and the mountains beneath the sun in the south and the heat over Africa, and we will probably still do that for a long time afterward when we sit together in our new surroundings [at our next assignment]."[47] Seven days later, while helping to carry out Operation Barbarossa — Hitler's surprise attack on the Soviet Union — Euler's squadron lost contact with his plane over the Azov Sea, near the Crimea.

Aided by Euler's mother and sister, Heisenberg desperately tried to locate his lost assistant through military channels.[48] He even inquired of British physicist P. A. M. Blackett, whose sympathies with the Soviet Union were no secret and who may have had contacts there. The search continued after the war, but no trace was ever found of Euler or his plane.[49] Two months after Euler's disappearance, his friend Grönblom fell defending his homeland. Deeply disturbed by both losses to the war, Heisenberg wrote a moving memorial for his student in the proceedings of the Finnish Academy: "The more outstanding his first achievements were in the field of science, the more reason we have to mourn the loss of a young man who was suddenly torn from us and his work because of a higher duty."[50]

CHAPTER 23

A Copenhagen Interpretation

While awaiting the arrival of more uranium and heavy water in 1940, Heisenberg's Leipzig and Berlin research teams tested paraffin and regular water as possible moderators, substances that would slow the fission neutrons enough to escape capture by U-238 and thus fission U-235. Each fission, set off by the absorption of a single neutron, produces on average two or more neutrons. Each of these released neutrons, if sufficiently slowed, could go on to produce more fissions of the rare U-235 isotope in a piece of natural uranium, each of which would in turn produce more neutrons; an energy-producing chain reaction would occur in a self-sustaining critical reactor.

Heisenberg's research teams tested the two moderator candidates in a nuclear pile, alternating layers of moderator and small amounts of natural uranium oxide — what they called for security reasons preparation 38, U_3O_8.[1] With only a general notion of the best layer configuration, the scientists had to find the exact geometry and amounts of material through trial and error. The Berlin team tried alternating horizontal layers of powdered preparation 38 and moderator in a cylindrical aluminum tank, 1.4 meters in height and diameter, that was immersed in water in the "Brunnengrube" (well hole) — a water pit dug inside the Virus House. Completed in October 1940, the Virus House stood under cherry trees, an insulated wood-frame house containing water and electrical connections to the nearby Kaiser Wilhelm Institute for Biology. The water in the 2-meter-deep Brunnengrube absorbed and reflected escaping neutrons from the containers lowered into it on chains by a crane spanning the 3-meter-wide pit.[2]

Through mid-1942, under the able direction of Wirtz and Fischer, the Berlin (B) group performed five experiments using paraffin and uranium oxide or metal powder. Heisenberg and his Berlin coworkers reported their findings in detailed technical reports to army research: none of the models worked. Most of the neutrons emitted by a small source in the center of the pile were absorbed by the contraptions rather than multiplying in fission.[3]

In the spring of 1942, the four Leipzig (L) experiments yielded quite a different result: the Germans' first positive neutron multiplication.[4] But after this promising start, they never managed to achieve a chain reaction. Fermi's Chicago group surpassed the Germans within months.

After first trying paraffin and water as moderators in early 1941 (model L-1), Döpel, who was closer to the power source than was the Berlin team, switched to precious heavy water, arranged in concentric spherical aluminum-lined shells, alternating with uranium-powder shells. After Paschen and Döpel placed the concentric shells in two aluminum hemispheres, they bolted the ball (radius about 40 centimeters) shut, then winched it into a water tank in the basement of the Leipzig Physics Institute. On inserting a weak neutron source through a connecting tube into the center of the ball, they measured the neutron flux as a function of radius. Model L-2 proved a dud. The source neutrons were all absorbed within the sphere, mainly, Heisenberg calculated, by the considerable aluminum.[5]

To confirm this result and the findings by others that powdered uranium metal oxide, 38-metal, was superior to preparation 38, Heisenberg and his Leipzig team turned to metal powder and heavy water late in 1941. But only enough metal powder was available in Leipzig to make one spherical shell surrounded on both sides by heavy water. Careful measurements of this model, L-3, reported in early 1942, indicated a much smaller loss of source neutrons.[6] Heisenberg and Döpel were convinced that one more layer of uranium metal, a total of 755 kilograms, with 164 kilograms of heavy water, would yield a genuine multiplication — proof at last that Heisenberg's "machine" really would work.[7]

But the dangers of 38-metal and water had been overlooked. When water (heavy or not) and uranium meet, they produce flammable hydrogen gas. According to a formal report filed by Döpel, one day in December 1941 institute technician Paschen and his apprentice, F. Zumkeller, were pouring the powder into one of the hemispherical shells when heavy water somehow leaked in. An enormous flame suddenly shot out of the sphere, singeing the ceiling and burning Paschen's hand so severely that he could not work for nearly a month.[8] Döpel himself, his wife standing by with a fire extinguisher in hand, gingerly poured the remaining powder into the sphere for the L-3 measurements.

More 38-metal finally reached Leipzig in early 1942, and sometime in late spring, pile model L-4 began multiplying neutrons at the rate of 13 percent. "A simple expansion of the layer arrangement described here would thus lead to a uranium burner," the ecstatic Heisenberg team coolly reported to Army Ordnance.[9] "With that," wrote Heisenberg and Wirtz, "[we] proved the possibility of an independently working, energy-producing uranium burner."[10] Nuclear fission research was no longer just a politically useful or theoretically interesting exercise; the likelihood of controlled — and even uncontrolled — fission suddenly became very real indeed.

Years later, Heisenberg reportedly recalled: "It was from September 1941 that we saw an open road ahead of us, leading to the atomic bomb."[11] The project was at that time reporting on model L-2 and probably only just beginning model L-3. Moreover, every German effort to extract from natural uranium the fissionable isotope U-235, the explosive material needed for a bomb, had so far proved a failure. A reactor and a bomb were still far beyond reach.

But in August 1941 Fritz Houtermans of Manfred von Ardenne's independent Berlin research institute obtained an extremely important result: a theoretical confirmation of the plutonium alternative.[12] The implication for what Houtermans called "the theme of our work" was clear from Weizsäcker's and Heisenberg's earlier reports. Once a natural uranium reactor was finally up and running, the neutrons absorbed by the plentiful U-238 isotope would lead through beta decay of the nucleus to the fissionable, unstable daughter nucleus neptunium (Eka Re), which decays in a matter of days into the stable yet highly fissionable element plutonium. Since plutonium differs chemically from uranium, it can be easily extracted from a uranium reactor. Although Houtermans did not say so, a working reactor would produce fissionable material that could be readily separated and used either for a mobile energy-producing machine or for a new "irresistible" offensive weapon. The first of the two Allied bombs dropped on Japan in 1945 was fired by the separated uranium isotope U-235, the second by reactor-generated plutonium.

Although Heisenberg may not have met Houtermans at that time, he apparently learned of the new possibility in that period. A letter from Heisenberg to his professors' club colleague Hermann Heimpel, dated October 1, 1941, strongly suggests that Heisenberg did perceive an open road to a deadly explosive and that he was already mindful of the possible consequences.

Thanking Heimpel for a copy of his book *Deutsches Mittelalter (German Middle Ages)*, Heisenberg wrote: "I really liked the passage in your book about the mind-set of the middle ages in contrast to our epoch. In this connection it suddenly came to me that such a transformation could occur once again in the near future. For perhaps we humans will recognize one day that we actually possess the power to destroy the earth completely, that we could very well bring upon ourselves a 'last day' or something closely related to it."[13]

Heisenberg discussed the newly opened road with his trusted Berlin institute staff. Sometime in August or early September, they determined that Heisenberg should discuss the turn of events with Niels Bohr in German-occupied Denmark. A September lecture series on astrophysics at a German propaganda institute in Copenhagen, in which Heisenberg had agreed to participate, provided a splendid opportunity. On September 15, 1941, Heisenberg traveled to occupied Copenhagen for the official purpose of delivering a lecture at that institute, along with Carl Friedrich von Weizsäcker and other German scientists, and with the unofficial intention of meeting with Bohr.

Heisenberg probably met with Bohr on the evening of September 16.[14] Their meeting is still shrouded in controversy and questions. Although the official circumstances of the trip and some of its immediate consequences are well documented, the only indication of the content of their meeting comes from postwar accounts by the participants, their colleagues, and colleagues of their colleagues. Given the intense feelings and tensions of the early postwar period, the veracity of all these reports is questionable. Most of the German accounts were offered in defense of Carl Friedrich's father, an official of the German Foreign Office, whose subdivision administered the German cultural propaganda institute in Copenhagen and who was tried and convicted at Nuremberg in 1948. These accounts are obviously products of the aims for which they were written. Rightly suspecting Gestapo surveillance of their meeting, Bohr and Heisenberg themselves did not dare commit any of their discussions to paper at the time.

Given the setting of the meeting (German-occupied Denmark), the occasion (Heisenberg's lecture in a propaganda institute), and the topic (nuclear fission, controlled and otherwise), it may be little wonder that Heisenberg's visit greatly disturbed his former mentor and that Heisenberg felt he had failed to communicate with Bohr. But Bohr and Heisenberg had been close friends and colleagues for nearly twenty years. If

Bohr came away from their meeting in distress, it was because Heisenberg must have said something distressful.

In a carefully worded postwar statement about the visit, apparently an early draft prepared for the Weizsäcker trial (it differed from the less informative official defense exhibit that he submitted), Heisenberg again recalled his realization in 1941 that the production of an explosive, an atomic bomb, was now a real possibility.[15] Heisenberg remembered that his most important talk with Bohr occurred one evening as they strolled along a tree-lined path in the large and secluded Fælledpark, just behind Bohr's institute. The Danish professor and his former assistant had often talked together while walking along these quiet paths. Aside from its other attractions, the venue offered the advantage of escaping whatever bugs the Gestapo had installed in the institute — discussing secret nuclear research was treasonous for the German and life-threatening for the Dane. The boyish-looking Heisenberg recalled opening the discussion with the taller and more distinguished-looking Bohr by asking whether Bohr believed that "as a physicist one has the moral right to work on the practical exploitation of atomic energy."

An obviously startled Bohr responded by asking whether Heisenberg believed that atomic energy could be exploited in this war. "Yes, I know that," Heisenberg answered. However, he claimed that he meant only a machine. Because of the technical difficulties involved, he told Bohr, a bomb could not be produced before the war was over.

Unraveling Heisenberg's postwar account of this meeting, his intentions, and Bohr's reactions requires a much fuller appreciation of the background of the meeting than was available then or since.[16] Again, our sources allow only speculation on crucial points. Heisenberg's remembered question on morality is a case in point, for available sources give no evidence that he had ever raised it before. But it was also not quite so certain until then that a reactor could actually be constructed, that it would soon be within reach, and that, once working, it would easily provide the material for an atomic bomb. Since Bohr had served for years as a father figure to his youthful charges, especially to Werner, and had often debated philosophical and ethical concerns with his younger colleague, it seems reasonable that Heisenberg might have turned to Bohr when faced with an ethical dilemma in his research. But during the past few years, particularly during the SS affair, Werner had sought ethical advice on science in the political arena not from Bohr but from his German academic elders, Max Planck and (perhaps) Max von Laue, both of whom were in Berlin and more accessible than Bohr. Though in semiretirement, Laue was still vice director of the Berlin Kaiser Wilhelm

Institute, as he had been since the days of Einstein. Yet among the scant surviving records there is no indication that Heisenberg or his colleagues approached Planck or Laue about the morals of nuclear research or that they even fretted over them.

Aside from any advice that he may or may not have been seeking, Heisenberg probably had other aims in seeing Bohr. In a letter to B. L. van der Waerden after the war, Heisenberg seems to suggest that he was attempting to stave off an Allied crash program on nuclear research.[17] However untenable and naive such an aim may have been — if indeed that is what his letter was implying — it is not ruled out by a consideration of Heisenberg's prewar trip to the United States two years earlier and especially by the circumstances surrounding his visit to Denmark.

Heisenberg's 1939 trip to the United States occurred within several months of the publication of the basic Bohr-Wheeler theory and the French confirmation of neutron multiplication in atomic fission. As indicated earlier, on at least two later occasions Heisenberg recalled discussing the possibility of nuclear explosives with Fermi in the United States — a discussion that may have occurred with others as well.[18] As reported years later by Heisenberg, both participants had expressed a ready willingness at the time to engage in fission research for their respective governments. Looking back on this conversation in 1942, when his uranium pile first went critical, Fermi could have felt no assurance that Heisenberg was not working as hard as possible on nuclear energy — which, in fact, he was.

Nor, for his part, could Heisenberg be sure in September 1941 that the Allies were not doing just the same. According to Heisenberg's recollection, Bohr apparently told him matter of factly that "the physicists of all nations would be put to work on war research."[19] Elisabeth Heisenberg claims that throughout the war her husband "constantly tortured himself" with the thought that the better supplied Allies might develop and use the bomb against Germany.[20]

By the fall of 1941, when Heisenberg visited Bohr, the German Reich had reached its zenith. Most of continental Europe was under Nazi occupation, the German army was plunging into Soviet Russia, and an end of the war may have seemed in sight. If the war ended soon with the German army in place, or if it bogged down at that point — as had World War I in the trenches of France — it was easy to suppose that the United States, which had not yet entered the war, would have enough time and resources to build a weapon, which the Allies would surely use on Germany. At least one secret German report in early 1942 indicated that the Germans somehow knew about secret American pile research —and they now knew where that research could lead.[21]

After the war, Heisenberg wrote that he learned just after the September visit that Bohr was in contact with the Allied scientists. The Gestapo had intercepted a secret message from Bohr to British scientists and had delivered it to Heisenberg. Heisenberg probably suspected as much even before his visit. Perhaps, as his postwar letter to van der Waerden suggests, Heisenberg was trying to avert an Allied crash program and an ultimate nuclear attack on Germany by letting the Allies know through Bohr that the Germans — who believed throughout the war that they were ahead of the Allies — were still a long way from constructing an explosive.

Whatever Heisenberg's aims and intentions, his understanding of Bohr's frame of mind in German-occupied Copenhagen and of how he himself would be perceived in Denmark was woefully incorrect and misguided. He had last seen Bohr in 1938. Two years later, the German army overran the frightened Danish kingdom practically without firing a shot. With the German Reich well entrenched in Denmark and across most of Europe by September 1941, it must have been cold comfort for Bohr to hear Heisenberg's remembered amoral qualifier about the prospect for nuclear weapons: "At this point, it is certainly only a question of the exploitation of energy in machines; the production of bombs would probably require such an enormous effort that the war would be at an end before they could be made."[22] Even as Heisenberg himself remembered it, only time and effort stood between him and the bomb, and Bohr could regard neither to be insurmountable in September 1941. Nor could he have been very pleased even with the much sooner prospect of a new energy source to power the German economy and to drive German ships and submarines around the world.

Until the autumn of 1943, the German occupation forces maintained the fiction that they had no intention of nazifying Denmark — they did not want a rebellious populace that would drain their forces from other invasions. For the most part German commanders left Danish Jews alone, and the German army and occupation authorities were under strict orders to avoid offending the Danes as much as possible.[23] Bohr, his institute, and his cyclotron were also undisturbed, apparently, according to Bohr, on encouragement from Carl Friedrich's father in the Foreign Office.[24] The Germans allowed Bohr's institute to function as normally as possible, with continued Rockefeller Foundation funding.[25] But such niceties could not disguise the fact that proud Denmark had been reduced to a colony of the Nazi Reich and that it was being subjected to incessant propaganda by its occupiers. Even a German propaganda expert had few illusions about Danish resentment: "A feeling of quiet rage prevails here, which only comes to the fore when the Danes believe themselves alone and unobserved."[26]

Several months before Heisenberg arrived in Denmark, just days before Carl Friedrich and a party of German scientists arrived for a visit, Danish communists and other anti-German Danes had been summarily arrested and deported to Germany, an action that incensed the Danes further toward any Germans. Meanwhile, Carl Friedrich himself had already made the German scientists thoroughly unwelcome. During a visit to Bohr's institute the previous March, Weizsäcker had reportedly insulted Bohr by bringing the head of the local German Culture Institute to meet him.[27]

The newly opened Culture Institute was a propaganda arm of the Culture Division in the German Foreign Office.[28] In March 1941, Carl Friedrich had spoken there and elsewhere in Copenhagen to Germans and local sympathizers. According to surviving Reich Education Ministry records, Weizsäcker's trip was so successful that afterwards the German occupation office in Denmark requested a return visit by Weizsäcker in the fall, this time accompanied by Professor Dr. Heisenberg.[29] Weizsäcker was in contact with Heisenberg right after the visit and probably informed Heisenberg of the plan — if he was not already privy to it.[30] Certainly by mid-July Heisenberg knew that Carl Friedrich and the Culture Institute were planning to demonstrate support of German sympathizers among Danish scientists by organizing a conference on astrophysics to be held at the end of September 1941.

On July 22 Weizsäcker wrote to the German Academic Exchange Service to confirm himself, Heisenberg, and several other German scientists as invited speakers and — aware of the lingering opposition to Heisenberg in the government — to argue for Heisenberg's participation. After conferring with Heisenberg and Weizsäcker in early August, the Reich Education Ministry scheduled the conference for September 18–24. Heisenberg, claiming personal commitments, would be in Copenhagen September 15–21.[31] But whether naive or malevolent, Carl Friedrich again insulted the Danes by cordially inviting them to attend his and Heisenberg's lectures on solar physics and cosmic rays — in the odious German Culture Institute.[32] Already imbued with "quiet rage," Bohr and his colleagues did not appreciate the scientists' participation in a crass propaganda campaign. Heisenberg later surmised that his meeting with Bohr "did not have the intended effect [because] Bohr evidently disapproved of my taking part in an astrophysics conference at the 'German Culture Institute.'"[33]

Although it is not clear from the available documents which came first—the plan to speak with Bohr about fission or the plan to speak at the cultural propaganda institute—the timing of the latter seems to precede Houtermans's results and the perceived open road to nuclear weapons. If so, this raises a question: What was Heisenberg doing in occupied Copenhagen in the first place? If he did decide to go there even before he contemplated approaching Bohr about nuclear research, it might be argued that he wanted to assure himself that Bohr and his institute were unmolested by the German occupation. But surely Weizsäcker had already determined that, and others could have kept Heisenberg apprised of the situation. Instead, the most likely answer seems to be that Heisenberg was indeed joining his friend and colleague Carl Friedrich in a conscious or unconscious propaganda effort instigated by the Foreign Office subdivision under Carl Friedrich's father. Heisenberg either readily joined Carl Friedrich in carrying out the effort or allowed himself to be drawn into it, probably as a way of proving his reliability to regime officials. There does not seem to be any other compelling reason for him to have visited occupied Copenhagen before or after the dilemmas of nuclear research became acute.

This interpretation is supported by further Reich Education Ministry documents. In his postwar affidavit, Heisenberg argued (as he would in other such cases) that his lecture at the Culture Institute was the smallest compromise possible to gain permission to visit Copenhagen. Although some opposition to his touring abroad did exist, it was not insurmountable. In particular, the REM office of Dr. Wilhelm Führer, earlier prominent in blocking Heisenberg's Munich appointment, had to approve all foreign travel, and Führer had no intention of approving Heisenberg's travel anywhere unless pressured. During a hastily arranged meeting with the REM in early September, Führer demanded that party headquarters pass final judgment on whether or not to allow the Leipzig professor out of the country. The party quickly consented after the senior Weizsäcker's Foreign Office suggested that the trip could be used as a test case of Heisenberg's suitability for future propaganda lectures.[34] For Heisenberg, his lecture trip to Copenhagen could be seen as a minor personal victory; for the regime, it would be a test of the professor's reliability as a precondition for future exploitation; for the Danes, it was nothing more than gross propaganda. Bohr's wife, Margrethe, never wavered in *her* opinion of the episode: "No matter what anyone says, that was a hostile visit!"[35] Bohr and Heisenberg were never as close thereafter as they had been before the war.

What of the "intended effect" that Heisenberg had hoped to achieve by his meeting with Bohr? If the effect was an Allied moratorium, or even a joint boycott of applied nuclear research, the visit could not have achieved any such thing, even if Heisenberg had been well received in Copenhagen. Isolated in occupied Denmark, Bohr had only little inkling of the progress in nuclear research on either side of the war, and because of this, Allied scientists probably would not have accepted his assessment of a boycott offer from German scientists. Weizsäcker, for instance, reported in March 1941 that "concerning the more technical questions [Bohr] knew a great deal less than we." Perhaps this and the German scientists' actions on their September trip are the origin of the view expressed in postwar Copenhagen that the Germans were actually on a spying mission.[36] No wonder Bohr was so thoroughly disturbed to learn from Heisenberg six months after Weizsäcker's first visit that German scientists already saw an open road to an atomic bomb.

Nevertheless, he still seemed skeptical two years later. Responding in 1943 to hints from British physicist James Chadwick (discoverer of the neutron and Bohr's secret liaison with the British nuclear research team), Bohr wrote (in halting and deliberately cryptic English): "Above all I have to the best of my judgment convinced myself that in spite of all future prospects any immediate use of the latest marvelous discoveries of atomic physics is impracticable."[37]

Several months after Bohr's letter to Chadwick, the German occupation turned uglier. In one of the most spectacular rescue operations of the war, the Danish underground conveyed the part-Jewish Bohr, his family, and virtually the entire Jewish population of Denmark to neutral Sweden, just ahead of a planned roundup and deportation of Jews to the German death camps. A British plane flew Bohr to England, where he met immediately with Chadwick. The little-known summary report of a British nuclear committee meeting shortly thereafter describes Chadwick's impressions: "Chadwick . . . says that Heisenberg has visited Bohr in Copenhagen. He also says that he himself has been in communication with Bohr within a month or so, and that Bohr believes that there are no military possibilities. He thinks that perhaps Bohr has been sold this idea by Heisenberg."[38] If he felt he had been led astray by Heisenberg, he would have thereafter resented Heisenberg's wartime visit all the more. For his part, Heisenberg was pleased to receive the imprimatur of the Foreign Office as a traveling spokesman for Germany.

During the months following his meeting with Bohr, Heisenberg revealed his own reactions to the lately proven potentialities of fission research. This occurred in the context of a shift in the institutional and political framework of the uranium project in the wake of Germany's changing fortunes at the front. By the end of 1941 the German Blitz had run its course. The invasion of Russia, begun in June, was bogged down outside Leningrad and Moscow by December. Previously confident of total victory by Christmas, Germany had squandered most of her raw materials. As the predicament became a crisis during that bitterly cold winter, Hitler took over as operational commander. For the first time, he ordered the full mobilization of the German economy in support of the war effort, along with the full exploitation of occupied territories. Germany would now use every means at her disposal to wage "total war." By April 1942 the Reichstag would vote Hitler absolute dictatorial powers. As a result of this new state of affairs, Erich Schumann, head of army research, informed the uranium research directors in December 1941 that henceforth the Army Ordnance Office could support their efforts only "if a certainty exists of attaining an application in the foreseeable future."[39]

Ironically, just as the United States was entering the war and was launching a crash program to build the bomb, Schumann was calling the German scientists to Berlin for a meeting to decide whether their efforts were worth continuing at all. During that meeting, on December 16, 1941, which Heisenberg probably attended, the uranium researchers agreed to prepare a comprehensive report on their progress and on the prospects of their research for General Emil Leeb, the head of Army Ordnance, and Schumann agreed to call a conference of all uranium researchers for the end of February 1942 to evaluate the status of the project.

The only available copy of the 144-page memorandum to General Leeb, dated simply "February 1942," lacks a title page and authors' names. Heisenberg probably did not help to write it, but some of the wording and ideas appear in his reports and lectures at that time. Apparently well aware of Allied research, the scientists' recommendations are clear: "In the present situation preparations should be made for the technical development and utilization of atomic energy. The enormous significance that it has for the energy economy in general and for the Wehrmacht in particular justifies such preliminary research, all the more in that this problem is also being worked on intensively in the enemy nations, especially in America."[40] The authors considered time no longer a problem, for recent reactor experiments in Leipzig and

Berlin implied that "success can be expected shortly." But the building of a nuclear weapon for the Wehrmacht depended on the development of new isotope-separation techniques or the generation of the new element plutonium in the first working reactor. The report concluded that progress toward achieving a working reactor was being hindered less by scientific problems than "by problems pertaining to the acquisition of materials" — problems that, the researchers felt sure, the military sponsors would not find difficult to solve.[41]

The army chose to ignore the scientists' qualified optimism. Leeb and Schumann slashed uranium research funding, reduced activities to Diebner's army laboratory in the Gottow suburb of Berlin, and abandoned research altogether at the Kaiser Wilhelm Institute for Physics, returning the institute to its sponsoring society. The scientists clung nevertheless to their three-day conference on nuclear technology at the institute, scheduled to start on February 26.

Recognizing a sudden opportunity to regain control of uranium research after two years, Abraham Esau of the rival Reich Research Council scheduled a separate, conflicting series of nontechnical lectures for the opening day of the Army Ordnance conference at the Council's House of German Research in the neighboring suburb of Berlin-Steglitz. With the army withdrawing from research and leading nuclear researchers eager for a new sponsor, Esau lined up an impressive panel of scientists — Heisenberg, Hahn, Bothe, Geiger, Harteck, and Clusius — to deliver short lay lectures on nuclear energy development. Their audience was to be the top echelon of army, government, and SS officers — Himmler, Göring, Bormann, Speer, Keitel, Raeder, and others. But fate was unkind to Esau. Apparently through error Esau's secretary enclosed the wrong list of lectures with invitations to the dignitaries.[42] Instead of receiving Esau's list of eight nontechnical lectures, they received the list of reports for the rival Army Ordance conference and found themselves invited to hear twenty-five technical talks on such arcana as neutron diffusion lengths, enriched isotopes, and neutron multiplication and absorption factors in the latest Leipzig model.[43] Most of the dignitaries declined to attend.

With REM chief Rust as chair, military researcher Schumann opened the session in the cozy lecture hall of the former private mansion with "Nuclear physics as a weapon." Hahn followed with "The fission of the uranium nucleus," after which Heisenberg presented his favorite subject, "The theoretical foundations for energy acquisition from uranium fission."[44] While emphasizing reactor construction, Heisenberg mentioned the possibility of weapons development, but as in the Leeb memorandum, he neither discouraged government support for atomic research nor encouraged the government to expect a weapon in the foreseeable future. Heisenberg was still walking a fine line.

Heisenberg did attest, however, that a uranium "machine" could soon be built to generate enough power to drive battleships and submarines. Moreover, enough pure U-235 would constitute an "explosive of totally unimaginable power." But he hastened to add that separating U-235 from a block of raw uranium was difficult and required sophisticated techniques that were still unavailable. An alternative route to weapons lay through the uranium machine: "As soon as such a machine is running, the question of the acquisition of an explosive material . . . also receives a new significance." A working machine, he told his audience of minor bureaucrats, would produce the more easily separable plutonium, a substance as explosive as U-235.[45]

The implication of Heisenberg's argument seemed to be that if the regime would leave scientists alone, ideologically and professionally, while at the same time supporting and protecting them through reactor research, it would help itself progress toward a more distant but no less cherished goal — the development of a powerful new explosive. Whether or not Heisenberg believed that this goal could actually be reached before the war ended, at the very least the regime could be assured in a few years of a vast new energy source to power the German economy. As one historian writes: "Tailored to both his audience and the times, Heisenberg's talk illustrated clearly and vividly the warlike aspects of nuclear power."[46] While true, that was only half the story: from another perspective, it also illustrated the extent to which Heisenberg, the young man who had once thrilled at the precipices of mountain peaks, was willing to flirt with the catastrophic consequences of atomic research for the sake of what he believed would be beneficial to himself and to German science.

Heisenberg's renewed commitment to this dangerous strategy of enticing Nazi bureaucrats with the potentialities of nuclear energy in order to gain personal and professional advantages was not the effort of a loner. It coincided closely with a major campaign launched by the German Physical Society, aimed at Reich officials, to encourage more material support for physics education and research.[47] With Germany now headed for total war, the society's main argument was the familiar one: physics and physicists were making tangible contributions to the war effort. Heisenberg fully intended his February 1942 Berlin lecture as a contribution to this campaign to make warfare serve physics by demonstrating how physics could serve warfare.

Two months after he delivered his Berlin talk, Heisenberg's colleague, Wolfgang Finkelnburg, now vice president of the German Physical Society, congratulated him for his efforts: his lecture to the Reich Research Council and the press reports about it seemed to be having a "satisfactory effect. I have received various inquiries from party officials with questions about the war relevance of theoretical physics and especially about the relevance of your work."[48] Heisenberg concurred: "In general, the interest of the highest officials in modern physics now seems to have become quite great."[49] A month earlier, the authorities had suddenly confirmed their interest in Heisenberg and his physics. At the end of April 1942 one of Himmler's two promises to Heisenberg was at last fulfilled: Werner received a call to Berlin to succeed Debye in the directorship of the Kaiser Wilhelm Institute for Physics and to assume a concurrent professorship in theoretical physics at the University of Berlin.[50]

As the leaders of the Reich, unbeknownst to Heisenberg or to anyone else, prepared their final assault on humanity—the "final solution" agreed on at the Berlin-Wannsee conference in January 1942—Heisenberg and his theoretical physics basked in the prospect of imminent rehabilitation and full recognition by some of those same leaders. The fine line had become a tightrope.

CHAPTER 24

An Ordered Reality

As the uranium project headed down the road that could lead to a bomb and as the Reich extended its grip over most of Europe, Heisenberg's intellectual activities moved back toward nontechnical matters. He was developing in particular a concern for grand philosophical issues that went far beyond the philosophy of physics. As usual, he presented his nontechnical thoughts primarily through the medium of public lectures, both at home and abroad. Between May 1941 and the end of 1942, Heisenberg produced five lectures and one book-length manuscript on philosophical issues. The manuscript was published nearly 50 years later as *Ordnung der Wirklichkeit* (*The Order of Reality*).[1] The lectures were delivered to educated Hungarians in pro-German Budapest, to Zurich university students in neutral Switzerland, to Leipzig faculty and students, and to radio listeners and newspaper readers throughout the Reich. The lectures seem either preparatory to or derivative of the large manuscript.[2] Between early 1941 and the fall of 1942, Heisenberg did not produce a single nonnuclear scientific paper and wrote only a few surviving letters touching on scientific matters.

Heisenberg produced his manuscript and lectures at the time he was working to enhance both nuclear research and the recognition of his profession. Mrs. Heisenberg reports that his intensive work on the 161-page philosophical typescript occurred only during his brief vacations with the family in Urfeld in 1941 and 1942, when he had privacy and time away from his other work and duties. Mrs. Heisenberg typed at least one of the two existing versions of the untitled work, which they simply called "Philosophie," and they presented it to close, trusted friends as a present for Christmas 1942.[3] Despite the obvious peril of committing private opinions to paper at the time, Heisenberg included several passages that were somewhat critical of the regime. Heisenberg declined to publish the work after the war, declaring it to be too personal; apparently the conditions under which it was written and of which it was a part had rendered postwar publication anachronistic.

As noted earlier, Heisenberg never engaged in serious philosophical inquiry or exposition without a reason. Usually the reason derived from professional matters that were then reflected in his addresses to nonspecialist audiences. Most of his philosophical publications derived from these public lectures. However, the motivations for Heisenberg's 1942 philosophy manuscript seem much more personal in nature, relating to his age and to his situation. This was a man who had devoted his life to rapid scientific achievement, who was still engaged in a constant struggle to preserve what he valued, and who had spent most of his nonprofessional life in youthful settings and among younger men who glorified the virtues of youth. For such a man, his fortieth birthday, arriving in the middle of an intensifying war, was probably something of a trauma. Heisenberg's manuscript, written during the months before and after he turned 40 in December 1941, opens with the stated goal of discovering how his life's work "harmonizes with the whole," meaning the whole of what he would call reality: "He who has dedicated his life to the task of going after the individual connections of nature will be confronted over and over again with the question of how those individual connections fit harmoniously into the whole, other than the whole presented to us by [everyday] life or the world."[4]

In an earlier time of chaos and confusion, Heisenberg had felt a desperate need for an order and harmony that could provide the basis for stability in his life. He believed that he had experienced some sort of stabilizing central order during one of his early youth-movement meetings. Now, as he headed into middle age in the midst of war and dictatorship, Heisenberg seemed to require a new type of order, one that would reassure him of the significance of his life's efforts by giving him and his science meaning in the context of a larger, transcendent scheme. The mature man established this new order himself under the influence of his readings. Writing in 1941 and 1942, he argued that a whole does exist, that the reality that it represents is organized in a hierarchical order, and that it is the task of the scientist, especially himself, to comprehend his science in the context of this hierarchy. "In every period the attempt has been undertaken to submit our knowledge of reality to a general order."[5]

Heisenberg adopted a hierarchical order from his favorite author, the romantic Goethe. In his Budapest lecture, delivered in April 1941, Heisenberg described Newton's and Goethe's incompatible theories of colors emitted by a prism irradiated with white light as being not right or wrong in themselves but as referring to two complementary types or layers of reality: the physical and the spiritual. At the end of the lecture and at the outset of his manuscript, he expanded the layers of reality to include the nine vertically ordered layers presented, from bottom to top, by Goethe: accidental, mechanical, physical, chemical, organic, psychic, ethical, religious, genial.[6] The order of these layers moves from what Heisenberg calls the objective to the subjective, from areas of reality that lie outside of us to those that lie wholly within human experience, the genial or creative powers capping the hierarchy. In his elaboration of each level in his manuscript, Heisenberg placed his own life's work, quantum physics, just below the organic level in the realm of the chemical, because it dealt with atoms. The man of knowledge who wants to comprehend more fully what he called the "grand connections" must climb this ladder of realities — much as a mountain climber struggles alone to reach the top of an enormous mountain, much as a yogi struggles to attain a knowledge of nirvana.[7]

Sitting alone in his Urfeld nest — alone even when surrounded by his large family — the physicist passed his fortieth birthday with a new understanding of the place of his life's work in the broader scheme of things. Faced again, as he had been several times in the past, with the stark contrast between the beautiful, genial world of his understanding of nature and the horrible death, destruction, and ugliness of the Nazi dictatorship at war, he found an order that reached far above that ugly world. Science and all that he treasured in it and in life were layers within a grander, transcendent hierarchy that made them all worthwhile.

Despite the escapist quality of his exercise, it may seem admirable to find Werner Heisenberg, the great twentieth-century scientist, contemplating the limits of his science, acknowledging that alternative ways of looking at nature and our place in it may be as valid as the cool, often exploitive, rationality of those who, since Bacon, have ascribed power to knowledge. Yet there is also a disturbing quality to an order of reality that is conceived as a vertical hierarchy with science, rationality, and the individual somewhere far below the top. In a way, Heisenberg's hierarchy of reality layers seems overwhelming, even depressing, to contemplate. Given the years in which the work was written, one cannot help wondering to what extent the hierarchical situation in which he lived had any impact on these views.

Perhaps unconsciously acknowledging the connection, in the closing pages of his long essay Heisenberg considered the immediate reality of the war and the role of the individual — himself — in it. As described in these pages, probably written in 1942, the war for him was not merely a struggle for power and territory initiated and pursued by a ruthless dictator, but an expression of more fundamental "movements in the foundations of human thought": a shifting of the layers of reality over the heads of individuals in such a way that the evil side of the irrational, the "dark demons" now loose upon the world, took on a greater role.[8] It is not clear whom he considered the dark demons to be, nor whether he expected their role to be permanent or merely temporary.

Heisenberg also made several statements that would have gotten him into deep trouble had they become known at the time. For instance, he lumped National Socialism together with Bolshevism as a "strange sort of this worldly [that is, nontranscendent] religion." He admired the Anglo-Saxon enemy for producing "the first great figures of the early modern period" that led to modern science and to the knowledge of objective reality (at the lower end of the hierarchy, however).[9] Contemplating the contemporary situation, he even provided a moral imperative for his readers: "We must make it clear to ourselves over and over again that it is more important to treat others humanely than to fulfill any sort of professional, national, or political duties."[10]

Nevertheless, to fulfill his humane duties to others, Heisenberg the individual at the same time had to sustain the role of his public persona —the "active opposition" that fostered the illusion of living at once in two separate worlds. In his essay, he reaffirms this by separating the larger public world from the smaller world of the private being. The grand "movements of thought" occurring in this war far transcended the actions and influence of mere individuals such as himself. The individual "can contribute nothing to this, other than to prepare himself internally for the changes that will occur without his action."

The helplessness of the individual before the forces of national and international struggle thus established, Heisenberg closed his essay with his own personal recommendation: while striving to help others, he wrote, one can do little more than to accept one's fate within the broader circumstances of one's life. The individual is conveniently relieved of any responsibility for what is going on outside himself: "For us there remains nothing but to turn to the simple things: we should conscientiously fulfill the duties and tasks that life presents to us without asking much about the why or the wherefore. We should transfer to the next generation that which still seems beautiful to us, build up that which is destroyed, and have faith in other people above the noise and passions. And then we should wait for what happens. . . . Reality is transformed by itself without our influence."[11]

And that is what Heisenberg did after moving to Berlin in 1942. Even when playing a leading role in a research project integrated into the hierarchy of the German war effort, even when serving as a cultural representative to occupied and oppressed territories, even when acceding to the demands of Nazi functionaries, he assured himself that his actions really made no difference on the grand scale of reality. On the other hand, in the smaller world of himself, his colleagues, and his family and friends, his actions might help to preserve something of that which "still seems beautiful to us." In other words, he had convinced himself of what he wanted to believe all along: that he could live and work as a subject of this system, but not be a part of it.

Heisenberg's transfer to Berlin in 1942 signaled shifts in the setting and status of German uranium research as the authorities subordinated Germany's economy and science even more directly to the war effort. Germany's waning fortunes on the Russian front and her rapidly diminishing supply of finished goods demanded a radical change of priorities, with production taking precedence. During the terrible winter of 1941 – 1942, Hitler appointed Albert Speer, his astute architect, to succeed the late Fritz Todt as head of arms production in the four-year plan, which was administered by Reich Marshal Hermann Göring, the commander of the Luftwaffe. In March 1942, Speer had Hitler decree that the needs of the German economy were subordinate to those of arms production —thus to Speer himself.[12] Among the potential elements of arms production at that time was a recently orphaned research project on exploring the possible exploitation of nuclear energy.

After the Army Ordnance Office decided to relinquish most of its control over nuclear research in early 1942, a hodgepodge of government bureaucracies grabbed for the plum. Reich Education Minister Bernhard Rust sought to enhance his standing by immediately assigning the project to his own Reich Research Council (RFR), now headed by SS Colonel Rudolf Mentzel. Dr. Mentzel turned it over to the physics section of the RFR, still headed by Professor Esau. But the Education Ministry and its Research Council collided with the Kaiser Wilhelm Society (KWG), which administered a network of government research institutes. Until its confiscation by Army Ordnance, one of these institutes had been Debye's Kaiser Wilhelm Institute for Physics in Berlin-Dahlem. The KWG vigorously sought to regain control of Debye's institute from Army Ordnance. When KWG president Albert Vögler informed Albert Speer of the potentialities of nuclear research, military and otherwise, Speer, already appreciating the importance of science for the war effort, began to take a personal interest in nuclear matters. In 1942 he had Hitler name Göring head of the new Reich Research Council to run under the four-year plan and to foster war-related research. By the end of the year, Göring had appointed Mentzel head of the managing committee of the RFR; at Mentzel's suggestion, he gave Esau the dual titles of head of the physics section of the RFR and Reich Plenipotentiary for Nuclear Research.[13] Although this left the German nuclear project at the end of 1942 temporarily suspended between competing bureaucracies—Speer and the KWG on the one hand and Göring, Esau, and the RFR on the other—the nuclear project was nevertheless now fully integrated into the newly established hierarchy. Research was to serve the war effort and would be dominated by two men—Speer and Göring.

While Vögler and his staff schemed at the beginning of 1942 to diminish the influence of the Education Ministry in KWG affairs, scientists in Debye's old institute were maneuvering to keep Heidelberg physicist Hans Bothe from replacing Kurt Diebner as head of the Kaiser Wilhelm Institute. The institute's staff still wanted Heisenberg. After discussing the conditions of work and residence with Vögler and Ernst Telschow, the KWG manager, Heisenberg finally signed a contract to head the institute in June 1942.[14]

Because the status of the institute was still uncertain in 1942, as was Debye's association with it, Heisenberg took a leave from his Leipzig chair rather than resigning outright.[15] Not until March 1943 did Army Ordnance, the RFR, and the KWG finally reach a formal agreement. While Speer's office controlled priority ratings and resources, henceforth the KWG would administer the Kaiser Wilhelm institutes, but the RFR would fund nuclear research. Army Ordnance would transfer most of its equipment to the KWG but would continue to fund nuclear research in the Gottow research station, just south of Berlin, to which Diebner now repaired, and in Harteck's and Clusius's Hamburg and Munich institutes.[16] Bureaucratic niceties prevented the KWG from naming Heisenberg as director of the physics institute—officially, Debye was still on leave. The KWG fell back on semantics and named Heisenberg not director *of* the institute but director *at* the institute. Until the summer of 1943, Heisenberg, who commuted weekly from his Leipzig home, registered his official residence as a rented room in the Schumacher home and listed his official job description as director at, not of, the institute.[17] Dualities could exist on more than one level of reality.

Since Einstein's day, an appointment to the directorship at or of the Kaiser Wilhelm Institute for Physics automatically meant a simultaneous appointment to the faculty of the University of Berlin (but with no teaching obligations). This brought the REM into the game, and both of Heisenberg's nominations—to the Kaiser Wilhelm Institute and to the Berlin faculty—again raised the specter of political opposition. The SS offered no objection; it regarded the appointments as fulfillment of Himmler's promise to have Heisenberg suitably placed somewhere other than Munich. But both Hess's former party bureau and Rosenberg's ideology office remained strongholds of anti-Heisenberg "deutsche" physics. The party's University Teachers League and Rosenberg's Main Office for Scholarship still claimed the right to evaluate and approve every academic and institute appointment.[18] In the spring of 1942, when the KWG and REM offered him the dual Berlin appointment, Heisenberg and his physics were still suspect to both political offices.

Heisenberg and his supporters could regard approval of his dual appointment in the heart of the German Reich as a major victory for theoretical physics over its ideological opponents. Heisenberg had earlier regarded politically motivated professional attacks to be personal in nature. Now he could regard his personal and professional success as a triumph for his entire profession. The identity of himself with his science under the wartime dictatorship was complete.[19] As Mrs. Heisenberg put it years later: "The only thing that gave Heisenberg any satisfaction in this matter was the fact that his summons to Berlin had to be viewed as a clear victory of modern physics over 'German physics,' . . . I know how important this was to him."[20]

Heisenberg's colleagues did all they could to ensure his appointment to Berlin as part of their broader, ongoing campaign to achieve greater recognition of the value of theoretical physics. To demonstrate the Physical Society's support, Ramsauer and Finkelnburg, supported by Prandtl, submitted a memo to the REM describing the decline of German physics compared with Anglo-Saxon physics and arguing the importance of increased support for research, especially nuclear research, for the war effort.[21] Coming on top of the February 1942 rounds of lectures on nuclear energy, just as the tide of war was beginning to turn against the Reich, the arguments of the physicists began to have an effect. Nearly every top-ranking official expressed a new appreciation of the contribution of scientific research to the war effort. Dr. Erxleben, who was the head of Scholarship Observation and Evaluation in Rosenberg's Scholarship Office, declared in September 1942: "It appears to us urgently necessary to do everything to effect an upswing of atomic physics research" — a 180-degree turn from the sentiment of even a year earlier.[22]

The more diplomatic of Heisenberg's colleagues managed to neutralize the remaining pockets of ideological opposition through direct negotiations, what they called "Religionsgespräche" ("religious debates"). One such debate had already occurred in 1940; Finkelnburg, a member of the Teachers League, organized a second in Munich on June 25, 1942, in support of Heisenberg's call to Berlin. In light of the REM's positive reaction to the Physical Society's memo, Finkelnburg easily routed REM functionary Dr. Wilhelm Führer and convinced Dr. Gustav Borger, the moderator of the meeting and head of the Teachers League Office for Scholarship, of the value of Heisenberg and his science to the German cause.[23]

Correspondence that summer between Borger's office, Erxleben's office, and party headquarters indicates complete support for Heisenberg. Reporting to the party on the Munich meeting and citing the evaluations of some of Heisenberg's former SS investigators, now on his own staff, Borger echoed the earlier SS conclusions that Heisenberg's personal behavior was exemplary and that "his political position is in no way to be designated as argumentative. He is doubtless the unpolitical academic type."[24] Erxleben delivered in turn his recommendation that the party withdraw from scientific conflicts. With a war raging, the competitive status of German physics was suddenly far more important: "Under no circumstances can we allow atomic physics research in Germany to remain inferior to work being done abroad. Professor Heisenberg's accomplishments in this field doubtless justify his call to the Kaiser Wilhelm Institute."[25] Heisenberg would have his institute, but his appointment to the teaching chair, from which he began lecturing in the fall of 1942, still needed official approval.

To put the matter to rest, Heisenberg's supporters held a final "religious debate" with representatives of "deutsche" physics in November 1942 in the Tyrolean mountain village of Seefeld. Thanks to the persistent efforts of Finkelnburg and others since 1940, the representatives of "deutsche" physics were already in retreat; they could not carry their argument on either scientific grounds or the practical grounds of the war effort. During the meeting, which Heisenberg attended as an observer, the parties adopted five points of agreement drafted at the previous meeting, effectively defusing the political influence of "deutsche" physics, but they also elicited several further compromises from the theoreticians, especially regarding the avoidance of Einstein's name in public.[26] Three months later, the REM officially appointed Heisenberg to a chair for theoretical physics at the University of Berlin and to the directorship at the Kaiser Wilhelm Institute for Physics. Friedrich Hund soon replaced Heisenberg as Leipzig professor of theoretical physics.[27]

Heisenberg had at last attained one of the most important and visible positions in all of German physics, and after nearly a decade of struggle, German physicists had at last achieved a victory over their ideological opponents. They regarded these events as outstanding achievements, which at first glance they were. They had lifted recognition for theoretical physics from the depths of the "white Jew" affair of 1937, in which a theorist could land in a concentration camp merely for teaching modern physics, to nearly universal appreciation in 1942 by leaders of the most influential power centers in the Reich. Deft diplomacy, combined with the practical needs of the war and a decline in the need for ideology as a weapon of control and political advance, had made this possible.

In many ways the victory was in the end too little too late. Irreparable damage had been done to German physics during the previous decade; ideology had been silenced but racism had not. Moreover, as one historian points out, the victory simply meant that the regime no longer saw the struggle over modern physics as a conflict between camps loyal and disloyal to the regime, as it had before the war, but saw it as a conflict between two loyal factions within the system, each representing a different approach to scientific research. If physicists regarded Heisenberg's appointment as a monumental victory for the recognition of modern or "Jewish" physics, the regime saw it merely as acceptance of a controversial approach that promised greater practical benefits for the war effort.[28]

Heisenberg and his colleagues painstakingly encouraged regime officials to regard nuclear research as an important potential source of practical applications. This was evident in Heisenberg's February 1942 lecture to regime officials. In it, he emphasized the potential benefits of nuclear fission, both controlled and uncontrolled, while echoing the caution of the army's Leeb report concerning technical hurdles. It was evident again when, on June 4, 1942, Heisenberg and other nuclear scientists met in the Kaiser Wilhelm Society's Harnack House to brief the newly appointed Speer and three military heads of weapons production on their work. The meeting occurred just before Hitler transferred the Reich Research Council and its nuclear subdivision to Göring's four-year plan, over which Speer exercised considerable influence.[29] Since only postwar recollections of the meeting are available, the details are sketchy. Apparently repeating much of his February lecture, Heisenberg emphasized the need for a cyclotron and requested more funding for isotope separation.

Although the Leipzig L-4 experiment was at that moment multiplying neutrons at the rate of 13 percent, in contrast to the February lecture the plutonium alternative to isotope separation did not receive much notice. Heisenberg apparently did not want to awaken undue optimism, nor, he claimed, did he want to be ordered to build a bomb, since it could not be achieved without enormous effort.[30] There is no indication that, as he later claimed, he withheld information on moral principle. Even though the German army was scoring successes at that very moment in North Africa and on the Russian front, the Battle of Britain was all but lost and Allied bombs were already striking northern German industrial regions. A new type of bomb would certainly have interested the weapons procurers among Heisenberg's listeners. The available version of Speer's office journal for 1942 contains no mention of a bomb discussed at that meeting. It states simply: "That evening there was a lecture in Harnack House on atom smashing and the development of the uranium machine and the cyclotron."[31] Speer recalled that Heisenberg seemed most disturbed by the lack of technical and financial support given to nuclear research and by the more massive backing to this field probably provided by the United States.[32]

During the discussion after the lecture, Speer surely did ask Heisenberg about a bomb. "His answer was by no means encouraging," Speer claimed long after the war, perhaps in an attempt to defend himself against accusations from some quarters that he had not pushed hard enough for a bomb to save the Reich. The scientific solutions had been found, Heisenberg seemed to indicate, but technical difficulties prevented their realization in the foreseeable future.[33] Field Marshal Milch recalled asking during the meeting how large a bomb would have to be to destroy a large city, such as London. To the astonishment of his audience, Heisenberg reportedly replied, "About the size of a pineapple"[34] — perhaps referring only to the U-235 content. The incredulous officials must have thought the scientist insane.

But Speer was still intrigued enough after the meeting to ask Heisenberg for a tour of his new institute, only blocks away. During their walk to the institute on the cool early summer evening, Heisenberg reportedly requested only moderate increases in support of the nuclear project: the construction of a radiation- and bomb-proof bunker, a cyclotron, and high-priority ratings for the acquisition of materials. Speer apparently became convinced that the project would make only a modest contri-

bution to the war effort.[35] Accordingly, Speer gave other projects—
Wernher von Braun's rocket research in particular—top priority, while
his report on the conference to Hitler was less than enthusiastic. Point
15 on his list of topics discussed with Hitler on June 23, 1942, states
only: "Reported briefly to the Führer on the conference on splitting the
atom and on the backing we have given the project."[36] Speer continued
to support the project on a modest scale for the rest of the war, hoping
to gain at least a new energy-producing machine to power ships and the
German economy. Ironically, at almost the same moment, American
science administrators Vannevar Bush and James Conant were informing President Roosevelt of their conviction that an Allied bomb would
be ready just in time to be used in the war.

On the day Speer reported to Hitler on the nuclear conference, just
one week before Heisenberg took up his duties in Berlin, the last Leipzig
model, still submerged in its water tank, began to emit a tiny stream of
bubbles. Robert Döpel, suspecting a leak somewhere in the sphere, had
it hoisted out of the water. When the hapless technician Paschen gingerly opened one of the inlet valves, air rushed in, then suddenly reversed, showering him with bits of burning radioactive uranium powder.
This spurt was followed by an intense flame that melted the aluminum
and set more uranium on fire. Döpel, Paschen, and apprentice F. Zumkeller managed to get the fire temporarily under control and lowered
the sphere back into the water to cool.

Heisenberg, summoned to the scene by his staff, appeared briefly, saw
that all was under control, and left to direct a seminar. He was summoned again soon afterward when the sphere began to heat up in the
tank. Watching the sphere in the water, Heisenberg and Döpel saw it
suddenly begin to swell. Both men leaped for the door, escaping from
the room just as the ball exploded in a burst of flame and smoke,
destroying the laboratory and raining burning uranium powder over the
entire area. The two hemispheres of the reactor, held together with
hundreds of bolts, had been literally ripped apart. Summoned by the
dazed staff, the fire department soon arrived to put out the numerous
fires ignited by the powder. After managing this task with enormous
difficulty, the fire chief thanked the shaken professors for their fine
display of explosive "atomic fission."[37] The incident marked the obvious end of the successful Leipzig series of reactor experiments—and
just in time for Heisenberg's move to Berlin.

Assuming his Kaiser Wilhelm Institute duties on July 1, 1942, while
his wife and family remained behind in Leipzig and Urfeld, Heisenberg
turned again to the Berlin experiments. At the end of the month, he

outlined a large semitechnical experiment in which the dangerous uranium metal powder was to be replaced for the first time by safer and
easier-to-handle uranium metal plates.[38] Cadmium control rods would
also be used for the first time, and a jacket of carbon would be wrapped
around the cylindrical tank in order to slow down and reflect any
neutrons escaping from the contraption. Henceforth, under Heisenberg's direction the Berlin project would combine the more difficult
theories of cylindrical reactor design with the results of different
amounts and geometries of horizontal metal plates separated by the
institute's 1.5 metric tons of heavy water in the hope of at last obtaining
a self-sustaining reaction by semiempirical means.

But Heisenberg's ambitious experiments depended on the industrial
production of the metal plates from metallic uranium oxide and the
completion of the underground bunker laboratory to house the
experiments — and neither was available when Heisenberg took over in
July 1942. An upgraded priority classification could have speeded plate
production, but despite Heisenberg's request to Speer during their meeting the previous month, Speer had given the project the lowest classification, *kriegswichtig* (important for the war), that still allowed it to function.[39] In addition, the cost of producing finished uranium plates from
raw uranium ore was so great that it consumed most of the entire RFR
uranium research budget for fiscal years 1943 and 1944.[40] Purifying and
casting the metal into plates apparently presented little difficulty, but
cutting them with specially designed hardened machine tools was a
problem, exacerbated by the incessant bombing of German industries
and the shortages of supplies and resources. Because of the cost and the
delay in plate production, full-scale reactor experiments did not begin
again until late in 1943 — just as conditions worsened with the Allied
bombing of Berlin. It wasn't until nearly a year afterward that Heisenberg and his coworkers, having realized that they were on the wrong
track, switched from metal plates to more efficient metal cubes.

During the summer of 1942 the Berlin team completed the last of the
B-5 series of experiments, using less suitable paraffin as a moderator to
slow the neutrons. While waiting for the plates for the B-6 experiments
to arrive, Heisenberg gradually turned to other matters.[41] The uranium
project had served its initial purpose for Heisenberg. His call to Berlin
and the personal support accorded the project by Speer and Göring, the
controllers of the economy, constituted concrete acknowledgment of
the practical importance of theoretical physics and of his own leadership
role in it. He had also thoroughly explored the scientific potential of
fission — in theory if not yet in practice. Heisenberg could now safely

turn to less dire scientific matters and to his role as an important German professor living in war-torn Berlin.

Commuting to Leipzig for weekends with his family, Heisenberg found his new life as a prominent Berlin professor to be a busy one. He lectured on theoretical topics twice a week at the university, directed graduate research, and oversaw a lecture series at the institute on his favorite subject, cosmic radiation. He edited and published the lectures in a monograph completed in June 1943.[42] In addition, as he had in Leipzig, he moved in the established circles of the cultural and intellectual elite of Berlin. On September 1, 1942, Luftwaffe commander Göring made him a corresponding member of the German Academy for Aeronautical Research.[43] REM officers approved his election to the science class of the Prussian Academy of Sciences in April 1943 — exactly ten years after the same class had expelled Albert Einstein.[44] And in November 1942, at the invitation of Prussian finance minister Johannes Popitz, he attended a meeting of the famous Wednesday Society, one of Berlin's socially elite men's clubs since the Wilhelmine era.[45]

The Wednesday Society selected its members from among the leaders of Berlin's cultural, academic, administrative, and military life. The group met every several weeks (on a Wednesday, naturally) in the home of one of its members. In addition to refreshments, the host provided a general lecture on his work. In 1942, the society's 28 members included such noted figures as the surgeon Ferdinand von Sauerbruch; General Ludwig Beck, chief of the army general staff; Ambassador Ulrich von Hassel; and professors Eduard Spranger, Wolfgang Schadewaldt, and Jens Jessen. As these names suggest, the Wednesday Society also served during the Third Reich as a meeting place for many members of the conservative Prussian military and professional opposition to Hitler and as a breeding ground for the hapless conspirators of the failed coup d'état of July 20, 1944. Members were chosen for their sympathy with the views of the non-Nazi German and Prussian cultural elite, which though patriotic and nationalistic, insisted on moral rectitude. Heisenberg, whose sentiments harmonized to an extent with the members of the society, was already well known to Spranger, a former member of the Leipzig professors' club, and to the future conspirators (but not club members) Carl Goerdeler, former Leipzig mayor now in Berlin, and Adolf Reichwein, a former youth-movement member. Werner attended his first meeting as a member in December 1942.[46]

Mrs. Schadewaldt, whose husband met his end in the wake of the failed coup d'état, later recalled a meeting of the society in their home on March 17, 1943. After the close of the meeting, Heisenberg, Hassel, and Sauerbruch lingered in conversation. "And then Heisenberg, in somewhat subdued terms, and Sauerbruch, in his spirited manner, grumbled about 'Schimpanski,' that was the code name for Hitler."[47]

According to most studies of the coup attempt, most of the conspirators had opposed National Socialist policies, in word if not in deed, since the early years of the regime, but they did not object to the German national and military revival. By the outbreak of war, their Prussian scruples regarding Nazism had led them to conspire against the Reich. But events did not favor a coup d'état until the demoralizing defeats of the winter of 1942–1943. As the German army went into captivity and retreat on the Russian front and in North Africa and as the Allies intensified their bombing campaign, the general populace began to realize that the promised military triumph was turning into another long and humiliating defeat. Members of the professional classes were emboldened to allow their anti-Nazi scruples free rein at last and to consider alternatives to Nazism. Many wanted a restoration of the monarchy, and some, especially the courageous Count Klaus von Stauffenberg, saw assassination of Hitler as the only way to gain military backing for a coup d'état.[48] Even if the coup attempt failed, they reasoned, it would demonstrate to the world that despite Hitler and his evil Reich, a decent Germany still survived. Everyone in the Wednesday Society knew of the society's anti-Hitler orientation and most knew of the festering conspiracy. It is uncertain, however, how many members actually took part in it as words turned to deeds in 1944. Mrs. Heisenberg claims that Reichwein, who was not a member of the society, came to Heisenberg's institute in 1944 to ask him to join the conspiracy, but Heisenberg, unwilling to conspire, respectfully declined.[49]

According to the published records of the Wednesday Society, Heisenberg attended most of its meetings and hosted two of them, including its last, on July 12, 1944, in the KWG's Harnack House.[50] The institute's gardens, near the Virus House, provided berries for the occasion, and the KWG supplied wine. Heisenberg's lecture that evening on the constitution of stars included an explanation of the process of nuclear fusion.[51] Ten members attended, among them the conspirators Beck, Jessen, Sauerbruch, and Ludwig Diels. Diels wrote in his diary: "The mood is subdued. Jessen, who is a defeatist, contributes especially to this."[52]

Heisenberg submitted the minutes of the meeting and a copy of his lecture to Popitz on July 19, then left for southern Germany and a visit with his family in Urfeld. He arrived in Urfeld the next day to learn of the failed assassination attempt. Stauffenberg, a wounded war veteran who had occasional access to Hitler's war room, had entered the room with a suitcase containing a bomb. He placed it next to Hitler, who was standing near a strategy table, then left the room. One of the officers, seeing the suitcase, moved it to a corner out of the way, where it exploded moments later—wounding a now enraged Hitler.

In the reign of terror that ensued, most of the members of the Wednesday Society were rounded up, summarily tried, and executed (or allowed to commit suicide because of their status). Even those in or out of the society who knew of the plot but did not participate were tried and consigned to the guillotine. Max Planck lost his eldest son, Erwin. With less than a year of his reign remaining, Hitler had destroyed many of the last leaders of the Wilhelmine upper cultural and social bourgeoisie. Heisenberg again miraculously escaped death. There is no indication that he was even interrogated. The high-ranking regime officials—Speer, Göring, Himmler—who supported him may have protected him from suspicion. His loyalty to the Reich was by then beyond suspicion.[53]

Even before they had any hint of opposition or of a conspiracy among the elite, the authorities in Berlin held Heisenberg and his coworkers, along with everyone else in such positions, under constant surveillance, lest they stray from the fold. Immediately upon moving to Berlin in July 1942, Heisenberg had named two party members at the institute to serve as party representatives.[54] A short while later, the SS security service representative in charge of "political questions" at independent research institutes insisted on inspecting Heisenberg's institute.[55] Heisenberg used the inspection to his advantage. When the two SS physicists who were to keep regular watch over the institute showed up one day, Heisenberg treated them cordially, showed them some interesting experiments, and impressed them with his usual charm. Already positively disposed toward their surveillance target, one of them later claimed that they offered to inform Heisenberg of any governmental actions that might interest him.[56]

Heisenberg's university institute also harbored Mathias Jules. An assistant in theoretical physics, Jules had played a crucial role in Himmler's investigation and exoneration of Heisenberg. Heisenberg's penciled notes to his institute secretary on certain government letters he had received indicate that Heisenberg made use of the man whenever he was engaged in delicate dealings with SS and party officials. But Jules demanded certain concessions, including the avoidance of Einstein's name. In the fall of 1942 Heisenberg, upon receiving a complaint from Jules, who had apparently received a complaint from Sommerfeld's publisher, wrote to Sommerfeld requesting that he delete as many references to Einstein as possible from his forthcoming book. Sommerfeld, aware of Heisenberg's situation, politely complied—approval of Heisenberg's Berlin university appointment had not yet been received.[57] Ten years into the Reich and in the midst of the Holocaust, the political battle over anti-Semitism was long over and names had been compromised away. By contrast, when Max von Laue was accused by Mentzel of using Einstein's name to excess in a lecture delivered in neutral Stockholm, the physicist absolutely refused to recant.[58]

Until the Allied bombing of German cities began in earnest in 1943, Werner's wife and children stayed in their Leipzig home while Heisenberg spent most of his time in Berlin. The two-and-a-half-hour commute by car or train to Leipzig allowed for weekend visits. This casual commuting ended after one of the Allied thousand-bomber raids practically leveled Berlin in early March 1943. The Heisenberg twins, Wolfgang and Maria, were with their father in Berlin to celebrate Grandfather Schumacher's seventy-fifth birthday. Heisenberg was at one of his social gatherings one evening when the bombs began falling. As Werner drove frantically from the city center to the Steglitz suburb, hellish walls of flame on both sides of the road etched the scene forever in his mind. He arrived to find the roof of the Schumacher house, hit by two incendiary bombs, entirely engulfed in flames. Having assured himself that his children and in-laws were safe, he ran next door, where a young woman was frantically crying for help. Her elderly father was fighting a losing battle against the flames of their burning home. The athleticism of the middle-aged professor served him well in this crisis—he was actually wearing a leotard track suit that night. Heisenberg managed to scale the outside wall to the roof of the burning house, leapt across a pool of flames, and led the old man to safety.

Everyone in the Schumacher household survived, but the children, temporarily sent to stay with the former Notgemeinschaft head Friedrich Schmidt-Ott, were so terrified by the experience that Heisenberg decided to move his family permanently to their Urfeld summer cottage. After renting the Leipzig house in April 1943 and shipping the furniture to Berlin for storage, Mrs. Heisenberg, the children, a woman friend whose non-Aryan husband had "disappeared," and her child moved together into the tiny Urfeld cottage for the duration.[59] After the Schumachers moved south that summer, Werner, completely alone in Berlin, moved into the bachelor quarters of Harnack House, which he shared with an influential member of the Foreign Office. It was here that he twice entertained his Wednesday Society friends.[60]

Eight months after the move to Urfeld, the Heisenbergs' Leipzig house on the quiet tree-lined street, with its beautiful flower garden and shady back yard, fell victim to another heavy bombing raid that also damaged the physics institute. The same series of raids bombed Heisenberg's sickly mother out of her Munich apartment on Ainmillerstrasse. With Werner's help, she found new rooms in the Bavarian border town of Mittenwald and arranged for a truck to move her furniture.[61] She might have moved to the cottage with Heisenberg's wife and children and their friend and her son, but though they might have had room for one more, they had just then all come down with scarlet fever.[62]

Until long after the end of the war, Elisabeth Heisenberg and the children would not see their husband and father again, except for brief vacations and holidays. They would have to face the last, most difficult years of the war alone in their mountain hideaway.

Although Heisenberg regarded his appointment to the prominent positions of head of the Kaiser Wilhelm Institute and professor of theoretical physics in Berlin as an enormous victory for German physics, Himmler's second promise, the publication of Heisenberg's views in a Nazi science journal, was not yet fulfilled. To Heisenberg, Himmler's 1938 letter served as an exoneration of the accusations of treason against him, while together the Berlin appointment and publication of the article would signify to him the ultimate defeat of "deutsche" physics and the rehabilitation of theoretical physics.

Despite the victory his appointment signified, Heisenberg used every conceivable strategy to make Himmler's second promise a reality. The task was not easy, for the journal that was to publish the article was the *Zeitschrift für die gesamte Naturwissenschaft* (*Journal for the Entirety of Science*), the official organ of the Reich Students League, which, like the University Teachers League, remained a stronghold of anti-Heisenberg sentiment in 1942. Working through Jules, who was a former member of the Students League, and probably through his SS security service (SD) connections, Heisenberg managed to have Himmler order SD chief Turowski to accept an article from Heisenberg at about the same time Heisenberg was summoned to Berlin.[63] Heisenberg had drafted the article two years earlier and submitted it to Himmler's infamous Reich Main Security Office for transmittal to the journal's editor, Bruno Thüring. Himmler's office also ordered SS officer Dr. Fritz Kubach, head of the science section in the Students League, to publish the article. But Thüring, who had earlier opposed Heisenberg's call to Vienna, was not in the SS and would not budge without further pressure and extracted compromise.

Heisenberg confronted Thüring during the Seefeld "religious debate" in November 1942. After the encounter, Thüring wrote to Heisenberg that he had the distinct impression that publication of the article "is supposed to be a type of rehabilitation for the . . . attacks directed against you"—a rehabilitation that Thüring strongly opposed.[64] Another letter from Heisenberg to Himmler's office (Himmler himself was "away on business"), in which he complained of the journal's intransigence, brought results. Heisenberg's article, "The evaluation of 'modern theoretical physics,'" finally appeared in the October 1943 issue of the Nazi journal and was eagerly read by physicists throughout the Reich.[65] The article upheld the compromises achieved earlier, especially those regarding Einstein and the separation of the physicist from his physics. Heisenberg reassured a concerned Theodor Vahlen, now president of the Prussian Academy of Sciences, which would appoint Heisenberg to membership several months later, that in this article he presented the viewpoint that the special theory of relativity "would have arisen even without Einstein."[66]

Now, with the restoration of personal honor and recognition, Heisenberg seemed to take every token of recognition received thereafter as further assurance of the rehabilitation of himself and his profession. It was as if recognition and respect, which he always required, had been withheld for so painfully long and, when finally given, were so tenuous that no amount of honor accorded him could ever make up for this or reassure him that his status had not somehow declined once again.

There were many opportunities for reassurance in 1943 and 1944. When he received the prestigious Copernicus prize of the Reich University of Königsberg, Heisenberg wrote to Dr. Borger, the head of the Teachers League, who had moderated the Seefeld conference, that he was especially pleased with this award, because it could be regarded as even further official rehabilitation of theoretical physics. "Hopefully this development . . . will continue in the future."[67] The development had already begun. In 1942, Heisenberg protégé Carl Friedrich von Weizsäcker was asked to accept the theoretical physics chair in German-occupied Strasbourg. In March 1943, the Nazi party newspaper, *Völkischer Beobachter,* asked Heisenberg for an article celebrating Planck's eighty-fifth birthday.[68] In October Göring recommended Heisenberg to Hitler for the War Service Cross, First Class.[69] In 1943 and 1944 Planck and Heisenberg were featured in front-page stories in *Das Reich,* Goebbels's cultural propaganda newspaper.[70]

Reassurance of his standing also seems to have been one of the personal motives behind Heisenberg's many cultural tours abroad throughout the war. Travel by professors beyond the Reich's original boundaries, especially on professional business, required explicit approval of the Education Ministry's foreign office and of the German

Congress Center and Foreign Currency Bureau, an arm of Goebbels's propaganda ministry. In granting approval, the Education Ministry relied on the evaluation of its own member in charge of foreign exchange and on the party's Teachers League representatives at the professor's university. Every permission to travel on official business could thus be seen by the recipient as concrete evidence that the authorities held him in high regard.

But of course the authorities saw such permission in a quite different light. Germany had long practiced propaganda abroad using noted scholars who participated either wittingly or unwittingly. While the Teachers League strongly preferred those of the proper ideological persuasion, the Congress Center was primarily concerned that professors could be exploited for the German cause; it did not demand ideological purity.[71] Philosopher Hans Gadamer, Heisenberg's Leipzig colleague from the professors' club, later commented on his own wartime trips abroad, one of which was to Portugal in the company of Carl Friedrich von Weizsäcker: "I did not fully recognize that thereby one was being used for purposes of foreign propaganda, for which a political innocent was sometimes suitable. Such instances were an escape with mixed feelings."[72]

Weizsäcker also accompanied Heisenberg, by now no innocent, on many of his foreign travels. Their 1941 trip together to German-occupied Copenhagen had been a test case of Heisenberg's suitability for future exploitation. It was regarded as such a success that the elder Weizsäcker's Foreign Office advised the REM that from the "political culture point of view" a further presentation of Heisenberg was extremely desirable.[73] Heisenberg probably knew of this recommendation. If he did, the co-optation seems complete: just as the German nuclear project made a case for its members and their science by pointing out their utility to the war effort, the Foreign Office argued for Heisenberg's unhindered travel throughout the Reich and even outside the Reich by pointing out his usefulness to the German propaganda cause. At the same time, Heisenberg could see such permission as both approval of himself and opportunity to pursue his own purposes during these largely voluntary visits, while entering into horrible compromises with the authorities that he regarded as of little significance. Naturally, those whom he visited often took quite a different view.

After 1941, most of Heisenberg's travels resulted from solicited or unsolicited invitations, usually asking him to lecture, as he did in Copenhagen, at one of the local cultural propaganda centers, the infamous German culture institutes. As noted earlier, the culture institutes, set up

on occupied soil by the political culture division in Ernst von Weizsäcker's Foreign Office, were created for the express purpose of extolling the virtue of German rule through the example of German culture.
Fritz von Twardowski, head of the political culture division through
1943, later admitted this function.[74] In many instances, nationals in
occupied countries identified the cultural visits with imperialism. No
wonder some felt insulted by an invitation to attend Heisenberg's lectures in the local propaganda institute; acceptance would have signified
collaboration. At least one foreigner would later accuse Heisenberg
among the several German visitors to his homeland of possessing "the
least understanding of the situation."[75] Yet Heisenberg was also a man
who could write in his private papers: "It is more important to treat
others humanely than to fulfill any sort of professional, national, or
political duties."[76] At least he seemed to know what was right, even if he
could not always carry this out.

Between the SS attack on Heisenberg in 1937 and the Teachers
League's retreat in the summer of 1942, Heisenberg undertook five
recorded trips abroad: north to England and south to Geneva and
Bologna in 1938, to the United States in the summer of 1939, to
Budapest in German-allied Hungary in 1941, and to Copenhagen in the
same year. Heisenberg was able to make these trips largely because of
the pressure exerted on party bureaucrats by Weizsäcker's Foreign Office.[77] After Heisenberg moved to Berlin in 1942, he had little trouble
traveling as he wished, but he still perceived every approved trip as a
victory, in addition to being a means of proving himself and maintaining
contacts with foreign physicists.[78]

None of his trips after July 1942 seems to have been self-motivated.
Every trip originated with an "invitation" communicated directly to
Heisenberg from the REM or through his new Berlin dean, Ludwig
Bieberbach, who was a leading proponent of the Nazi "deutsche"
mathematics movement. The surviving examples of invitations are
worded in a way that strongly suggests acceptance. Heisenberg could
have refused, with some difficulty, but duty and the opportunities it
presented required no urging, even when he would be traveling into an
unsavory situation and even when his visit was being used to bolster
faltering support for the Reich. Heisenberg readily accepted an invitation conveyed on May 25, 1943, from Dr. Hans Frank, General Governor of occupied Polish territories, to lecture in a Cracow cultural
propaganda institute and to stay as a guest in Frank's Schloss Warten-
berg.[79] Just two months before Frank offered his invitation, Frank and
his henchmen had liquidated the Cracow Jewish ghetto; only days

earlier, in mid-May, they had annihilated the heroic Warsaw ghetto. The Lodz ghetto was still in existence when Heisenberg arrived that December.[80] If Heisenberg did not know of these events before his arrival, he must have heard rumors about them during his stay.[81]

Between July 1942 and the end of the war, Heisenberg made another seven recorded foreign trips, in addition to the December 1943 visit to Cracow. Two were to neutral Switzerland in November 1942 and December 1944. He made a return trip to Budapest in the company of Planck and Weizsäcker at a time (November–December 1942) when Hungary's allegiance to the Reich had begun to waver. He was in Pressburg (Bratislava) in Reich-aligned Slovakia also at a time (March–April 1943) when pro-German sentiment had begun to falter. He made two trips to Copenhagen, in January and April 1944, after Bohr's departure. The seventh trip was to the occupied Netherlands in October 1943.[82]

Reich Education Ministry guidelines strictly controlled the activities of its emissaries. A professor had to submit lists of everything he carried into and out of a country; he was required to report to local occupation authorities (so he could be watched); he had to prepare a written report of his activities, including his observations of the political situation.[83] Compliance ensured control.

Three visits for which considerable documentation, as well as controversy, is extant are Heisenberg's trips to Denmark in 1944 and the trip to the Netherlands in 1943. Documentation is, however, no guarantor of truth. No matter how thorough the written record, how vivid the memories of witnesses, they are overshadowed by the terrible circumstances of the times. Matters of life and death, murder and torture, resistance and collaboration, manipulation and self-delusion are all at play against an extraordinary background of war, plunder, oppression, and genocide. It is difficult enough to reconstruct past events occurring in normal, peaceful times with some degree of accuracy. How much more do the horrors, feelings, and fears of those terrible times hinder our confidence in any source describing events of that period.

Nevertheless, the available sources do provide a partial account of Heisenberg's travels and the circumstances in which they occurred. His January trip to Denmark came at a particularly dark moment in Danish history. Having subjected the Danes to occupation since 1940, the German authorities suddenly dropped their facade of benign intentions in the fall of 1942 after the monarch fled and Hitler decided to incorporate Denmark fully into the Reich.[84] The German command replaced the civilian governor with SS officer Dr. Werner Best. As resistance and sabotage increased in reaction to nazification, Best declared a state of emergency, imposed martial law, and in cooperation with Berlin prepared to round up Danish Jews on October 1, 1943, for transport to the German death camps.[85] Thanks to a German industrialist, the resistance learned of the plan, and with the help of the entire population and the cooperation of the Swedish government, within several days Niels Bohr, his family, and nearly all of the estimated 8000 Danish Jews were ferried in private boats across the narrow Øresund Straits to neutral Sweden.[86]

Soon after this spectacular escape, the resistance inquired of Bohr whether his institute should be blown to bits to prevent its use by the Germans. Bohr declined the offer and, as expected, German troops occupied the building on December 6, 1943, on the pretext that Bohr and his institute were conducting anti-German propaganda and contacting the Allies.[87] The troops immediately imprisoned the institute's only resident, Professor Hans Bøggild, a cosmic-ray researcher, subjected his colleague, a Dr. Olsen, to five days of Gestapo interrogation, and barred the remaining members from the premises. The Danes informed Dr. Hans Suess, a German heavy-water expert who was passing through Copenhagen on his way home from the Norwegian plant, of the takeover. Suess informed Heisenberg of the Copenhagen events after returning from Christmas vacation.[88] At the same time, Heisenberg learned of Reich Research Council plans to strip Bohr's institute of its cyclotron and other equipment and to ship them to Germany.[89]

Working through Walther Gerlach, Esau's successor as head of Research Council physics, and Dr. E. Six, Twardowski's successor as head of cultural propaganda, Heisenberg obtained permission to fly to Copenhagen on January 24, 1944, accompanied by Kurt Diebner, to help arrange for the disposition of Bohr's institute.[90] Diebner could throw the weight of the Army Ordnance Office behind any plan worked out with the institute staff. Officially, the two were traveling as members of a committee to investigate the charges against Bohr's institute. Heisenberg's uneasiness is evident in a letter to his mother written the night before he left. "I am not at all happy about the trip . . . [but] it is probably necessary that I try to determine what is right and, if possible, to rectify the situation."[91]

Several sources exist—all open to question—describing what transpired on this journey. One is a lengthy report written in Danish by the institute members soon after the institute was released from Nazi control in 1944.[92] Its purpose is unclear, but it casts Heisenberg in an ambiguous light at best. Other accounts include anecdotal postwar recollections casting Heisenberg in a much harsher light and a January 8, 1944, appeal to Heisenberg concerning Bohr's institute from Swedish physicist Hans von Euler with emendations at the end in Heisenberg's hand.[93]

According to the report prepared by the institute members, Heisenberg wrote to them on January 10 informing them of his arrival in 14 days. Shortly after their arrival, Heisenberg and Diebner met at the institute with Bohr's colleagues, physics professors Christian Møller and Hans Jacobsen, whom Heisenberg knew through their work on nuclear forces and cosmic rays, and with the Gestapo officer in charge of the institute. Committee member Heisenberg demonstrated to the officer that Bohr was not engaged in underground activities by pulling out one of his own letters from among Bohr's voluminous correspondence.[94]

As intermediary, Heisenberg then discussed three proposals for the institute, which are outlined in the report. They are compatible with four proposals that Heisenberg abbreviated, with pros and cons, at the foot of the letter from Hans von Euler. They were written perhaps in preparation for the meeting. The three proposed scenarios were (1) German scientists would take over and run the institute for war research; (2) the German forces would return the institute to the university under the conditions that no war research would be carried out in the institute and all research would be published; and (3) the institute would be returned to the university, but the Germans would strip the institute of the apparatus they needed. According to the report, the Danes naturally rejected all three proposals, refused any conditions at all, and demanded Bøggild's immediate release from prison before discussing anything further. Heisenberg warned them that under the current circumstances, unless conditions were imposed the authorities might not release the institute at all.[95]

An impasse was avoided the next day when the senior Weizsäcker's Foreign Office, probably notified of the situation by Carl Friedrich, informed the university rector that the institute would be released without condition on February 3. Bøggild was released from jail that afternoon, and after filing a report on the harmless nature of the institute, Heisenberg flew back to his duties in Berlin.[96]

The Copenhagen authorities would not let the matter rest so easily. Two months later, Heisenberg received a strong request from the Reich Education Ministry to accept an invitation from the German occupation administrators in Denmark to lecture at the newly reopened German Culture Institute in Copenhagen.[97] He could surely have refused without much difficulty but readily decided to accept. Heisenberg returned to occupied Copenhagen, delivered his lecture as requested, and, to the dismay of the Danes, even dined publicly with the brutal Reich commissar, SS-Oberstürmbannführer Dr. Best. This return trip, his warm thanks to Best, and a positive report to the REM on the culture institute were, he apparently somehow convinced himself, payments demanded by the Reich in exchange for its reduced interference in the affairs of Bohr's institute.[98] One 1949 report, based on Danish assessments at the time, declared: "It is thought that Heisenberg is not a Nazi but is an intense nationalist with the characteristic deference to the authorities in control of the nation."[99]

Heisenberg's trip to the occupied Netherlands is even more questionable and certainly more difficult to interpret. Nazi racists considered the non-Jewish Dutch, like the non-Jewish Danes, to be an "Aryan" people and therefore suitable for full incorporation into Hitler's Reich. But while the Danish government remained in place until 1943, the Dutch monarch and government had fled to London immediately upon the country's occupation in 1940, which provoked instant, large-scale resistance to the German occupation.

The Reich commissar for the Netherlands was the infamous Dr. Artur Seyss-Inquart—an Austrian Nazi who had earlier presided over Hitler's annexation of Austria and then served as second-in-command to Hans Frank in the Polish territories. He was later tried and executed at Nuremberg. Like the first governor of occupied Denmark, Seyss-Inquart at first attempted to co-opt what he called "obliging circles in the field of economy, especially [in the fields of] agriculture, culture, art, and science."[100] The Dutch responded with grudging acquiescence and apparent collaboration in some quarters. But attempts to nazify Dutch society and to introduce anti-Semitic measures as early as 1941 provoked strikes and riots, which were met, on orders from Himmler, with brutal suppression.

Mass executions, the internment of Dutch army officers, and the start of Jewish deportations to the death camps in the summer of 1942 provoked open resistance, sabotage of Nazi targets, and the organized concealing of Jews and other endangered persons. According to historian Werner Warmbrunn, students and faculty members at Dutch universities "kindled and spearheaded ideological resistance to national socialism." He estimates that students, subject to being sent into forced labor in Germany, made up as much as one-third of the persons executed during the Nazi occupation.[101]

German policies met with particularly strong resistance at the University of Leiden, where Heisenberg would have his most remembered encounters. Unlike the essentially passive reaction in Germany and elsewhere under German rule, Leiden students struck their university upon the dismissal of Jewish professors in November 1940. A year later the majority of the faculty responded to a German attempt to replace an outspoken faculty member by handing in their resignations and refusing to cooperate further with the authorities. Scores were subsequently taken hostage, and many were executed in reprisals for underground actions. Hendrik B. G. Casimir, then a researcher at the Philips Company in Leiden, reports that at the university's famous Kamerlingh Onnes low-temperature laboratory, where at least one Jew was concealed, "things went on much as usual." As for the rest, writes Warmbrunn, "Leiden ceased to exist as a university."[102]

Both university protests and repressive measures to quash them spread across the Netherlands in 1943 as the deportations to the death camps continued. By September they were complete, with the remainder of Dutch Jews holding out in hiding. A month later Werner Heisenberg, successor in Berlin to Dutch national Peter Debye, arrived in the Netherlands for what must have been an extremely tense encounter with his Dutch colleagues.

Considerable official correspondence and documentation pertaining to Heisenberg's visit are available, as is some ostensibly private correspondence between Heisenberg and his colleagues in the Netherlands. Letters between Heisenberg and correspondents in occupied territories were never written as private communications, for the authors knew that the Gestapo could and very likely would read the contents before they were delivered. The disturbed conditions in the Netherlands at the time make it difficult today to interpret these sources. With the end of the deportations apparently in sight, on June 15, 1943, the REM forwarded an invitation to Heisenberg from the Dutch Education Ministry and unnamed Dutch scientists to visit the Netherlands in the fall. The REM urged acceptance, and Heisenberg did so but inquired which scientists were involved.[103] Two months later, he received a letter from his old Copenhagen colleague, H. A. Kramers, now in Leiden, with an explanation of the origin of the invitation—or as much of an explanation as he cared to put into writing.[104]

According to Kramers's account, the chief of education in the Dutch ministry wanted to improve the conditions of Dutch professors by arranging for personal contacts with foreign colleagues. Kramers, apparently in close touch with the education chief, discussed this with his physics colleagues, de Haas, Casimir, Kronig, and others. They agreed to invite Swiss experimentalist Paul Scherrer and German theoreticians

Heisenberg and Richard Becker to visit in the fall.[105] Taking this account at face value, aside from the intellectual stimulus of foreign colleagues, the Dutch physicists apparently felt that if the newly rehabilitated German theorists took a personal interest in Dutch science, they would use their influence with the German authorities to protect Dutch laboratory equipment and to improve the Dutch scientists' working conditions.[106] At any rate, an official in the Main Department for Science, Education, and Promotion of Culture in the office of the Reich commissar made little effort to conceal his reasons for approving the invitation. In a letter to Heisenberg in September, he wrote: "We desire that in this way the rather unstable relationships we have in the scientific field may be strengthened once again. You will probably have heard that the majority of Dutch university professors reject or mistrust our political views and ideas, but on the other hand they have no desire to sever connections in the professional sphere."[107]

Heisenberg's Dutch travels from October 18 to 26, 1943, took him to the Education Ministry in Apeldoorn; to the Kamerlingh Onnes Laboratory in Leiden, where he met with Kramers, Casimir, and Keesom and delivered a colloquium lecture on elementary particles; to Utrecht, where he stayed with Léon Rosenfeld; to the universities of Delft and Amsterdam, where he lectured; and to a meeting with Seyss-Inquart at occupation headquarters, where he argued for the survival of the Kamerlingh Onnes Laboratory.[108]

Back in Berlin, Heisenberg continued to work with his contacts to try to prevent German interference in Dutch research and reported to the REM his official observations. According to his report, the Dutch expressed "a blunt rejection of the German viewpoint. . . . However, on a purely scientific basis, a collaboration with Dutch colleagues is perfectly possible."[109] Although this statement may express Heisenberg's actual views, it seems equally an echo of the views expressed by the occupation official in the Main Department for Science. Dutch interest in scientific collaboration had been strengthened by the visit — precisely the outcome the official had desired. At the same time, the stated possibility of scientific collaboration was a good argument for the preservation of Dutch laboratories and research.

During the war, two physicists whom Heisenberg had visited in the Netherlands warmly thanked him for his interventions on behalf of themselves and their science.[110] Nevertheless, immediately after the war and in subsequent writings, several scientists recalled distressing private encounters with Heisenberg during his visit. According to a June 1945 report filed by G. P. Kuiper, a Dutch-American member of the Alsos mission, an Allied science intelligence unit, Casimir had told Kuiper that during a private walk with Heisenberg in Leiden, Casimir learned that Heisenberg knew about the German concentration camps and about Germany's plunder of occupied territories, yet "he wanted Germany to rule." As Kuiper reported it, Heisenberg said: "Democracy cannot develop sufficient energy to rule Europe. There are, therefore, only two possibilities: Germany and Russia. And then a Europe under German leadership would perhaps be the lesser evil."[111]

These statements circulated widely among Allied scientists after the war and contributed to their negative assessment of Heisenberg in those years. Like other evidence, they require closer scrutiny than face-value acceptance. Human frailty being what it is, recollections can be distorted by painful memories of wartime experiences; occasionally they are distorted by efforts to protect those who acted less than nobly in a difficult situation by accusing others of even worse behavior. In any case, if Heisenberg did make this statement, then, as Casimir has pointed out, he showed an appalling lack of sensitivity, to say the least. In view of the bitter hostility harbored by the Dutch toward Germany at that time (especially in Leiden), the expression of such a sentiment would, as one writer puts it, have forever poisoned Heisenberg's relations with his Dutch colleagues, however friendly and grateful for favors they may appear in their wartime letters.[112] And whether or not Heisenberg actually said those things, the very fact that Casimir and Kuiper claimed he did could hardly have helped Heisenberg's case during the postwar evaluation of his actions during the Third Reich.

But what were Heisenberg's true feelings about the war? We have already seen that he wanted to protect Germany from defeat and that he regarded the war as a shift of power far beyond the control of mere mortals like himself. In October 1943, when he visited the Netherlands, the Western Allies were still only a weak presence on the European continent, while the Russian army, after the crushing defeat of the German Sixth Army at Stalingrad, was already advancing toward Poland. Heisenberg at that moment may well have seen the war as a struggle over which dictatorial system would ultimately rule Europe, and it looked very much to him, as it did to many Germans, as if Germany would lose. Ever since the trauma of the Bavarian soviet republic, Heisenberg could imagine nothing worse than the conquest of Germany by Soviet troops —no matter that others were already suffering the terrible consequences of conquest by Hitler's troops.

Heisenberg's choice of regime as the lesser evil became even more obvious after the war: having refused to leave Germany during the 12 years of Nazi rule, he reacted to early cold-war fears that Soviet forces might overrun Western Europe by making contingency plans to emigrate immediately to the United States.[113] Heisenberg's lately elevated status in Nazi Germany apparently evoked compromises in his sensitivities and perceptions in addition to those in his life and work.

CHAPTER 25

Target Number One

A breakthrough in Heisenberg's nonnuclear research paralleled the culmination of his efforts at professional recognition in his appointment to Berlin in 1942. Just two months after he arrived in Berlin, Heisenberg submitted the first part of what, by 1945, was a four-part work on a new and widely studied fundamental theory of elementary particles. The move to Berlin did not in itself inspire this productivity; rather, he wrote to the ailing Hans Geiger, to whom he dedicated the first paper, he had been working on these new ideas for quite some time.[1] But Heisenberg's bachelor life in Berlin and his return to academic activities did enable him to formulate the first two parts of the work so rapidly that they were completed by the end of October 1942.[2] The third and fourth parts were developed over the next two years in collaboration with his colleagues in Denmark and in the occupied Netherlands.[3]

Heisenberg's prewar researches in quantum field theory, undertaken in part with Pauli, had led him into the study of cosmic rays. When an extremely high-energy cosmic ray strikes the earth's atmosphere or a more solid piece of matter, it induces a shower of newly created particles and photons. This effect was to be explained on the basis of quantum field theory. Heisenberg's researches had convinced him of the inadequacy of available field theories for this task; infinities and divergences plagued all three of the available theories — quantum electrodynamics, the Fermi field, and Yukawa's meson theory.

The small size of elementary particles and the close approach of the particles to each other in a cosmic-ray collision — which triggered the particle shower—indicated to Heisenberg during the late 1930s that the difficulties in quantum field theory could be resolved only if a universal minium length, a new fundamental constant, were introduced into the theory. But every attempt to build such a constant into available field theories had failed. In a 1938 paper and in his section of a recently published report with Pauli intended for the ill-fated 1939 Solvay Congress, Heisenberg argued that the new fundamental length not only should occupy a central place in the future field theory—accounting for the existence of particles as stationary states—but also should mark the "limits of applicability of present quantum theory."[4] For Heisenberg and Bohr over a decade earlier, classical mechanics had lost validity when applied to the atom, and this required a new quantum mechanics; now, quantum mechanics broke down when applied to events occurring in regions smaller than a fundamental length about the size of an elementary particle, 10^{-13} centimeter.[5]

A great many calculations and equally numerous letters exchanged with Heisenberg throughout the 1930s had never really convinced Pauli of Heisenberg's program for the fundamental length and field theories. As in the 1920s, something entirely new was needed. Pauli had already suggested that Heisenberg, as he did when formulating the 1925 breakthrough to quantum mechanics, should focus only on observables and attempt to exclude all unobservable variables from the theory. Heisenberg did so by iterating the Hamiltonian, but he did not get very far before bogging down in mathematics.[6] Heisenberg tried another approach involving exclusive reliance on observables, one that seemed more promising: it led to his widely studied S-matrix theory of elementary particles.

At the height of the war, in 1942, Italian physicist Gian Carlo Wick, who had previously worked in Leipzig, returned to Germany for a visit. The visit stimulated Heisenberg's researches at a time when the uranium project had reached a hiatus. After stopping in Munich to see Sommerfeld in June, Wick spoke to Heisenberg's last four-student Leipzig seminar, then accompanied Heisenberg to his new job in Berlin before returning to Rome in early July.[7] According to a footnote to Heisenberg's first publication, Wick brought with him reports on his own work as well as on that being conducted by Breit, Bohr, Placzek, and Peierls along similar lines. Heisenberg rushed his first S-matrix paper, entitled "The 'observable quantities' in the theory of elementary particles," to the *Zeitschrift für Physik* by early September.[8]

In his new approach, Heisenberg used his hypothesized fundamental length to limit the possible changes in momentum and energy of two colliding high-speed particles.[9] This limitation could then help identify observables in present theories that, because they are measurable in laboratory experiments, must reappear in any future theory.[10] He considered events on this side of the limits set by the fundamental length to be observable, those beyond to be unobservable. For two colliding particles, this yielded four sets of observables with which to work: the properties of two particles as seen in the laboratory long before and long after the particles collide. During their collision, they approach within a distance of the universal length and are thus unobservable.[11] When far away from the collision, the particles act like undisturbed plane waves, and their momentum and energy can be expressed by a four-dimensional wave vector. Before collision, the two particles, or plane waves, possessed the vectors $\mathbf{k}'_1$ and $\mathbf{k}'_2$. After collision they possess vectors $\mathbf{k}''_1$ and $\mathbf{k}''_2$. Linking these observable plane waves without losing any particles and without entering the unknown region of interaction, Heisenberg represented the entire collision as a matrix transformation from the before to the after plane waves. This could be done by multiplying the waves before collision by a two-by-two scattering or S-matrix, represented by $S(\mathbf{k}'_i; \mathbf{k}''_i)$, where $i = 1$ or 2.

Although Heisenberg could not actually specify the elements of the S-matrix, it contained in principle all of the information about the collision including, he demonstrated, the expected relationships between spin and statistics. He also showed that the imposition of certain requirements on the interaction, such as conservation laws and symmetry properties, restricted the arbitrariness of S. In particular, by requiring a total scattering probability of 1 (the particles *will* collide and scatter), he proved that S must be unitary; that is, it commutes and multiplies with its Hermitian conjugate to form the unit matrix 1: $S^\dagger S = SS^\dagger = 1$. From this he obtained the important result that S can be expressed as $S = e^{i\eta}$, where η is a special (Hermitian) matrix containing only observable elements. In his second paper, completed in October 1942, Heisenberg further showed that several simple examples of relativistic η-matrices yielded expected scattering cross sections and even the possibility of an old chestnut — his favorite phenomenon of cosmic-ray explosion showers, an instanteous burst of shower particles in a single collision.[12]

As it had in 1925, Heisenberg's retreat to observables had made for an entirely new approach to problems at the subatomic level. But unlike 1925, no new physics emerged. In 1925 Heisenberg could use the correspondence principle and the quantum conditions to link the observed oscillator frequencies and amplitudes to the equations of classical mechanics. In 1942 not even the correspondence principle could be used to link the forbidden realm with the laboratory, if field theory failed completely in that region. In other words, Heisenberg's S-matrix was a theoretical orphan; it could be derived neither from any present theory nor, thought Heisenberg, even from a future theory. Thus, as Pauli, now in Princeton, put it in 1943, the S-matrix was an empty concept.[13] But for Heisenberg, who could leap to solutions without bothering with the intervening steps or without even articulating an underlying theory, it was, like the fundamental length, a concept that would reappear of necessity in the future theory, of whatever type. He was convinced "that in the future theory the matrix η or the unitary matrix S belonging to it will play an important role, even when in this theory there is no longer . . . a Hamilton function and a Schrödinger equation."[14]

Quantum field theory rested on the notion of a field that carried the forces between particles. A hypothesized field was built into a field theory by postulating a Hamiltonian function related to the energy, then formulating the quantum field theory equivalent of the Schrödinger wave equation. By introducing the S-matrix, observables, and a forbidden inner realm of interaction, Heisenberg had replaced the futile search for a suitable Hamiltonian formulation of field theory with a search for a suitable representation and manipulation of the S and η matrices as clues to the future theory—which might not be a field theory at all. Despite the war and its devastating effects, the tantalizing prospects of a future theory inspired Heisenberg and many other physicists into the early postwar years—and have continued to motivate scientists ever since.[15]

The most ardent postwar supporters of Heisenberg's S-matrix theory had learned of it directly from Heisenberg during his wartime travels. Wentzel and Stückelberg, who made extensive use of the theory, heard about it during Heisenberg's two wartime visits to Switzerland. Christian Møller, who brought the theory to the attention of Allied scientists after the war, corresponded with Heisenberg and discussed the theory extensively with him during Heisenberg's visit to Copenhagen in April 1944. And one evening in October 1943 Heisenberg presented his new theory to an informal colloquium in the Kramers home near Leiden in the German-occupied Netherlands. The university was closed at the time, and official colloquia were banned. Unlike the more unpleasant aspects of this visit, records and memories of the talk and dinner abound. During the discussion of Heisenberg's talk, Kramers made the insightful remark that if the actual elements of the η-matrix could ever be determined without a complete theory, they would yield a so-called analytic function in the complex plane of k variables — that is, a function containing so-called real and imaginary parts in a plane, the axes of which are the real and imaginary parts of the complex momentum variable k.[16]

After the discussion — and after the dinner and the inevitable musical interlude that followed — Kramers worked through most of the night on Heisenberg's views. Back in Berlin, Heisenberg wrote immediately that he had grown "more and more enthusiastic" about Kramers's remark "because I believe that with it one can really arrive at a complete model of a theory of elementary particles."[17] Heisenberg had discovered that the zeros of the analytic S-matrix, the values of k that satisfy $S(k) = 0$, fall along the positive imaginary k-axis (in the complex k-plane) and that these zeros correspond to the stationary states of bound systems or to the energies of individual elementary particles.[18] Heisenberg suggested to Kramers that they collaborate on a paper — a suggestion he made at the same time he was reporting to the Reich Education Ministry that scientific collaboration with the Dutch would be quite possible.

Kramers declined to write the suggested paper with Heisenberg. He argued that the post was too slow between Leiden and Berlin for such work — a letter either way took about a month to arrive — but he did ask that his suggestions be acknowledged.[19] Apparently shaken out of an intellectual depression by Heisenberg's visit, Kramers now required the intensity of direct collaboration.[20] But Kramers perhaps also felt that it would be inappropriate to collaborate publicly with the German physicist. Heisenberg obviously did not agree. As he progressed with Kramers's idea during the following months, he repeated his inquiry about collaboration at least three times.[21] Finally, as Heisenberg prepared to publish in March 1944, Kramers responded with all sorts of excuses for not writing a paper with him. His ambiguously worded conclusion can be understood to express either an unwillingness or a real inability to collaborate with Heisenberg: "I feel that the moment is not right for a joint publication — the moment is indeed right for joint work, but I would have to give up my scientific program entirely for the next months."[22]

Heisenberg submitted his third S-matrix paper — with a grateful acknowledgment to Kramers — soon after receiving Kramers's letter. About a month earlier, he had had long discussions about the theory with Christian Møller and Niels Arley in Copenhagen.[23] Just two days after declining to publish with Heisenberg, Kramers presented his thoughts on the theory to a Utrecht symposium.[24] Kramers would publish, but not with Heisenberg.

Late in 1944 Heisenberg prepared a fourth installment of his theory —a paper dealing not just with two colliding particles but with many interacting particles, which gave rise to an even more complicated η-matrix. He presented its essentials in Zurich at the end of 1944, but the paper could not be published before the presses stopped at war's end. There was a sudden spurt of interest in the analytic S-matrix during the early postwar years, thanks to the work of Møller, Kronig, Wentzel, Wergeland, and Heisenberg's own postwar summaries of his wartime S-matrix research.[25] Heisenberg's travels apparently did stimulate those whom he had visited.

The enthusiasm for the S-matrix soon died, however, after Pauli's Princeton student, S. T. Ma, discovered additional S-matrix zeros that did not correspond to stationary states or to elementary particles.[26] Pauli renounced the S-matrix during the 1946 Cambridge meeting of the British Physical Society, and interest soon returned to field theories when so-called renormalization rehabilitated quantum electrodynamics in the late 1940s. The infinities in the theory, such as self-energy and charge, could be defined away, rendering quantum field theories usable after all to very high energies and distances much shorter than the fundamental length. The hitherto mysterious elements of the S-matrix could be calculated harmlessly from the old quantum field theory. Although there were many difficulties in applying the renormalization technique to other types of field theories, Heisenberg's original nonfield theoretic S-matrix program languished in the corners of interest until the early 1960s.[27]

In the throes of war in 1943, while Heisenberg prepared his S-matrix papers and ingratiated himself and his physics with the authorities, Carl Ramsauer and the German Physical Society launched a new and even more ambitious campaign in 1943 for even greater acknowledgment and appreciation of physics by the authorities. Ramsauer's strategy, which one historian calls "the self-mobilization of science," involved all of the usual tactics employed by the scientists: memos to bureaucrats, personal diplomacy, and lectures to high-ranking audiences.[28] Göring's German Academy for Aeronautical Research provided a convenient forum for the last of these devices, and academy member Heisenberg was a useful and willing participant in this forum.

Ramsauer's new plan, as he explained it to Heisenberg, consisted of a three-part "logic": the premise that German physics would be a decisive factor in Germany's future; the assertion that German physics "was quantitatively and qualitatively far surpassed by American physics"; and the conclusion that new measures had to be undertaken that would require far more than the current financial and organizational support available for research.[29] Ramsauer presented his syllogism in a widely disseminated address to the aeronautical academy in early April 1943. One of his concluding recommendations called for "the most rational use of the existing physicists, particularly the setting up of new guidelines for the military use of physicists."[30] One month later Heisenberg, together with Hahn, Bothe, and Clusius, made the case for nuclear physics in particular.

Organizing the lectures by the nuclear scientists required some diplomacy. By agreement among Vögler, Mentzel, and Esau, nuclear fission research had been split between the Kaiser Wilhelm Society (KWG), the Reich Research Council (RFR), and the Army Ordnance Office. Speer's Ministry for Armaments and War Production and the KWG largely supported the work of the KWG institutes. But late in 1942, Göring had placed Esau in charge of all nuclear research outside Army Ordnance and had transferred Speer's KWG nuclear research to Esau's subsection of the RFR. Esau certainly did not appreciate KWG scientists Heisenberg and his colleagues acting as independent spokesmen for nuclear research.

For his part, by March 1943 Heisenberg was equally resenting Esau's attempts to direct his research. That month, Army Ordnance finally withdrew entirely from nuclear research and transferred control of Diebner's Gottow research team to the Reich Physical-Technical Institute (the German bureau of standards), also headed by Professor Esau.[31] Within days of Diebner's transfer, Esau confiscated for Diebner's use about 600 liters of heavy water that Heisenberg had stored in his institute's air-raid shelter for the B-6 series of large-scale experiments.[32] Esau still had to be placated, however, and to neutralize his objections to the scientists' lectures before Göring's aeronautical academy, Heisenberg proposed that Esau chair the meeting, which he did.[33] Then, to prevent his further meddling, Heisenberg pressured Speer to have Esau removed. Speer in turn pressured Göring and Mentzel, who replaced Esau at the end of the year with a man more to Heisenberg's liking, the Munich experimentalist Walther Gerlach.[34] Heisenberg and the physicists had clearly regained their influence.

Heisenberg's lecture to Göring's aeronautical academy several months earlier, on May 6, 1943, bore the neutral title "The acquisition of energy from nuclear fission."[35] As in his 1942 talk to Speer, Heisenberg performed a balancing act before the scientific academy and the larger audience of regime officials who would hear or read his talk. On the one hand, Heisenberg revealed privately, he wanted to emphasize the practical significance of nuclear research "in order to help the work achieve a favorable status as well as practical support."[36] On the other hand, after the shocking debacle at Stalingrad and amidst the Allied bombing raids that were systematically reducing German cities to smoldering rubble, he did not want the unlikely prospect of a new bomb in the foreseeable future to result in either the project being taken away from the scientists or the scientists being ordered to produce what they could not soon deliver.

Heisenberg deftly steered between the two extremes. Without mentioning the plutonium alternative, he argued that fission energy could be obtained only by using natural uranium with a moderator or by enriching the U-235 content of uranium by complicated—and expensive—processes. He admitted that if enough U-235 could be separated from natural uranium and concentrated into a small enough ball, fission would take place almost instantaneously, whereby "a correspondingly large amount of energy is released explosively." This comment was the closest he came to reminding his audience of the bomb. The remainder of his talk focused on the development of a uranium "burner," but his conclusion covered both possibilities—a burner and a bomb—while again dampening any undue optimism about the latter by noting the enormous technical and practical difficulties yet to be overcome. Atomic energy in large amounts is technically feasible, he declared, but "on the other hand the practical realization of this goal would naturally face great difficulties in this tense wartime economy."[37]

Heisenberg's lecture and the physicists' campaign had the desired effects. Steady funding and a sufficient "urgency classification" to keep the nuclear project going under the severest wartime conditions were ensured to the end of the war.[38] At the same time, as defeat loomed larger and Hitler began touting the imminent use of Wunderwaffen, secret weapons that would turn the tide of war, pressure for immediate production of a bomb and efforts to wrest control of the project from the scientists had been discouraged. The scientists even achieved further favors from the regime: better coordination of all war-related research through a new planning office in the Reich Research Council and temporary exemption from military service for as many as 5000 scientists, engineers, and students to allow them to pursue war-related research and studies.[39]

The preservation of young German scientists for the future was one reason offered after the war by Max von Laue to explain the willingness of German scientists to continue research on the project under Hitler.[40] The opportunity to use his influence to protect young scientists was also one reason Mrs. Heisenberg gave for her husband's decision to accept the prominent Berlin posts.[41] After the war, however, criticism surfaced regarding Heisenberg's apparent failure to use his influence to the fullest to rescue or to protect as many people as he might have. One does indeed wonder why Heisenberg and his colleagues did not turn their influential associations with such powerful Nazi figures as Himmler, Speer, and Göring to broad-scale rescue. One postwar critic said of Heisenberg years later: "He saved physics; he did not save physicists."[42] Kramers's biographer writes that Kramers asked Heisenberg several times to intercede for friends interned in concentration camps but that "there is no indication that this had any effect."[43]

Throughout his life, Heisenberg saw himself as primarily responsible only for his own circle of friends, colleagues, and students, and he believed himself incapable of really helping anyone beyond that circle, let alone the general public. This trait is evident in his perception of himself as leader of his youth group, in his reaction to the dismissal policy of the early Nazi regime, in his desire to preserve an island of students and assistants in Leipzig, and in his efforts to extract from the authorities military exemptions for physicists (but not nonphysicists) during the war. This trait reached its extreme in a wartime "order of reality" that, in his view, rendered the individual a helpless pawn of historical forces. The ideological and political blinders of the Third Reich had continually encouraged such a distorted perception, to which Heisenberg was already prone. At the same time, with rehabilitation achieved at last, they encouraged Heisenberg and his colleagues indeed to emphasize professional matters above all else, as if these were somehow independent of the fates of individual practitioners.

Although the available written record is again of varying insight and reliability, when Heisenberg did act, he did so as he had in past situations — offering too little, too late. With concerted, large-scale efforts out of the question, records do indicate at least five gestures by Heisenberg to save individuals from threatening situations. Two postwar affidavits by Heisenberg refer to four other cases, and Mrs. Heisenberg recalls two more instances — all in addition to the Edwin Gora episode discussed in Chapter 22. All of the cases not involving institute members appear to have occurred during or after 1943, when Heisenberg felt he had gained the full confidence of the regime. Whether for lack of sensitivity to the dire predicaments of those outside his circle or for lack of comprehension of his options or both, his efforts were pitifully weak as responses to the life-and-death situations of those whom he sought to help.

The records show Heisenberg making a variety of brief approaches to different authorities. In March 1943, he forwarded a letter to Himmler from a Münster mathematician on behalf of the French mathematician Elie Cartan.[44] He also encouraged an SS supporter in his institute to act on behalf of a Dr. Wetzel imprisoned in Stuttgart for incautious remarks; he appealed to Army Command Headquarters via the Kaiser Wilhelm Society to have returned from the front a promising physics student who was also an infantry officer; he worked through his Foreign Office roommate in the KWG's Harnack House to try to obtain the release of a foreign scientist imprisoned by the Gestapo.[45] In the last of his S-matrix letters to Heisenberg in 1944, Kramers requested Heisenberg's help for a young jurist friend of the family who had been imprisoned in Buchenwald. Heisenberg responded: "Whether I can do something for Herr D. is unfortunately very doubtful, but I will try."[46] There is no indication of what he attempted on the young prisoner's behalf, nor of what became of him.

The most tragic and far-reaching of the situations in which Heisenberg attempted to help involved Samuel A. Goudsmit's aged parents. Goudsmit, one year younger than Heisenberg and the Dutch coinventor of electron spin, was still an awestruck physics student when he first met the great Heisenberg during the period in 1925 when Heisenberg was formulating the foundations of matrix mechanics. Since the two physicists worked along similar lines, they occasionally corresponded afterwards. After receiving his doctorate under Paul Ehrenfest in 1927, Goudsmit left the Netherlands for the University of Michigan. He and Heisenberg often met there when Heisenberg went to Michigan to visit or to lecture at the Ann Arbor summer school in physics. They had met there last in the summer of 1939.

Although Goudsmit had moved permanently to the United States, his Jewish parents had remained behind in The Hague. Goudsmit had already obtained American visas for them to go to the United States, and they had just received their travel papers when tragedy struck. During the Nazi deportation of the Dutch Jews in 1943, the Goudsmits were taken out of their home, loaded onto a cattle car, and transported to the Auschwitz death camp. Goudsmit's friend and colleague Dirk Coster, who had been instrumental in rescuing Lise Meitner after the annexation of Austria in 1938, appealed for help to Heisenberg by letter. Heisenberg responded by writing a letter to Coster on February 16, 1943, to be shown to the authorities. In it, he described Goudsmit's friendly hospitality to visiting Germans and his own anxiety regarding the safety of Goudsmit's parents.[47] It is uncertain what effect Heisenberg expected this letter to Coster in the Netherlands to have in rescuing the Goudsmits from Auschwitz. Perhaps he did not know that they had already been deported. There is no indication that Heisenberg tried any other avenue. Possibly due to the notoriously slow post between the Netherlands and Germany, Heisenberg wrote his letter too late. Five days before the date of Heisenberg's letter, Goudsmit's father and his blind mother had died in the Auschwitz gas chamber on his father's seventieth birthday.[48]

Goudsmit returned to the Netherlands less than a year later. He was now the scientific head of the Alsos mission, the secret Allied science intelligence unit that, by then, was after the German nuclear energy project and its leading scientist — Werner Heisenberg. Standing in tears in the wreck of what had been his boyhood home, he wrote several years later, "I was gripped by that shattering emotion all of us have felt who have lost family and relatives and friends at the hands of the murderous Nazis — a terrible feeling of guilt."[49] It was also a feeling of rage, rage at the Germans and surely rage at Heisenberg for having failed to help. He did not mention Heisenberg's letter in this passage. It is uncertain when he learned of it, but even if he did know at the time, it would not have altered his influential assessment of Heisenberg after the war as a great physicist with a deep and tragic character flaw. Even years later, Goudsmit's opinion of the episode had not changed much. In an obituary for Heisenberg, written in 1976 for the American Philosophical Society, Goudsmit was much kinder but still angry with the late physicist: "Heisenberg was asked in 1943 to intercede in the case of acquaintances who were being sent to a concentration camp. He responded merely with a vague letter. I doubt that he could have done anything else. I doubt that I or most of the physicists I know would have done better under the same circumstances."[50]

While the restored recognition accorded Heisenberg and other nuclear scientists did not affect their rescue of threatened persons, it did noticeably affect the German nuclear effort. Heisenberg and his colleagues were now confident both of their nuclear science and of their control of their profession. While Allied scientists, with the much greater material and organizational support of the United States government and military, were now settled in the Nevada desert constructing an atomic bomb, Heisenberg and the German scientists believed that the Allies could not have progressed any further in the utilization of nuclear energy than the Germans had. In a report to Mentzel and Göring shortly after the aeronautical academy lectures in May 1943, Abraham Esau, in his last days as head of nuclear research in the Reich Research Council, drew that conclusion from the scientists' papers and lectures. Transferring Esau's report to Göring, Mentzel echoed Esau in declaring that, while nuclear research for technical reasons would not lead soon to a machine or an explosive, as Heisenberg had argued, "it is certain that the enemy powers could not present us with surprises in this field."[51] The continual loss of physics personnel following Hitler's rise to power and Ramsauer's case for the overall superiority of Anglo-American physics notwithstanding, this report and various subsequent events make it clear that German physicists could not believe themselves inferior in the field where they had predominated for so long—especially now that they had won new respect from their superiors. And, by the same token, the "enemy powers" believed that German science had advanced at least as much as Allied research. For the Germans, the long struggle to preserve German physics could not have been for nothing.

Heisenberg's new status as Germany's premier nuclear physicist apparently blinded him in another way. In 1943 he seemed less than eager to acknowledge the achievement of a superior uranium pile arrangement by a longtime nuclear competitor (and perceived inferior), Kurt Diebner. Although Heisenberg did not devote as much attention to nuclear technology after moving to Berlin as he had before, as head of the Berlin pile research team he did set the research program. There seemed no theoretical reason for the superiority of Diebner's pile. Even more important, Heisenberg had already settled in July 1942 on a new large-scale experiment: 1.5 metric tons of heavy water and 3 metric tons of uranium metal plates arranged in horizontal layers within a cylindrical metal tank.[52] Heisenberg had based the experiment's design on the results of his Leipzig and Berlin experiments; he also wanted to compare the results of this arrangement with the theoretically derived properties of such a complicated pile.[53] Heisenberg's research program remained on course for nearly two and a half years.

Working at the army's Gottow weapons research station, the inventive Diebner had hit on the alternative idea of arranging the uranium metal not in plates but in cubes suspended in a cylindrical tank of heavy water. This allowed more contact between the uranium and the heavy water; presumably, more of the neutrons released in uranium fission would thus be slowed by the heavy water to energies enabling further fission rather than capture. Diebner's first attempt, using frozen heavy water to support the metal cubes, yielded about 36 percent neutron multiplication—well above the best Leipzig result. (Any multiplication indicates that more neutrons are produced than absorbed; the greater the multiplication, the closer to a chain reaction.) In his second attempt, undertaken in 1943 under the auspices of Esau's Reich Physical-Technical Institute, Diebner suspended his uranium cubes on thin wires in liquid heavy water—some of which had been commandeered from Heisenberg's institute. This contraption yielded nearly 110 percent multiplication.[54] More cubes and water might go critical. But Diebner's experimental destiny collided again with Heisenberg's when Allied bombers leveled the Degussa Company, the manufacturer of Diebner's uranium cubes. Degussa's subsidiary, the Auer Company, was the sole remaining uranium metal producer, and Auer already had a long-standing contract to deliver Heisenberg's metal plates.

Heisenberg had actually made a brief evaluation of Diebner's Gottow experiments in early 1943 and had grudgingly admitted during his aeronautical academy lecture in May that, compared with the Leipzig layer arrangement, the Gottow group offered "a somewhat improved apparatus . . . in which the neutron multiplication was somewhat higher."[55] But he remained confident of his planned plate experiments. Apparently he even promised Vögler, the head of the Kaiser Wilhelm Society, that the "uranium machine" would be up and running by the end of the year—if the plates were delivered soon.[56] They were not. Mentzel of the Reich Research Council, an Auer representative, and a now lame-duck Esau were still arguing over production priorities at the end of the year. They finally agreed that "production of cubes for Gottow should not interfere with plate production."[57] Heisenberg would have his plates, but because of his low war priority rating, they did not actually arrive until January 1944. Heisenberg did not give up the less efficient plates for Diebner's more fissionable cube arrangement until nearly a year after that.

By the time the uranium plates finally arrived in Berlin, Vögler was beginning to wonder what had become of the promised machine. Karl Wirtz, the head of pile construction in Berlin, had only just begun assembling the first of the B-6 experiments in the Berlin institute's newly completed underground bunker. He and his team were assisted by several members of Bothe's Heidelberg group, who moved to Berlin after Bothe turned to cyclotron construction. The elaborately equipped bomb- and radiation-proof bunker in which they now worked had 2-meter-thick walls of iron-reinforced concrete. Inside were a main laboratory with a water-filled pit, rapid air and water pumps in case of accident, a workshop, remote-control apparatus for handling radioactive materials, and heavy-water tanks with a purification system.[58]

Wirtz and his team arranged the 1-centimeter-thick uranium plates in the magnesium-alloy cylinder (1.24 meters wide by 1.64 meters high) and lowered the vessel into the pool of water, which acted as an absorber and reflector of neutrons. Then they filled the vessel with heavy water and measured the neutron flux as a function of radius when a small constant neutron source was lowered into the center of the device. After four trials, they found an optimal separation of 26 centimeters between the five uranium plates, which resulted in an encouraging 206 percent multiplication.[59] Wirtz and Heisenberg repeated the configuration as experiment B-7 in December 1944. This time, however, instead of using water as the main reflector and absorber of neutrons, they wrapped graphite around the container — the first time graphite was used in a German experiment. The neutron multiplication factor rose even higher, but not enough to hope that similar pile configurations would ever lead to a critical state.[60] The plate geometry had failed to achieve a chain reaction, but Heisenberg and Wirtz could console themselves: "With the layer experiments a satisfactory agreement was achieved with the theory, which was ever more refined in the course of time (Heisenberg, Weizsäcker, Höcker)."[61]

Meanwhile, the intensified bombing of Berlin was having an increasing impact on Heisenberg and his project. By the summer of 1943 the Allies were conducting round-the-clock raids. Max von Laue wrote to Heisenberg from Pomerania describing the "uncanny experience" of hearing bombers flying overhead for three solid hours — "uncanny mainly because nothing was done against them."[62] In July, Speer ordered all war research institutes to identify places to which they could move should the air raids make work impossible. Although the nuclear project was safely nestled in its new concrete bunker, there was no guarantee of a steady flow of electricity, water, and supplies to keep the project and its members going, nor were there living quarters within the bunker for all the project members. Heisenberg looked for a safe haven to the south and west in order to be closer to his family in Urfeld and because he wanted to be closer to the Western Allies than to the Soviet east when the war finally ended.[63] Physicist Walther Gerlach, who replaced Esau as administrator of nuclear research, had earlier taught in the university town of Tübingen to the southwest. Apparently on a tip from Gerlach, Heisenberg learned of a group of peaceful little villages nestled among the hills of the Swabian Alps region of the Black Forest, just south of Stuttgart and Tübingen.

As conditions worsened daily in the summer and fall of 1943, Heisenberg decided to dispatch to the south all personnel who would not be needed for the forthcoming B-6 series of experiments. By the end of the year, about a third of his 55-member Kaiser Wilhelm Institute staff, including assistant director Max von Laue, were settled in the Black Forest.[64] They moved into a large and nearly vacant textile factory at Weiherstrasse 1 in the picturesque town of Hechingen, where they set up offices and rooms for measurement of materials and construction of apparatus.[65]

The institute members were not alone for long, however. Otto Hahn, the codiscoverer of fission, and his staff moved to nearby Tailfingen after Hahn's Berlin institute suffered a direct hit and burned to the ground. Weizsäcker and his family escaped to Hechingen just ahead of the Allied bombing and capture of Strasbourg. But not until January 1945 would Heisenberg, Wirtz, and the remaining staff move permanently to their Black Forest retreat.

The splitting of his institute and the uprooting of his family put a new strain on Heisenberg. When not traveling to occupied countries or writing S-matrix papers, Heisenberg shuttled for the next year and a half by train and auto between his three widely dispersed homes in Berlin, Hechingen, and Urfeld.[66] The strain increased with the difficulties his family encountered in their new quarters. No sooner had Heisenberg's wife, six children, and their live-in friends recovered from their December bout with scarlet fever in Urfeld than they had to contend with a partially collapsed roof caused by a heavy snowfall. The mayor of Kochel, the nearest large town, was unable to release materials for the repairs, so Heisenberg had to obtain permission from the mayor of Leipzig to transport roofing tiles from the ruin of their Leipzig house to Urfeld.[67] In addition to the distrust of the Berlin professor's family by the local Bavarian peasants, Mrs. Heisenberg recalled the constant hunger and illness of their Urfeld existence, the latter caused by the cold (the house was intended only as a summer residence) and the former by the impossibility of vegetable farming (the house was situated on a rocky slope near Lake Walchen, at the foot of the mountains).

At first Heisenberg had intended to move his family to Hechingen, as other institute staff members had done, but the Berlin researchers who had descended on the small town ahead of him had snapped up the available family housing, leaving Heisenberg himself with nowhere to live but a rented room in the home of a befriended family. The Heisenberg family furniture, which had been trucked to storage in Berlin, now landed in the basement of the Hechingen family's house.[68] Isolated from the institute families and denied Hechingen's agricultural advantages, Mrs. Heisenberg came to resent her husband's "Swabian idyll," which, she wrote, "was always a slight cause of dispute between us." Heisenberg's musical concerts for the Hechingen residents surely did not help matters.[69] Nevertheless, as life became daily more dangerous and as Germany careened toward her inevitable defeat, their moments alone together in their Urfeld "eagle's nest" were ever more precious. Two weeks before the not-unexpected D-day invasion, Heisenberg wrote his mother from Berlin: "We experience each beautiful day that is given to us as a gift from the Good Lord, for which we are thankful," and when they were able to sit together for a few days there in Urfeld and listen to the play of the children, "we don't want to think of anything but this happiness, for these could indeed be our last moments together."[70]

Back in Berlin briefly, Heisenberg welcomed the news of the D-day invasion not for its likely end of the Reich but for its hastening the end of the war "one way or the other."[71] Later, in Hechingen, he waxed romantic in his thoughts about the future. After the war ended, he opined, "the sun will continue to shine as it has before, we will be able to make music and to do science, and whether or not we live richly or modestly, it will make no great difference."[72] But many hardships would be endured before the war ended, and the Allies were not likely to allow Germany's premier physicist to steal quietly away into an idyllic existence in some mountain retreat.

As British, French, and American troops smashed their way inland from the Normandy beachhead, and as British and American scientists worked feverishly at Los Alamos to assemble the first atomic bombs, Allied intelligence agents scrambled for reliable information about the progress of German nuclear research. Through reports received from Debye and refugee scientists as well as perhaps from inside sources, Allied scientists were only too well aware of the secret research under way at the Kaiser Wilhelm Institute in Berlin and of Heisenberg's role as head of the main reactor project.[73] The danger that Germany might actually succeed in building a bomb seemed so great that in 1942 two of Heisenberg's former colleagues suggested that the Allies kidnap him when he came to Zurich to lecture at the end of the year.[74] At the very least, they suggested, Wentzel in Zurich or Wick in Rome should "interview" Heisenberg in Zurich to extract whatever information he could about the German nuclear effort. Nothing had come of either suggestion by the time Heisenberg made his 1942 visit; the Allies had only just begun to establish an international intelligence network. But the suggestions did apparently inspire special interest in Axis science, especially regarding nuclear fission.

In 1942, soon after establishing the Office of Strategic Services (OSS), the precursor of the CIA, General William (Wild Bill) Donovan began recruiting agents for atomic intelligence. His top atomic spy was Morris (Moe) Berg, a man of many talents. An erudite, multilingual Princeton graduate who was familiar with physics, Berg had played until 1942 as a catcher for the Boston Red Sox baseball team. He came to Donovan's attention through his propaganda broadcasts to Japan — in Japanese — and through his diplomatic efforts in Latin America to counter anti-American Nazi influence. Donovan recruited the catcher to catch European intelligence.[75] Berg's first assignment was in Yugoslavia, then in Italy in 1944, where he was to operate behind enemy lines to capture nuclear physicists Wick and Amaldi in order to determine the progress of Axis science. Capturing the two proved unnecessary after the Allies broke through the Gustav Line and liberated Rome just before D-day. Berg was in Rome within days interrogating his subjects about German research.

Moe Berg's experience and abilities made him a natural to focus on German affairs. But OSS activities conflicted with those of other agencies of the armed forces, and Berg's mission to Italy conflicted with that of the so-called Alsos mission. General Leslie Groves, the military chief of the Manhattan Project, had already dispatched the small Alsos mission (*alsos* is Greek for *grove*) to Italy to capture Italian atomic scientists and materials as the Allied front lines advanced and to obtain from the Italians as much information as possible about an Axis bomb.[76]

In order to minimize conflicts, Groves's superiors placed him in command of all American nuclear intelligence. Late in 1944, Groves assigned Berg to Allen Dulles's OSS office in neutral Switzerland, and he dispatched a reconstituted Alsos mission to London for the Normandy invasion. As it had been in the past, Groves's mission was under the field command of Colonel Boris T. Pash, a Russian-American veteran of anti-Soviet battles known for his often ill-considered bravado. But the new Alsos unit included a scientific section for the first time — headed by Samuel A. Goudsmit. Goudsmit was chosen for his familiarity with European physicists, physics, and languages and for his lack of familiarity with the Manhattan Project, should he be captured. Accountable only to Groves, Goudsmit and Pash and their small entourage rolled across northern Europe with the Allied armies, confiscating and examining every scrap of material even vaguely related to German science and even vaguely hinting at the whereabouts of the premier German physicist who had failed to help when needed.[77]

Moe Berg, working in Zurich independently of Goudsmit and Pash, had already obtained a lead on Heisenberg's location. Because Switzerland, while neutral, was surrounded by Axis countries, spies of every stripe flooded the country during the war. According to one count, Germany alone had 23 organizations operating in northern Switzerland.[78] Berg established a liaison with Paul Scherrer, professor of experimental physics at the Zurich Polytechnic, who, as a fervent anti-Nazi, was eager to help Berg in every way he could.[79] From the postmark on a letter from Heisenberg to Wentzel in 1944, Berg learned of the Hechingen outpost and relayed the information to Goudsmit. But Berg's greatest service to his country was his role in an operation even more radical than the earlier proposed kidnapping. In a later account to a friend—no doubt embellished—Berg claimed that he was ordered to have Scherrer invite Heisenberg to lecture in Zurich. At the slightest hint that Heisenberg was constructing an atomic bomb, Berg, standing ready with a loaded pistol, was to assassinate the scientist.[80]

Although Scherrer may not have known of Berg's intentions, the first part of the scheme went off as planned. Heisenberg knew Scherrer well from their common interests in cosmic rays and from an earlier wartime visit to Zurich at Scherrer's invitation. Heisenberg readily accepted a second invitation to speak in December 1944 but insisted that he would lecture only on a nonpolitical subject. He knew he would be carefully watched by spies on both sides of the war and in general avoided public lectures on political issues. He chose to lecture on the fourth installment of his S-matrix theory.

Carl Friedrich von Weizsäcker, whose wife was Swiss, accompanied Heisenberg to Zurich, where about 20 people, including Berg and several pro-German Swiss scientists, attended the lecture in the Polytechnic's physics institute. Although Heisenberg was now practiced in treading warily before a lecture audience, he apparently managed to get himself into trouble at a private dinner party in the Scherrer home. Berg sat next to Heisenberg with open ears and a loaded pistol but was disappointed: Heisenberg's main indiscretion, later reported all the way up to Roosevelt, was a defeatist remark about Germany's failing war fortunes. This time there was no talk of a future Europe "under German leadership," but according to Goudsmit, who apparently heard it from one of the Swiss scientists, Heisenberg supposedly made an equally incriminating remark: "How fine it would have been if we had won this war."[81]

Pro-German spies reported Heisenberg's defeatist remark to the Gestapo, which brought it to the attention of the Berlin SS. Under the conditions of German total war, defeatism of any sort was construed as treasonous and could lead to the same fate as that meted out to Munich student antiwar protesters — execution. Unbeknownst to Heisenberg's family in Urfeld, the SS planned a full investigation of both Heisenberg and Weizsäcker. SS officer Mentzel informed Gerlach, and Gerlach warned Heisenberg. Fortunately, Gerlach was able to defuse the issue. When an SS general appeared in Gerlach's office to lodge the complaint against Heisenberg and Weizsäcker, Gerlach feigned horror at the charge and promised that Heisenberg would receive a severe reprimand for his behavior. That apparently satisfied the general.[82] Heisenberg had once again eluded danger and death. But the baseball catcher in Zurich had caught enough to steer Goudsmit and Pash to the area of the Black Forest just south of Tübingen.

During the weeks following Heisenberg's Zurich trip in December 1944, the massive air raids on Berlin continued unceasingly, and near panic gripped the city as the Soviet army rolled relentlessly westward. Amidst constant bombing, mounting rubble, and frequent power failures, Gerlach, who had moved his offices from Munich to Heisenberg's Berlin institute, finally ordered all nuclear pile research transferred out of Berlin. During his year as Göring's so-called plenipotentiary for nuclear research, Willy Wien's successor as Munich's professor of experimental physics had supported both of the competing pile designs, Diebner's cubes and Heisenberg's plates, until Heisenberg finally acceded to the "more favorable arrangement."[83] He wanted to see which of the two designs would prove more likely to produce a chain reaction before the war was over.

With the war just three months from its conclusion, at the end of January 1945 Wirtz and the remainder of Heisenberg's Berlin team had assembled their largest pile experiment to date: hundreds of cubes cut from the B-7 plates and suspended on aluminum wires from the lid of the reactor cylinder, which was then filled with the institute's 1.5 metric tons of heavy water. The vessel lay, wrapped in a mantle of pure graphite, in the institute's bomb-shelter water pit, ready for neutron multiplication measurements that would constitute experiment B-8.[84]

No sooner had they assembled this appliance, however, than Gerlach ordered it and Diebner's contraption dismantled and shipped south — better to have a slight delay in research than to have men and materials fall into Russian hands. Gerlach, Wirtz, and Gerlach's newly appointed assistant, Kurt Diebner, in German army uniform and with a revolver strapped to his hip, left Berlin the next day accompanied by several trucks. They headed for Hechingen, where Heisenberg awaited them, but they got only as far as Diebner's new outpost in Stadtilm in the province of Thüringen, about halfway across Germany toward Hechingen in the southwest. Gerlach abruptly decided to stop there and have the apparatus reassembled under Diebner's direction in the desperate hope of achieving a chain reaction sooner. A worried telephone call from Wirtz to Heisenberg in Hechingen brought Heisenberg and Weizsäcker to Stadtilm as quickly as the Allied bombing and strafing of Germany's transportation system would allow. Bagge's arrival from Hechingen with a convoy of trucks to transport Heisenberg's uranium and heavy water the rest of the way finally convinced the plenipotentiary of the preferability of a Heisenberg reactor to a Diebner reactor. Diebner had won Gerlach's favor for his innate experimental abilities, but Heisenberg was the more powerful. And with trucks at Heisenberg's disposal, Gerlach apparently did not want to risk further delay due to an unseemly fight over materials. Heisenberg's equipment and materials finally arrived in Hechingen at the end of February 1945, four weeks after they had left Berlin — and just two months before the end of the war.

During the final months of the war in Europe, Heisenberg's team, with Weizsäcker and members of Bothe's group, worked feverishly on what would be their last attempt to achieve a critical reactor. The work took place in the nearby picturesque village of Haigerloch, which, as Gerlach knew from his earlier days in Tübingen, offered ideal protection for the experiment. A huge rock formation, topped by a Renaissance church and monastery, dominated the center of the town. At the base of the rock a small cave had been dug horizontally into the side of the rock to serve as a wine cellar for the local innkeeper. It now served the scientists as an air-raid-proof "atom cellar."[85] Advance teams from Berlin had already disposed of the innkeeper's wine (one way or another), enlarged the cave, dug the water pit, attached water and power cables, and assembled the winches and heavy equipment needed to handle Germany's last pile attempt.

Even as Germany disintegrated in devastation and chaos, the scientists worked calmly and steadily at their task. The physicists' successful campaign in previous years to enhance the status of their science now worked to their advantage. When Hitler decreed total mobilization for all-out war in July 1944, a now less accommodating Speer ordered the cessation of all research that would not directly affect the outcome of the war.[86] Riding the success of the physicists' campaign, Gerlach managed to have the nuclear project exempted from the general ban and easily mobilized such previously unenthusiastic figures as Himmler and Bormann to help him do it. After they further encouraged Speer to allow the nuclear project to continue, Himmler ordered the release of another

14,600 scientists from active military duty, and Bormann excused nuclear researchers, including Heisenberg, from all but minimal participation in the Volkssturm, the people's army. This mobilization of the entire population was Hitler's last line of defense and a vehicle of his last desperate attempt to maintain control.[87]

Ironically, amid total war and facing imminent defeat, both the scientists and the Nazi leaders suddenly committed themselves totally to a small-scale technical research effort that at that stage seemed to offer no practical benefit to the war effort. At the beginning of the war, they had believed that nuclear engineering research could give them a new war machine and possibly even a new and immensely powerful explosive. Now, at the end of the war, they hoped only—but fervently—that the simple experimental model in Haigerloch would go critical before the collapse. The motives for this hope among both the scientists and the Nazi officials are difficult to pin down. In addition to scientific curiosity and the exhilaration that success would bring in the face of defeat, several indications suggest that both groups were operating on the mistaken belief that German research had advanced much further than had Allied research. No doubt the Nazi leaders hoped to use the secret of nuclear fission as a bargaining chip in negotiating a conditional surrender with the Allies. They were in for a nasty shock when the Allies expressed absolutely no interest in their offer.[88]

The scientists, especially Heisenberg, were also looking to the future in the last months of the war. They believed that Germany, as it had in the years following the defeat of World War I, would look again after this war to its leading science—and in particular to its leading scientists—as the remaining pillar of German competitive greatness, whatever defeats and humiliations might be suffered in other areas. Heisenberg wrote to a former student in April 1944: "It is indeed very important that after the war we take part once again in the competition of research."[89] Heisenberg also wanted badly to reassure himself that, as a successful scientist producing useful results—as would be proved by a working reactor—he and his profession would not be disregarded and abused, as they had been under Hitler, by whatever regime succeeded Hitler. These goals would be realized if they could achieve a chain reaction before the end of the war. In the end, however, the effort failed.

By early March 1945 Heisenberg and Wirtz and their team of technicians began the final assembly of experiment B-8 in the innkeeper's "atom cellar."[90] Two weeks later they winched the graphite-covered lid supporting the chains of uranium blocks into place over the cylindrical reactor vessel, then slowly filled the vessel with heavy water. As the pumping progressed, the neutron multiplication rate increased. It seemed to Heisenberg and Wirtz that at long last the pile might go critical! Then in the middle of their excitement came the sudden realization of their extreme peril—they had ignored all but the most rudimentary safety precautions. A block of neutron-absorbent cadmium was at hand, ready to be tossed into the tank should the reaction get out of control. Only now did the scientists begin to wonder seriously if that would be enough to halt the reaction in time. It is testimony to their determination—and desperation—that no one tried to stop the experiment. Everyone was determined that Germany should be the first to achieve a sustained chain reaction, regardless of the danger to themselves.

They watched nervously as the remaining heavy water flowed unchecked into the tank—but, alas, it was not enough. The experiment had yielded the highest multiplication rate yet, 670 percent, but Heisenberg quickly calculated that they still needed nearly 50 percent more uranium and heavy water for the reaction to be self-sustaining.[91] Perhaps more of both could be found at Diebner's Stadtilm outpost—but it was too late. American troops were already advancing through Thüringen in east-central Germany; by early April they were within miles of Stadtilm. Diebner finally abandoned his outpost on April 8 and headed south to join his mentor, Walther Gerlach, who had resettled in Munich.

On the same day, Secretary of War Stimson met with General Groves in Washington to decide what to do about the German scientists in the south.[92] Since entering northern Europe in 1944, the Alsos mission had absorbed every available bit of information about the German project from Joliot's Paris laboratory and from the papers left behind in Weizsäcker's hastily evacuated Strasbourg institute.[93] Goudsmit, Pash, and the Alsos team then crossed the Rhine with the Allied armies in February 1945. At the end of March, as experiment B-8 lay in its cave-protected water pit, the Alsos mission entered the old university town of Heidelberg, situated northwest of Stuttgart. As the U. S. Army set up a forward command post, the Alsos team established its "Advance Base, South."[94]

After seizing and interrogating Bothe and Gentner in Heidelberg, Goudsmit and his staff contemplated their next move. By then they knew the locations of all target scientists and laboratories and had relayed to Washington their conclusion that Hitler's promised secret weapons did not include an atomic bomb.[95] Groves demanded that they be absolutely certain and that they capture all the remaining project members. Unfortunately, most of their targets were within the areas earmarked at the Yalta meeting of Allied leaders for invasion and occupation by French and Russian forces. Mission priorities suddenly shifted as a result: instead of gathering intelligence in an attempt to thwart the German bomb effort, the Alsos team was now bent on snatching up German scientists, papers, and equipment before the Russians and the French could take them into custody.

The three main Allies had agreed that the entire region south of Stuttgart should be occupied by the French, but Colonel Lansdale, attached to the Alsos mission, had different ideas, as he reported several weeks later: "Our feeling was that the individuals and materials down there should be seized by the Americans in advance of the French, or if that were impossible, destroyed to the fullest extent."[96] Atomic scientists and equipment were simply too valuable to allow to fall into the hands of any other nation. Groves and Stimson, meeting in Washington, considered a full-scale American invasion of the south. An army operations commander, approached in Heidelberg by Colonel Pash, recommended an airborne assault or, at the very least, the carpet bombing of the entire region. But the French were advancing too rapidly for either plan to be enacted. In the end a local Heidelberg commander assigned a combat engineer battalion to the zealous Colonel Pash, who took off immediately with a convoy of jeeps and armored cars and arrived in Haigerloch and Hechingen less than an hour after the French forward line swept through the area on April 23.

Pash and his men promptly set to work. They arrested Wirtz, Bagge, Weizsäcker, and Laue, found and confiscated their papers and equipment, began dismantling the Haigerloch pile, and blew up the alloy containment vessel of the last German reactor experiment. They then moved on to Tailfingen to arrest Otto Hahn. By the time the French commanders realized what was happening, the Germans' heavy water and uranium were on their way to Alsos mission headquarters in Paris, and the prisoners and their papers were on their way to Heidelberg for interrogation and study by Goudsmit and company.[97]

Three important targets remained at large: Gerlach and Diebner in Munich, and "target number one" — Heisenberg — who, interrogations revealed, had left Hechingen for Urfeld shortly before Pash's arrival. In the last weeks of the war, with a nearly critical reactor, no way to obtain more materials, and defeat and occupation at hand, Heisenberg's first priority remained what it had been for the past year, even above family and personal safety: to ensure the survival of his staff and his equipment for the future. After burying the uranium cubes, to be retrieved later, he waged what he called a never-ending battle for the lives of his institute members. They were apparently endangered both by a lack of food and by the zeal of the local populace to fulfill Hitler's order to fight to the last German.[98] There were many instances across Germany of small-town lynch mobs going after anyone who counseled surrender to invading Allied troops.

As the French lines advanced toward Hechingen on April 19, Heisenberg installed his staff, along with whatever food supplies remained, in the textile factory basement for protection against bombing and artillery, then set out on the only transportation available, a bicycle, to attend to his second priority, his family. The Nobel laureate bicycled first to nearby Kleintissen, where his brother and family, whom he had seen only rarely since the outbreak of war, had settled for the duration. After staying with his brother for a few days, he then embarked on an incredible marathon bicycle trip all the way across war-torn southern Germany to Urfeld, a distance of about 250 kilometers (150 miles). Pedaling only at night to avoid marauding German army units and low-flying Allied aircraft, both of which shot at anything that moved, he made it to Urfeld in an amazing three days.

There the situation was desperate. Eisenhower had ordered U.S. Army units to turn from their advance toward Berlin and head south in a vain search for the purported "Bavarian redoubt"—a rumored stronghold where Hitler's most fanatical followers would make a last stand. With the complete breakdown of civilian and military order, Waffen-SS units retreating from the American advance were rampaging through the Lake Walchen area in a last frenzy of pillage and murder. One night they hanged 17 soldiers from a German recuperation company in the woods near the Heisenberg home for "desertion."

The Heisenberg family situation was also dire. Mrs. Heisenberg was still fighting a losing battle with family illnesses, lack of food, and house repairs when her husband suddenly reappeared. Soon after he arrived, one of his sons became so ill with what appeared to be appendicitis that Heisenberg had to drive him over snow-covered, bombed-out roads to the nearby military hospital that had just lost 17 of its patients. The doctors determined that the ailment was not appendicitis.[99] Heisenberg also managed to move his aged mother to Urfeld from her apartment in Mittenwald. Scouring the nearby village, Heisenberg gathered a stockpile of groceries and fuel for the family before settling in to await the end of the war and the arrival of the American Seventh Army.

The end for Heisenberg was different from what he expected. On April 30, the same day Hitler and Eva Braun, his new bride and former mistress, killed themselves in their Berlin bunker as Russian troops closed in on them, U.S. Army headquarters in Heidelberg dispatched two teams to Bavaria in search of the remaining scientific targets. One team, led by the Heidelberg Alsos commander, was to locate and capture Gerlach and Diebner in Munich; the other, in an "alpine operation" under the command of the indomitable Colonel Pash, was to capture Heisenberg in Urfeld, which was still under enemy control.

In a report to Washington and later in an action-packed monograph, Pash described the execution of what he regarded as "the most important single intelligence mission of the war."[100] Pash's task force of ten men and four vehicles arrived in Bavaria on May 1 and advanced to the town of Kochel, which lay on the opposite side of a mountain from Urfeld. A reconnaisance the next day revealed that a bridge on the road around the mountain to Urfeld had been destroyed, cutting off the road to vehicles. Determined that "nothing was going to stop me from getting to Urfeld that day," Pash led a foot patrol over the snow-covered mountain. Exhausted from the climb, they arrived in town by late afternoon and promptly engaged in a shooting match with a small German force, killing two Germans and scaring off the rest.

While Pash and his patrol were holding their positions in Urfeld, two high-ranking German officers rode into town on motor scooters and attempted to surrender their battalion-sized unit to them. Obviously outnumbered, Pash bluffed his way out by demanding that the officers bring their entire force into town the next day; then he beat a hasty retreat to Kochel. That night the combat engineer unit repaired the bridge to Urfeld, and at 6:00 A.M. on May 3 Pash's team rolled into Urfeld, followed later that day by an infantry battalion from the Kochel area to take the German forces prisoner. Arriving in town and deploying his team, Pash and two of his men climbed the hill to Heisenberg's cabin and found their target sitting calmly on his veranda overlooking the lake. Heisenberg politely invited them in, introduced them to his stunned wife and curious children, who had obviously not expected their husband's and father's arrest, or at least not before the Americans had captured the Urfeld area.

As Heisenberg quietly gathered his belongings and papers, the sound of small-arms fire sent Pash rushing out the door and down the hill, waving a pistol in the air as he ran. A small German unit had attacked and quickly retreated. Fearing imminent attack by a much larger force, possibly the surrendering battalion, Pash loaded his prize, along with his papers and belongings, into an armored car and hastened with his men back to Kochel. The next morning Heisenberg commenced a bone-crushing trip to Heidelberg in the back of Pash's jeep. Three days later, on May 7, General Alfred Jodl, chief of operations of the German High Command, and Admiral Hans-Georg von Friedeburg, German U-boat commander, signed the instruments of unconditional surrender at Reims, ending the war, the Third Reich, and German wartime uranium research. Heisenberg would not see his family again for over eight months.

26

Reconstructing

The perceptions and rationale that Heisenberg and many other Germans accepted before and during the war had enabled them to continue their work and daily lives under the circumstances of the German Reich at war. The rationale and perceptions suddenly ceased to function once the circumstances no longer obtained. They were replaced with a new structure based on the past as reconstructed for the world in which they now lived. The Allies, sweeping across Germany and the rest of Europe in the first half of 1945, brought the long-awaited collapse of the Nazi dictatorship. Their arrival also brought an end to German nuclear research and captivity to the German nuclear scientists. VE day in May was followed three months later by the capitulation of Japan under the shadows of the mushroom clouds of Hiroshima and Nagasaki.

The postwar era brought with it the realization of two terrible truths. The first came with the disclosure to the world of the utter depravity of the Nazi regime, exemplified by the unspeakable horrors of the Nazi death camps. The second truth to be faced was the awesomely destructive fury of nuclear weapons, a fury unleashed through scientific research. Both of these truths had ramifications that went far beyond the immediate experience of the war, changing forever our perceptions of human progress and human potential. They taught us to be skeptical of so-called modern, enlightened societies, however cultured, and to be wary of modern science, however promising. While each person, and especially each nuclear scientist, struggled after the war to come to terms in his or her own way with one or both of these terrible lessons,

Heisenberg having the most to explain, took a leading role in publicly articulating the reactions of leading German scientists.

As newspapers around the world blazoned reports of Nazi atrocities uncovered by Allied troops during the spring of 1945, the public remained unaware of nuclear weapons until they were used on Japan in August. The German nuclear scientists, who thought their research equal to, if not ahead of, Allied research, supposed that the Alsos mission that captured them in early May was merely an attempt by the Allies to tap Germany's superior knowledge. The day after Heisenberg arrived at the Alsos outpost in Heidelberg, he was ushered to an interrogation on his work by his erstwhile colleague, Samuel A. Goudsmit, the scientific head of the Alsos mission.

Heisenberg and Goudsmit had last seen each other in Ann Arbor shortly before the war. Much had happened since. Now, as Goudsmit faced the man he had looked up to as a young physicist but who had apparently made little or no effort to rescue his parents from a Nazi death camp, the German physicist seemed to him despicably haughty and self-involved. Heisenberg, for his part, seemed to welcome the attention the Allies accorded him for his wealth of nuclear knowledge. The extraordinary efforts that Pash had made to arrest him had no doubt reinforced such pretensions. When asked about his nuclear research, Heisenberg was so confident of its significance that he offered to instruct the Americans on uranium fission.[1] Goudsmit, knowing of Allied progress, though not about an imminent bomb, politely thanked him for the offer.

On impulse, Goudsmit repeated his question of six years earlier: "Wouldn't you want to come to America now and work with us?" Heisenberg repeated the answer he had given earlier: "No, I don't want to leave. Germany needs me."[2] To Goudsmit, this seemed further evidence of Heisenberg's overweening self-importance. But with most of Germany in ruins and her economy near collapse, Goudsmit could hardly have expected any other response at that point from a man so attached to his country.

Heisenberg was interned. Of the fourteen leading nuclear physicists rounded up in the flurry of Alsos strikes, four were sent—willingly or not—to the United States to help with American research. Goudsmit remanded the rest—including Hahn, Laue, Weizsäcker, Bothe, and Harteck—to American military authorities, who held them incommunicado for two months in a series of prisoner-of-war camps in France and Belgium. Their families had to fend for themselves. Heisenberg, Diebner, and Gerlach joined their seven fellow prisoners at a camp near

Versailles, known as "Dustbin." Despite this inauspicious omen, the Allied military treated its scientific prisoners astonishingly well, providing them with adequate food, English-language newspapers, weekly physics colloquia, and a jogging track.[3] Nevertheless, the reason for their internment in the first place — to allow the Allies "to catch up" — seemed to the scientists hardly sufficient to keep them so long. Laue in particular could not understand why he, who had not worked on fission, should be held against his will. To their inquiries, a British officer replied only that they were "detained for His Majesty's pleasure."[4] Thereafter they called themselves the "detainees."

Scottish physicist R. V. Jones, professor of natural philosophy in Aberdeen and head of intelligence for the British Air Staff, had been following German science since the start of the war and had assisted the Alsos mission on both occasions when it landed in Europe — the first in Italy, the second in Britain. But America's irksome decision in the last months of the war to exclude the other Allies from sharing the mission's nuclear booty inspired Jones and his staff to begin looking out for their own interests. When an American general reportedly expressed the opinion that the best solution to the problem of German nuclear physics was to shoot all the German nuclear physicists, Jones took action.[5] Not only were executions, or even war-crimes trials, out of the question, but the British seemed in awe of the prestigious detainees. Jones graciously offered to relieve the Americans of responsibility for the physicists. Apparently wanting not the physicists themselves but only their silence, the Americans agreed — on condition that the scientists be kept out of Russian or French control. The Russians and especially the French were already diverging from the British and Americans on postwar policy toward occupied Germany.

Fearful that the prisoners would be captured by the other Allies or sent to the United States if they remained on the continent, Professor Jones decided to move them to Britain. As an intelligence chief, he knew of a country safe house in the tiny village of Godmanchester near Cambridge and a large Allied air base. The house had been used by MI-6 agents as a staging area for parachuting into German-occupied territory. In early July, after outfitting the house with secret microphones, Jones had his ten German scientists flown under heavy military guard from their camp in Belgium to their new home in England. There they remained until Jones figured out what to do with them.

The British knew they could not hold the scientists forever, but they did not want to turn them loose in England for fear that they might learn too much about British research from less security-conscious colleagues. By the end of the year, the British had decided that only a revival of the German economy and a measure of political and cultural autonomy in the British zone of occupation were consistent with British commitments and German social and political stability. Science and technology were envisoned as crucial elements of the intended revival. With the dust of the atomic blasts now settled and the British zone firmly under British control, on January 3, 1946, a British transport plane flew the detainees to less restricted detention in a northern German town in the British occupation zone. There they could move about during the day, but they had to return to British quarters at night. Within several months, they were all released. Most, including Heisenberg, settled in the undisturbed university town of Göttingen, intended by the British authorities to be a crystallization point for the revival of West German science.

The detainees could hardly complain of their Godmanchester prison. The jail actually consisted of a sumptuous English country manor, known as Farm Hall, which was located on a rolling grassy estate surrounded by flowering hedges, large trees, and an unobtrusive fence. For their leisure-time pleasure, several tennis courts were located in the rear; a well-tuned grand piano stood in the parlor; and they had books, newspapers, game paraphernalia, and a radio—even the *Physical Review*. The prisoners whiled away their hours with relaxation, lectures to each other on their nonnuclear work, and walks about the grounds. Two British officers, carefully chosen for their ignorance of Allied research, watched over the prisoners, providing them with new clothes, shoes, and hearty English meals. The royal treatment prompted one ungracious officer to comment that the prisoners were living better than the average English family—to say nothing of the average German family or the average family across most of war-torn Europe.[6]

The only real complaint of the detainees was that they were not permitted to communicate with their wives and families. All contact with the outside world was prohibited until the evening of August 6, 1945, when they were abruptly made aware of the reason for His Majesty's pleasure. On that evening, the British officers joined the scientists as usual in the manor dining room where the evening meal was served punctually at 7:45 P.M. Dinner conversation, however, was far from usual. The British officers calmly told the scientists that the Americans had reportedly dropped what they called an atom bomb on Japan. A brief BBC radio message seemed to confirm this. Pandemonium reigned. The shocked and disbelieving scientists huddled around the radio at 9:00 to hear a more detailed report from the BBC—but those crumbs of information only deepened their perplexity. If the Allied scientists had really been successful—and it seemed they had—then German nuclear superiority had been a mere fantasy.

The news was devastating. Each man responded to it in his own way. Walther Gerlach, the last Reich physics head, behaved like a routed general and apparently suffered a nervous breakdown of sorts. Heisenberg and Weizsäcker, who shared a bedroom next to Gerlach's, feared he might attempt suicide and looked in on him that night to assure themselves of his safety. The angered younger physicists, long chafing at the bottom of the power hierachy, accused their elders of mismanagement; Hahn and Laue, initially shaken by the news, washed their hands of the whole affair; Heisenberg set to work calculating.[7] According to Jones, he concluded that the bomb would have to have contained several tons of uranium—quite a load for a plane to deliver. Heisenberg thought that the bomb might in fact have been a hydrogen bomb.[8] Apparently, he had never determined in detail the actual size of an explosive.

The news of a successful Allied bomb turned a glaring public spotlight on two painful issues for the Germans—one acutely embarrassing, the other more profoundly distressing. Transcripts of conversations recorded that evening and the next day by Jones's hidden microphones—and hitherto released only in very brief excerpt and faulty translation—quote Hahn at the dinner table succinctly summing up the first problem: "If the Americans have a uranium bomb, then you're all second-raters. Poor old Heisenberg."[9] As Laue expressed it to his son the next day: "The main question is naturally why we in Germany did not achieve a bomb."[10] Why was the German achievement, whether or not they were ultimately aiming for a bomb, so slight in comparison? The second question has been asked ever since: In view of the incredible death and destruction wrought by both the bomb and the Hitler regime, what moral scruples, if any, did German nuclear scientists bring to their wartime work? It is a question that can be asked of both sides of the war.[11]

British news reporters were already formulating their own anti-German answers to these questions, which the detainees found appalling both for their inaccuracies and for the damage they were doing to German reputations. The scientists spent the entire day after Hiroshima pondering and debating their position—an indication that they had never jointly articulated one during the war. Some of their conversations were recorded, others were not. Heisenberg put together his response with the diplomatic Weizsäcker, his closest and most trusted colleague among the scientists, while strolling unmonitored on the grounds after lunch. By evening they had settled on their positions, and in view of negative press reports, the older scientists decided to present their side of the story in a press release. Gerlach and Wirtz assisted Heisenberg in drafting the statement, an early version of which, in Heisenberg's hand, survives on the pages of an English military school exercise book.[12] The following day, a final draft was formulated, typed, signed by all, and handed to their captors. It was probably never released to the public.

One of the younger physicists, Erich Bagge, wrote in his diary that "the story [in the press statement] found wide-ranging but not complete acceptance." It had been signed by all only after difficulties with the younger physicists had been resolved.[13] Apparently some, including Bagge and his superior, Diebner, objected to the statement's special pleading and facile exoneration of the older scientists. Nevertheless, Max von Laue, who remained aloof from fission research, apparently joined the other elders in endorsing the statement. He repeated the gist of the argument in a letter on August 7 to his son, who was sequestered in Princeton to avoid the German draft. In late September, he forwarded a copy of the statement to his son for distribution in the United States.[14] The Farm Hall statement has served ever since as the foundation of the German scientists' position regarding their wartime work, what one historian has called a self-serving apologia.[15]

To explain why Germany never achieved even a chain reaction, let alone an atomic bomb, Heisenberg and the Farm Hall elders argued that under wartime conditions in Germany, their economic, technical, and material resources were woefully insufficient and that no one wanted to demand more aid since they did not believe they could achieve their goal before the war ended. Under the conditions of total war, the regime would invest massive resources only if it were assured of immediate success. When the German effort in the context of the uneasy relationship between science and government in Germany is compared with the enormous Allied effort and the close collaboration between scientists and the military on the Manhattan Project, the argument is a relatively accurate statement of the reason the German project was so unsuccessful. It would have been enhanced by acknowledgment of the damage done to German science by Nazi policies — as the physicists themselves had argued during the war.

The German scientists were not satisfied with this explanation. With moral issues being raised by British reporters and their competence at stake, they regarded short-sightedness as indismisable and material conditions as insufficient for their failure. They now added a moral dimension. According to General Groves's excerpts from the Farm Hall transcripts of August 7, Weizsäcker had insisted over Hahn's objections that they had been driven by moral scruples to concentrate on building a machine rather than a bomb. As quoted by Groves, Weizsäcker declared: "I believe the reason we didn't do it was because all the physicists didn't want to do it, on principles. If we had wanted Germany to win the war we could have succeeded."[16]

There is no evidence to support Weizsäcker's claim. Perhaps speaking only for himself, Max von Laue nevertheless echoed the ethical sentiment of the resulting Farm Hall statement in a letter to his son that same day: "All of our uranium research was directed toward the achievement of a uranium machine as an energy source, first because no one believed in the possibility of a bomb in the foreseeable future, and second because fundamentally no one of us wanted to put such a weapon in Hitler's hands."[17] The Farm Hall statement is far less explicit on the second issue, but the syntax implies the influence of scruples nonetheless. After describing the difficulties of 1942 that led them to give up any hope of a bomb, the next sentence of the statement simply reads: "The further work therefore concentrated on the problem of the machine, for which in addition to uranium 'heavy' water is necessary."[18] There was no mention of efforts at large-scale isotope separation or of the plutonium alternative.

To comprehend more fully the position Heisenberg and his compatriots presented in their Farm Hall statement and its derivatives — "so violently debated in all scientific circles ever since," as Groves put it — one needs the perspective afforded by hindsight.[19] First of all, whatever their failings as scientists and as citizens, German scientists were not solely responsible for the moral character of their country. It is true that, as noted earlier, the mere fact that these world-renowned scientists continued to live and work in Germany after the moral and political affronts of early years left them already politically and morally compromised and lent the regime an unwarranted and false credibility. They compounded their failing by continuing to seek out and to accept collaborative accommodations with the regime — a regime that continually demonstrated its utter contempt for decency of any sort from the very beginning and clearly held the scientists and their science in outright contempt. Nevertheless, Heisenberg and his colleagues were not any more — or any less — responsible than other Germans for the unspeakable atrocities of the Third Reich.

Nor were they alone in their unprincipled eagerness to prove their value to their government by creating the weapons of war. Ever since Archimedes built catapults for the king of Syracuse, ever since Bacon declared that knowledge is power, science has been the handmaiden of every nation's economic, military, and political interests. Not until after World War II — after the advent of weapons of mass destruction and after the example of German scientists so willingly working for their government — have moral scruples really played any role in the willingness of scientists to arm their respective nations. Recent studies of the ways in which American physicists allowed themselves to be manipulated to serve cold-war military aims suggest that fundamental moral or ethical issues — those beyond feelings of patriotism and the desire to defend one's culture — still do not play a significant role in contemporary weapons research.[20]

The postwar quandary of Allied scientists points up the dilemma. Some of the Allied scientists who had devoted their best efforts to the creation of not one but two atomic bombs were profoundly shocked by the destruction they had wrought, and they suffered great moral and emotional anguish because of it. For instance, like many others, Heisenberg's alter ego, J. Robert Oppenheimer, the scientific head of the Manhattan Project, was driven as much by desires to succeed in building the bomb and proving the value of physics to his government as he was in defending the world against Nazism. Even after he and his collaborators had built the weapon and proved the utility of physics and even after Germany had been totally defeated, he willingly helped select the Japanese cities to be targeted. But no sooner had the radioactive dust settled on Hiroshima and Nagasaki than he and others began to realize what they had done and to experience moral pangs about it. Many of the scientists immediately joined in an effort to control nuclear weapons, and Oppenheimer now joined with several others in an effort to oppose the building of an even more destructive weapon, the hydrogen bomb, largely on moral grounds.[21]

A number of Allied scientists consoled themselves with the reflection that they had built the bomb to counter the far greater evil of a bomb in Hitler's hands. While certainly valid, this rationale could not disguise the fact that their efforts had resulted in the deaths of hundreds of thousands of civilians. The Smythe report, the official account submitted to Congress on the Allied nuclear effort, attempted to counter congressional criticism by stating the obvious: "This weapon has been created not by the devilish inspiration of some warped genius but by the arduous labor of thousands of normal men and women for the safety of their country."[22]

Amidst the throes of their own moral anguish, many of those men and women were simply appalled to perceive a conspicuous lack of any similar soul-searching among those on the other side. They perceived a shocking failure on the part of the German scientists to acknowledge that they too had been working as well as they could on nuclear fission for their country and, like scientists everywhere, had been willingly exploited by their leaders.[23] On the contrary, the Germans had the audacity to congratulate themselves for the fact that they had not built the bomb! They were even beginning to claim that they had avoided building the bomb out of principle, apparently unconcerned by the fact that, apart from the question of a bomb, they had been working nonetheless for one of the most depraved regimes that ever existed. American physicist Philip Morrison probably spoke for many of his colleagues in 1947: "No different from their Allied counterparts, the German scientists worked for the military as best their circumstances allowed. But the difference, which will be never possible to forgive, is that they worked for the cause of Himmler and Auschwitz, for the burners of books and the takers of hostages."[24]

With the help of his fellow detainees, Heisenberg expounded the Farm Hall position in the months and years following their return to Germany. The story became sharper as it spread farther. In December 1946 British authorities allowed Heisenberg to publish a summary of German nuclear research. An article appeared in the German journal *Naturwissenschaften*, with a partial translation in the British journal *Nature*. Beginning in 1947, Heisenberg sat for a series of interviews by German newspapers and by the science editor of the *New York Times* in response to Goudsmit's reports on German war research and the Alsos mission.[25]

In each of Heisenberg's accounts, early 1942 is depicted as the turning point. During the previous year, 1941, Heisenberg's Leipzig team had proved that an atomic bomb was possible in principle, and his Berlin team had learned that a reactor could breed plutonium for a bomb. But the technical hurdles to be overcome were still enormous and costly. By 1942, as the war situation worsened, the Army Ordnance Office had decided to forgo most of its nuclear research effort, since the office could not be sure that the effort would lead soon to a weapon. Because of that decision, as well as the reduced capacities of German industry and the technical obstacles still to be surmounted, "all hope of making bombs was given up," Heisenberg told the *New York Times*.[26] Although Heisenberg had tantalized Reich officials in early 1942 by hinting that a running reactor would produce equally fissionable plutonium, after the war he insisted that he only wanted to ensure their continued support. He dampened any expectation of an imminent weapon by stressing the technical difficulties. The strategy worked, he said — even though such a strategy was unnecessary, since he seems to have believed that the difficulties really were enormous.

For Heisenberg, the decisive meeting occurred with Speer on June 6, 1942. After the meeting, Speer ordered that the project be continued only on a modest scale but that the researchers should work, in Heisenberg's postwar words, for "the only attainable goal": "to build an energy-producing uranium burner for powering machines."[27] Of course, the production of a burner to power, say, submarines would have been no small contribution to the war effort. But in his 1946–1947 accounts, Heisenberg also seems to imply the Farm Hall ethic — that the scientists had made a conscious decision not to build a bomb for Hitler and to deter the regime from ordering them or anyone else to do so. As he put it in the pages of *Nature:* "The German physicists had consciously worked from the very beginning toward maintaining control over the project, and they used . . . their influence to direct the work in the sense depicted in this report."[28] Beyond that, even if they had not decided against building a bomb, they were still immune from postwar moralism. Since the project had never progressed much beyond its status in 1942, Heisenberg wrote in 1946, he and his colleagues were therefore conveniently spared "the difficult moral decision" of whether or not to build atom bombs for Hitler.[29] Scruples, whether invoked or not, were simply unnecessary.

Heisenberg's preposterous account parallels but overinterprets actual events. He especially did try to maintain scientific control over the project. He was also aware of the theoretical possibility of a nuclear explosive by late 1941, he did not demand a crash research and development program to build one, and he did seem content to work for the rest of the war on the more modest goal of building a reactor. It is difficult to assess his intentions and motives beyond that. But from what we know of his activities and research, there is nothing to support the notion that Heisenberg actually hindered the project in any way to keep an explosive out of Hitler's hands or even that he himself had that much control of the situation. Moreover, the feverish effort at the end of the war to construct a reactor must be attributed to both curiosity and the plain desire to succeed. Had they succeeded earlier in producing a chain reaction, Heisenberg and his coworkers would surely have advanced to the next step: extracting weapons-grade plutonium from their machine.[30]

The perceived technological and industrial difficulties, and Heisenberg's assessment that the war would be over before they could develop a bomb, seem far more decisive than any scruples he may or may not have had. If he did not inform the authorities in 1943 of the possibility that a working reactor would produce weapons-grade plutonium, it seems to have been solely to protect himself and his project from orders to build the bomb immediately. Failure to do so could have been construed as sabotage of the war effort — with obvious consequences.

In a draft version of his 1946 report for the German and British journals, Heisenberg implied more overtly that the control exercised by his group over the project was intended to keep it out of the hands of "other less scrupulous physicists" — as publicist Robert Jungk later called them — who "might in different circumstances make the attempt to construct atom bombs for Hitler."[31] This is the courage that many of Heisenberg's supporters saw in his taking control of the project. No doubt Heisenberg was referring to his chief rivals in fission research, Kurt Diebner and Diebner's supporters: Esau, Schumann, and the Army Ordnance Office (HWA). As indicated in Chapter 24, an administrative turning point of the project also occurred in 1942 when the Kaiser Wilhelm Society regained de facto control of the Kaiser Wilhelm Institute for Physics from Diebner and the HWA, and Esau's days as head of physics research in the Reich Research Council (RFR) were numbered.[32]

If these were the "less scrupulous" physicists Heisenberg had in mind, the argument does not cohere. If Heisenberg had concluded that a bomb was unattainable before the end of the war, then what did he have to fear from his rivals? Even if the more technically adept Diebner and his supporters had somehow managed to exclude Heisenberg entirely from fission research, it is difficult to imagine them moving significantly faster than Heisenberg and his collaborators to develop a reactor, much less a bomb. Before 1942, Heisenberg perhaps did have more cause to fear his rivals, but competition concerned more general administrative and ideological issues, not the direction of one particular project. At that time, he was pursuing his strategy of using uranium research to enhance the prestige of modern physics and himself in the eyes of the regime; losing control of his research teams would defeat that strategy. But his control over Reich uranium research was never actually as complete as he could have wished. Before 1942 the HWA exercised its military authority through Diebner, who was its representative. Later, it was Mentzel and the RFR who oversaw uranium research through their representative, Gerlach, and right up to the end, Gerlach chose to give equal support to the Diebner and Heisenberg research teams.

After the war, Heisenberg and his colleagues had good reason to portray their project as they did. To reestablish German science, to ensure that scientists could never again be disregarded and abused by their government, and to counter public criticism of their wartime behavior, it was essential that they acquire as much influence as possible, first in the British zone, then within the emerging West German government. Emphasizing again the prestige and utility of nuclear research and technology was the surest means of establishing themselves as vital to Germany's science and to her economic revival. To realize their goal, suspicions of their having worked to arm Hitler with nuclear weapons had to be addressed. The Allied occupation forces had placed denazification and the control of nuclear energy at the top of their priorities list. Heisenberg, not a party member and ideologically victimized by Stark, sought and gained the confidence of the occupation authorities in the matter of denazification.

In addition, after the war Heisenberg and his supporters took great pains to distance themselves from former Army Ordnance researchers and from anyone else who openly admitted working toward the goal of an atomic bomb under Hitler — even if actual work never proceeded beyond the prereactor stage. During the late 1940s, Heisenberg's circle also began an intensive public campaign to establish a German nuclear power program, a campaign that continued until the rescinding of Allied science control laws and the granting of sovereignty to the West German Federal Republic in 1955 as part of the NATO alliance. With West German self-rule imminent after 1950, Heisenberg and the nuclear scientists pushed for the establishment of a cabinet-level ministry for nuclear energy policy. At the same time, they now acted to mobilize public opinion against the acceptance by the German government of NATO plans to equip the West German army with tactical nuclear weapons.

The scientists succeeded on both scores. While ensuring that the West German army would remain nonnuclear, they successfully negotiated with Washington for permission to begin a full-scale nuclear reactor program, a program that by the late 1960s was the most successful in the world. West Germany was then the leading exporter of nuclear technology. Heisenberg later wrote with satisfaction in his memoirs: "The fact that in wartime no attempt was made in Germany to construct atom bombs, although knowledge of the principles existed, probably had a favorable effect on these [Washington] negotiations."[33]

If no actual attempt was made to construct an atom bomb in Germany (regardless of whether a reactor was or was not intended as the first step in that direction), a strong difference of opinion emerged between German and American scientists as to why the attempt was not made. The loudest and most divisive debate occurred between Heisenberg and the former Alsos science head Samuel A. Goudsmit, then professor of physics at Northwestern University. Goudsmit offered his highly influential views in a series of articles and in a monograph, widely read among American scientists, entitled *Alsos*. Their debate raged through the pages of the *New York Times* and in an exchange of long and fascinating letters.[34]

In many ways Goudsmit was bitterly disillusioned concerning Germany, German science, and one German scientist in particular, Werner Heisenberg. Moreover, the broader concerns that he and his colleagues faced regarding science in the United States were quite different from those the Germans were facing. As the cold war deepened, the paramount issues for American scientists were those of secrecy, administration, and the relationship between science and the military. Goudsmit expressly intended his account of the failed German project — "failed" apparently because it did not produce an atomic bomb — as a case study of what can go wrong, an example of "how incompetent control (which is not restricted to totalitarian countries) can kill scientific progress in a short time."[35] If Heisenberg was arguing the competence and success of the German scientists in perserving their science and their scruples under Hitler, Goudsmit was arguing just the opposite — each, in part, for his own contemporary audience. And indeed each audience has tended ever since to subscribe to the respective views Goudsmit and Heisenberg offered.

According to Goudsmit, a variety of factors caused the death of science in Nazi Germany. Nazi racial doctrine removed essential personnel from the laboratory and the classroom and weakened the scientists' adherence to fundamental scientific theories. The organization of German science and its support systems was disastrous in its lack of coherence and cooperation. The scientists themselves, who had grown accustomed to leading the world in modern science, became convinced that their superiority was absolute and therefore grew complacent: if they could not make an explosive uranium-235 bomb, neither could the Allies. And finally, said Goudsmit, the German scientists indulged in an excess of hero worship, such as that practiced by "the smug Heisenberg clique," that overlooked less heroic but more practical-minded technicians, such as Diebner or the self-made Manfred von Ardenne.[36]

The German researchers had concentrated on a reactor because they believed that, uncontrolled, it would eventually explode. But even then, they believed that the Allies were far behind them. In Goudsmit's opinion, the Germans had completely missed both fast-neutron fission and the plutonium alternative. If they had seen them, they, like the American scientists, would have pressured their government for more support. Thinking themselves far ahead, wrote Goudsmit, in actuality German scientists had only the vaguest notions of how a uranium bomb or even a reactor actually works, as shown by the lack of control rods in their experiments. They were obviously far behind the Allies in such technical efforts as isotope separation and moderator testing and production.

Heisenberg vehemently objected to Goudsmit's account on nearly every score. In long exchanges with Goudsmit, in letters to and interviews with the *New York Times*, and through C. F. von Weizsäcker and B. L. van der Waerden, then in the United States, Heisenberg vigorously maintained the advanced state of German war research.[37] Possibly through his American uncle Karl, who had lived in New York, Heisenberg gained the backing of Waldemar Kaempffert, the German-American science editor of the *New York Times*. In an interview by Kaempffert in response to Goudsmit's *Alsos*, Heisenberg, speaking "with an objectivity that is convincing," insisted that the destruction of German industry and unresolved technical problems forced the German scientists to give up "the idea of devising an atomic bomb and to concentrate on the development of atomic power for industry."[38] Three days after the interview appeared, Goudsmit wrote a letter to the *Times* taking issue with Heisenberg's account. "Heisenberg stresses the lack of industrial resources during the second half of the war. The book, 'Alsos,' points at the lack of vision of the German scientists." Kaempffert angrily replied that "liars do not win the Nobel prize" — a remark that prompted Goudsmit's publisher to inquire of Einstein whether in fact Nobel laureates do lie.[39]

Of course, there were glaring errors in Goudsmit's sometimes angry and sometimes oversimplified account of the German wartime research effort, but Heisenberg in particular was concerned that the research effort should be seen not only as ethically untainted but also as greatly successful. He had made it his personal mission to preserve the high quality of modern physics in Germany despite the adverse conditions. He and his colleagues could not afford to appear to be incompetent fools if they were to be influential in West German science affairs; the more they were thought to have known about atomic bombs, the more noble they would seem to contemporaries for not having attempted to build them. He defended the obvious hero worship, the formation of a clique around himself, as a means of excluding "unscrupulous persons" from influence on the course of uranium research. Heisenberg's 1941 visit with Bohr was now described as an effort to convey to the Allies that the Germans knew about the bomb but would not pursue it. They would work on nuclear energy only to gain funding and recognition and to save young physicists from the draft.[40]

Bohr's views on that visit now became crucial to Heisenberg's case, and Uncle Karl again assisted his nephew. He had earlier befriended Bohr during one of Bohr's many fund-raising trips to the United States. With his uncle's help, Heisenberg managed to reestablish contact with Bohr and received permission — from both Bohr and the British — to travel to Copenhagen in 1947. With his wartime motives and behavior in question abroad, Heisenberg apparently wanted to discuss the situation with the influential Bohr, but more importantly, to learn what Bohr remembered of their controversial 1941 encounter. Accompanied by an Allied control officer, Heisenberg made the trip just as Goudsmit's book appeared in the United States.[41]

After all that had occurred in Denmark during the war, and probably still angry that Heisenberg had led him astray during the 1941 visit, a cordial Bohr proved less supportive than Heisenberg had expected. Bohr flatly refused to discuss the details of the visit, and Heisenberg did not report much of what transpired during this encounter. Bohr had taken the 1941 meeting, Heisenberg reported, merely as an indication of German progress on nuclear fission research.[42] Brushing Heisenberg off, Bohr told him to get in touch with Goudsmit; Heisenberg would have to refute Goudsmit without Bohr's backing. A year later, the diplomatic B. L. van der Waerden, Heisenberg's selfappointed "attorney," composed an English aide-mémoire on the German position and presented it to Bohr.[43] There is no record of a response. Bohr's relationship with Heisenberg was civil but strained thereafter.

Heisenberg did get in touch with Goudsmit. Although it is unclear how much, if any, Heisenberg read of the book, soon after *Alsos* was published Heisenberg wrote to Goudsmit attempting to explain the difficult psychological situation the Germans had to face during the war.[44] It was a terrible moral dilemma. On the one hand, he claimed, the German scientists were well aware of the "horrible consequences" that a German victory would mean for Europe; on the other hand, they did not wish to see Germany defeated — not because of patriotism but because of "the hate that National Socialism had sown" — a compromising statement at best. The dilemma led the scientists to pursue "a more passive and modest posture," he argued. This was a reference to the position outlined in his earlier essays "Active and passive resistance" and "The order of reality"—to help on a small scale where it is possible and otherwise to do work that will perhaps prove useful later.[45]

Heisenberg's letter elicited an angry five-page, single-spaced typed response from Goudsmit, repeating many of the arguments in his book. Active opposition, Goudsmit told the German physicist without ado, was simply a self-serving rationalization, fabricated ostensibly for the pursuit of an impossible goal—the mere preservation of relativity and quantum theory under Hitler. "How could you ever hope to be successful? How could you ever think that these were important issues?"[46] The two argued back and forth in public and private exchanges over the following year. Heisenberg consistently maintained Germany's scientific success, despite Nazi policies, and the German scientists' moral dilemmas, while Goudsmit was unrelenting on Germany's scientific failure and the scientists' compromising position toward the Hitler regime.

In 1948, at the request of American occupation authorities, Heisenberg and Wirtz published a technical account of the German project in a series of U.S. Army reports on German science and technology.[47] They argued, of course, that Germany had been well advanced in reactor engineering, but they did not even hint at any broader issues. The report enabled Heisenberg to reexamine available research reports, and at Heisenberg's insistence Goudsmit reexamined copies of the captured reports in Washington.

As a result of this exercise, Goudsmit corrected his most obvious errors, conceding that the Germans had, in fact, been aware that a bomb differed from a reactor and that they had also been cognizant of the plutonium alternative. But, Goudsmit wrote in the *New York Times*, Heisenberg's claims of advanced theoretical knowledge notwithstanding, the reports "show clearly that their scientists had only a very vague notion of the working of the atomic bomb, and their ideas about a uranium pile were in a very preliminary stage." The reason for their meager progress was, again, their lack of vision. And their lack of vision was the direct result of "the stifling atmosphere in which scientists work under a totalitarian regime."[48]

This, of course, had been Goudsmit's fundamental point all along. Again and again Goudsmit made the same point to Heisenberg and his emissaries: what he really wanted to see from Heisenberg, Hahn, and other leading scientists was articles about the frustration of scientific progress under a totalitarian system of government.[49] He insisted that they should stop extolling the greatness of German science and acknowledge its decimation by the Nazis—a demand that they were hardly in a position to fulfill. In fact, their position was almost impossible to maintain under any circumstances: trying to distance themselves from the Nazi regime while at the same time claiming that they had done great but harmless work under it.

That Heisenberg would even attempt to defend the pursuit of decent science under the Nazi regime, or believe such were possible, seemed outrageous to many American scientists. Goudsmit had already declared of Heisenberg: "He fought the Nazis not because they were bad, but because they were bad for Germany, or at least for German science."[50] Carl Friedrich von Weizsäcker learned of this outrage firsthand when, in 1949, he met in Chicago with émigré physicists James Franck and Maria Goeppert-Mayer, both of whom—in view of the increasing secrecy, military control of nuclear research, and the decision to build the hydrogen bomb—were very concerned with ethical issues. Franck was especially critical of the German scientists, Weizsäcker reported to Heisenberg. In Franck's view, even the defense of decent physics and the acquisition of support and draft deferments could not justify the compromises Heisenberg had made with the Nazi regime.[51]

Goudsmit and Heisenberg never did settle their quarrel. Years later the two men, now old, met one last time in an attempt to heal old wounds. During his last trip to the United States, in the spring of 1973, Heisenberg lectured at the Smithsonian Institution in Washington, D.C., and Goudsmit, who had long since moved to the Brookhaven National Laboratory on Long Island, traveled to Washington to meet him.[52] Goudsmit once again admitted the technical errors in his *Alsos* portrayal of a backward German project and apologized for any personal injury he had caused the German physicist. A complete reconciliation, however, was impossible. Heisenberg died three years later.

In his obituary of Heisenberg for the American Philosophical Society, Goudsmit offered his assessment of their controversy after decades of reflecting on it. In Goudsmit's view, Heisenberg had failed to realize that German physics was already in precipitous decline relative to physics in other nations even before the Nazis came to power. The United States in particular was rapidly outpacing Germany. The American research system of cooperative university departments, large-scale industrial research, and close collaboration between experimentalists and theorists was much more conducive to the progress of contemporary physics, especially nuclear physics, than was the German tradition. Thanks in part to American fascination with European science, which made possible Heisenberg's own many trips to the United States, American physics was already surpassing German physics when the Nazis began driving many of their best scientists from Germany. Heisenberg's efforts to maintain an illusory German lead in contemporary physics were thus completely misplaced. "If Heisenberg had realized this," wrote Goudsmit, "he would not have taken the German failures so personally."[53] Perhaps he would not have been so willing either to enter into the debilitating compromises he endured by convincing himself that he was indeed personally responsible for the preservation of leading German physics.

The concern of many scientists in the United States during the late 1940s over the issue of government control of research was soon settled to their disappointment in the tightened secrecy criteria of the cold-war era, and in the government-mandated H-bomb program.[54] Whatever the lessons of the past, some scientists would work nonetheless to fashion weapons of mass destruction that were becoming ever more prevalent and ever more powerful. The prospect of "mutually assured destruction" became more assured indeed with the invention of intercontinental ballistic missiles, which could carry the instruments of destruction to any spot on the globe within minutes. Concern for moral issues and the social responsibility of the scientist mounted everywhere, especially in the United States. Some American scientists felt satisfaction that at least they were building bombs not for a Hitler but for the protection of American democracy.

The satisfaction was soon challenged by the publication of Robert Jungk's history of the atom bomb, *Brighter Than a Thousand Suns*, which appeared in German in 1956 and in English in 1958.[55] Those years were ones of vehement debate in West Germany over nuclear weapons and reactor technology, and Jungk's book took up the German scientists' case. In accounting for the meager success of the German wartime project, Jungk went beyond the issues of inadequate support, insufficient industrial resources, and faulty management to "the actual personal attitudes of the German experts in atomic research." These, he claimed, had not been discussed hitherto for reasons of "discretion and tact."[56] He even quoted from a letter he had received from Heisenberg, in which Heisenberg reiterated the argument of "active and passive resistance," to show how close he was to his sources.[57]

According to Jungk's account, the German scientists — meaning those around Heisenberg — so distrusted the regime and other less

scrupulous physicists that, pursuing their active resistance, they continued to work on the project but secretly sought "to divert the minds of
the National Socialist service departments from the idea of so inhuman
a weapon."[58] They gladly welcomed the authorities' conclusion that an
atomic bomb could not be made under wartime conditions in Germany
and were content to concentrate on trying to build a reactor, while
awaiting the inevitable defeat. Meanwhile, Jungk continued, the Americans worked feverishly on an atomic bomb and succeeded in providing
their government with a weapon of awesome destruction that was
promptly used on Japan. The implication was clear: the Americans were
morally inferior to the upright Germans.[59] It was the old, apologetic
Farm Hall story carried to its ultimate conclusion.

Heisenberg and Weizsäcker, heavily involved at that time in the
scientists' dual campaign against weapons and for reactors, had helped
Jungk reach this conclusion and privately professed agreement "by and
large" with his book.[60] The book naturally provoked a new round of
debate about German wartime research. German physicist and editor
Paul Rosbaud, who by much later postwar accounts had supplied the
Allies with inside information on the German uranium project during
the war, wrote in his review of Jungk's book for *Discovery*: "Out of all
of the theory [of the A-bomb] emerges a strange picture in which it
sometimes appears that the German physicists alone have no actual or
moral guilt for the A-bomb."[61]

Max von Laue, who in the meantime had reconsidered the Farm Hall
story that he had once so strongly supported, wrote to Rosbaud denying
the moral element in the story and repudiating Jungk's portrayal of a
morally driven Heisenberg. Referring to the discussions on August 7,
1945, he wrote: "The version was developed that the German atomic
physicists really had not wanted the atomic bomb, either because it was
impossible to achieve it during the expected duration of the war or
because they simply did not want to have it at all. The leader in these
discussions was Weizsäcker. I did not hear the mention of any ethical
point of view. *Heisenberg was mostly silent* [his emphasis]."[62] Why they
would not want to have it at all is unclear, but ethics for Laue were an
ex post facto invention.

In 1964 science students at Cornell University, after reading Jungk's
book, asked their mentor, Hans Bethe, to give a talk on the social
responsibilities of scientists and engineers. In his talk, which was later
published in the newsletter of the Society for the Social Responsibility
of Science, Bethe reviewed the public record of the Goudsmit-Heisen-
berg controversy and noted its seeming lack of focus on moral issues:

"Neither Goudsmit nor Heisenberg indicated that conscience played any part in the German failure to develop the atomic bomb."[63] Their debate had revolved around the failure of German science.

Without access to the full record, Bethe had apparently overlooked the strong moral argument that Heisenberg was making, as indicated, for instance, in his letter to Goudsmit on the German scientists' dilemma. Bethe's talk brought an objection from Heisenberg, who again emphasized his moral scruples. In his letter to Bethe, Heisenberg acknowledged that German physicists were morally no better or worse than their American counterparts, but he still maintained that they did not build bombs because they did not want Hitler to win the war. On the other hand, certain that Germany would eventually lose, they did not "wish a total and obliterating defeat of Germany." Apparently this sentiment, voiced earlier, was their justification for working to build a reactor to power the German economy before and after defeat. Again Heisenberg claimed that the controversial 1941 meeting with Bohr proved their moral concern. With the technological and administrative turning points in 1942, they were thus relieved that they could concentrate on developing a reactor without fear that they would be ordered to build a bomb.[64] Jungk's overemphasis on morality apparently inspired all parties to stress moral scruples when telling their stories.

The 1967 publication of David Irving's account of the German project, published as *The Virus House* in Britain and as *The German Atomic Bomb* in the United States, again revived the debate.[65] Irving, a British author, offered the first historical account of the German project to take advantage of a wealth of available, though obscure, primary sources. A number had been only recently declassified. Irving's overall portrayal was sympathetic and unemotional, yet he too doubted that moral scruples played a role at any stage of research. He believed that curiosity drove the German scientists in their work and would have driven them all the way to a bomb, had circumstances allowed it.

This prompted Heisenberg to offer another round of interviews and book reviews, in Germany and the United States, on the German project.[66] Heisenberg expressed satisfaction that "Irving's investigation confirms the German report in all important points," in the pages of the *Bulletin of the Atomic Scientists* — the magazine of concerned American physicists — and in the influential *Frankfurter Allgemeine Zeitung*. But he believed that Irving's interpretation of motives was faulty. Irving "does not recognize sufficiently how deep a mistrust could exist, yes, often has to exist, between human beings in a totalitarian state, even between those who work closely together." Because of this, since the war situation prevented large-scale projects, "German physicists did not insist on pursuing, by means of practical measures, a path which could not have led to success during the war."[67]

In a preface to Heisenberg's review of Irving's book, *Bulletin* co-founder Eugene Rabinowitz took direct issue with Heisenberg's contention that distrust, hence scruples about the use of their work, had determined the scientists' behavior. He conceded that distrust of the regime probably did exist and probably did make the German scientists reluctant to provide weapons to Hitler. But the possibility that Germany might lose the war and that defeat would mean the end of the great revival of the German nation begun by the Nazi regime at first made the scientists much less reluctant. This changed after 1942. "As the war dragged on, and the likelihood of German defeat loomed more and more ominously for all who retained a modicum of rationality, the scruples of the leading German physicists became stronger, and the alibi of developing not an atom bomb, but a postwar reactor, actually became reassuring."[68]

The controversy continues in full force to this day.[69] To a large extent, Goudsmit, Rabinowitz, Irving, and others are correct. We (in the American tradition, at least) do feel there was a profound failure by Heisenberg and others to be completely candid about their attitudes during Hitler's rule and especially during the war—a failure to explore their errors as well as their successes, to point out the human frailty as well as the human resilience of the scientist and the citizen in this encounter with the nightmare world of genocidal dictatorship, to debate not merely the stifling of science by governments but the stifling of the human spirit itself. Certainly it was unacceptable for Heisenberg and colleagues to claim that they had consciously delayed the project because of moral scruples; it was not much better for Heisenberg to say that he might have built the bomb, had it been attainable during the war, but otherwise to absolve himself of any moral failing. After all, had he and they not worked on reactors to power the German war machine? Had he and they not allowed themselves to be exploited by a monstrous regime? How this all came about, how these highly educated scientists, blessed with the best culture and learning and the highest ideals of scientific inquiry, could find themselves in this situation, are questions that they themselves, through candid reflection, could perhaps have helped us to answer better than could any postwar biographer. Perhaps it was too much to expect a soul-searching confession from Heisenberg rather than an apology carefully tailored to the changing postwar situation. How many American scientists who have worked and who continue to work today on instruments of mass destruction have bared their souls to us?

What is remarkable in Heisenberg's case is that despite their frustration with him, many of Heisenberg's severest American critics remained sympathetic and more than politely cordial toward him, even while publishing the most devastating repudiations. It was as if they recognized how much they shared his difficulties, as if they only wanted him to admit what has become common knowledge since World War II: that scientists everywhere, no matter how devoted they may be to the search for truth and universal understanding, will work for their governments, whether worthy or loathsome, and that many will serve their governments by fashioning the weapons of war and destruction.

The closing paragraph of Samuel Goudsmit's obituary of the man he had so admired and so reviled expresses the frustration as well as the pity that many others must have experienced. "Heisenberg was a very great physicist, a deep thinker, a fine human being, and also a courageous person," he wrote. "He was one of the greatest physicists of our time, but he suffered severely under the unwarranted attacks by fanatical colleagues. In my opinion he must be considered to have been in some respects a victim of the Nazi regime."[70]

CHAPTER 27

Candles at the Door

Heisenberg returned in January 1946 to a Germany on its knees. Bombs and artillery had reduced nearly every city and town to heaps of rubble. Roads, railway lines, river passages, and bridges had been severed or destroyed altogether; gas, water, and electric lines had been cut. Nearly a quarter of German housing was lost, and agricultural production was at a standstill. Severe shortages of food, clothing, and shelter engendered fears of starvation and epidemic. Everywhere children begged for food while their parents rummaged through garbage for whatever they could find.

As Germany's economy and infrastructure lay in ruins, her population increased dramatically. Millions had been lost to death or were in Allied captivity, but many millions more had arrived in Germany—former slave laborers and those who had fled or been expelled from eastern countries under Soviet domination. "[The Germans] are immeasurably depressed," Max von Laue wrote to his son in Princeton, after Laue had returned to Germany. "The complete suffering of war makes itself felt only now."[1] The psychological trauma of defeat, coupled with the shock of acknowledging that the man who had vowed to lead Germany to greatness was an unspeakably depraved criminal, added moral devastation to the physical destruction suffered by the once proud German nation.

Death stalked Heisenberg's family and friends as it did many others. So many of the older generation died, it was as if fate did not intend that generation to survive into the postwar era. The conspiratorial members of the Wednesday Society, which had embodied the strengths and weaknesses of the Wilhelmine and Weimar eras, were only the most well known to perish in the last months of the war. In the only letter he received from home while in British captivity, Heisenberg learned of the sudden death of his 74-year-old mother in July 1945. She had broken her hip in a fall in Urfeld and died of complications several days later in a hospital in the nearby town of Bad Tölz. Werner had been close to his mother right up to the end and had brought her to Urfeld in the last days before he was captured. In his lonely captivity, the news of her death was a heart-wrenching blow. "It was difficult to get over it as I was so alone."[2]

Fearing that the physicist might be kidnapped by one of the other Allied powers, the British did not allow Heisenberg to travel to Bavaria, then in the American zone of occupation, until several months after his return to Germany. After heading for Urfeld, he made a tearful pilgrimage to his mother's grave. Under the dire postwar conditions, the family had not been able to provide a proper funeral. "It was very sad," he wrote his Uncle Karl after visiting the grave, "for it was nothing more than a heap of earth beneath a small wooden cross without a name."[3]

Within a year after the death of Heisenberg's mother, her sister died; so did Heisenberg's Osnabrück aunt, Guste, and his Uncle Karl and Aunt Helen, who both died in the same month.[4] In the science world, the aged Geiger and Hoffmann died in the days following the capitulation, and the 90-year-old Planck died in Göttingen after losing all his children to the war, sickness, and the executioner. Robert Döpel's wife and an institute technician were killed in the last bombing raid on Leipzig, which also destroyed Heisenberg's already damaged house and most of the physics institute.[5] Kaiser Wilhelm Society head Vögel and Reich Education Ministry chief Rust committed suicide; Mentzel, Stark, Lenard, and the senior Weizsäcker faced criminal trials. Every former member of the Nazi party faced a denazification tribunal; Himmler and Göring committed suicide; Rosenberg was executed along with other war criminals at Nuremberg; and Speer received a 20-year prison sentence.

With the IG Farben chemical complex broken up as a criminal enterprise, Heisenberg's brother, who had worked for the Agfa film and chemical company, his wife, and their four children lived meagerly after the war as Black Forest pottery makers.[6] Elisabeth's brother, Fritz Schumacher, returned to Germany with the British forces to help begin Germany's economic recovery. Her sister and elderly parents waited patiently for better times while they prepared to welcome home their captured relatives at the south German country house to which they had fled during the war.[7]

The Heisenbergs in Osnabrück had also fled to the countryside to escape the bombing. Heisenberg returned to the demolished city in early 1946 to find the house he had visited so often as a child in ruins. Sadly, he picked through the ashes of the once warm and cozy kitchen for a memento of his Osnabrück childhood, collecting a piece of tile from the old wood-burning stove near which he had once contentedly played. In the ashes of this safe and happy place to which he could never return, he saw the end of an era, both in his own life and in the life of the German people.[8]

With his mother and father gone, Heisenberg had just turned 44 when he arrived from England in early 1946 in the north German town of Alswede, near Lübeck—a village left undamaged by the ravages of war. He was still a detainee of the British Crown. There he and the other detainees were housed in a confiscated confectionery store to which they had to report every evening until moved to their permanent locations.[9] A month after Heisenberg arrived at the confectionery store, his determined wife negotiated the chaotic railway and border system to travel across two occupation zones and nearly the entire length of Germany, finally reaching the husband she had not seen for nearly nine months.[10]

The weeks and months following her husband's capture in Urfeld had been dire ones for Elisabeth. While Werner feasted on British officer's rations, sported on the tennis courts of the Farm Hall mansion, and formulated the motives of his wartime work, his wife, who had already borne six of his seven children and whom he had left at the height of the war for bachelor quarters in Berlin, had to cope alone for months. Heisenberg openly worried about the "Urfelder," but there was little more he could do for them.[11] Colonel Pash, who had taken Werner away in May, had returned to the house several times with groceries, but these had run out by the summer of 1945.

Within a few weeks of Heisenberg's mother's death in July, Elisabeth's friend who shared their Urfeld home also died. Elisabeth was left alone in the lakeside cottage to care for her friend's son as well as her own six children. She no longer had her husband's Berlin salary and, in any case, there was little on which to spend it in the mountainous Urfeld area.[12] The Sommerfeld family, still living in Munich, helped her as best they could, but it would be nearly 18 months before the Heisenbergs could be reunited under better circumstances. A photograph of Heisenberg and his wife, taken shortly after their reunion, shows a gaunt and determined Elisabeth. Heisenberg is energized, ready to face the postwar world with his warm smile, radiant eyes, and still blond hair that made him look much younger than his years.

During his captivity in Farm Hall, Otto Hahn learned that he had been awarded the Nobel prize for the discovery of fission; Meitner and Strassmann were not included. Of the ten detainees, the three Nobel laureates — Hahn, Heisenberg, and Laue — emerged as the leading spokesmen for German science. During their captivity, they had met twice in London with British scientists and had once driven into the German hills with their British control officer to discuss the future of German science. With the permission of the United States, the British had decided to settle the scientists permanently in the British occupation zone that stretched across the northwestern section of Germany.[13] Following the Cambridge model, the scientists emphasized the necessity of reestablishing their institutes in a university town. Göttingen was the obvious choice for its academic tradition and many research institutes, and it had come through the war nearly unscathed. Only its proximity to the Soviet occupation zone caused Heisenberg some worry.[14] Göttingen was to become for the British a scientific and technological center that would eventually bring about the revival of German science, first in its own zone, then in the emerging West German state. Not wishing to subsidize or colonize its occupation zone indefinitely, the British more than any other Allied power intended the scientific revival to occur hand in hand with an economic and political revival that would eventually lead to German autonomy.

Some of the nonlaureate detainees were less enthusiastic about the British plan. During the weeks following their arrival in Germany, Harteck and Diebner returned to their old Hamburg institute, and Harteck eventually emigrated to the United States. Gerlach went to Bonn and later returned to his chair for experimental physics in Munich. At the end of February 1946, Heisenberg and Hahn were the first to arrive in Göttingen. They were joined soon after by Laue, Weizsäcker, and several other former detainees.[15]

The British arranged for Heisenberg and Hahn to reestablish their old Kaiser Wilhelm institutes for physics and chemistry, respectively, in the empty rooms of Prandtl's former Aerodynamics Experimental Institute. The institutions were still part of the state-run Kaiser Wilhelm Society and thus independent of any university. The plan nearly foundered at the start. In Hechingen, the French, still irked at the American seizure of nuclear equipment and scientists from their occupation zone, were holding instruments and technicians from the two institutes and were unwilling to release them to the British. With only three coworkers, no laboratory materials, and no association with the University of Göttingen, Heisenberg was miserable. To add to his troubles, the flood of refugees in the small town made it impossible for him to find decent living quarters; he was reduced to sleeping on a straw sack in a dingy tenement.[16] For the moment, the Urfelder would have to remain in Urfeld.

In the midst of Heisenberg's Göttingen misery, the aged Sommerfeld revived the old question of who should be his successor in the Munich theoretical physics chair. He had submitted yet another list of three candidates to the Munich faculty: Heisenberg, Weizsäcker, and Hund.[17] Heisenberg once again gratefully accepted, but he again needed the consent of higher authorities — this time the Allies — and their approval was uncertain. They had refused initially to allow Gerlach to return to his Munich chair.

Within a year it was settled: Heisenberg stayed in Göttingen but by his own choice. His institute was finally thriving, but more important, in the summer of 1946 British troops had vacated a beautiful mansion that they had confiscated on the Hainberg, the hill just outside Göttingen where Heisenberg had strolled with Bohr so many years ago.[18] British science officers arranged for the house to go to Heisenberg. Elisabeth, in another display of almost superhuman determination, managed by herself to arrange for trucks to cross through three different zones to gather their furniture from Hechingen and Urfeld and to take it, along with the six children, to Göttingen. The friend's child was apparently sent to live with relatives. Together at last, the Heisenbergs moved into the Hainberg mansion in September 1946. It was so large (and they had so little money) that they rented out two of the rooms to Göttingen students, one the son of Popitz, the executed Wednesday Society conspirator, the other a Weizsäcker family friend.[19]

Once the Heisenbergs had settled in Göttingen and with Werner's institute flourishing, a move to Munich lost any remaining attraction. They still owned the Munich house they had bought in 1937, but since it was undamaged by the war American officers had confiscated it for their use. In addition, American authorities were not then nearly as supportive of German science as were the British. Fritz Bopp, a former Sommerfeld pupil and a collaborator on the uranium project, eventually succeeded the great Sommerfeld in Munich in 1947.[20] Over a decade later, amidst a reorganization of West German nuclear energy policy, Heisenberg finally moved, along with his institute, to his beloved Munich, where he remained for the rest of his life. Long before the move, his institute was renamed the Max Planck Institute for Physics and Astrophysics, an institute within the Max Planck Society, the network of federal research institutes that replaced the Kaiser Wilhelm Society in the early postwar years. Heisenberg was eventually named an adjunct professor at the University of Munich, but he never again held an academic teaching chair. He remained director of the nonacademic state-supported institute until he was forced by illness to retire in 1970.

Despite the pleasant radiance that emerges from early postwar photographs of Heisenberg, several of Heisenberg's colleagues throughout the later years observed that he seemed to suffer from a perpetual depression.[21] Aside from the toll taken by advancing age, he seemed particularly depressed by several factors: that his actions during the Third Reich were not understood abroad; that German physics, for which he had worked and suffered so long, was now indeed eclipsed by American physics; and that his own research was not as successful or as well received as it once had been. His response to depression was as it always had been — work and ever more work.

When not coping with Goudsmit and other critics of German war research, Heisenberg focused his energies after the war on two major concerns: science and science policy. Until the western Allies ended the formal occupation in 1955 and granted sovereignty to West Germany as part of the NATO alliance, science policy issues took precedence and so absorbed him during the late 1940s and early 1950s that it is hard to believe he had much time, or emotional energy, for anything else. His most well-known new physics, a proposed unified field theory, did not fully emerge until the late 1950s.

Heisenberg regarded his efforts to influence science policy again as a service to German science, and again he had convinced himself that he was the most qualified to assume this burden. His experiences with the Third Reich had deepened rather than mitigated his perceptions of duty. He seemed determined that German science should again reach world-class standing and that German scientists, himself naturally included, should never again be disregarded or abused by government authorities. In the international arena, he avidly lobbied for the establishment of a European accelerator facility, the Centre Européen pour la Recherche Nucléaire (CERN) in Geneva, to rival American machines, and he served actively on its governing policy commitee.[22] In 1952 he became president of the Alexander von Humboldt Foundation, a federal agency that brought foreign postdoctorates to West Germany to perform research and in the process to broaden themselves and reestablish their countries' contacts with German science. As a one-time foreign postdoctoral student in Denmark, Heisenberg knew the value of work abroad and came to cherish the Humboldt Foundation presidency above all the many offices he held. It was the last official post from which he resigned when illness set in during the years before his death.[23]

Domestic policy issues consumed even more of Heisenberg's time. Until preparations began to end the military occupation and establish the West German state in the former French, British, and American zones, Heisenberg focused his efforts on occupation problems. The occupation itself was coordinated in Berlin through the Allied Control Authority, which was headed by the military governors of the four occupation zones. After the military occupation ended in 1949, the civilian Allied High Commission maintained control until 1955. The four Allies did not fully relinquish their control rights until German reunification over 35 year later.

At least two Allied policies directly affected science: denazification and Allied Control Law 25, the control of scientific research.[24] Denazification, declared to be a dominant objective of the occupation, was intended to rid German public life of Nazi influence. It was enforced to widely varying degrees in the different zones, but all Germans, in whatever zone, were required to fill out a questionnaire about their political affiliations. Former members of the Nazi party or party organizations had to appear before a military tribunal to explain their activities. In the British and American zones, the tribunals were soon turned over to local courts. For many, an appearance before a tribunal was likely to result in the loss of a job, since former party members were generally excluded from civil service posts, which included all teaching positions. Antiparty feeling ran high, and many German academics found it impossible to work. After one science teacher lost his job due to party membership, Max von Laue complained angrily to his son that denazification as practiced by the Americans in particular made "every use of reason impossible."[25] Realizing finally that the western zones could not be turned into a power vacuum and that their economy and society could not be revived if every single party member was excluded, the U.S. occupation command practically ended denazification as a broad-ranging policy after 1947.[26]

People called before a denazification tribunal were permitted to submit affidavits from prominent persons testifying to their behavior. The testimonials were appropriately nicknamed "Persilscheine" (whitewash certificates), after the popular laundry soap Persil, whose slogan is *nicht nur sauber sondern rein* (not just clean but pure). Heisenberg was frequently sought out as a writer of such whitewash certificates by his friends and coworkers and by those to whom he owed a debt — those who had helped him during the SS affair, as well as receptionists and lower functionaries in Rust's and Himmler's offices who had enabled access to their bosses. But he refused to supply testimonials to those whom he did not know or did not care to support, and he wrote an evaluation of Stark on behalf of the prosecution at Stark's trial.[27]

Although it is uncertain how effective Heisenberg's certificates were, the occupation authorities seem to have greatly valued Heisenberg's judgment. The British and American authorities had already decided, or had become convinced, that Heisenberg and the other atomic scientists were far too valuable to the revival of German science to be held in any but the highest regard. Heisenberg himself could claim that he had been victimized by the Nazis, and having never joined the party or its affiliations, his questionnaire was both clean and pure.[28] Moreover, most of the denazification proceedings were completed before the controversy in American scientific circles over Heisenberg's wartime activities. Most local authorities in the British and American zones probably took little note of the squabble.

Countering the Allied law controlling science was a more difficult matter. At the top of the projects prohibited by the law was research in applied nuclear physics, which included every form of nuclear reactor research and isotope studies, as well as cyclotron construction and experimental high-energy physics. All research had to be cleared in advance by a science officer, and all results and publications had to be submitted to the officer for review. The prohibited nuclear research included the very subjects to which Heisenberg had intended to return. In particular, unlike the Allied emphasis on product research and development, Heisenberg had planned that nuclear reactor technology would be the foundation of the revival of German physical science, and he argued that progress in this area would lead to the revival of the entire German economy. He believed too that the prestige and momentous impact of nuclear energy must not be denied to West Germany or to West German scientists. Heisenberg, Hahn, Gerlach, Weizsäcker, and others campaigned vigorously for the promotion of reactor research, taking their message to the public as well as to the occupation authorities.[29]

Control Law 25 was finally relaxed in 1949, but nontheoretical applied nuclear research was not permitted in Germany until as late as 1955 — the Allies did not want the Germans playing with fission under any guise. Once it did become legal, however, both the government and Heisenberg's institute took off with it, and within a decade West Germany was the world's leading exporter of nuclear technology. At the same time, Heisenberg and his closest colleagues vigorously opposed a 1955 NATO plan to equip the West German army with battlefield nuclear weapons. The scientists mobilized the German populace in ways they had never considered previously, and the plan was so soundly defeated in 1958 that the West German army has remained nonnuclear ever since. The prospect of nuclear weapons in German hands had already caused enough problems for the scientists.[30]

Heisenberg's domestic governmental policy efforts spoke directly to a century-old conflict in German culture that resurfaced at the end of the war: the sometimes virulent competition between the federalist states, or Länder, and the central government, or Bund, over responsibility for supporting scientific research. Financial support was tantamount to administrative control. Heisenberg and his supporters allied themselves with the centralist faction, and Chancellor Konrad Adenauer's establishment in 1955 of a cabinet-level "federal ministry for atomic questions" to promote nuclear energy development was their most striking achievement. The ministry was the predecessor of the two present-day cabinet ministries for university research and industrial technology. But Heisenberg and the centralists were much less successful in forging a strong national authority over other branches of scientific research. The delicate balance today between Bund and Länder regarding research funding is the hard-won result of decades of careful negotiations and contractual agreements.[31]

For more than a century, the regulation and support of education and culture, including science, had been the prerogative of each Land, and since Bismarck the prerogative had been law. Most research was done at universities, and universities were under the authority of the cultural ministry in each Land. Two interstate organizations had challenged this arrangement in the early decades of the twentieth century. In the throes of industrial revolution, Kaiser Wilhelm II expressed his personal interest in promoting science and technology by establishing the Kaiser Wilhelm Society, the network of pure research institutes in various scholarly and scientific fields supported directly by the Reich. Second, as discussed in Chapter 9, during the early years of the Weimar Republic German scientists had established the nationwide Notgemeinschaft der Deutschen Wissenschaft (Emergency Association of German Scholarship) to support impoverished laboratories and researchers. It bypassed the federal government by funneling public and private funds directly into science through Notgemeinschaft-appointed committees of scientists and administrators. Grants were awarded on the basis of peer review of project applications, a process they called the self-administration of science. Max Born had earlier supported the young Heisenberg with several Notgemeinschaft grants.

Most of the original Notgemeinschaft administrators enjoyed close ties to the division for higher education in the large and powerful Prussian Culture Ministry. But soon after coming to power in 1933, Hitler had reduced the semiautonomous states to powerless administrative districts. When he elevated the Prussian Culture Ministry to the status of Reich Education Ministry, Notgemeinschaft administrators became minor functionaries controlled by party and SS bureaucrats. Rudolf Mentzel's science section, which oversaw the Notgemeinschaft, grew out of the Prussian division for higher education.

After the war, most members of the former higher education division migrated to the Culture Ministry of Lower Saxony in the British occupation zone. The able Prussian bureaucrats promptly founded a new Notgemeinschaft; they were determined to rebuild German university science and to avoid the errors of the past in the democracy of the future.[32] In their view, the major error had been neither their own disdain for a decentralized democracy nor the blatant support for Hitler exhibited by some of their leaders. Their problems had their roots in the failure of the Weimar constitution to prevent the creation of the Reich ministries, which encroached on state prerogatives and ultimately on the administrators' personal spheres of influence (as though Hitler would have paid the slightest attention to any constitutional checks on his power).

This time the bureaucrats were determined to keep the federal government off their cultural and scientific turf, and self-administration served as their chief ideological weapon. Their position found powerful support in the Christian Democratic Union, which was emerging as the dominant party in the British zone and in West Germany.[33] They also found themselves supported by the state cultural ministers and university rectors and by the policies of the occupation forces. A 1948 aide-mémoire from the military governors to the German Parliamentary Council, which was deliberating the constitution of the new government, declared: "The powers of the federal government shall be limited to those expressly enumerated in the constitution and, in any case, shall not include education, cultural and religious affairs."[34]

The Allied attitude, especially that of the Americans, who were inherently more federalist than the British and who oversaw the most federalist state of Bavaria, was hailed by the rabidly federalist-minded states and by Kurt Zierold, the administrator of the higher education department in the Culture Ministry of Lower Saxony. Zierold was a jurist and former Notgemeinschaft official.[35] In 1946, with advice and influence from Heisenberg and Hahn, the British reincarnated the Kaiser Wilhelm Society with Otto Hahn as president. Within two years, the society was operating on an interzonal basis with interzonal funding and a new name: the Max Planck Society. This revival served to induce Zierold and the western culture ministers to reincarnate the Notgemeinschaft for the repair and support of university laboratories. The plan naturally met with little objection from the increasingly powerful conference of West German university rectors, of which Zierold happened to be president.

The rectors and culture ministers officially recognized the new Notgemeinschaft in 1949, just as the Parliamentary Council laid the foundations for the West German Federal Republic in its new constitution, or Basic Law. According to its bylaws, the Notgemeinschaft once again called for the self-administration of science. Its supporters even believed this notion to be legally sanctioned by Article 5 of the Basic Law, which declared: "Art and science, research and teaching are free." This article was interpreted to mean free from interference by federal bureaucrats. Not surprisingly, the new organization located its headquarters in Hannover—in the department for higher education in the Culture Ministry of Lower Saxony.

The administrative territory staked out by the Notgemeinschaft was challenged just two months later by the founding of the very different Deutscher Forschungsrat (German Research Council), or DFR. Directed by Heisenberg, the DFR was headquartered in Göttingen—a town also located in Lower Saxony. Its roots, like those of the Notgemeinschaft, lay deep in Germany's past, but it inclined in the opposite direction: toward federal authority, elite scientists, and research policy and overall planning more than direct funding. Like the Notgemeinschaft, the DFR enjoyed powerful patrons among the occupation authorities— specifically in the Research Branch of the British Control Commission, with which Heisenberg and Hahn were closely associated.[36] It also had the backing of the more centralist Social Democratic Party (SPD), re-emerging in West Germany, and of Konrad Adenauer, the new federal chancellor.

In December 1948, Carlo Schmid, the head of the SPD delegation to the Parliamentary Council, read into the record a letter received from Heisenberg and three other members of an advisory science council established by the British Research Branch. It urged that the organization and promotion of scientific endeavors be assigned to the Bund rather than to the Länder in the new Basic Law.[37] The arguments of the scientists are not surprising. "The individual state cannot bear the responsibility for and the financing of German scientific research, which has long since outgrown the boundaries of individual states." The needs of large-scale projects, such as nuclear reactor development, or large networks of institutes, such as the Max Planck Society, could not be met by the meager resources of individual Länder. They also declared, "We must look with horror at where an attempt to limit the life-sustaining field of science has already led in the last years." In other words, local bureaucratic control, not the creation of the Reich, had enabled the nazification of German science.

The scientists' letter and the parliamentary debate had two effects: first, the phrase "promotion of scientific research" was inserted in article 74 of the West German Basic Law, which enumerates the concurrent powers exercised by Bund and Länder; second, the DFR was formally created in March 1949 by the Max Planck Society and the western academies of science, who placed it under the direction of their favorite physicist: Werner Heisenberg. The original 15 self-appointed members of this council were all well-known academic scientists close to Heisenberg and his Göttingen circle, and most had been associated in some way with Heisenberg and his nuclear fission project during the war.[38] But not one of these council members was associated with technology, industrial research, or education at technical colleges and universities — the very fields the DFR intended to manage. This lack of experience and apparent condescension toward technology naturally displeased leading figures in those fields.

In his address to the inaugural convention of the DFR, Heisenberg went so far as to proclaim the new council "the sole representative of all German science" and the professional representative of science and technology in Germany and abroad.[39] The council's bylaws accorded it the power not only to advise federal and state governments and to represent German science at international gatherings but also to participate "in the financing of scientific research, in particular in the solicitation and distribution of public funds for research purposes." Having thus completely ignored its competitor, the council proceeded to make clear in its bylaws the inferior status of the Notgemeinschaft: "In the fulfillment of these tasks the German Research Council will rely foremost upon the Emergency Association of German Scholarship."[40]

Needless to say, the Notgemeinschaft did not appreciate being cast as a supporting player in the DFR's show — it was all too reminiscent of the attempted usurpation of power by the Reich Research Council. Zierold complained privately to British authorities that the DFR was a superfluous organization, consisting wholly of effete snobs who had no intention of reviving German science where it belonged — at the universities and technical colleges.[41] Democracy itself — a word to reckon with during this era — became a pawn in the game. Zierold argued that a self-appointed committee of elite academics could have little acquaintance with democratic principles, to which Heisenberg replied that the regional culture ministers and university rectors could hardly set themselves up as the guarantors of democracy after their "shameful" behavior in the Third Reich: "an all-too-eager submission to the state authorities."[42] No matter that the scientists had hardly distinguished themselves for their resistance to the Nazi regime or that Heisenberg's organization now made no secret of its eagerness to submit to federal authority.

State and federal authorities immediately came to the rescue of their respective organizations. Within a month of the founding of the DFR, the western culture ministers issued a joint declaration, the Königstein Agreement, the first sentence of which struck to the heart of the matter: "The Länder of the three western zones consider the promotion of scientific research as fundamentally a task of the Länder."[43] Article 74 of the Basic Law notwithstanding, the declaration remained in force for over a decade and has been the basis of all state-federal negotiations ever since.

Not to be trumped by the culture ministers, Chancellor Adenauer requested and received a memo from the DFR on its intentions.[44] The memo, written by Heisenberg, called for recognition of the DFR as the sole science advisory panel to the chancellor and for the establishment of a small service bureau in the chancellor's own office — presumably as close to his desk as possible — to coordinate direct advice from the scientists and to sort out the research policies of the Bund vis-à-vis the Länder. Heisenberg discussed his strategy in a letter to a DFR colleague: "The Federal Chancellor is the only strong personality," he wrote, who was in a position "to carry out our wishes for the centralized direction of research."[45] No stranger to realpolitik, Adenauer deftly played both ends against the middle. While refusing to supply federal funds or even to grant official recognition to the DFR, he gladly welcomed Heisenberg's service bureau as an opportunity to control these elite scientists while outmaneuvering the culture ministers. By appearing to listen to the scientists' wishes, he could co-opt their allegiance, while pointing to their organization as an excuse to ignore the pressure exerted by the culture ministers.

Despite their competition, both policy groups managed to register notable successes during the two years of their joint existence, a crucial period for the Federal Republic. While the Notgemeinschaft raised over a million marks for university and local research labs through its fund-raising efforts in German industry, Heisenberg and the DFR gained U.S. Marshall Plan money for German science, even though the Marshall Plan had originally excluded science from support. He also arranged for the admission of the Federal Republic to UNESCO's International Union of Scientific Councils and applied for a further relaxation of new science control laws issued by the Allied High Commission in 1950. He presented several detailed memos to Adenauer on the organization and funding of research in the present and future West German state. Heisenberg told his man in the chancellor's office: "It is . . . a funda-mental question of the international competitive ability of German research and thus, ultimately, of the German economy."[46] As might be expected, in each memo the foremost concern was that Germany be prepared for immediate development of nuclear energy as soon as the Allies permitted it.[47]

Yet pressure was also mounting on the DFR to "fuse" with the Notgemeinschaft. German industry, academies, and scientists realized that the embarrassing competition for funds and control was hindering rather than helping the recovery of German science and technology. The culture ministers at first categorically demanded that the DFR be joined with the Notgemeinschaft, but the general membership of the Notge-meinschaft rejected a temporary agreement proposed in 1949 because it did not go far enough—they really wanted the total subordination of the DFR.[48] By 1951 the pressure was unbearable: industrial donors especially refused to contribute to one organization, the Notgemein-schaft, for purposes claimed by another organization, the DFR. Ade-nauer, for his part, allowed the controversy to continue as a way of diverting the scientists and administrators from the increasing power of his chancellery.

Thanks to the intervention of Heisenberg's friend and colleague Walther Gerlach, who was now rector of the University of Munich as well as a leading figure in the Notgemeinschaft, both sides were gradu-ally reaching an agreement on fusion. Just as fusion seemed imminent, however, Adenauer suddenly tendered his official recognition of the DFR — but only if it remained an independent organization. Heisenberg responded with a new memo on the organization of German science in which the DFR's independence was assured.[49] Only Gerlach's renewed diplomacy could save the fusion, set for early August 1951. Again Adenauer blocked the effort with an eleventh-hour promise of new research funds, and again only outside pressure on the reluctant Heisenberg, exerted by scientists, industry, and the academies of science, forced his acquiescence.[50] The two bodies finally joined in August to form the German Research Association (DFG), the present-day equivalent of the U.S. National Science Foundation and National Endowment for the Humanities, combined.

According to prior agreements, the DFR, upon joining with the Notgemeinschaft, became the distinguished "senate" of the new German Research Association. Heisenberg was elected to the presidium of the DFG and to the chairmanship of its influential committee on nuclear research, with the understanding that the senate would continue to pursue the goals of the now defunct DFR. In practice, however, the power of the states could not be broken by the Bund, nor could Adenauer overcome the predominance of federalist ideology in his own party, the Christian Democratic Union, and its coalition partner, the Bavarian Christian Social Union. As a result, for nearly two decades the organization and promotion of most of German scientific research remained in the hands of the state culture ministers, and funding was coordinated primarily through the self-administered distribution of individual research grants.

Until as late as 1969, the Bund supported only such interstate research organizations as the Max Planck Society and only those fields that most contributed to national prestige and influence — namely, nuclear energy, space research, and computer technology. Matters changed in the late 1960s when, in the midst of a supposed technology gap, marked by a "brain drain" and fears of being flooded with high-technology American products, the more centrally oriented Social Democrats came to power for the first time in the federal government. The new government began to pursue a more broad-based effort to stimulate German science and technology through federal initiatives and closer federal-state-industry cooperation.[51]

Heisenberg himself regarded the whole affair as a defeat both for himself and for German democracy — and as further cause for despair. In the 1951 closing report of the DFR, he described what he saw as the cause of the defeat, which, in fact, applied equally to both the DFR and the Notgemeinschaft. "We do not have an old democratic tradition," he wrote, "and we Germans are in general grateful when we can turn over the responsibility for public life to our superior authorities."[52]

Heisenberg's intense concern with science policy issues throughout the postwar era did not entirely hinder his scientific concerns, which even flourished simultaneously with political events. The simultaneity is evident at the start: Heisenberg's Farm Hall interlude initiated a flurry of postdetention publications that set the stage for his later research. The contemplative leisure afforded by Farm Hall and the shared detention with Max von Laue and Carl Friedrich von Weizsäcker stimulated Heisenberg's continued work in two areas: the phenomenon of super-conductivity, which he discussed with Laue, and the problem of hydro-dynamic turbulence, which he explored with Weizsäcker. His papers on superconductivity during the late 1940s were the less successful of the two efforts. Although he succeeded in demonstrating that superconduc-tivity can be seen as a type of phase transition, similar to the condensa-tion of a liquid from vapor, he failed to account for the transition. Current theory accounts for superconductivity — the reduction of elec-trical resistance to near zero in some bodies at low temperature — by the correlation of particles in pairs, which sets up a weakly attractive force that enables the phase transition at very low temperatures.[53]

Pursuing his interest in astrophysics during and after the war, Weiz-säcker examined the properties of turbulent rotating masses of hot gases as a model of spiral nebulae and as an account of the formation of planets in the solar system. Weizsäcker's work apparently encouraged Heisenberg's Farm Hall return to hydrodynamic turbulence, as well as a joint paper on spiral nebulae. During the late 1940s, Heisenberg pub-lished several studies and talks on a new statistical theory of turbu-lence.[54] Nevertheless, the main impetus for Heisenberg's return to tur-bulence was probably an even more fundamental concern: quantum field theory.

Just before the war, Heisenberg had believed he could treat explosion showers in cosmic rays as a spray of "droplets" unleashed by a turbulent matter field confined to a region of particle collision that is smaller than a universal minimum length (Chapter 21). In order to initiate the explo-sion, a nonlinear interaction between the high-speed particles had to occur, and the result would be a burst of Yukawa mesons and Pauli neutrinos that could penetrate large blocks of lead. Unfortunately, detection of Heisenberg's predicted showers could not be confirmed, and most physicists, especially those in the United States, doubted Heisenberg's program for a future theory of the fundamental length.

Soon after the war, two events revitalized quantum research. First, a number of theoreticians in the United States invented the process of so-called renormalization. The infinities and divergences plaguing quan-tum electrodynamics could be defined away, rendering the theory appli-cable to all energies and to even the very smallest distances of approach between particles. Again, it seemed, there was no need for any new theories.[55]

Second, in 1947 Cecil Powell and coworkers finally unraveled the puzzle of cosmic-ray mesons and in the process confirmed the existence of Heisenberg's explosion showers, now called multiple processes. Heisenberg's showers were predicated on a so-called strong coupling in Yukawa's theory of nuclear forces. The force between protons and neutrons involved the exchange of a heavy particle—the pi-meson, or pion. Powell discovered that the meson usually observed in cosmic-ray experiments is not Yukawa's pion at all but rather the mu-meson, or muon, belonging to Fermi's weak-coupling theory of beta decay. Pions and muons are produced in multiple processes, and they seem to arise from field theories that were, unlike quantum electrodynamics (QED), not renormalizable. Infinities, hence explosion showers, were so inherent to the field that they could not be defined away.

Powell's discoveries as well as difficulties with the wartime S-matrix revitalized Heisenberg's quantum field physics. This led to the development of strikingly original theories during the following years, culminating in his 1958 proposal of a new unified field theory that, though beset by difficulties and never accepted by most physicists, implied for Heisenberg a new revolution in physics. As Heisenberg struggled for his policy preferences during the late 1940s and early 1950s, in a series of papers presented to the Göttingen Academy and to the newly established *Zeitschrift für Naturforschung (Journal for Natural Research)* he evaluated the situation in quantum field theory and set his program for future advance.[56]

In field theory, he observed, particles and their interactions are represented by fields that satisfy one or more relativistically invariant differential equations. But whenever these equations are quantized, the physical properties of the fields diverge to infinity. These infinities seemed to arise from so-called local interactions, interactions occurring at arbitrarily small distances. They could be avoided either by a cutoff of short wavelengths or by Heisenberg's preferred method—the introduction of a minimum length as a lower boundary. But, he discovered in 1951, the latter proposal allowed for violations of causality arising from special relativity theory. A minimum distance, Δr, implied a minimum time, $\Delta t = \Delta r/c$, c being the speed of light, between a cause and its effect. Inversely, a minimum time implied a certain distance between the impact of a cause and the appearance of its effect.[57] Local causality seemed incompatible with the quantization of relativistic equations. The S-matrix was one way to avoid this dilemma, and the use of dispersion relations to represent the matrix helped to ensure relativistic causality, but the S-matrix could hardly be considered satisfactory. It connected events long before and long after a collision, without illuminating anything about the actual collision. The S-matrix, Heisenberg declared in his first postwar paper, was meant only to provide the "mathematical framework of quantum field theory."[58]

To Heisenberg, the situation was similar to that of the early 1920s, the years before quantum mechanics, and explosion showers and turbulence provided clues for the correct model of a future field theory. Previous field theories treated elementary particles and the forces between them as distinct field entities. The dissipation of the turbulent field into a literal shower of myriads of particles at short distances suggested to Heisenberg that the new field theory should deal not with individual particles and fields at all but with one general overall matter field, which he symbolized by $\psi(x)$. Distinct elementary particles would appear as stationary energy states of this general matter field. Like a liquid, the field would turn turbulent when confined to a "bottle" smaller than the universal minimum length. Unlike QED, the differential equation of this field must be nonlinear in order to generate the turbulence. A power of the universal length, l, must control the nonlinearity; the length would determine the masses of the elementary particles condensed from the matter field; and selection rules and conservation laws would emerge from the symmetry properties of the field equations.

For the general matter field itself, Heisenberg chose Dirac's spinor wave function. For its nonlinear, relativistic differential equation, Heisenberg, drawing on the earlier work of Born and Infeld during the 1930s, chose the simplest form,

$$\gamma_v \frac{\partial \psi}{\partial x_v} - l^2 \psi \, (\psi^\dagger \psi) = 0$$

where γ_v is the relativistic gamma-matrix, $\psi^\dagger$ is the Hermitian conjugate of ψ, and repeated indices are summed. This field theory was not only nonrenormalizable, but because it was nonlinear it was not susceptible to the old method of perturbation theory. Even renormalizable theories such as QED, Heisenberg believed, were merely low-energy approximations to the nonlinear unified theory.

As so many times before, the ingenious Heisenberg had incorporated the main difficulties of current theories into a new and potentially revolutionary theory, a theory that would be truly unified because all of matter and its interactions could be reduced to one simple set of equations for one unified field representing every form of matter and force. It was indeed a revolutionary notion.

Heisenberg spent over eight years exploring the problems and possibilities of his new nonlinear theory. The properties of various formulations of nonlinear equations, among the most difficult differential equations to solve, and their relationship to various models and properties of elementary particles had to be explored: the Lee model; preservation of the unitarity of the S-matrix, which coincided with the probability interpretation of quantum mechanics; various approximation methods, such as the new Tamm-Dancoff method; and finally the utilization of symmetry properties via group theoretical methods. In 1956, the last problem was rendered more difficult when T. D. Lee and C. N. Yang discovered that parity is not conserved — that is, symmetry under space reflection is violated — in beta decay.[59]

By 1957, Heisenberg had modified his matter field to form an eight-component type of spinor field known as an isospinor. But what kind of field equation would it satisfy? As in the early 1920s, Heisenberg in Göttingen turned increasingly to his old friend and colleague Wolfgang Pauli, now back in Zurich, for advice and criticism. Their correspondence attained even greater intensity than before. This was especially the case when in February 1957, amidst the scientists' opposition to German nuclear weapons, Heisenberg fell ill and retreated with Elisabeth to the town of Ascona on Lake Maggiore in northern Italy. A mathematical "battle of Ascona" broke out between Heisenberg and Pauli over the technical arcana of relativistic unified field equations. Heisenberg insisted on invoking an earlier proposal, apparently made by Dirac, to extend the so-called Hilbert space of wave functions by allowing the introduction of an indefinite space metric. Pauli just as resolutely refused.[60] After six weeks of what Heisenberg described as a painful battle, Pauli finally capitulated. On his return home, Heisenberg stopped off in Zurich for a medical checkup and a mopping-up operation on Pauli, who finally conceded "boring unanimity" of opinions. Heisenberg returned to Göttingen to continue the work and to tangle in public with Adenauer over weapons and reactors.

Nine months later, Heisenberg returned to Switzerland, this time for policy meetings at the CERN accelerator in Geneva, and stopped in Zurich for a stimulating visit with Pauli. Within a few weeks of the visit, Heisenberg recalled, he happened on a very simple field equation that seemed to satisfy every symmetry property demanded of it — both the relativistic Lorentz group and the isospin group. Pauli was elated; a joint paper on the Heisenberg-Pauli equation, the basis of a unified field theory, seemed in order as soon as the mathematical consequences could be worked out. Fittingly, it would be their first joint publication since they had laid the foundations of field theory nearly 30 years earlier.[61]

The paper was never written. As mathematical difficulties mounted, Pauli left for a prearranged two-month visit to the United States, while Heisenberg, his family, and his institute prepared to move from Göttingen to their new quarters on the northern outskirts of Munich. With the mathematics still unsolved, the two physicists decided to publish the equation only as a so-called preprint, a preliminary communication of results to be sent to selected physicists. The distribution was set for February 27, 1958. A 14-page typescript was prepared in English and duplicated on a mimeograph.[62]

Three days before the preprint was to be distributed, Heisenberg announced the new formula in a lecture at the University of Göttingen physics institute. An eager reporter in the audience relayed word of a sensational new "world formula" around the world. One enthused press agent proclaimed, "Professor Heisenberg and his assistant, W. Pauli, have discovered the basic equation of the cosmos!"[63]

Two months later, more than 1800 listeners turned out to hear Heisenberg reveal the secret of the cosmos in the same auditorium on the occasion of Max Planck's one-hundredth birthday. During his highly technical talk, Heisenberg carefully wrote his new equation on the overhead projector in the darkened room:

$$\gamma_\mu \frac{\partial \psi}{\partial x_\mu} \pm l^2 \; \gamma_\mu \gamma_5 \; \psi \; (\overline{\psi} \gamma_\mu \gamma_5 \psi) = 0$$

As the two-foot-high symbols slowly appeared on the huge screen, flashbulbs popped all over the hall. Just as an equation-filled page from one of Einstein's field-theory manuscripts made it to the front page of the *New York Times* in 1949 under the heading "New Einstein theory gives a master key to the universe," Heisenberg's so-called world formula found its way to front pages throughout Germany.[64] In both instances, the more incomprehensible the purported key to the cosmos and the more public the physicist, the greater the public's fascination with both the physics and the physicist.

The public quickly regained its senses, however, after Pauli's sudden renunciation of Heisenberg's world formula. Ever the critic, Pauli had grown increasingly doubtful until, two weeks before Heisenberg offered the formula to the eager Planck celebration, he refused any further support of the theory in a strong letter to Heisenberg and in a two-paragraph statement in English that he distributed to 67 leading physicists.[65]

The letter and renunciation did not deter Heisenberg from presenting his formula to receptive audiences all over West and East Germany.

A July conference on elementary particles at CERN in Geneva brought Heisenberg face to face with his critic for the first time since Pauli's renunciation. Pauli was chairman of the section at the CERN conference in which Heisenberg was scheduled to present again his new field equation. Pauli opened the session on fundamental ideas in field theory with the remark, "What you will hear today is only a substitute for fundamental ideas."[66] Heisenberg's work was "mathematically objectionable," he stated in the discussion, then proceeded to tear it apart.[67]

Over a decade later, Heisenberg was still smarting. "Wolfgang's attitude to me was almost hostile," he wrote. "He criticized many details of my analysis, some, I thought, quite unreasonably."[68] Most physicists, especially those in the United States, were already doubtful of Heisenberg and did not consider the theory further. Nevertheless, it has been pursued, improved, and modified ever since by Heisenberg and his disciples in Munich, especially by Hans-Peter Dürr, Heisenberg's able successor as institute director. A scientific and historical evaluation of its place in the history of physics is still outstanding.

The CERN meeting was the last that Heisenberg would see of the man who had worked so closely with him throughout his career and who had so greatly influenced his many contributions to physics. Pauli returned to Zurich after the CERN conference. Four months later he died suddenly of cancer at the age of 58.

Heisenberg's move at last to Munich in September 1958 as director of the Max Planck Institute, 22 years after his initial call to succeed Arnold Sommerfeld at the university, and now 56 years of age, marked the beginning of the last phase of his life. It was a phase characterized by extensive travel, continued work on field theory, gradual withdrawal from science policy affairs, greater involvement with his institute and family, increasing concern with placing his work in philosophical perspective, and the sad deaths of many of his teachers, colleagues, and competitors: Max von Laue, Erwin Schrödinger, Niels Bohr, Otto Hahn, Lise Meitner, Paul Scherrer, Max Born, and Hans Kienle.[69] Sommerfeld had died in 1951, Einstein four years later.

Faced with his own mortality and the likelihood that he would never regain the stature he had once enjoyed, Heisenberg now increasingly attempted to place his life's work in a permanent intellectual tradition. His fiftieth, fifty-fifth, and sixtieth birthdays brought him renewed concerns about his advancing age and his ability to continue first-rate physics.[70] Under the influence of his long-time friend and colleague Carl Friedrich von Weizsächer, his preferred intellectual tradition derived from ancient Greek philosophy. One former student recalled that Weizsächer and Heisenberg began every lecture course, no matter what the subject, with a reference to Greek philosophy.[71]

By the winter of 1955–1956, when Heisenberg delivered the Gifford Lectures on physics and philosophy at the University of St. Andrews in Scotland, he had already distinguished contemporary elementary particle physics from nineteenth-century atomism. For him, the latter was a form of repugnant mechanistic materialism derived from the atomic theories of Democritus and Leucippus; the former held closest affinity to the work of the sagacious Aristotle. The underlying matter field of Heisenberg's unified field theory bore similarities to the notion of substance in Aristotelianism, an intermediate type of reality. Measuring the properties of elementary particles seemed closest to the Aristotelian notion of potentia, since the particle comes into being only in the act of measurement.[72]

By the 1960s, particle qualities had succumbed to the symmetry properties of field equations, and Aristotle had succumbed to Plato. The Platonic atoms of his remembered youth were now fundamental. "The particles of modern physics are representations of symmetry groups and to that extent they resemble the symmetrical bodies of Plato's philosophy," he declared in one of his last publications.[73] In his 1969 memoirs, written as a Platonic dialogue, he claimed that Platonism had dominated his thinking throughout his career. Toward the end of the memoir he wrote of his happy days in the old Urfeld cottage during the 1960s when—with Colonel Pash and the war far behind him—"we could once again meditate peacefully about the great questions Plato had once asked, questions that had perhaps found their answer in the contemporary physics of elementary particles," a physics that found its meaning in the ancient idealism and transcendent philosophy of Plato.[74]

The Platonic contentment was interrupted more and more by illness. A liver condition caused increasing weakness, dizziness, and depression. The slender young man who had dazzled audiences with his scientific bravado and friends with his physical courage had turned by the late 1960s into an obviously aging physicist to whom life and physics were no longer as kind as they had once been. As he closed his memoirs in 1969, he knew that the end would not be long in coming. In his last sentences, he turned to what had been the one sustaining force in his life—the wondrous beauty and harmony of classical music. As he recounted listening to two of his sons and a colleague play Beethoven's youthful Serenade in D Major one sunny afternoon in an institute in the beautiful Bavarian countryside of the land he so loved, he closed by extolling how the Serenade "brims over with vital force and joy. . . . Faith in the central order keeps casting out faintheartedness and weariness. And as I listened, I grew firm in the conviction that, measured on the human scale, life, music and science would always go on, even though we ourselves are no more than transient visitors or, in Niels' words, both spectators and actors in the great drama of life."[75]

A few years later, Heisenberg fell ill again and was hospitalized. Exploratory surgery indicated advanced cancer of the kidneys and gall bladder for which little could be done. Chemotherapy helped delay the inevitable. In 1975 his condition worsened. He was hospitalized again and returned home too weakened to recuperate. Werner Karl Heisenberg died peacefully at home in Munich on Sunday, February 1, 1976. The following evening, his many assistants, colleagues, friends, and coworkers sadly gathered at the institute to walk in a candlelight procession to his nearby home, where each one silently placed a glowing candle on the steps before his door.

Appendix A

CHRONOLOGY

1901.	5 Dec.	Werner Karl Heisenberg born in Würzburg, Germany, to August and Anna (née Wecklein) Heisenberg
1906.	Sep.	Enters primary school in Würzburg
1909.	30 Dec.	August Heisenberg appointed professor of middle and modern Greek studies at University of Munich, begins teaching in Jan.
1910.	June	Heisenberg family moves from Würzburg to Munich
	Sep.	Attends Elisabethenschule in Munich for last year of primary education
1911.	Sep.	Begins nine-year course of study at the humanistic Maximilians-Gymnasium in Munich, then under the rectorship of his grandfather, Nikolaus Wecklein
1914.	1 Aug.	Outbreak of World War I
1916–1918		Member of Military Preparedness Association (Wehrkraftverein) of Max-Gymnasium
1918.	Aug.–Sep.	Participates in Bavarian agricultural service
	8 Nov.	Bavarian socialist republic declared
	11 Nov.	Armistice
1919.	1 May–June	Participates in military suppression of Bavarian Soviet Republic
	July	Becomes leader of youth movement group of boys at Max-Gymnasium
	1–3 Aug.	Attends Schloss Prunn meeting of Neupfadfinder, a youth movement organization
1920.	15 July	Graduates from Max-Gymnasium; selected for support by Maximilianeum Foundation, Munich
	Aug.	Nearly dies of typhoid after tour with youth group
	Oct.	Enters University of Munich as student of Arnold Sommerfeld; first meeting with Wolfgang Pauli
1921.	17 Dec.	Submits his first paper for publication
1922.	12–22 June	Attends Göttingen lectures by Niels Bohr; first meeting with Bohr
	Sep.	Delivers invited paper to hydrodynamics conference in Innsbruck
1922–23.	Oct.–May	Studies in Göttingen with Max Born and others while Sommerfeld visits United States
1923.	11 May	Receipt of paper by Born and WH on excited helium
	July	Completes requirements for doctorate in physics from University of Munich
	Aug.	Leads youth group on trip to Finland

	Oct.	Returns to Göttingen as privately funded assistant to Max Born
1924.	15–27 Mar.	Visits Bohr's institute in Copenhagen
	7 June	Meets Einstein in Göttingen for first time
	28 June	Delivers habilitation lecture in Göttingen; *venia legendi* in physics conferred by University of Göttingen in Oct.
1924–25.	17 Sep.–1 May	International Education Board (Rockefeller Foundation) fellow in Bohr's institute
1925.	29 June	Receipt of WH's paper from Göttingen providing break-through to quantum mechanics
	16 Nov.	"Three-man paper" by Born, Heisenberg, and Jordan on foundations of quantum mechanics received from Göttingen
1926.	1 May	Begins appointment as university lector and assistant to Niels Bohr in Copenhagen
1927.	23 Mar.	Paper on uncertainty principle received from Copenhagen
	11–20 Sep.	Attends Volta Conference, Lake Como, where Bohr presents principle of complementarity
	24–29 Oct.	Attends 5th Solvay Congress in Brussels
1928.	1 Feb.	Delivers inaugural lecture as professor of theoretical physics, University of Leipzig
	20 May	Presents a quantum theory of ferromagnetism
1929.	19 Mar.	First of two papers (with Pauli) laying the foundations of relativistic quantum field theory
	Mar–Nov.	Travels to United States, Japan, China, and India
1930.	22 Nov.	Death of his father
1932.	7 June	Receipt of his first paper on neutron-proton model of nucleus
1933.	30 Jan.	Hitler comes to power
	22–29 Oct.	Attends and addresses 7th Solvay Congress on nuclei
	3 Nov.	Receives Max Planck Medal of German Physical Society
	11 Dec.	Receives Nobel Prize for Physics for 1932
1934.	21 June	Presents theory of the positron
1935.	Jan.	Sommerfeld recommends WH as his successor in Munich
	8 May	Faculty protest meeting fails in effort to rescind dismissals of faculty at Universiy of Leipzig
1936.	29 Jan.	Attack on WH and theoretical physics in a Nazi party newspaper
	28 Feb.	Responds to attack in same newspaper; circulates petition in protest
	8 June	Presents theory of cosmic-ray showers based on Fermi field and critical minimum length

1937.	Jan.	Meets Elisabeth Schumacher at private music recital
	29 Apr.	Marries Elisabeth Schumacher in Berlin
	15 July	WH and others viciously attacked in SS periodical
1938.	Jan.	Birth of twins, the first of seven children
	Mar.	Travels to Cambridge and Manchester, England
	21 July	Himmler exonerates WH of SS charges, but appointment to succeed Sommerfeld blocked
	Aug.–Sep.	Serves in reserve mountain infantry unit during Sudeten Crisis
	9–10 Nov.	Witnesses violence of Nazi Kristallnacht
1939.	June	Buys Urfeld summer home as retreat for family from unrest
	June–July	Travels to United States for lectures; refuses to immigrate
	1 Sep.	Start of World War II
	26 Sep.	Joins German uranium fission research as leading theoretician
	6 Dec.	Presents first of secret theoretical surveys of fission to German Army Ordnance Office
1941.	Spring	Leipzig uranium pile first multiplies neutrons
	15–22 Sep.	Travels to German occupied Copenhagen to lecture at a propaganda institute and discuss nuclear research with Bohr
1942.	26 Feb.	Presents lecture to Reich officials on energy acquisition from uranium fission
	4 June	Briefs Albert Speer, Minister for Armaments and War Production, on nuclear research
	1 July	Appointed director at the Kaiser Wilhelm Institute for Physics in Berlin, one of the main pile research laboratories; commutes from Leipzig
	8 Sep.	Presents the first of a multipart paper on the S-matrix in elementary particle physics
	9 Dec.	Attends his first meeting of the Wednesday Society in Berlin
1943.	Feb.	Appointed by Reich Education Ministry to chair for theoretical physics at University of Berlin, in addition to his post at the Kaiser Wilhelm Institute
	Apr.	Election to Prussian Academy of Sciences approved; family moves permanently to Urfeld as Allied bombing increases
	6 May	Lectures on nuclear fission before Göring's Aeronautical Academy
	Summer	Dispatches most of his institute staff to the Black Forest to escape bombing of Berlin

	18–26 Oct.	Travels to German occupied Netherlands
1944.	24 Jan.–4 Feb.	Travels to German occupied Copenhagen after German Army confiscates Bohr institute, brief return in Apr
	12 July	Hosts last full meeting of Wednesday Society
	20 July	Failed assassination attempt on Hitler involving some members of Wednesday Society
	Dec.	Lectures in neutral Switzerland
1945.	Jan.	Vacates Berlin with remainder of research team to continue pile research in the south
	3 May	Arrested in Urfeld by a U.S. colonel, brought to Heidelberg and held in France and Belgium
	7–8 May	German surrender
	July	Detained along with nine other leading German atomic scientists at an English country estate; mother dies in Bad Tölz
	6 and 9 Aug.	Atomic bombs dropped on Hiroshima and Nagasaki
1946.	3 Jan.	Transported with other German scientists to Alswede in British occupation zone
	Mar.	Settles in Göttingen
	July	Named director of the Kaiser Wilhelm Institute for Physics in Göttingen (later Max Planck Institute)
1947.	Dec.	Lectures in Cambridge, Edinburgh, and Bristol
1949.	9 Mar.	German Research Council established by Max Planck Society and West German academies; appointed president of council
1950.	23 Feb.	Proposes nonlinear spinor field theory containing universal length
	30 Aug.–6 Sep.	Attends International Congress of Mathematicians in Cambridge, Mass.
1951.	Aug.	Research Council fused with Notgemeinschaft to form German Research Association; appointed to presidium
1952.	Feb.	Chairman of Commission for Atomic Physics in German Research Association
	Mar.	Heads German delegation to European Council for Nuclear Research
1953.	10 Dec.	Appointed by Adenauer president of Alexander von Humboldt Foundation
1954		West German delegate to conference on Atoms for Peace in Geneva
1955–1956.	Winter	Delivers Gifford Lectures on "Physics and Philosophy" at University of St. Andrews
1957.	12 Apr.	Issues declaration with 17 other Göttingen scientists against West German nuclear weapons

1958.	27 Feb.	Issues preprint (with Pauli) on a proposed unified field theory of elementary particles, later renunciated by Pauli
	Sep.	Moves with his family and institure to Munich
1970.	31 Dec.	Resigns directorship of Max Planck Institute for Physics
1975.	Oct.	Resigns presidency of Alexander von Humboldt Foundation
1976.	1 Feb.	Dies of cancer at his home in Munich

Appendix B

HEISENBERG'S TEXTBOOKS AND READINGS
AT THE MAXIMILIANS-GYMNASIUM

Based on the *Jahresberichte* and *Schulordnung*. Class years are given on the right.

Mathematics and Physics

Franz Dicknether, *Lehrbuch der Arithmetik*, part 1	1–2
Ibid., part 2	3–4
Jos. Lengauer, *Die Grundlehren der ebenen Geometrie*	4–6
Lengauer, *Dr. E. Bardeys Aufgabensammlung*	4–8
Lengauer, *Die Grundlehren der ebenen Trigonometrie*	7–8
Lengauer, *Die Grundlehren der Stereometrie*	8
Johann Kleiber and Adalbert Grüttner, *Kleiber-Nath: Physik für die Oberstufe*, 7th ed. (1916); 8th ed. (1918)	7–9

Greek

Grammar and readers	4–5
Xenophon, *Anabasis*, I, III, IV	5–6
Homer, *Odyssey* (selection)	6–7
Herodotus, *Persian War*, VIII	7
Lucian, *Sharon*	7
Demosthenes, I, III	8
Olynthian speech	8
Sophocles, *Aias*	8
Sophocles, *Oedipus*	9
Plato, *Apology*	9
Plato, *Symposium* (selection)	9
Plato, *Phaedo* (selection)	9

Latin

Grammar and readers	1–5
Caesar, *De bello Gallico*	4–5
Ovid, *Metamorphoses*	6
Livius, XXI	6
Vergil, *Aeneid*, I, II, VI	7
Cicero, *Cato maior*	7
Horace, *Oden* (selections)	8
Tacitus, *Germania* (selection)	8
Horace, *Satires* (selections)	9
Tacitus, *Annals*, I, II	9

German

Readers	1–5, 9
Nibelungenlied (Hochdeutsch)	4
Ibid., (Mittelhochdeutsch)	7
Schwab, *Der gehörnte Siegfried*	4
Letters from the front	4
Reports from Army Command Headquarters	4
Reports from war correspondents	4
Theodor Storm, *Pole Poppenspäler*	5
Schmitthenner, *Die Frühglocke*	5
Detlev von Liliencron, *Kriegsnovellen*	5
Goethe, *Hermann und Dorothea*	6
Goethe, *Iphigenie*	8
Goethe, *Götz von Berlichingen*	8
Goethe, *Tasso*	9
Goethe, *Egmont*	9
Schiller, *Das Lied von der Glocke*	6
Schiller, *Tell*	7
Schiller, translation of *Aeneid*	7
Schiller, *Wallensteins Tod*	9
Schiller, *Die Braut von Messina*	9
Schiller, *Wallensteins Lager*	9
Schiller, *Piccolomini*	9
G. Keller, *Das Fähnlein der sieben Aufrechten*	6
Das Waltharilied	6
A. Stern, *Die Flut des Lebens*	6
Riehl, *Der Stadtpfeifer*	6
Ratzl, *Aus dem Kriege mit Frankreich*	7
Lessing, *Laokoon*	8
Lessing, *Minna von Barnhelm*	8
Grillparzer, *Libussa*	8
Keller, *Kleider machen Leute*	8
Gustav Freytag, *Die Journalisten*	8
Shakespeare, *Macbeth* (translation)	9
Hebbel, *Agnes Bernauer*	9
Mörike, *Mozart auf der Reise nach Prag*	9
Auerbach, *Diethelm von Buchenberg*	9

Notes

Abbreviations of frequently cited references and source locations are listed below. Archival references in the notes are followed by call numbers.

Papers reprinted in easily accessible collected works are cited by year and location in the collected works. The original sources may be found there. Unless otherwise indicated, published versions of all letters cited from and to Wolfgang Pauli may be found in the appropriate volume of PWB, the contents of which are arranged chronologically. All correspondence cited between WH and his family is in HP.

The wording of quotations not originally in English may be found in the corresponding translation of this book. Most translations of unpublished German and French quotations are by the author. Published English translations have been compared with the originals and edited as necessary. References to writings by WH are to HCW. Translations used for English quotations, as well as republications of the original works in HCW, may be found in WH, *Bibl.*

Abbreviations Used in Notes

ADJ	Archiv der Deutschen Jugendbewegung, Burg Ludwigstein
AHQP x, y	Archive for History of Quantum Physics, Berkeley, New York, and elsewhere; x = microfilm number, y = section
AIP	American Institute of Physics, New York (later Maryland)
Alsos	Samuel A. Goudsmit, *Alsos* (New York: Henry Schuman, 1947; reprinted Los Angeles: Tomash Publishers, 1983)
AP	*Annalen der Physik*
AS, *Atombau*	Arnold Sommerfeld, *Atombau und Spektrallinien*, various editions (Braunschweig: F. Vieweg und Sohn, 1919 ff.)
BA Koblenz	Bundesarchiv Koblenz
BAW	Bayerische Akademie der Wissenschaften, Archiv, Munich
BCW x, y	Niels Bohr, *Collected Works*, 9 vols., ed. Léon Rosenfeld et al. (Amsterdam: North-Holland); x = volume, y = pages
BDC	Berlin Document Center, U.S. Army, Berlin (later BA Koblenz)
Beyerchen	Alan D. Beyerchen, *Scientists under Hitler: Politics and the physics community in the Third Reich* (New Haven: Yale University Press, 1977)
BGC	Bohr General Correspondence, NBA
BHStA	Bayerisches Hauptstaatsarchiv, Allgemeines Archiv, Munich
BMS	Bohr Manuscripts, NBA, and on microfilm in AHQP
BP	Hans Bethe Papers, Archives, Cornell University, Ithaca, New York
BPC	Bohr Private Correspondence, NBA
BSB	Bayerische Staatsbibliothek, Handschriftenabteilung, Munich

BSC *x*, *y*   Bohr Scientific Correspondence, microfilms in AHQP; *x* = microfilm number, *y* = section

DC, "Cosmic rays"   D. Cassidy, "Cosmic ray showers, high energy physics and quantum field physics: Programmatic interactions in the 1930s," *HSPS, 12* (1981), 1–39

DC, "TH Berlin"   D. Cassidy, "Gustav Hertz, Hans Geiger und das Physikalische Institut der Technischen Hochschule Berlin in den Jahren 1933 bis 1945," in *Wissenschaft und Gesellschaft,* ed. R. Rürup, vol. 1 (Berlin: Springer-Verlag, 1979), pp. 373–387

DC, "WH"   "Werner Heisenberg and the crisis in quantum theory, 1920–1925," Ph.D. dissertation, Purdue University, 1976

DJ *x*, *y*   David Irving, comp., *Third Reich Documents,* Group 11, *German Atomic Research,* microfilms of photocopied primary sources assembled from various archives and private sources (Wakefield, England: Microform Academic Publishers, 1966); *x* = microfilm number, *y* = frame numbers

DP   Peter Debye Papers, at MPG

Dresden   Max Dresden, *H. A. Kramers: Between tradition and revolution* (New York: Springer-Verlag, 1987)

EA   Albert Einstein Archive, The Hebrew University, Jerusalem

EB, *Briefw.*   Albert Einstein, Hedwig Born, and Max Born, *Briefwechsel 1916–1955,* with commentary by Max Born (Munich: Nymphenburger Verlagshandlung, 1969). The English translation of this work is not complete.

EH, *Recoll.*   Elisabeth Heisenberg, *Inner exil: Recollections of a life with Werner Heisenberg,* trans. S. Cappellari and C. Morris (Boston: Birkhäuser, 1984)

EHR   Paul Ehrenfest Papers, microfilms in AHQP

ES, *Briefw.*   Albert Einstein and Arnold Sommerfeld, *Briefwechsel,* ed. Armin Hermann (Basel: Schwabe & Co., 1968)

ETH   Eidgenössische Technische Hochschule, Handschriftenabteilung, Zurich

GP   Samuel A. Goudsmit Papers, AIP

HA   Heisenberg Archive, Max Planck-Institut für Physik und Astrophysik, Munich. To be transferred to MPG.

HCW *xy*, *z*   WH, *Gesammelte Werke/Collected Works,* multiple vols., ed. W. Blum et al., (Berlin: Springer-Verlag, and Munich: Piper-Verlag, 1985—); *x* = series (A, B, C), *y* = volume, *z* = pages

Hendry, "BKS"   John Hendry, "Bohr-Kramers-Slater: A virtual theory of virtual oscillators and its role in the history of quantum mechanics," *Centaurus,* 25 (1981), 189–221

Hendry, *Creation*   John Hendry, *The creation of quantum mechanics and the Bohr-Pauli dialogue* (Hingham, Mass.: D. Reidel, 1984)

Hermann, *Jahrhundert*	Armin Hermann, *Die Jahrhundertwissenschaft: Werner Heisenberg und die Physik seiner Zeit* (Stuttgart: Deutsche Verlags-Anstalt, 1977)
Hermann, *WH*	Armin Hermann, *Werner Heisenberg: In Selbstzeugnissen und Bilddokumenten* (Reinbek: Rowohlt, 1976)
HP	Heisenberg private papers, in care of the Heisenberg family
HSPS	*Historical studies in the physical and biologial sciences*
IZ	Institut für Zeitgeschichte, Munich
Irving	David Irving, *The German atomic bomb: The history of nuclear research in Nazi Germany* (reprint New York: Da Capo Press, 1967)
Jammer, *Conceptual*	Max Jammer, *The conceptual development of quantum mechanics* (New York: McGraw-Hill, 1966)
Jammer, *Philosophy*	Max Jammer, *The philosophy of quantum mechanics: The interpretations of quantum mechanics in historical perspective* (New York: John Wiley & Sons, 1974)
JM	Christa Jungnickel and Russell McCormmach, *Intellectual mastery of nature: Theoretical physics from Ohm to Einstein*, 2 vols. (Chicago: University of Chicago Press, 1986)
KFZ Karlsruhe	Kernforschungszentrum, Bibliothek, Karlsruhe
LC	Library of Congress, Manuscript Division, Washington, D.C., microfilm collection of captured German war documents
LNN	*Leipziger Neueste Nachrichten*
Ludwig	Karl-Heinz Ludwig, *Technik und Ingenieure im Dritten Reich* (Königstein/Ts.: Athenäum-Verlag, 1974/1979)
MG	Maximilians-Gymnasium, or Max-Gymnasium
MNN	*Münchner Neueste Nachrichten*
MPG	Archiv zur Geschichte der Max Planck-Gesellschaft, Berlin-Dahlem
MR x, y, z	Jagdish Mehra and Helmut Rechenberg, *The historical development of quantum theory*, multiple volumes and parts (New York: Springer-Verlag, 1982 ff.); x = volume, y = part, z = pages
NARA	National Archives and Records Administration, Washington, D.C.
NBA	Niels Bohr Archive, Niels Bohr Institute, Copenhagen
Nwn	*Die Naturwissenschaften*
Pais, *Subtle*	Abraham Pais, '*Subtle is the Lord . . .*': *The science and the life of Albert Einstein* (Oxford: Clarendon Press, 1982)
Pais, *Inward*	Abraham Pais, *Inward bound: Of matter and forces in the physical world* (Oxford: Clarendon Press, 1986)
PBl	*Physikalische Blätter*

PCSP *x, y*	Wolfgang Pauli, *Collected scientific papers*, 2 vols., ed. R. Kronig and V. F. Weisskopf (New York: Interscience, 1964); x = volume, y = pages
PR	*Physical Review*
PRS	*Proceedings of the Royal Society of London*, series A
PWB	Wolfgang Pauli, *Wissenschaftlicher Briefwechsel mit Bohr, Einstein, Heisenberg u. A.*; vol. 1: *1919–1929*, ed. A. Hermann et al. (New York: Springer-Verlag, 1979); vol. 2: *1930–1939*, ed. Karl von Meyenn (Berlin: Springer-Verlag, 1985)
PZ	*Physikalische Zeitschrift*
RAC	Rockefeller Archive Center, North Tarrytown, New York
Ringer	Fritz K. Ringer, *The decline of the German mandarins: The German academic community, 1890–1933* (Cambridge, Mass.: Harvard University Press, 1969)
Robertson	Peter Robertson, *The early years: The Niels Bohr Institute 1921–1930* (Copenhagen: Akademisk Forlag, 1979)
Röseberg	Ulrich Röseberg, *Niels Bohr: Leben und Werk eines Atomphysikers 1885–1962* (Berlin: Akademie-Verlag, 1987)
SAM	Stadtarchiv Munich
SAW, MPK	Sächsische Akademie der Wissenschaften, mathematisch-physikalische Klasse
Serwer	Daniel Serwer, *"Unmechanischer Zwang*: Pauli, Heisenberg, and the rejection of the mechanical atom," *HSPS*, 8 (1977), 189–256
SN	Sommerfeld Nachlass, Deutsches Museum, Munich
SPK	Staatsbibliothek Preussischer Kulturbesitz, Berlin.
SStA	Sächsisches Staatsarchiv Dresden
StAM	Staatsarchiv Munich
UA	Universitätsarchiv or University Archive
UB	Universitätsbibliothek or University Library
Waerden	B. L. van der Waerden, ed. and trans., *Sources of quantum mechanics* (New York: Dover, 1967)
Walker	Mark Walker, *German National Socialism and the quest for nuclear power 1939–1949* (New York: Cambridge University Press, 1989)
Weart	Spencer Weart, *Scientists in power* (Cambridge, Mass.: Harvard University Press, 1979)
WH, *Bibl.*	*Werner Heisenberg: A bibliography of his writings*, comp. D. Cassidy and M. Baker, Berkeley Papers in History of Science, 9 (Berkeley, Calif: Office for History of Science and Technology, 1984)
WH, *PB*	WH, *Physics and Beyond: Encounters and Conversations*, trans. Arnold J. Pomerans (New York: Harper & Row,

1971). Quoted translations are edited to conform to the original German.

WH-Wirtz   WH and Karl Wirtz, "Grossversuche zur Vorbereitung der Konstruktion eines Uranbrenners," in *Naturforschung und Medizin in Deutschland 1939–1946*, FIAT review of German Science, vol. 14, ed. W. Bothe and S. Flügge (Weinheim: Verlag Chemie, 1953), pp. 142–165

ZP   *Zeitschrift für Physik*

ZStA Potsdam   Zentraler Staatsarchiv Potsdam

Chapter 1

1. A. Heisenberg, autobiographical sketch, in *Geistiges und künstlerisches München in Selbstbiographien*, ed. W. Zils (Munich: Max Kellerers Verlag, 1913), pp. 156–160.

2. Ibid.

3. A. Heisenberg to Erich Petzet, 8 Jan 1913 (BSB, E. Petzetiana IVb).

4. Hartmut Kaelble, "Sozialer Aufstieg in Deutschland 1850–1914," *Vierteljahrsschrift für Soziologic und Wirtschaftsgeschichte*, 60 (1973), 41–71.

5. This has been argued in Ringer.

6. A. Heisenberg, "Studien zur Textgeschichte des Georgios Akropolites," Ph.D. dissertation, Munich, 1893.

7. Surmised from A. Heisenberg to Erich Petzet, 26 Oct 1895 (BSB, E. Petzetiana IVb). August's moves from and to other schools are not included here.

8. A. Heisenberg to his father, 13 Apr 1897 (HP).

9. Report of Staatsministerium, 17 Apr 1893, in "Acta des Königlichen Staats-Ministerium des Inneren für Kirchen und Schul-Angelegenheiten. Dr. Heisenberg August" (BHStA, MK 17732).

10. N. Wecklein, "Zeising: Adolf Z.," *Allgemeine deutsche Biographie*, vol. 55 (1910), pp. 404–411. Zeising was best known for his recovery of the law of the golden mean in mathematical aesthetics: ancient and renaissance artists had chosen the sizes and proportions in their works according to certain explicit ratios. Otto Hagenmaier, *Der Goldene Schnitt. Ein Harmoniegesetz und seine Anwendung* (Munich: Heinz Moos, 1963).

11. An outline of Wecklein's life was provided by his school colleague J. Melber in "Geheimer Hofrat Dr. Nik. Wecklein," *Bayerische Blätter für das Gymnasialschulwesen*, 69 (1927), 88–102.

12. Ibid., and Wecklein to Halm, 27 Sep 1875 and 19 Jan 1877 (BSB, Halmiana IX).

13. A. Heisenberg-Krumbacher correspondence (BSB, Krumbacheriana I).

14. Heisenberg, note 1.

15. M. Schanz, "Auszug des Berichts des Koreferats," 6 Jun 1901, ("Acta," note 9).

16. "Bericht des academischen Senats der Königl. Universität Würzburg," 30 Nov 1901 ("Acta," note 9). His lecture was published as "Die bisherige Entwicklung der byzantinischen Philologie und ihre künftige Aufgaben," *Allgemeine Zeitung* (Munich), no. 274 (28 November 1901), Beilage, pp. 1–5.

17. "Frequenz der humanistischen Gymnasien . . . am Schlusse des Schuljahres 1901/02," *Blätter für das Gymnasial-Schulwesen*, 38 (1902), 661–663.

18. Dieter Schäfer, *Der Weg der Industrie in Unterfranken* (Würzburg: Stürtz, 1970).

19. Klaus Schönhoven, "Arbeiterschaft, Gewerkschaften und Sozialdemokratie in

Würzburg 1848–1914," in Hans W. Loew and Klaus Schönhoven, eds. *Würzburgs Sozialdemokraten: Vom Arbeiterverein zur Sozialdemokratischen Volkspartei 1868–1978* (Würzburg: Stürtz, 1978), pp. 1–39, on pp. 22–24.

20. Records in August Heisenberg's personnel file ("Acta," note 9). Inflation was minimal over those four years.

21. Franz Dölger, "Die Byzantinisten der Akademie. Jakob Philipp Fallmerayer, Karl Krumbacher, August Heisenberg," in *Geist und Gestalt. Biographische Beiträge zur Geschichte der Bayerischen Akademie der Wissenschaften vornehmlich im zweiten Jahrhundert ihres Bestehens* (Munich: Beck, 1959), pp. 139–157, on p. 153; Ernest Reisinger, recollection of August Heisenberg as teacher, in *Meine Jugend in Alt-Schwabing* (Munich: Franzis-Verlag, 1952), pp. 68–71.

22. "Qualifikationsliste. K. Lateinschule Lindau," 3 May 1898 ("Acta," note 9).

23. For instance, Geneviève Bianquis, "La femme allemande à l'époque moderne et contemporaine (du XVIIe au XXe siècle)," in *Histoire mondiale de la femme*, ed. Pierre Grimal, vol. 4 (Paris: Nouvelle Librairie de France, c. 1980), pp. 253–290; and Annette Kuhn und Gerhard Schneider, eds., *Frauen in der Geschichte* (Düsseldorf: Schwann, 1979).

24. *Jahres-Bericht über das Kgl. Alte Gymnasium zu Würzburg* for 1901–1902 to 1909–1910.

25. Bibliography in Franz Dölger, "August Heisenberg. Geboren 13. November 1869, gestorben 22. November 1930," *Jahresbericht über die Fortschritte der klassischen Altertumswissenschaft*, 241 (1933), 25–55, on 43–55.

26. A. Heisenberg, *Grabeskirche und Apostelkirche: Zwei Basiliken Konstantins*, 2 vols. (Leipzig, 1908); and "Acta des K. Acad. Senats der Ludwig-Max.-Universität München. Betreffend Dr. August Heisenberg" (UA Munich).

27. Kaspar Hammer, "Qualifikationsliste. Altes Gymnasium, 1902" ("Acta," note 9).

28. Interview with C. F. von Weizsäcker, Starnberg, 30 Apr 1982.

29. George L. Mosse, *Nationalism and sexuality: Respectability and abnormal sexuality in modern Europe* (New York: Howard Fertig, 1985).

30. Reports in "Acta," note 9.

31. J. L. Heilbron, *The dilemmas of an upright man: Max Planck as spokesman for German science* (Berkeley: University of California Press, 1986), p. 4.

32. One need only think of Freud's patients in nearby Vienna or of the hypocrisy of Victorian morality exposed by British literary figures of the period. See, for example, Carl Schorske, *Fin-de-Siécle Vienna: Politics and culture* (New York: Knopf, 1979).

33. WH to his parents, 11 Jan [1928].

34. "Double dialogue with Werner Heisenberg" (1974), HCW C3, 464–486, on 475.

35. WH, "Ordnung der Wirklichkeit" (c. 1942), HCW C1, 218–306; WH, "Naturwissenschaftliche und religiöse Wahrheit" (1973), HCW C3, 422–439.

36. Interview with Elisabeth Heisenberg, Göttingen, Feb 1982.

37. C. F. von Weizsäcker, "Heisenbergs Entwicklung seit 1927," in Weizsäcker and B. L. van der Waerden, *Werner Heisenberg* (Munich: Hanser, 1977), pp. 25–40, on p. 40.

38. This seems in accord with Mosse's assertion (note 29) that the lately developed notion of the nuclear family was such that most members of the upper middle class sharply distinguished between the world of public affairs and the very private world of family and household.

39. EH, note 36.

40. Ibid.

41. Interview of WH by T. S. Kuhn, 30 Nov 1962 (AHQP).

42. "Zeugnisnoten-Protokoll des K. Maximilians-Gymnasiums in München, 1913/14, Klasse IIIA," quoted in Hermann, WH, p. 8.

43. Interview with Fr. Gottfried Simmerding, former youth comrade, Munich, 1982.

44. EH, Recoll., p. 13.

45. K. Maximilians-Gymnasium München, "Zensur, Werner Heisenberg," class 4A, 1914–1915, and class 5A, 1915–1916 (Archive, Maximilians-Gymnasium, Munich).

46. WH to Koji Kigoshi, 5 Mar 1971 (HA).

47. EH, note 36.

48. WH, "Theory, criticism and a philosophy," (1968), HCW C2, 423–438, on 438.

49. Melber, note 11.

50. August Heisenberg's very active involvement in gymnasium political affairs is indicated by "Mitteilung über die Gymnasiallehrer-Vereinigung München," Blätter für das Gymnasial-Schulwesen, 41 (1905), 554–556, and by his speeches to teachers' meetings in Bericht über die XXIV. Generalversammlung des Bay. Gymnasiallehrervereins, München, 4. April bis 6. April 1907, pp. 18–20, on p. 18, supplement to Blätter für das Gymnasial-Schulwesen , 43 (1907); and in Bericht über die XXV. Generalversammlung des Bayerischen Gymnasiallehrervereins, Augsburg, 15. April bis 17. April 1909, supplement to Blätter für das Gymnasial-Schulwesen, 45 (1909). The issues of the period are described in Eugen Brand, "Die Entwicklung des Gymnasiallehrerstandes in Bayern von 1773–1904," Blätter für das Gymnasial-Schulwesen, 40 (1904), 433–588; and Johannes Guthmann, Der Bayerische Lehrer- und Lehrerinnenverein: Seine Geschichte, vol. 2 (Munich: Oldenbourg, 1959).

51. "Rede von Professor Werner Heisenberg," delivered to Werner Heisenberg-Gymnasium, Garching, (1972), HCW C5, 433–435, on 433.

52. Theodore Preger, "Karl Krumbacher, geb. 23. September 1856, gest. 12. Dezember 1909," Blätter für das Gymanasial-Schulwesen, 46 (1910), 78–79.

53. Dr. O. Crusius to Akademischer Senat, 21 Dec 1909 ("Acta," note 26).

54. Ibid.

55. A. Heisenberg to Akad. Senat, 9 Jan 1910 ("Acta," note 26).

Chapter 2

1. This according to a file card on August Heisenberg (SAM) and Adressbuch für München und Umgebung (Munich, 1911). Congenial neighbors, an army lieutenant and the widow of a gymnasium teacher, occupied the other two apartments on the floor.

2. Robert Eben Sackett, Popular entertainment, class, and politics in Munich, 1900–1923 (Cambridge, Mass.: Harvard University Press, 1982) and Allan Mitchell, Revolution in Bavaria, 1918–1919: The Eisner regime and the soviet republic (Princeton: Princeton University Press, 1965).

3. Oscar Brunn, Plan von München, 1911 (Graphiksammlung, Stadtmuseum, Munich).

4. Alois Wagner, Zu meiner Zeit. Ein Bubenleben in Schwabing 1904 bis 1918 (Munich: Süddeutscher Verlag, 1980).

5. The Heisenberg family also acquired an automobile in this period. A photograph survives of August, the boys, and the W021ins seated in what must have been an early model. Published in Hermann, WH, p. 8.

6. Heisenberg's primary school records, once in his private papers, could not be found. The school he attended is surmised from information in card file and Addressbuch, note 1, and "Bericht über die Städtischen Volks- und Mittelschulen

Münchens für das Geschäftsjahr 1911. Erstattet von Stadtschulrat Dr. Sg. Kerschenstein" (SAM, Schulamt 2236).

7. Hans Scharold, *100 Jahre Maximilians-Gymnasium. Ein Beitrag zur Geschichte des Gymnasiums in Bayern* (Munich, 1949).

8. "Erste Klasse A," K. Maximilians-Gymnasium [hereafter MG] in München, *Jahresbericht für das Schuljahr 1911/12* (Munich, 1912), pp. 39–40.

9. Johannes Melber, Hsgb., *Die Schulordnung an den höheren Lehranstalten Bayerns nach der Königlichen Verordnung vom 30. Mai 1914* (Munich: Lindauersche Universitätsbuchhandlung, 1914).

10. A classical caricature of an authoritarian gymnasium teacher and his class, which was not far from the truth in many cases, is Heinrich Mann's *Professor Unrat* (Frankfurt, 1905). It was the basis of the film *Der blaue Engel*.

11. MG, *Jahresbericht 1911/12*, p. 47.

12. During the middle years, 1914–1917, Herr Wolff was at the front, then assigned to the lower grades.

13. The texts for each subject are listed in the *Jahresberichte* and in Melber, note 9.

14. MG, "Zeugnisnoten-Protokoll," quoted by Hermann, *WH*.

15. MG, *Jahresbericht 1912/13*, p. 46, which includes the poem. It is not known whether Erwin turned down the honor or whether he was even asked.

16. "München am 1. Mobilmachungstag," *MNN*, 67: 393 (3 Aug 1914), Morgenblatt, 3 (dated 2 Aug)

17. Wagner, note 4, p. 43.

18. WH, "Es war einmal" (1964), HCW C4, 21.

19. For example, Ringer, and Konrad H. Jarnasch, *Students, society, and politics in imperial Germany: The rise of academic illiberalism* (Princeton: Princeton University Press, 1983), and other works surveyed in D. Cassidy, "Recent German perspectives on German technical education," *HSPS*, 14 (1983), 187–200.

20. A. Heisenberg, "Die jüngste Entwicklung der Sprachfrage in Griechenland," *Internationale Wochenschrift für Wissenschaft, Kunst und Technik*, 5 (1911), 685–702; Karl Krumbacher, *Das Problem der neugriechischen Schriftsprache* (Munich, 1902).

21. Klaus Schwabe, *Wissenschaft und Kriegsmoral. Die deutschen Hochschullehrer und die politischen Grundfragen des Ersten Weltkrieges* (Göttingen: Musterschmidt, 1969). Only three professors are known to have refused, one of whom was Albert Einstein.

22. See, for example, "Aufrufe und Aeusserungen der 'Intellektuellen'," *Die Eiche*, vol. 3, no. 2 (Apr 1915), 94–196; and Hermann Kellermann, *Der Krieg der Geister. Eine Auslese deutscher und ausländischer Stimmen zum Weltkriege 1914* (Weimar, 1915).

23. A. Heisenberg, *Der Philhellenismus einst und jetzt* (Munich: Beck, 1913). Founding address to Deutsch-Griechische Gesellschaft, 8 Dec 1912.

24. A. Heisenberg, "Die Zukunft Griechenlands," *Süddeutsche Monatshefte*, 12 (1915), 939–947; "Griechenland und die Mittelmeerfrage," *Panther*, 5 (1917), 349–356; *Neugriechenland* (Leipzig: Teubner, 1919).

25. A. Heisenberg, note 24, "Zukunft," p. 942.

26. "Kriegserlebnisse eines Münchner Universitätsprofessors [Heisenberg]," *MNN*, 68: 283 (6 Jun 1915), 4, and 68: 291 (10 Jun 1915), 3.

27. A. Heisenberg to Otto Crusius, 2 Feb 1915 (BSB, Crusiana). A powerful, though fictional, portrayal of the horrors of trench warfare was provided by the Osnabrück native Erich Maria Remarque, *All quiet on the western front* (New York: Fawcett Crest, 1928/1956).

28. A. Heisenberg to Crusius, 24 Apr 1915 (BSB, Crusiana).

29. WH, interview by T. S. Kuhn, 30 Nov 1962 (AHQP), p. 1. No indication of wounds or even of a relapse of rheumatism is available.

30. Based on MG, *Jahresberichte*.

31. For example, Karl Theodor Heigel (President, Bavarian Academy of Sciences), "An die akademische Jugend!" *Süddeutsche Monatshefte*, 11 (1914), 776–779.

32. [Ernst Kemmer], "Krieg und Schule," MG, *Jahresbericht 1914/15*, p. 36.

33. Ernst Kemmer, "Schule, militärische Jugenderziehung und vaterländischer Hilfsdienst," MG, *Jahresbericht 1916/17*, pp. 26–29, on p. 27.

34. "Kriegsgefahr und Lebensmittelmarkt," *MNN*, 67: 389 (1 Aug 1914), Vorabend-Ausgabe, 3.

35. Wagner, note 4, p. 48.

36. J. Melber, "Geheimer Hofrat Dr. Nik. Wecklein," *Bayerische Blätter für das Gymnasialschulwesen*, 69 (1927), 88–102.

37. Kemmer, note 33, p. 28.

38. "Jungmannen auf Land," *MNN*, 71: 192 (17 Apr 1918), Morgen-Ausgabe, 2; "Mittelschüler im landwirtschaftlichen Hilfsdienst," *MNN*, 71: 238 (12 May 1918), 3.

39. MG, "Jahreszeugnis Werner Heisenberg, 1917/18, 7. Klasse" (HP).

40. According to surviving Jugendwehr records, Heisenberg was appointed Gruppenführer in September 1918 (communication from Dr. Heyl, Archivdirektor, BHStA Kriegsarchiv).

41. WH, note 29, p. 3.

42. WH to his father, 15 May 1918.

43. WH to his parents, [May 1918].

44. WH, *PB*, p. 2.

45. This is based on reports of events at the front in *MNN*. Optimism among the Munich educated at the end of 1915 is displayed by physicist Arnold Sommerfeld, then touring the front, in a letter to his colleague Wilhelm Wien, 5 Dec 1915 (SN): "The war situation now appears to me to justify the more subtle hopes, not for an end in the near future but for a fundamental victory."

Chapter 3

1. WH, "Zur Quantentheorie der Linienstruktur und der anomalen Zeemaneffekte" (1922, rec. 17 Dec 1921), HCW A1, 134–158.

2. Yearly grade reports for WH (MG Archive, Munich), their emphasis.

3. "Reifezeugnis" for WH, 15 Jul 1920 (HP).

4. Johannes Melber, report on examination of WH (Maximilianeum-Stiftung, Archive, Munich).

5. Interview of WH by T. S. Kuhn, 30 Nov 1962 (AHQP); WH, "Rede zur 100-Jahrfeier des Max-Gymnasiums in München am 13.7.1949," HCW C5, 395–408.

6. Ibid.

7. Ibid.

8. A. Einstein, *Über die spezielle und die allgemeine Relativitätstheorie. Gemeinverständlich* (Braunschweig: Vieweg, 1917). Recalled by WH, "Begegnungen und Gespräche mit Albert Einstein" (1974), HCW C4, 202–216.

9. Johannes Melber, ed., *Die Schulordnung an den höheren Lehranstalten Bayerns nach der Königlichen Verordnung vom 30. Mai 1914* (Munich: Lindauersche Universitätsbuchhandlung, 1914), p. 92.

10. WH, note 8, p. 202.

11. WH, note 5, p. 2. The "Einstein relation" may be the equation $E = mc^2$.

12. WH, note 5, p. 2.

13. Melber, note 4.

14. The professor made the most of the opportunity. A. Heisenberg, "Dialekte und Umgangssprache im Neugriechischen," Bay. Akad. Wiss., *Jahrbuch 1918*, pp. 1–26, Festrede on 29 May 1918; "Phonographische Fixierung von Sprach- und Gesangsproben bei griechischen (kriegsgefangenen) Truppen in Görlitz ausgeführt von Prof. August Heisenberg 1917/1919" (BAW VII, 466).

15. Grade report for 1915–1916, note 2.

16. Fermat claimed that he could prove the theorem. He died without revealing the secret.

17. L. Kronecker, "Über die Auflösung der Pellschen Gleichung mittels elliptischer Functionen" (1863), and "Zur Theorie der elliptischen Functionen" (1883–1889), *Leopold Kronecker's Werke*, ed. K. Hensel, vol. 4 (Leipzig: Teubner, 1929), pp. 219–225 and 345–495.

18. WH, note 5, p. 2; WH, "Die Bedeutung des Schönen in der exakten Naturwissenschaft" (1970), HCW C1, 369–384, on 369.

19. WH, notebook (HA).

20. Paul Bachmann, *Zahlentheorie*, 5 vols. (Leipzig: Teubner, 1892 ff.).

21. Acquisition list, Library, MG, Munich. It is probable, however, that only the teachers had direct access to the library, in which case Wolff could have borrowed the books for Werner.

22. See Appendix B.

23. WH, *PB*, p. 2.

24. Plato, *Timaeus*, trans. Desmond Lee (New York: Penguin Books, 1977), par. 55–56.

25. WH, *PB*, p. 8.

26. WH, note 5, "Rede," p. 402.

27. WH to Pauli, 24 Nov 1925.

28. Interview with C. F. von Weizsäcker, Starnberg, Apr 1982.

29. MG, *Jahresbericht 1919/1920*.

30. Foundation documents, in *125 Jahre Maximilianeum 1852 bis 1977* (Munich: UNI-Druck, 1977).

31. Melber, note 4.

32. "Die Mitglieder des Maximilianeums von 1852 bis 1977," in *125 Jahre*, note 30, pp. 55–111.

33. *125 Jahre Maximiliansgymnasium München. Rückblick-Ausblick. Eine Dokumentation* (Munich: Max-Gymnasium, 1974), p. 113.

34. Recalled in Felix Andreas Wittmann to Dr. Riedl, 1972 (Maximilianeum-Stiftung, Archive, Munich).

Chapter 4

1. A. Heisenberg to Karl Krumbacher, 26 June 1892 (BSB Krumbacheriana I).

2. Data provided in Allan Mitchell, *Revolution in Bavaria, 1918–1919: The Eisner Regime and the soviet republic* (Princeton: Princeton University Press, 1965), p. 17. Political alliances, the Bavarian revolt, and Bavarian history throughout this period are described by Mitchell; Richard Grunberger, *Red rising in Bavaria* (New York: St. Martin's Press, 1973); Michael Doeberl, *Entwicklungsgeschichte Bayerns*, vol. 3

(Munich, 1931); and Karl Schwend, *Bayern zwischen Monarchie und Diktatur* (Munich, 1954).

3. This is argued by Mitchell, note 2.

4. Mitchell, note 2, p. 218; Wolfgang Treue, *Die deutschen Parteien: Vom 19. Jahrhundert bis zur Gegenwart* (Frankfurt am Main: Ullstein, 1975).

5. Dr. Gustav Wyneken, "An die Schüler der höheren Schulen!" *MNN*, 72: 100 (3 Mar 1919), 4.

6. Dresler, "An die Schüler," *MNN*, 72: 168 (12–13 Apr 1919), 4–5, and in other newspapers.

7. Account of Heisenberg's reminiscences in *Die neue Seite, 3. MPZ, Stammesmitteilungen*, 2 (1962), 2–3.

8. Interview with Fr. Gottfried Simmerding, Munich, Mar 1982. Simmerding was a former youth-group comrade of Heisenberg's and was a resident of the city at the time.

9. WH, *PB*, p. 7.

10. Robert G. L. Waite, *Vanguard of Nazism: The free corps movement in postwar Germany, 1918–1923* (Cambridge, Mass.: Harvard University Press, 1952). For example, Rudolf Höss, *Kommandant in Auschwitz: Autobiographische Aufzeichnungen*, ed. M. Broszat (Munich: DTV, 1963).

11. The most detailed account of the invasion and its planning is provided in *Darstellungen aus den Nachkriegskämpfen deutscher Truppen und Freikorps*, vol. 4: *Die Niederwerfung der Räteherrschaft in Bayern 1919*, ed. Kriegsgeschichtliche Forschungsanstalt des Heeres (Berlin, 1939). Notice is taken, however, of the period and of the audience for which this work was written.

12. Ibid., and "Die Beteiligung der II. Marine-Brigade," *MNN*, 72: 171 (5 May 1919), Abend-Ausgabe, 3.

13. Oven's order is printed in *Darstellung*, note 11, pp. 218–219.

14. Grunberger, note 2, p. 144.

15. "Verluste," note 11, pp. 209–212. Toller eventually emigrated to the United States and committed suicide in New York after Franco's victory in the Spanish Civil War. He wrote an illuminating autobiography, *Eine Jugend in Deutschland* (Amsterdam, 1933; reprinted Reinbek: Rowohlt, 1983).

16. Interview of WH by T. S. Kuhn, 30 Nov 1962 (AHQP); also WH, "Rede zur 100-Jahrfeier des Max-Gymnasiums in München am 13.7.1949," HCW C5, 395–408, on 401.

17. *MNN* for the period.

18. "Allgemeine Studentenversammlung an der Universität," *MNN*, 72: 181 (10–11 May 1919); also Aufruf der Universität zum Eintritt in die Freikorps," *MNN*, 72: 177 (8 May 1919), Abend-Ausgabe, 3.

19. The Jewish heritage of most of the revolutionaries was often used as a pretext for engaging in and fomenting violent anti-Semitism. One overview of the rise and public political expression of anti-Semitism in Germany is provided in Lucy S. Dawidowicz, *The war against the Jews 1933–1945* (New York: Holt, Rinehart and Winston, 1975).

20. WH, note 16, "Rede," p. 397.

21. WH, note 16, interview.

22. WH, *PB*, p. 7.

23. EH, *Recoll.*, ch. 1.

24. WH, "Wissenschaft als Mittel zur Verständigung unter den Völkern" (1946), HCW C5, 385.

Chapter 5

1. WH, *PB*, p. 1.

2. Walter Z. Laqueur, *Young Germany: A history of the German youth movement* (London: Routledge and Kegan Paul, 1962).

3. Willibald Karl, *Jugend, Gesellschaft und Politik im Zeitraum des Ersten Weltkriegs*, Miscellanea Bavarica Monacensia, vol. 48 (Munich: SAM, 1973).

4. Interview with Rev. Wolfgang Rüdel, Ansbach, Mar 1982; and Rüdel, "Erinnerungen 27.1.1962" (poem), *Die Neue Seite. 3. MPZ, Stammesmitteilungen*, 2 (1962), between pp. 20 and 21.

5. Histories of the German Pathfinders, including the Neupfadfinder, are provided in Karl, note 3, Günther Bandick, "Ursprung und geistige Entwicklung der deutchen Pfadfinderbewegung bis 1933," doctoral dissertation, Universität Hamburg, 1955; and Karl Seidelmann (a leader of the Neupfadfinder), *Die Pfadfinder in der deutschen Jugendgeschichte*, vol. 1: *Darstellung* (Hannover: Hermann Schroedel, 1977).

6. Seidelmann, note 5.

7. Rüdel, note 4.

8. This is surmised from lists of those in the various activities. Schlenk had also served with Werner in the agricultural service, as recalled in Carl Zenker to WH, 6 Aug 1943 (HA).

9. Interview with Fr. Gottfried Simmerding, Munich, Mar 1982.

10. WH, *PB*, p. 10.

11. Ibid.

12. *Schloss Prunn. Der deutsche Pfadfindertag von 1919*, supplement to *Der Weisse Ritter*, 2 (1919), ed. Franz Ludwig Habbel and Ludwig Voggenreiter (Regensburg: Der Weisse Ritter-Verlag, 1919).

13. Such views are described by numerous authors, among them Laqueur, note 2; George L. Mosse, *The crisis of German ideology: Intellectual origins of the Third Reich* (New York: Schocken Books, 1964, 1981); Peter Gay, *Weimar Culture* (New York: Harper, 1968); Paul Forman, "Weimar culture, causality and quantum theory, 1918–1927," *HSPS*, 3 (1971), 1–115.

14. F. L. Habbel, "Vorrede zu der Zeitschrift 'Der Aufbau,' 1918," reprinted in *Die Pfadfinder in der deutschen Jugendgeschichte*, vol. 2,1: *Quellen und Dokumente aus der Zeit bis 1945*, ed. Karl Seidelmann (Hannover: Hermann Schroedel, 1980), pp. 51–52, his emphasis.

15. Ernest Kemmer, "Windsbach," *Der Weisse Ritter*, 3 (1921), 189–191, on 189. Kemmer's reemergence in the youth movement supports Mosse's characterization (note 13, p. 266) of the movement's activists, regardless of age, as rejecting bourgeois institutions for having brought Germany to its current state, while affirming reactionary youthful activism to restore Germany's greatness.

16. F. L. Habbel, "Unser Pfadfindertum (1919)," *Pfadfinder*, note 14, pp. 52–55, on p. 54.

17. Karl Ettinger, statement, note 12, pp. 39–40.

18. Steidle, statement, note 12, pp. 14–18.

19. F. L. Habbel, Karl Sonntag, Ludwig Voggenreiter, "Ein Geleitwort," *Der Weisse Ritter*, 2 (1919), 4–6, on 6. The journal started with volume 2. Its predecessor, *Der Aufbau*, was designated volume 1.

20. This is also noted in George L. Mosse, *Nationalism and sexuality: Respectability and abnormal sexuality in modern Europe* (New York: Howard Fertig, 1985), p. 47.

21. This is argued in Mosse, note 13, pp. 173–174.

22. Michael H. Kater, *Studentenschaft und Rechtsradikalismus in Deutschland 1918–1933* (Hamburg: Hoffmann und Campe, 1975).

23. WH to Kurt Pflügel, 21 Oct 1923 (Pflügel papers).

24. Simmerding, note 9.

25. Indicated by WH, *PB*, p. 54.

26. EH, *Recoll.*, pp. 25–26.

27. Mosse, note 13, pp. 207 ff.; Bandick, note 5.

28. Völkel, "Der Bund," *Der Weisse Ritter*, 5, Sonderheft: "Der Naumburger Bund," Pfingsten 1925, 35–40.

29. "Die Freideutsche Jugend will nach eigener Bestimmung, vor eigener Verantwortung, in innerer Wahrhaftigkeit ihr Leben gestalten! Für diese Freiheit tritt sie unter allen Umständen geschlossen ein." ("The Free German Youth will form their lives in self determination, responsible only to themselves, and with inner rectitude! They will stand up for this freedom under any circumstances.")

30. *Schloss Prunn*, note 12, p. 32. The oath was written by the Austrians.

31. This is argued by Mosse, note 13, and Harry Pross, *Jugend-Eros-Politik: Die Geschichte der deutschen Jugendverbände* (Bern, 1964).

32. Mosse, note 13, p. 176. According to Mosse, note 20, p. 49, the rediscovery of the human body and male relationships did not imply a revival of sensuality but a longing for the "genuine."

33. Schürnbrand, statements, note 12, pp. 13 and 20.

34. Women students constituted only about 10 percent of the total student population in the 1920s. According to one count, of a total of 10,596 women who received doctorates in Germany beween 1918 and 1933, only 24 became professors. J. Kein and D. Cassidy, "The history of women in science," *Women's studies international forum*, 7 (1984), 313–317.

35. Geneviève Bianquis, "La femme allemande à l'époque moderne et contemporaine (du XVIIe au XXe siècle)," in *Histoire mondiale de la femme*, ed. Pierre Grimal, vol. 4. (Paris: Nouvelle Librairie de France, c. 1980), pp. 253–290.

36. Colonel Seisser was one of the three officials who at first supported Hitler during his 1923 Beer Hall Putsch, then reneged that night.

37. WH, *PB*, p. 11.

38. *Schloss Prunn*, note 12, p. 41.

39. WH, *PB*, p. 11.

40. WH to his father, 16 Aug 1919.

41. Based on materials in Nachlass Karl Seidelmann (ADJ).

42. Wolfgang Rüdel to WH, 15 Feb 1923 (HP).

43. Mosse, note 13, p. 181, quotes one source reporting in 1926 that the Neupfadfinder admitted "only those Jews . . . who are more Nordic than the Nordic Aryans. Secretly they are made fun of all the same."

44. Hans Blüher, *Secessio Judaica: Philosophische Grundlegung der historischen Situation des Judentums und der antisemitischen Bewegung* (Berlin: Der Weisse Ritter-Verlag, 1922). Blüher's influence is discussed by Mosse, note 13, and Pross, note 31.

45. Kater, note 22. Although a direct connection cannot be established, such passages as the following from Hitler's *Mein Kampf*, written while Hitler and Hess were in prison, parallel and may have been influenced by Voggenreiter's readings: *Mein Kampf*, trans. Ralph Manheim (Boston: Houghton Mifflin, 1971), pp. 150, 297–299, 344–346, 440–441.

46. Wolfgang Hurt to Martin Völkel, 24 Apr [1921] (Nachlass Seidelmann). However, Walter's brother Franz did not help matters.

47. Karl Sonntag, recollections, in *Die Neue Seite. 3. MPZ, Stammesmitteilungen*, 4 (1967), 77–86.

48. Mosse, note 13, p. 185.

49. Sonntag to Habbel, 19 Dec 21; Martin Völkel, "Rundbrief," 29 Mar 1922 (both Nachlass Seidelmann). Homosexual activity was a crime under the Weimar constitution.

50. Völkel to Habbel, 26 May 1922; Sonntag to Voggenreiter, 9 Apr 1925; Völkel-Sonntag correspondence (all Nachlass Seidelmann).

51. Plans for camp activities in Karl Sonntag, Gaubrief, Apr 1924; and "Sommerlager, Lagerordnung," 1925 (Nachlass Seidelmann).

52. For example, [M. Völkel], "Der grosse Häuptling spricht," *Die Spur in ein deutsches Jugendland*, 2 (Mar 1924), 155.

53. Rüdel, note 4.

54. Interview with Friedrich Hund, Feb 1982.

55. Mrs. Heisenberg to Kurt Pflügel, 4 Aug [1922], in *Die Neue Seite*, 2 (1962), between pp. 20 and 21.

56. Diary of trip (Simmerding private papers); WH, *PB*, p. 52.

57. Interview with Elisabeth Heisenberg, Feb 1982; the Rüdel brothers to WH, 22 Aug 1920 (HP).

58. WH to his father, 15 May 1918.

59. WH, interview session 1, 1962 (AHQP).

60. Simmerding, note 9. Bandick, note 5, notes that in this respect the Neupfadfinder did not follow the Puritan ethic of ambition for worldly success.

61. Rolf Wägele, "Werner und der jüngere Jungstamm," *Die Neue Seite*, 2 (1962), 27–31; Karl Sonntag, remarks, *Die Neue Seite*, 2 (1962), 8.

62. WH, "Wissenschaft als Mittel zur Verständigung unter den Völkern" (1946), HCW C5, 384–394, on 385.

63. WH, "Alte Werte in neuen Formen" (1961), HCW C4, 19.

64. WH, *PB*, pp. 18–19.

65. For example, WH to Renée Weber and WH to Fritjof Capra, 1970s (HA). WH, "Ordnung der Wirklichkeit," 1942, published for the first time in HCW C1, 218–306.

66. Ruth Nanda Anshen, "World perspectives: What this series means," in WH, *PB*, pp. 249–260, on pp. 249 and 252.

67. Seidelmann, sunrise lecture to the Jungstamm on Easter Sunday, 1931 (Nachlass Seidelmann).

68. Simmerding, note 9.

69. WH, fragment of autobiographical sketch, c. 1934 (Archiv für Geschichte der Naturforschung und Medizin, Deutsche Akademie der Naturforscher, Leopoldina, Halle).

Chapter 6

1. Born to Sommerfeld, 5 Mar 1920 (SN); Born and Heisenberg, "La Mécanique des quanta," 1927 Solvay Congress report, HCW B, 58–96.

2. William Carr, *A history of Germany 1815–1945* (New York: St. Martin's Press, 1979), p. 283.

3. In November 1921 Wilhelm Brüchner, university student, founded the *Sturmabteilung* (SA), or storm troopers, of the Nazi Party and populated it with

many of his classmates. Among them was Rudolf Hess, a guest auditor who commanded his own SA company. Hitler's party experienced a sharp rise after the Kapp putsch, gaining heavy support from student groups during his Beer Hall Putsch three years later. Karl Schwend, *Bayern zwischen Monarchie und Diktatur* (Munich, 1954).

4. Arno Seifert, "In den Kriegen und Krisen des 20. Jahrhunderts," *Ludwig-Maximilians-Universität. Ingolstadt-Landshut-München 1472–1972*, ed. Laetitia Böhm and Johannes Spörl (Berlin: Duncker und Humblot, 1972), pp. 315–362.

5. "Studentenversammlung in der Universität," *MNN*, 73: 23 (17–18 Jan 1920), 2; "Münchner Studentenschaft und Arco-Prozess," *MNN*, 73: 24 (19 Jan 1920), 4; "Eine verhinderte Vorlesung," *MNN*, 73: 29 (22 Jan 1920), Morgen, 3.

6. Based on *Chronik der Ludwig-Maximilians-Universität* for the period.

7. Aloys Fischer, *Die Wirtschaftliche Lage der Studenschaft Münchens und die Bedeutung für die Studentenfürsorge* (Munich: Verlag des Vereins Studentenhaus München, 1921).

8. In 1920, August Heisenberg's total salary was 35,096 marks per year; by October 1921 it was 101,426 marks per year. Based on salaries and extras as given in "Beamtenbesoldungsgesetz," *Gesetz- und Verordnungsblatt für den Freistaat Bayern*, 1920, 275–322; and "Gesetz zur Abänderung des Beamtenbesoldungsgesetzes," *Gesetz- und Verordnungsblatt für den Freistatt Bayern*, 1923, pp. 1–3.

9. Erich Simon, "Der Haushalt eines höheren Beamten," *Jahrbücher für Nationalökonomie und Statistik*, 119 (1922), 425–432.

10. WH to his mother and WH to his father, both Jena, 23 Sep [1921].

11. WH to his mother, Berlin, 28 Sep 1921.

12. Richard Willstätter, *Aus meinem Leben. Von Arbeit, Musse und Freunden*, 2nd ed. (Weinheim: Verlag Chemie, 1949, 1958), p. 302.

13. Max Born, *My Life: Recollections of a Nobel laureate* (New York: Charles Scribner's Sons, 1978), p. 194.

14. WH, "Begegnungen und Gespräche mit Albert Einstein" (1974), HCW C4, 202–216. Most of the letters are published in ES, *Briefw.*

15. Report in *PZ*, 21 (1920), 649–668.

16. The hatred directed at relativity theory and Einstein has not yet been fully analyzed or explained. Summaries of the backgrounds and attitudes of Nazi Nobel laureates Philipp Lenard and Johannes Stark are offered in Beyerchen, chs. 5 and 6.

17. An excellent history of the rise of German theoretical physics is offered in JM.

18. JM, vol. 2, pp. 286–287.

19. Ibid., recounting an episode in 1912.

20. WH to his mother, Jena, 21 Sep 1921.

21. The article, published in *Die Weltbühne*, is reprinted in ES, *Briefw.*, pp. 89–90. Einstein canceled his appearance in Einstein to Somerfeld, 27 Sep 21 (ES, *Briefw.*).

22. Einstein to Sommerfeld, 9 Oct 1921 (ES, *Briefw.*).

23. For example, Italian physicist E. Amaldi writes of the 1930s: "We worked with incredible stubbornness. We would begin at eight o'clock in the morning and take measurements almost without a break until six or seven in the evening, and often later. . . . 'Physics as soma' was the phrase we used to refer to our work performed while the general situation in Italy grew more and more bleak." E. Amaldi, recollection, in Enrico Fermi, *Collected papers (note e memorie)*, vol. 1 (Chicago: University of Chicago Press, 1962), p. 811.

24. Pauli, "An Hermann Weyl zum 6. Nov. 1955," PWB, vol. 1, p. 33, n. 6.

25. WH, *PB*, p. 27.

26. A discussion of the function of the mathematical-physical seminar is provided in JM, vol. 1, pp. 78–107.

27. Based on *Personalstand der Ludwig-Maximilians-Universität München, Winterhalbjahr 1920/21.*

28. WH, *PB*, p. 16.

29. *Personalstand*, note 27.

30. WH, interview session 1 (AHQP).

31. Ulrich Benz, *Arnold Sommerfeld. Eine wissenschaftliche Biographie* (Stuttgart: Wissen. Verlagsgesellschaft, 1975); Michael Eckert et al. *Geheimrat Sommerfeld—Theoretischer Physiker* (Munich: Deutsches Museum, 1984).

32. Klein's strategy is described in Karl-Heinz Manegold, *Universität, Technische Hochscule und Industrie* (Berlin, 1970).

33. Kultusminister Matt to Univ. Senat, 3 Aug 1920 (Acta des K. Akad. Senats. Das physicalische Cabinet betr. UA München); Personalakt Wilhelm Wien (Littera E, Abt. II, Fascikel 698, UA München); W. Wien, "Ein Rückblick," in Wien, *Aus dem Leben und Wirken eines Physikers* (Leipzig: Barth, 1930), pp. 1–50.

34. A. Sommerfeld, "Das Institut für theoretische Physik," in *Die wissenschaftlichen Anstalten der Ludwig-Maximilians-Universität zu München*, ed. K. A. von Müller (Munich: Oldenbourg und Wolf, 1926), pp. 290–292; "Academie der Wissenschaften. Mathematisch-physikalisches Cabinett—Sammlung, nun seit 1909 Institut für theoretische Physik," vol. II: 1853–1927 (Acta des K. Staatsministeriums des Innern für Kirchen und Schul-Angelegenheiten. BHStA MK11317).

35. Röntgen, "Bericht der Kommission für die Wiederbesetzung der ordentlichen Professur für theoretische Physik," 20 Jul 1905 (Arnold Sommerfeld, Personalakt, E II-N, UA München). Röntgen and the foundation of the Munich Institute for Theoretical Physics are discussed in greater detail in JM, vol. 2, pp. 274–287.

36. Einstein to Sommerfeld, 14 Jan 1922 (ES, *Briefw.*).

37. Quoted in JM, vol. 2, p. 284.

38. Ludwig-Maximilians-Universität zu München, *Vorlesungsverzeichnis*, for entire period. An account of Heisenberg's Munich studies is also offered in MR 2, ch. 1.

39. Born to Sommerfeld, 13 May 1922 (SN); A. Sommerfeld, "Vorwort," in AS, *Atombau*, 1st ed.

40. WH, note 30.

41. P. P. Ewald, "Sommerfeld als Mensch, Lehrer und Freund," *Physics of the One- and Two-Electron Atoms*, ed. F. Bopp and H. Kleinpoppen (Amsterdam: North-Holland, 1969), pp. 8–16.

42. WH to his mother, 23 Sep 1921.

43. Pauli, "Relativitätstheorie" (1921), PCSP 1, pp. 1–237.

44. A Sommerfeld, "Zwanzig Jahre spectroscopischer Theorie in München," *Scientia*, Nov–Dec 1942, pp. 123–130.

45. Matriculation Verzeichnis for WH, winter semester 1920–1921 and summer semester 1921 (UA Munich). The rest were lost in World War II.

46. Matriculation Verzeichnis for Pauli, winter semester 1918–1919 (UA Munich).

47. WH, note 30.

48. JM, vol. 2, pp. 355–356.

49. Paul Forman, "The Financial Support and Political Alignment of Physicists in Weimar Germany," *Minerva*, 12 (1974), 49–66.

50. Detlev Richardt, report of WH's remarks at Lindau meeting of Nobel Prize winners, 1974, in *PBl*, 30 (1974), 30–37.

51. A. Heisenberg to WH, 2 Nov 1922.

52. Born to Sommerfeld, 5 Jan 1923 (SN).

53. The Heisenberg-Pauli relationship is also described, with some differences, in Dresden, pp. 259–261.

54. WH, note 30. Forman and Serwer (Serwer) raise the possibility of tacit (or even explicit) conflict between the two physicists arising from Pauli's Jewish heritage and Heisenberg's participation in an anti-Semitic youth movement. Heisenberg gives no overt indication of anti-Semitism when writing privately of Pauli, Einstein, and other Jewish physicists and acquaintances.

55. WH, *PB*, pp. 25–26.

56. Born, note 52.

Chapter 7

1. Much of this chapter derives from my earlier works: DC, "WH," and "Heisenberg's first core model of the atom: The development of a professional style," *HSPS*, 10 (1979), 187–224. It has greatly benefited from the following studies: Serwer; MR 2; Paul Forman, "The environment and practice of atomic physics in Weimar Germany: A study in the history of science," Ph.D. thesis, University of California, Berkeley, 1967; and Paul Forman, "Alfred Landé and the anomalous Zeeman effect, 1919–1921," *HSPS*, 2 (1970), 153 261. Readers are referred to these works for elaboration and for technical details concerning the topics treated in this chapter.

2. J. L. Heilbron and T. S. Kuhn, "The genesis of Bohr's atom," *HSPS*, 1 (1969), 211 290. Even more technical detail is provided in MR 1, 1.

3. Max Born, remark, EB, *Briefw.*, p. 114.

4. Niels Bohr, "On the sprectrum of hydrogen," address delivered 20 December 1913, translated and published in English in 1922 (BCW 2, 283–301).

5. A. Sommerfeld, *AP*, 51 (1916), 1–94, 125–167.

6. The "quantum conditions" may be expressed by the equation $\oint p_i dq_i = n_i h$, where $i = 1, 2, 3, \ldots, f$; f is the number of degrees of freedom; p_i and q_i are conjugate momentum and position variables; and the integral is taken over a closed orbit.

7. A. Sommerfeld, *PZ*, 17 (1916), 491–507; Peter Debye, *PZ*, 17 (1916), 507–512; Sommerfeld to Bohr, 4 Sep 1913 (BSC 7, 3).

8. AS, *Atombau*, 1st ed., p. vii; also Sommerfeld, "Ein Zahlenmysterium in der Theorie des Zeemaneffektes," *Nwn*, 8 (1920), 61–64.

9. W. Wien, "Vergangenheit, Gegenwart und Zukunft der Physik," in Wien, *Aus dem Leben und Wirken eines Physikers* (Leipzig: Barth, 1930), pp. 120–135.

10. A. Sommerfeld, *PZ*, 21 (1920), 619–621.

11. A. Sommerfeld, *AP*, 63 (1920), 221–263, on 231.

12. Forman, "Environment," note 1; Univ. Munich, *Jahresberichte* for the years in question.

13. Based on unpublished data gathered by Inventory of Sources for History of Twentieth Century Physics, Office for History of Science and Technology, University of California, Berkeley.

14. T. S. Kuhn et al., *Sources for history of quantum physics: An inventory and report* (Philadelphia: American Philosophical Society, 1967), pp. 2–3.

15. The main contributors were Bohr, Born, Dirac, Heisenberg, Jordan, Kramers, Landé, Pauli, Sommerfeld, and possibly Van Vleck. This list does not include the main contributors to and supporters of wave mechanics: Schrödinger, Einstein, L. de Broglie, and Planck. Nor does it include experimentalists.

16. This is based on P. Forman, J. L. Heilbron, and S. Weart, "Physics *circa* 1900: Personnel, funding, and productivity of the academic establishments," *HSPS*, 5 (1975), 1–185, tables on pp. 12 and 31. While Germany and the United States possessed about the same numbers of academic physicists in 1900—103 and 99, respectively—Germany touted 16 theorists to America's 3.

17. WH, *PB*, p. 35. Letters from Paschen to Sommerfeld (AHQP 33, 1) and Back to Sommerfeld (AHQP 29, 5) indicate that they regularly supplied him with data, but none of the available letters contains the data in question. They were probably on enclosures that have since been lost.

18. WH, notebook (HA).

19. WH, interview session 1 (AHQP).

20. WH, *PB*, p. 35.

21. Sommerfeld to Landé, [31] Mar 1921 (AHQP 32, 1). Sommerfeld may have discouraged Heisenberg from publishing in order to protect Back, who was submitting the Tübingen data for his habilitation (MR 2, 31).

22. A. Landé, *ZP*, 5 (1921), 231–241, rec. 16 Apr 1921.

23. For singlets and triplets, $m = 0, \pm 1, \pm 2, \ldots, \pm j$. For doublets, m took on the even number $(2j)$ of values $m = \pm 1/2, \pm 3/2, \pm 5/2, \ldots, \pm(j\text{-}1/2)$.

24. AS, *Atombau*, 2nd ed. (1921), p. ix; W. Voigt, *AP*, 41 (1913), 403–440; W. Voigt, *AP*, 42 (1913), 210–230; A. Sommerfeld, *ZP*, 8 (1922), 257–272, rec. 12 Dec 1921.

25. Bohr, "On the series spectra of the elements" (1920), BCW 3, 241–282.

26. Sommerfeld, note 24, second work, p. 257.

27. Heisenberg's version of the complicated Sommerfeld-Voigt equation for doublets may be written:

$$E = \bar{E} + h\nu_L(m^* \pm 1/2 \cdot \sqrt{1 + (2m^*/k^*)\tau + \tau^2}).$$

Here E is the energy of the Zeeman term; $\bar{E}$ is the average of the unperturbed doublet energies, $\tau = \Delta\nu/\nu_L \propto 1/H$, H being the magnetic field intensity; $\Delta\nu$ is the doublet separation; and ν_L is the Larmor frequency. An asterisk on a number reduces it by $1/2$, thus yielding half-integers and Sommerfeld's number mystery.

28. The model, however, did not address the so-called magneto-mechanical effects, the effects of Barnett and Einstein-de Haas.

29. For example, Hans A. Bethe and Edwin S. Salpeter, *Quantum mechanics of one- and two-electron atoms* (Berlin, 1957), p. 211; discussed in MR 3.

30. Pauli to Landé, 5 Oct 1921.

31. WH to his mother, Jena, 21 Sep 1921.

32. WH to his father, Jean, 23 Sep [1921].

33. Sommerfeld to Einstein, 11 Jan 1922 (ES, *Briefw.*).

34. WH to Landé, 29 Oct 1921 (AHQP 6, 2).

35. Landé to Bohr, 12 May 1922 (BSC 4, 3). "Ether" in this case refers to the electromagnetic field.

36. WH to Landé, 19 Nov 1921 (AHQP 6, 2).

37. Pauli, "Über das Modell des Wasserstoffmolekülions," *AP*, 68 (1922), 177–204.

38. WH to Pauli, 19 Nov [1921].

39. AS, *Atombau*, 3rd ed. (1922), pp. 53–54 and 310–324.

40. WH to Landé, 28 Nov 1921 (AHQP 6, 2).

41. WH, "Zur Quantentheorie der Linienstruktur und der anomalen Zeemaneffekte" (1922), HCW A1, 134–158; Sommerfeld, note 24, second work.

42. Sommerfeld to Bohr, 25 Mar 1922 (BSC 7, 3).

43. Bohr to Landé, 15 May 1922 (BSC 4, 3).

44. Bohr, "On the application of quantum theory to atomic structure: Part I—The fundamental postulates" (1922), BCW 3, 455–500.

45. See, among others, Forman, "Environment," note 1, and Forman, "Scientific internationalism and the Weimar physicists: The ideology and its manipulation in Germany after World War I," *Isis*, 64 (1973), 151–180.

46. For example, Sommerfeld to Carlsberg Foundation, 26 Oct 1919 (Robertson, pp. 34–35).

47. Funding for the lectures derived from interest on prize money donated by Wolfskehl for the proof or disproof of Fermat's last theorem.

48. Bohr to Sommerfeld, 30 Apr 1922 (BCW 3, 691–692).

49. Bohr, "Sieben Vorträge über die Theorie des Atombaus" (BMS 10), English translation in BCW 4, 341–419. The lectures were delivered Jun 12–22, 1922.

50. WH to Karl and Helen Heisenberg, 15 Jun 1922 (HP).

51. Bohr, lecture 3, Jun 14, 1922 (BCW 4, 372–387, on 370–371). Kramers's paper is *ZP*, 3 (1920), 199–223.

52. Interview with Friedrich Hund, who was in attendance, Göttingen, Apr 1982. Kramers's work and Heisenberg's objection are also discussed in MR 2 and in Dresden, pp. 124–132.

53. WH, *PB*, pp. 37–42; WH, "Quantum theory and its interpretation" (1967), HCW C2, 345–346.

54. WH, note 50.

55. WH to his parents, 15 Jun 1922. Heisenberg was more generous to Bohr when, writing Bohr in 1933, shortly after he learned that he was to receive the Nobel Prize, he recalled that "you talked me out of a false theory of the Zeeman effects." WH to Bohr, 27 Nov [1933] (BSC 20, 2).

56. Bohr, lecture 5, June 20, 1922 (BCW 4, p. 391).

57. WH, note 55.

Chapter 8

1. Indicated in "Akten des Rektorats der Universität München. Dr. Arnold Sommerfeld" (UA Munich, Personalakt E II-N).

2. A. Sommerfeld and WH, "Eine Bemerkung über relativistischen Röntgendubletts und Linienschärfe" (rec. 3 Aug 1922), HCW A1, 159–164; "Die Intensität der Mehrfachlinien und ihrer Zeemankomponenten" (rec. 26 Aug 1922), HCW A1, 165–188. The plan nearly collapsed when Wilhelm Lenz tried more than once to get Heisenberg to replace Pauli as his assistant in Hamburg. Heisenberg reassured a disturbed Sommerfeld that he would not accept, "despite the large money bags of the Hamburg people" and his own wish for financial independence. WH to Sommerfeld, 15 Jan [1923] (AHQP 83, H).

3. WH, "Nichtlaminare Lösungen der Differentialgleichungen für reibende Flüssigkeiten" (1922), HCW B, 23–26.

4. Franz Ludwig Habbel, "Die Aussenpolitik der deutschen Pfadfinderbewegung," *Der Weisse Ritter*, 7 (1927), 60–80, on 66. For the physicists, see Paul Forman, "Scientific internationalism and the Weimar physicists: The ideology and its manipulation in Germany after the First World War," *Isis*, 64 (1973), 150–180.

5. Habbel, note 4, p. 61.

6. WH to his father, 11 Nov [1922].

7. M. Völkel, draft of circular letter marked *streng vertraulich* (top secret), 28 Feb 1922 (Nachlass Karl Seidelmann, ADJ).

8. Karl Sonntag, circular letter, Bavaria District, 11 Mar 1922 (Nachlass Seidelmann).

9. Ibid.

10. Habbel, note 4; Kurt Pflügel, "Finnlandfahrt," *Die neue Seite, 3. MPZ Stammesmittelungen,* 2 (1962), 13–18.

11. "Der Kampf um die Überfahrt," *Die Spur in ein deutsches Jugendland,* 3 (1924), 37–39 (HCW C4, 8–10).

12. *Die neue Seite, 3. MPZ Stammesmitteilungen,* 2 (1962), 18–22, reprint and remembrance of note 11.

13. Kurtei, "Wie wir 3 Enten schossen und nur eine erbeuteten," *Die Spur in ein deutsches Jugendland,* 4 (1925–1926), 134–135.

14. Indicated in Elis J. Huetin to WH, Helsingfors, 1 Oct 1923 (HA).

15. WH to his mother, 13 Nov [1923].

16. When Karl Sonntag criticized the Mannen, the young men of Werner's age, for getting up too late during the winter camp over Christmas and New Year's 1922–1923, Werner wrote a long defense, indicating how adolescent he really was. For instance: "I only remember that in Kurtei's hut (the upper one), in which most of the Mannen lived, by sunrise the cabin was always clean and the cocoa was always ready." WH, affidavit, winter camp 1922–1923 (HP).

17. WH to Niels Bohr, Zell am Ziller, 31 Aug [1925] (BSC 11, 2).

18. WH to his parents, Leipzig, 17 Sep [1922].

19. Philipp Lenard, *Ueber Aether und Uräther. Mit einem Mahnwort an deutsche Naturforscher* (Leipzig: Hirzel, 1922). This episode is further discussed in Beyerchen, 91–93.

20. WH, note 18.

21. Handbill (Sammlung Darmstädter, F1e 1908(7): Einstein, SPK).

22. WH, *PB,* pp. 43–45.

23. WH to his parents, 19 Sep [1922]; WH to Pauli, 29 Sep [1922].

24. Ibid.

25. Heisenberg still referred to the speaker as "Einstein" in WH, *PB,* p. 44, written in 1969.

26. Max Born, *My life: Recollections of a Nobel laureate* (New York: Charles Scribner's Sons, 1975), p. 212.

27. [Felix Klein?], "Universität Göttingen—Philosophische Fakultät I. Mathematisch physikalisches Seminars" (UA Göttingen, "Zur Geschichte des Königlichen Math.-phys. Seminar"). Göttingen mathematics and physics during the earlier part of the century are discussed in Pyenson, *The young Einstein: The advent of relativity* (Bristol: Adam Hilger, 1985), pp. 101–193.

28. Published in EB, *Briefw.*

29. Born, note 26, pp. 199–200; "Personalakt Max Born" (UA Göttingen).

30. "Chronik der Georg August-Universität für die Rechnungsjahre 1921/23," Universitätsbund Göttingen, *Mitteilungen,* 6 (1923), issue 1 and 2; Born, note 26, p. 210; MR 1.

31. Born to Walther Gerlach, 11 May 1921 (AHQP 19, 1).

32. Records show that 1257 of 3263 students studied math or one of the sciences. *Amtliches Namenverzeichnis, Georg August-Universität zu Göttingen, Winterhalbjahr 1922/23.*

33. The others were Fischer, Ludloff, and Wessel. Born to Sommerfeld, 5 Jan 1923 (SN).

34. WH to Helen and Karl Heisenberg, 12 Jun [1922] (HP). Although the Göttingen town dialect is considered high German, the peasants did not speak it.

35. *The University of Goettingen* (Göttingen: Verein für Fremdenverkehr, c. 1926), pp. 18–19.

36. Born to Einstein, 7 Apr 1923 (EB, *Briefw.*).

37. WH to his brother, 6 Dec [1922] (HP).

38. Eberhard Rüdel to WH, 23 Nov 1922 (HP).

39. WH to his mother, 7 Dec [1922].

40. Götz von Selle, *Die Georg-August-Universität zu Göttingen 1737–1937* (Göttingen, 1937); August Tecklenburg, *Göttingen: Die Geschichte einer deutschen Stadt* (Göttingen, 1930); Albrecht Saathoff, *Geschichte der Stadt Göttingen seit der Gründung der Universität* (Göttingen, 1940).

41. WH to his father, 5 and 16 Nov [1922]; WH to his mother, 20 Nov, 1 and 7 Dec [1922].

42. WH to his mother, 1 Dec [1922].

43. Born, note 26, p. 211.

44. WH to his father, 16 Nov 1922.

45. WH to Sommerfeld, 15 Jan [1923] (AHQP 83, H).

46. WH to his brother, 6 Dec [1922] (HP); WH to his mother, 7 Dec [1922].

47. Born to Sommerfeld, 5 Jan 1923 (SN).

48. WH to his father, 5 Nov 1922.

49. Born, note 26, p. 211.

50. WH, note 44. The following discussion is based on DC, "WH." For technical details, the reader is referred to this and to MR 2.

51. As indicated earlier, these conditions may be written $\oint p_i dq_i = n_i h$, where $i = 1, 2, 3, \ldots f$, f is the number of degrees of freedom, and the n_i are positive integers.

52. Born and Pauli, "Über die Quantelung gestörter mechanischer Systeme" (1922), PCSP 2, 48–69.

53. Born to Landé, 29 May 22 (AHQP 4, 2).

54. This apparently motivated Pauli to undertake an independent, detailed calculation of the ground state of helium, but he dropped it after moving to Copenhagen in the fall of 1922. Pauli to Bohr, 7 Jul [1922] and 5 Sep 1922.

55. Bohr, "Über die Anwendung der Quantentheorie auf den Atombau. I. Die Grundpostulate der Quantentheorie," ZP, 13 (1923), 117–165, rec. 15 Nov 22 (English translation, 1924, BCW 3, 458–499).

56. Kramers, ZP, 13 (1923), 312–341, on p. 339, rec. 31 Dec 22.

57. WH to Landé, 15 Sep 1922 (AHQP 6, 2); WH to Sommerfeld, 28 Oct [1922] (AHQP 83, H).

58. WH to Sommerfeld, 28 Oct 1922 (AHQP 83, H).

59. A. Sommerfeld, Optical Society of America, *Journal*, 7 (1923), 509–515. Nevertheless, Sommerfeld had to withdraw the model a year later after it was proven unstable.

60. Born, *Nwn*, 11 (1922), 677–678, dated 27 Jun 1922.

61. Born, note 26, p. 202.

62. WH, note 44.

63. Born and WH, "Über Phasenbeziehungen bei den Bohrschen Modellen von Atomen und Molekeln" (rec. 16 Jan 23), HCW A1, 189–200. The hypothesis was replaced in 1925 by Pauli's exclusion principle.

64. WH to Pauli, 12 Dec 1922.

65. WH to Sommerfeld, 4 Jan 1923 (AHQP 83, H).

66. WH, note 45.

67. WH to Bohr, 2 Feb [1923] (BSC 9, 2).

68. Their results appeared in Born and WH, "Die Elektronenbahnen im angeregten Heliumatom" (rec. 11 May 23), HCW A1, 201–215.

69. Born to Bohr, 4 Mar 1923 (BCW 4, 669).

70. WH to Pauli, 19 Feb [1923]; also Born to Bohr, note 69. They reiterated this view in Born and WH, note 68, p. 215.

71. WH to Pauli, 26 Mar 1923.

72. Bohr to Born, 2 May 1923 (BCW 4, 673).

73. Born, Nwn, 11 (1923), 537–542, on 542.

74. Eberhard Rüdel to WH, 12 Jun 1923 (HP).

75. Reynolds's work is outlined in MR 2, 53.

76. WH, "Über Stabilität und Turbulenz von Flüssigkeitsströmen," doctoral thesis (UB Munich); revised published version (1924), HCW A1, 31–81. More detailed reviews of Heisenberg's work on the problem and its background are offered in MR 2, 49–63, and in Giovanni Battimelli, "The mathematician and the engineer: Statistical theories of turbulence in the '20s," Rivista di storia della scienza, 1 (1984), 73–94.

77. WH, note 3, and WH, "Die absoluten Dimensionen der Kármánschen Wirbelbewegung" (1922), HCW A1, 27–30. Prandtl, however, expressed objections to the second of these works in a note appended to it.

78. Promotions-Ordnung der Philosophischen Fakultät (II. Sektion) der Ludwig-Maximilians-Universität München (Munich, 1922).

79. "Protokoll, Promotion des Herrn Werner Heisenberg" (UA Munich, OCI 49p). Sommerfeld added by hand to the paragraph a concession to criticism: "Also stylistically [the thesis] is not entirely well written."

80. Ibid. It is possible, though unlikely, that the exams were held separately, as they are today.

81. "Protokoll, Promotion des Herrn Wolfgang Pauli" (UA Munich, OCI 49p).

82. WH, interview session 1 (AHQP).

83. Ibid. and "Protokoll," note 79.

84. Born, note 26, p. 213.

85. Jahrbuch der Ludwig-Maximilians-Universität München, 1919–1925 (Munich, 1925).

86. WH to his parents, 29 Nov [1923].

87. WH, note 82.

Chapter 9

1. Landé, "Das Versagen der Mechanik in der Quantentheorie," Nwn, 11 (1923), 725–726; Landé, PZ, 24 (1923), 441–444; Born, Nwn, 11 (1923), 537–542; Paschen, PZ, 24 (1923), 401–407.

2. Landé, PZ, 24 (1923), 441–444, on 441; and ZP, 19 (1923), 112–123 (rec. 16 Aug 23), on 123; Born, note 1, p. 542.

3. WH to Kurt Pflügel, 24 Nov [1923] (Pflügel Papers).

4. "Bekanntmachung über das Diensteinkommen der Hochschulprofessoren," Gesetz- und Verordnungsblatt für den Freistaat Bayern, 1923, p. 313; "Gesetz über

die Festsetzung der Teuerungszuschläge der Staatsbeamten," *Gesetz- und Verordnungsblatt für den Freistaat Bayern*, 1923, 7.

5. "Teuerungszahlen der Gemeinden vom Juli 1923 bis Januar 1925," *Statistisches Jahrbuch für das deutsche Reich*, 44 (1924–1925), 261.

6. Hans Guradze and Karl Freudenberg, "Existenzminimum des geistigen Arbeiters im März-August 1923," *Jahrbücher für Nationalökonomie und Statistik*, 121 (1923), 354–355.

7. "Teuerungszahlen," note 5; "Lebensmittelpreise im Kleinhandel in einigen deutschen Städten im Durchschnitt 1913/14 und vom Juli 1923 bis Januar 1925," *Statistisches Jährbuch für das deutsche Reich*, 44 (1924–1925), 262.

8. W. Wien, "Ein Rückblick," in Wien, *Aus dem Leben und Wirken eines Physikers* (Leipzig: Barth, 1930), pp. 1–50, on p. 47.

9. Born to Einstein, 7 Apr 1923 (EB, *Briefw.*).

10. Ibid.; "Personalakt Max Born" (UA Göttingen).

11. Brigitte Schröder-Gudehus, "The argument for the self-government and public support of science in Weimar Germany," *Minerva*, 10 (1972), 537–570; Paul Forman, "The financial support and political alignment of physicists in Weimar Germany," *Minerva*, 12 (1974), 39–66; and, for example, Friedrich Schmitt-Ott, "Die Kulturaufgaben und das Reich," *Internationale Monatsschrift für Wissenschaft, Kunst und Technik*, 11 (1919), 450–459; Georg Schreiber, *Die Not der deutschen Wissenschaft und der geistigen Arbeiter* (Leipzig: Quelle und Meyer, 1923).

12. This is argued in Forman, note 11.

13. Notgemeinschaft, *Bericht I* (1922), p. 38.

14. Ibid., pp. 38–39.

15. See Forman, note 11.

16. Records of Elektrophysikausschuss der Notgemeinschaft (BA Koblenz, R73), and Steffen Richter, *Forschungsförderung in Deutschland 1920–1936.* (Düsseldorf: VDI-Verlag, 1972).

17. WH to his parents, 20 Nov 1923. The stipend actually amounted to $12 per month, "Aufstellung. 27.10.23," in records, note 16.

18. Ibid.; Richter, note 16; Born to Universitätskurator, 18 Jul 1925, in "Personalakt Dr. Heisenberg" (UA Göttingen, 4/Vc 317).

19. Ibid., and Courant to Universitätskurator, 30 Oct 1925 in Personalakt, note 18. On September 1, 1924, the Rentenmark was replaced by the Reichsmark, of equal value and same abbreviation. It lasted until 1948.

20. Notgemeinschaft, *Bericht V* (1926), p. 85. The statement was probably written by Planck.

21. WH to Kurt Pflügel, 21 Oct 1923 (Pflügel Papers). An overview of Bavarian political history during the Weimar era is given in Karl Schwend, *Bayern zwischen Monarchie und Diktatur* (Munich, 1954).

22. WH to Pflügel, 21 Oct 1923 (Pflügel Papers).

23. Pflügel to WH, 24 Oct [1923] (HP).

24. WH to Pflügel, 31 Oct 1923 (Pflügel Papers).

25. The dramatic effect was lost on the farmers near Göttingen, who astutely observed when it was all over: "Hei lebet noch!" (He's still living!). WH to Pflügel, 24 Nov [1923] (Pflügel Papers).

26. Interview with Fr. Gottfried Simmerding, Munich, 17 Mar 1982. The story is unconfirmed.

27. Pflügel to WH, 12 Nov [1923] (HP). The influence of theology is evident in such statments.

28. WH to Pflügel, 24 Nov [1923] (Pflügel Papers) and WH to his mother, 13 Nov [1923]. In the letter to Kurt, he also referred to Hitler's "Operettennationalismus" (opera nationalism).

29. Wolfgang Rüdel to WH, 22 Jan 1924 (HP).

30. WH to Pflügel, 24 Nov [1923]. Kurt responded by reiterating the decision of the Munich Pathfinders: "We had discussed this and found this position to be correct. . . . In any case, neither words nor acts but work is now called for." Pflügel to WH, 6 Dec 1923 (HP).

31. Much of the physics and the history of physics in this chapter is discussed by numerous authors from varying historical perspectives. This chapter relies primarily on the following works, to which the reader is referred for details: DC, "WH"; Serwer; MR 2, 106–124; Hendry, "BKS."

32. A. Landé, ZP, 15 (1923), 189–205; E. Back, ZP, 15 (1923), 206–243. The formulas are equivalent to the modern formulas if one attributes the half-integral momentum to electron spin.

33. A. Sommerfeld, AP, 70 (1923), 32–62.

34. Thus, as before, the stationary-state energy of a Zeeman term is $E = E_0 + gmJ$, where E_0 is the energy of a multiplet state corresponding to J, and

$$g = 1 + \frac{J^2 - 1/4 + R^2 - K^2}{2(J^2 - 1/4)}$$

35. A. Landé, ZP, 11 (1922), 353–363, on 353–354.

36. Landé, note 32, p. 197.

37. Pauli to Landé, 26 Jul 1923.

38. WH to Pauli, 21 Feb [1923].

39. WH to Pauli, 26 Mar [1923].

40. There are $2J + 1$ m-values. If J increases by $1/2$, the number of m-values increases by $1:2 (J + 1/2) + 1 = 2J + 2$.

41. WH to Pauli, 26 Mar [1923].

42. Bohr to Landé, 3 Mar 1923 (AHQP 4, 1). A previously unpublished manuscript they produced is probably "Atomteoretiske Problemer" (1923, NBA), BCW 3, 569–574.

43. Bohr, AP, 71 (1923), 228–288, rec. 15 Mar 1923 (BCW 4, 550–610).

44. In the doublet core model, the total angular momentum quantum number j of each state is $j = k$ or $k - 1$. In Heisenberg's core model, there are $2j^* + 1$ magnetic states, where $j^* = j - 1/2$; thus $2(k - 1/2) + 1 + 2(k - 3/2) + 1 = 2(2k - 1)$. In Landé's vector model, there are $2j$ magnetic states; thus $2k + 2(k - 1) = 2(k - 1)$.

45. Pauli to Bohr, 21 Feb 1924.

46. Pauli to Sommerfeld, 6 Jun 1923.

47. Pauli, ZP, 16 (1923), 155–164, rec. 26 Apr 23 (PCSP 2, 151–160).

48. Ibid., p. 164, and Pauli to Sommerfeld, 6 Jun 1923.

49. Ibid.; Pauli to Landé, 17 Aug [1923]; Pauli, note 47, p. 155. Pauli's position is well summarized in Serwer.

50. Max Born and Friedrich Hund, Vorlesungen über Atommechanik, vol. 1 (Berlin: Springer-Verlag, 1925). The lectures were delivered during the 1923–1924 school year.

51. WH to Pauli, 9 Oct 1923.

52. Ibid., and WH to Bohr, 22 Dec [1923] (BSC 11, 2).

53. $\int H_{cl} dJ' - F(J)$, where the integral is taken from $J' = J - 1/2$ to $J' = J + 1/2$

so $H_{cl}(J) = \partial F(J)/\partial J$. Heisenberg then replaced $F(J)$ with H_{qu} and the differential with a difference.

54. This discussion derives from the letters in notes 51 and 52 and from Heisenberg's paper, "Über eine Abänderung der formalen Regeln der Quantentheorie beim Problem der anomalen Zeemaneffekte" (rec. 13 Jun 24), HCW A1, 289–305.

55. WH, note 51.

56. Landé, note 32, p. 200; also Pauli to Landé, 23 Sep 1923.

57. Hund, "Mathematisches Tagebuch," entry for 16 Apr 1923 (Hund papers, Göttingen).

58. Ibid., entry for 9 Oct 1923; and WH, "Über ein neues Quantenprinzip und dessen Anwendung auf die Theorie des anomalen Zeemaneffektes," draft manuscript (1923), HA and AHQP 45, 8.

59. WH to his mother, 7 Nov 1923; WH to his father, 29 Nov [1923]. Runge even lectured on Heisenberg's theory before Göttingen's Mathematische Gesellschaft.

60. WH to his father, 29 Nov [1923].

61. WH to Pauli, 7 Dec 1923; the manuscript was that in note 58. WH to Landé, 7 and 13 Dec 1923 (AHQP 6, 2). Heisenberg also wrote Sommerfeld of his new theory, 12 Dec 1923 (SN).

62. Indicated in WH to Bohr, 22 Dec [1923] (BSC 11, 2). Heisenberg's earlier letters to Bohr were all in Fraktur (Gothic) manuscript.

63. WH to Bohr, 22 Dec [1923] (BSC 11, 2). Heisenberg soon expressed the hope that even half integers would find a basis in the new principle. WH to Bohr, 3 Feb [1924] (BSC 11, 2).

64. Bohr to WH, 31 Jan 1924 (BSC 11, 2).

65. Pauli to Landé, 14 Dec [1923], to Bohr, 21 Feb 1924, and to Kramers, 19 Dec 1923.

66. A report of the meeting is given in "Gemeinschaftliche Tagung des Gauvereins Niedersachsen der Deutschen Physikalischen Gesellschaft in Braunschweig am 9. und 10. Februar 1924," Deutsche Physikalische Gesellschaft, Verh., 5 (1924), 7–11. The three lectures were by Hund ("Rydbergkorrektion und Grösse des Atomrumpfes"), WH ("Quantitatives über die Deformierbarkeit edelgasähnlicher Ionen"), and Born ("Der Einfluss der Ionendeformation auf physikalische und chemische Konstanten").

67. Pauli to Bohr, 11 Feb 1924.

68. An entry in Hund's diary at that time may be an example of this kind of reasoning: "The quantum effects are better understood with half-integral K than with whole integrals." Hund, note 57, entry for 17 Jan 1924.

69. Pauli, note 65.

70. WH to his parents, 15 Mar [1924]. He arrived that day.

71. Ibid., and Bohr to Wickliffe Rose, carbon copy, 16 Apr 1924 (BGC).

72. WH to his parents, 20 and 27 Mar [1924]. Bohr did not apply to Rose for Werner's funding until 16 Apr 1924 (BGC) and 15 May 1924 (RAC, IEB series 1, Box 50, W. Heisenberg file, 1924–1929).

73. WH to his parents, 23 Mar [1924].

74. Ibid.

75. WH to his parents, 27 Mar [1924].

76. Post-doctoral physicist Slater was on a Sheldon Traveling Fellowship. He stayed in Copenhagen until April 1924, according to data in Robertson, pp. 156–159; the other two Americans were probably F. C. Hoyt and H. C. Urey.

77. Bohr, Kramers, and Slater, Phil. Mag., 47 (1924), 785–802 dated January 1924 (BCW 5, 101–118; English version in Waerden, pp. 159–176); Kramers, Nature, 133

(1924), 673–676, dated 25 March 1924. Heisenberg reported (note 70) that on his arrival Bohr handed him two papers to read; they were most probably these.
78. WH, "Die Entwicklung der Quantentheorie 1918–1928" (1929), HCW B, 109–115.
79. This distinction and the BKS theory are described more fully in Hendry, "BKS," and in Dresden, pp. 159–215.
80. Ladenburg, ZP, 4 (1921), 451–468.
81. Pauli to Eddington, 20 Sep 1923.
82. Slater to Kramers, 8 Dec 1923 (AHQP 8, 10); Slater, *Nature*, 113 (1924), 307–308.
83. BKS, note 77 in Waerden, p. 165. Slater did not concur with the full theory. Slater, *Nature*, 116 (1925), 278; Slater, *Phys. Rev.*, 25 (1925), 395–428.
84. The contrast is indicated in WH to Pauli, 4 Mar [1924], and WH to Bohr, Munich, 16 Apr [1924] (BSC 11, 2).
85. WH to Pauli, 26 Mar 1924.
86. WH to Bohr, 15 May 1924 (BSC 11, 2).
87. WH, to his mother, 31 May [1924]. There is a play on words here. The usual notice reads "Hund entlaufen," Hund (dog) being Born's other assistant.
88. Born and WH, ZP, 23 (1924), 388–410 rec. 22 Mar 24, (HCW A1, 247–269); WH, ZP, 26 (1924), 196–204, rec. 4 Jun 24, (HCW A1, 280–288); Born and Hund, *Vorlesungen*, note 50, in which Born thanked Dr. Heisenberg for advice and calculations; and Born, "Über Quantenmechanik," ZP, 26 (1924), 379–395 (English translation in Waerden, pp. 181–198). Born again thanked Heisenberg for "much advice and help with the calculations" (Waerden, p. 182).
89. WH to Landé, 22 Dec 1923 (AHQP 6, 2).
90. The rule is described in more detail in DC, "WH," 296–302; and MR 2, 117–120.
91. WH to Landé, 6 May 1924 (AHQP 6, 2); Landé and WH, ZP, 25 (1924), 279–286, on 286, rec. 18 May 24 (HCW A1, 270–277); and WH, note 54, p. 301.
92. WH to Bohr, 15 May 1924 (BSC 11, 2).
93. Heisenberg wrote Landé of a difficulty for the new scheme posed by Landé's "absolute value rules." Bohr, Heisenberg claimed, "came here because of this matter." WH to Landé, 15 Jun [1924] (AHQP 6, 2).
94. Rose to Bohr, 2 Jun and 10 Jun 1924 (RAC, IEB series 1, Box 50, W. Heisenberg file); Bohr to Rose, 5 Jul 1924 (RAC, IEB Denmark, Box 2146, Folder 403).
95. Born to Bohr, 16 Apr 1924 (BCW 5, 299).
96. Kramers, note 77.
97. Born, note 88 (Waerden, p. 190).
98. Ibid., p. 191.
99. WH to Landé, 15 Jun [1924] (AHQP 6, 2).
100. Born, note 88 (Waerden, p. 198).
101. WH to his mother, 5 Jun 1924.
102. WH to his parents, 8 Jun [1924].
103. WH to Pauli, 8 Jun [1924]; WH to his parents, 8 Jun [1924].
104. Einstein to Born, 29 Apr 1924 (EB, *Briefw.*).
105. Born, note 88 (Waerden, p. 189).
106. WH to Landé, 6 Jul [1924] (AHQP 6, 2).
107. Bohr to WH, 5 Jul 1924 (BSC 11, 2).
108. Personalakt Dr. Heisenberg (UA Göttingen, 4/Vc 317). Heisenberg's venia legendi for the field of theoretical physics was not officially conferred until 10 Oct 1924 (Certificate in HA).
109. WH to Bohr, 5 Jul and 25 Aug [1924] (BSC 11, 2).

Chapter 10

1. WH, *ZP*, 33 (1925), 879–893 (HCW A1, 382–396; English translation in Waerden, pp. 261–276). All quoted translations are subject to slight editing to conform to original.
2. Ibid. (Waerden, p. 261).
3. Ibid. (Waerden, p. 262).
4. WH, "Die Entwicklung der Quantentheorie 1918–1928" (1929), HCW B, 109–115, on 111.
5. The founding of the institute is discussed in Robertson.
6. WH to his parents, 27 Mar [1924].
7. Robertson, pp. 41–49.
8. Ibid., pp. 94 and 106–109.
9. Ibid., table on pp. 156–159.
10. A Privatdozent lectured privately (that is, without state appointment) while awaiting a vacant teaching chair. Heisenberg was funded by lecture fees and Notgemeinschaft stipends.
11. Heisenberg received an IEB stipend of $1000 plus travel expenses for one year beginning 1 September 1924. Rose to Bohr, 10 June 1924 ("Rockefeller Foundation, 1923–25," BGC).
12. WH to his parents, 19 Sep [1924], and WH to his mother, 25 Sep [1924]; WH to Sommerfeld, 18 Nov 1924 (AHQP 83, H). The level of Heisenberg's active Danish is uncertain. Although his Copenhagen lecture notes are in Danish, Heisenberg always responded in German to Bohr's Danish letters.
13. Ibid.
14. MR 2, 149, and Dresden, p. 268. Dresden, pp. 252–276, is an excellent summary of the constellation of relationships among these prima donnas in Copenhagen.
15. Heisenberg apparently sought to replace Kramers as Bohr's personal favorite, not his permanent assistant. In a letter to Rose, 2 Mar 1925 ("Rockefeller Foundation, 1923–25," BGC), Bohr reported that Heisenberg indeed wished to stay longer in Copenhagen, "but on account of his future settling down in Germany he is adviced (sic) not to stay away for more than a year at the time, especially since he has been created Privatdozent in Göttingen just before he came here."
16. WH to Bohr, 5 Jul, 25 Aug, 4 Sep [1924] (BSC 11, 2).
17. WH to his parents, 19 Sep [1924]; WH to Wallace Lund, 27 Jan 25 (HA). This chapter relies on DC, "WH," and the large body of other secondary literature concerned with these issues, especially Serwer, MR 2, Dresden, and Hendry, *Creation*.
18. Kramers, *Nature*, 114 (1924), 310–311.
19. Adolf Smekal, *Nwn*, 11 (1923), 873–875.
20. Dresden, p. 274.
21. New to each other, their relationship seemed at first quite cordial. Heisenberg wrote Pauli on 30 Sep [1924]: "Kramers just said I should write you that one can discuss with me much better than with you!"
22. Bohr, *Nwn*, 12 (1924), 1115–1117, on 1115, dated 1 Nov 1924; (BCW 5, 145–147; English translation on 148–154).
23. Heisenberg was no stranger to correspondence, having appealed to it in defense of the core model in a letter to Landé, 29 Oct 1921 (AHQP 6, 2), and having used it in Sommerfeld and WH, *ZP*, 11 (1922), 131–154 (HCW A1, 165–188).
24. Sommerfeld to Kramers, 5 Jul 1924, discussed in MR 2, 155–156.

25. WH to Pauli, 30 Sep [1924].

26. WH to Sommerfeld, 18 Nov 1924 (AHQP 83, H).

27. WH to Pauli, 8 Oct [1924]; WH, "Über eine Anwendung des Korrespondenzprinzips auf die Frage nach der Polarisation des Fluoreszenzlichtes," ZP, 31 (1925), 617–626 (HCW A1, 344–353).

28. WH to Pauli, 8 Oct [1924].

29. WH, note 27, ZP, 622–623.

30. For details of these arguments, see DC, "WH," ch. 6, and MR 2, 154–169. Heisenberg's quantitative results were soon confirmed by Lothar Nordheim, ZP, 33 (1925), 729–740.

31. Kramers and WH, "Über die Streuung von Strahlung durch Atome," ZP, 31 (1925), 681–708, rec. 5 Jan 1925 (HCW A1, 354–381; English translation in Waerden, 223–251). The theory, its background, and its implications are discussed more extensively in Dresden; Hendry, "BKS"; MR 2; and DC, "WH," ch. 6.

32. Kramers and WH, note 31, ZP, p. 684.

33. WH, interview session 4, 13 Feb 1963 (AHQP).

34. WH to his parents, 6 Dec 1924.

35. Bohr to WH, 2 Jan 1925 (BSC 11, 2); WH to Kramers, 8 Jan 1925 (AHQP 8, 3).

36. WH, "Quantum theory and its interpretation" (1967) HCW C2, 345–361, on 351.

37. Kramers and WH, note 31 (Waerden, p. 234).

38. Serwer, and E. MacKinnon, "Heisenberg, models, and the rise of matrix mechanics," HSPS, 8 (1977), 137–188, reach opposite conclusions on this point.

39. WH to Landé, 15 Jun 1924 (AHQP 6, 2).

40. Pauli and Heisenberg met in Munich on 8 Jan 1925. Pauli's criticisms are indicated in WH to Bohr, 8 Jan 1925 (BSC 11, 2) and WH to Pauli, 28 Feb 1925. Pauli's position in the period 1924–1925 is well summarized in Serwer and in Hendry, Creation.

41. Pauli to Bohr, 2 Oct 1924.

42. Inferred from WH to Pauli, 8 Oct [1924].

43. Pauli to Landé, 24 Nov 1924; also Pauli to Landé, 10 Nov, 14 Nov, and 15 Dec 1924.

44. Pauli, ZP, 31 (1925), 373–385, on 385, rec. 2 Dec 1924 (PCSP 2, 201–213, on 213).

45. Pauli to Landé, 24 Nov 1924.

46. Pauli, note 44, p. 213; Pauli, ZP, 31 (1925), 765–783, on 768, rec. 16 Jan 1925 (PCSP 2, 214–232, on 217).

47. Ibid., and Pauli to Bohr, 12 Dec 1924.

48. For details, see J. L. Heilbron, "The origins of the exclusion principle," HSPS, 13 (1983), 261–310; Serwer; Karl von Meyenn, "Paulis Weg zum Ausschliessungs-prinzip," PB, 36 (1980), 293–298, 37 (1981), 13–19; and B. L. van der Waerden, "Exclusion principle and spin," Theoretical physics in the twentieth century, ed. M. Fierz and V. F. Weisskopf (New York: Interscience, 1960), pp. 199–244. The four numbers were the principal number n, the "magnetic number" m, and two "side quantum numbers" k_1 and k_2.

49. Pauli to Bohr, 12 Dec 1924.

50. WH to Pauli, 15 Dec [1924], and WH to Landé, 4 Jan 1925 (AHQP 6, 2).

51. Bohr to Pauli, 22 Dec 1924.

52. Ibid.

53. This play on words is also suggested by WH to Landé, 4 Jan and 18 Feb 1925 (AHQP 6, 2), and WH to Pauli, 26 Feb 1925.

54. WH to Landé, 4 Jan 1925 (AHQP 6, 2).

55. WH to Landé, 18 Feb [1925] (AHQP 6, 2).

56. WH, "Zur Quantentheorie der Multiplettstruktur und der anomalen Zeemaneffekte," ZP, 32 (1925), 841–860, rec. 10 Apr 1925 (HCW A1, 306–325).

57. Pauli to WH, 28 Feb 25.

58. Hans Geiger to Bohr, 17 Apr 25 (BCW 5, 352–353).

59. WH, note 56; WH to his parents, 3 Apr [1925]; WH to Bohr, 18 Apr 1925 (BSC 11, 2).

60. Heisenberg wrote his parents on 15 Mar [1925], a Sunday, that Pauli was to arrive at the end of the following week. Assuming that Monday is the first day of the week, as is usual in Germany, then Pauli arrived on Saturday or Sunday, 21 or 22 March. Rüdel visited from 20 March to shortly before 2 April 1925. WH to his parents, 20 Mar and 3 Apr [1925].

61. It was also later found to work in a third, intermediate Schema, as indicated in section 5 of WH, note 56.

62. Pauli to Sommerfeld, 6 Dec 1924.

63. J. H. Van Vleck, Bull. Nat. Res. Coun., 10 (1926), entire vol., on p. 287.

64. Bohr to WH, 18 Apr 1925 (BSC 11, 2).

65. Ibid. Kramers was devastated by the collapse of the BKS theory, as discussed by Dresden, 207–215.

66. WH to R. D. Kronig, 8 May 1925 (AHQP 16, 6).

67. Bohr, note 64.

68. Pauli to Bohr, 31 Dec [1924].

69. WH to Bohr, 21 Apr [1925] (BSC 11, 2).

70. He still believed hydrogen to be the least problematic for quantum theory. WH to Landé, 18 Feb [1925] (AHQP 6, 2).

71. Recalled by WH in "Erinnerungen an die Zeit der Entwicklung der Quantenmechanik" (1960), HCW C2, 263–270. To be sure, hydrogen did cause some difficulties, especially when subjected to crossed electric and magnetic fields. WH, note 27, "Anwendung," footnote on p. 353.

72. Indicated in WH to Bohr, 16 May [1925] (BSC 11, 2), and WH to Kronig, 8 and 20 May 1925 (AHQP 16, 6).

73. WH to Bohr, 16 May [1925] (BSC 11, 2).

74. Pauli to Kronig, 21 May 1925.

75. WH to Pauli, 24 Jun [1925].

76. WH to Pauli, 9 Jul 1925.

77. WH, note 75.

78. The last of these steps, stated explicitly in the letter to Pauli (note 76) accompanying his manuscript, appeared only implicitly in the manuscript and later published paper (note 1).

79. Born to Landé, 14 Jul 1925 (AHQP 4, 2).

80. WH to Bohr, 16 May [1925] (BSC 11, 2); WH to Kronig, 8 May 1925 (AHQP 16, 6); Born and Jordan, ZP, 33 (1925), 479–505, 11 June 1925.

81. Born and Jordan, note 80, p. 493, n. 1.

82. WH to his parents, 15 May [1925]. Heisenberg had already used the anharmonic oscillator as a simplification of complicated atomic behavior as early as WH to Pauli, 29 Sep [1922].

83. WH to Kronig, 5 Jun 1925 (AHQP 16, 6).

84. Ibid.

85. Heisenberg's struggle with this and other constants in his theory is followed in MR 2, ch. 4. Heisenberg's derivation of his new physics is discussed in more detail there and in DC, "WH."

86. WH, note 83, and WH to Bohr, 8 Jun [1925] (BSC 11, 2).
87. Later, he recalled that he also brought with him a copy of Goethe's *West-Östlicher Divan.*
88. WH to B. L. van der Waerden, private communication, recounted in Waerden, p. 25, and analyzed exhaustively in MR 2, 251–260. Without reference works on that barren island, it is not certain how Heisenberg was so sure that the result "agreed." Perhaps he had memorized the data during his intensive work on them.
89. WH to Pauli, 29 June 1925.
90. WH, note 1, HCW, pp. 388–389.
91. Indicated by WH to Pauli, 21 and 24 Jun [1925], 4 Jul [1925].
92. WH to Pauli, 24 Jun [1925].
93. Ibid.
94. WH to his father, 30 June [1925]. Nevertheless, Heisenberg apparently put on an optimistic display for Kramers, then visiting Göttingen. WH to Pauli, 29 June [1925].
95. This sequence according to Born, *My life: Recollections of a Nobel laureate* (New York: Charles Scribner's Sons, 1978), pp. 216–217, Born's comment in EB, *Briefw.,* p. 125, and an unpublished chronology prepared by B. L. van der Waerden (in Born folder, AHQP).

Chapter 11

1. Pauli to Kramers, 27 Jul 1925.
2. Pauli to Kronig, 9 Oct 1925. The development of matrix mechanics is summarized, along with a more extensive discussion of technical details, in Waerden; Jammer, *Conceptual*; and MR 3.
3. Born to Schrödinger, 16 May 1927 (AHQP 41, 7).
4. Born and Jordan, ZP, 34 (1925), 858–888, on 858.
5. Ibid. (English translation in Waerden, p. 296).
6. Dirac's background and the technical details of his work in this period are described in Helge Kragh, *Dirac: A scientific biography* (New York: Cambridge University Press, 1990).
7. Dirac's recollection is quoted in Waerden, p. 41; also Dirac interview session 1, 1 Apr 1962 (AHQP).
8. Dirac, PRS, 109 (1925), 642–653, rec. 7 Nov 25.
9. Dirac, PRS, 110 (1926), 561–579, rec. 22 Jan 26.
10. Born, WH, and Jordan, "Zur Quantenmechanik, II," ZP, 35 (1926), 557–615, rec. 16 Nov 1925 (HCW A1, 397–455). Again, for technical details, see the works in note 2.
11. WH-Jordan correspondence (AHQP 18, 2); Born to Bohr, 10 Oct 1925 (BCW 5, 311–312).
12. Franck substituted for Born as director of the theoretical physics institute, while Hund and Heisenberg split Born's teaching duties. Heisenberg's duties included a four-hour lecture, probably on the kinetic theory of matter, a recitation section (Übung), and a seminar. The lecture required only minimal preparation, since Born left his extensive lecture notes for Heisenberg. Gathered from Born to Bohr, 10 Oct 1925 (BCW 5, 311–312); Dekan Courant to University Kurator, 30 Oct 1925 (Personalakt Dr. Heisenberg, UA Göttingen, 4/Vc 317); and *Verzeichnis der Vorlesungen auf der Georg August-Universität zu Göttingen während des Wintersemesters 1925/26.*

13. The method required a "canonical transformation" — preserving the commutation relation — of the Hamiltonian matrix $H(p_0, q_0)$ for the initial momentum matrix (p_0) and position matrix (q_0) to diagonal form $H(p, q)$ by means of a "principle axis transformation" from conjugate coordinate matrices p_0, q_0 to p and q. This required finding an appropriate (unitary) matrix S and its inverse, S^{-1}, such that $SS^{-1} = 1$, to carry out the transformation of H and of each variable in H. The transformation could be expressed as a matrix multiplication:

$$H(p, q) = SH(p_0, q_0)S^{-1}$$

The initial Hamiltonian is transformed into the new Hamiltonian by multiplication from the left by S and from the right by S^{-1}. Perturbing degenerate atoms further complicated matters. A complete discussion is available in the original papers and in MR 4.

14. Pauli, ZP, 36 (1926), 336–363; Dirac, PRS, 110 (1926), 561–569.

15. R. Kronig, "The turning point," and B. L. van der Waerden, "Exclusion principle and spin," in *Theoretical physics in the twentieth century: A memorial volume to Wolfgang Pauli*, M. Fierz and V. F. Weisskopf, eds. (New York: Interscience, 1960), pp. 5–39, and 199–244. The invention of spin is discussed in Serwer; E. Rüdinger and K. Stolzenburg, "Introduction," BCW 5, 219–240; and MR 3.

16. Einstein, "Zum gegenwärtigen Stand des Strahlungsproblems" (1909) in *The Collected Papers of Albert Einstein*, J. Stachel et al., eds., vol. 2 (Princeton: Princeton University Press, 1989) on p. 549.

17. Bohr to WH, 20 Feb [1926] (BSC 11, 2).

18. WH to Pauli, 16 Nov 1925; Pauli to Bohr, 17 Nov 1925.

19. WH to Pauli, 21 Nov [1925].

20. Bohr, note attached to Uhlenbeck and Goudsmit, letter to the editor, *Nature*, 117 (1926), 265 (BCW 5, 289).

21. WH to Pauli, 24 Nov 1925, response to Pauli's "philosophical" letter.

22. Bohr to Ehrenfest, 22 Dec 1925 (BCW 5, 329–330).

23. WH to Jordan, 9 Apr 1926 (AHQP 18, 2); WH to Bohr, 26 Feb 1926 (BSC 11, 2).

24. WH to Pauli, 27 Jan 1926.

25. Pauli to Bohr, 12 Mar 1926; Pauli to Wentzel, 8 May 1926.

26. WH and Jordan, "Anwendung der Quantenmechanik auf das Problem der anomalen Zeemaneffekte," ZP, 37 (1926), 263–277, rec. 16 Mar 1926 (HCW A1, 516–530).

27. Ibid., p. 516.

28. Pauli to Bohr, 12 Mar 1926.

29. The uncertainty principle apparently remained for the Nobel Committee too controversial an issue.

30. Walter Moore, *Schrödinger: Life and Thought* (New York: Cambridge University Press, 1989).

31. Schrödinger's wave mechanics papers, which all appeared in *AP*, 1926, are reprinted in Schrödinger, *Gesammelte Abhandlungen*, vol. 3 (Vienna; Verlag der Österreichischen Akademie der Wissenschaften, 1984), and in English translation in, among others, Schrödinger, *Collected papers on wave mechanics*, 2nd ed. (New York: Chelsea Publishing Co., 1978). Most of the English quotations here are taken from this version.

Schrödinger's formulation of wave mechanics has been extensively examined, most notably in Moore, note 30; MR 5; Linda Wessels, "Schrödinger's route to wave

mechanics," *Stud. hist. phil. sci.*, 10 (1979), 311–340; Helge Kragh, "Erwin Schrödinger and the wave equation: The crucial phase," *Centaurus*, 26 (1982), 154–197.
32. The hypothesis was experimentally confirmed by electron diffraction experiments, interpreted as involving the diffraction of electron waves. C. J. Davisson and G. P. Thomason shared the 1937 Nobel prize in physics for these experiments.
33. Sommerfeld to Pauli, 3 Feb 1926 (SN); and Sommerfeld, *PZ*, 28 (1927), 231–239, on 238.
34. Pauli to Jordan, 12 Apr 1926.
35. Born and Wiener, *ZP*, 36 (1926), 177–187. His enthusiasm is recalled in Born to Schrödinger, 16 May 1927 (AHQP 41, 7).
36. Born, *ZP*, 37 (1926), 863–867, on 864, rec. 25 June 1926. Mara Beller, in "The genesis of interpretations of quantum physics, 1925–1927," Ph.D. thesis, University of Maryland, 1983, pp. 120–121, attributes part of Born's initial enthusiasm for Schrödinger to Born's irritation over the exclusive credit given to Heisenberg as the discoverer of matrix mechanics. Such irritation did not begin to surface until a year later (Born to Schrödinger, 16 May 1927 [AHQP 41, 7]). For instance, the opening sentence of Born's 1926 paper, cited at the beginning of the note, begins "The quantum mechanics founded by Heisenberg . . ."
37. WH to Jordan, 28 Jul [1926] (AHQP 18, 2).
38. WH to Jordan, 8 Apr 1926 (AHQP 18, 2).
39. Born to Schrödinger, 16 May 1927 (AHQP 41, 7). The next sentence reads: "Inzwischen habe ich mich aber wieder zu Heisenbergs Standpunkt zurückgefunden" ("In the meantime I have found my way back again to Heisenberg's standpoint').
40. WH to Pauli, 18 Sep 1925.
41. WH to Pauli, 16 Nov 1925.
42. Pauli to Kronig, 9 Oct 1925; WH to Pauli, 16 Nov 1925.
43. WH to Einstein, 30 Nov 1925 (EA).
44. Schrödinger, note 31, English translation, p. 45, his emphasis.
45. Schrödinger to Wien, 18 Jun 1926 excerpted in W. Wien, *Aus dem Leben und Wirken eines Physikers* (Leipzig, 1930), pp. 72–73.
46. Schrödinger, note 31, English translation, p. 59, slight changes.
47. Schrödinger, note 31, English translation, pp. 59 ff.
48. Schrödinger, *AP*, 79 (1926), 734–756, on 734.
49. WH to Pauli, 8 Jun 26.
50. WH, "Quantenmechanik," *Nwn*, 14 (1926), 989–994 (HCW B, 52–57). The designations are significant. Matrix mechanicians always called their physics quantum mechanics, while calling Schrödinger's work a mere method or theory. Schrödinger called his work a wave mechanics.
51. Pauli to Jordan, 12 Apr 1926, a letter intended for circulation. Heisenberg learned of the proof in Bohr to WH, 10 Apr 1926 (BSC 11, 2)
52. Beller, note 36, and "Matrix mechanics before Schrödinger: Philosophy, problems, consequences." *Isis*, 74 (1983), 469–491, and subsequent papers. The debate is also discussed in Hendry, *Creation*; MR 5, 2; A. I. Miller, "Redefining *Anschaulichkeit*," in *Physics as natural philosophy: Essays in honor of Laszlo Tisza on his 75th birthday*, ed. A. Shimony and H. Feshbach (Cambridge, Mass.: MIT Press, 1982), 376–411; Linda Wessels, "Schrödinger's interpretation of wave mechanics," Ph.D. dissertation, Indiana University, 1975; Edward MacKinnon, *Scientific explanation and atomic physics* (Chicago: University of Chicago Press, 1982), Ch. 7.
53. Bohr to WH, 18 Nov 1925; WH to Bohr, 20 Nov 1925 (BSC 11, 2).

54. Bohr to Born, 25 Nov 1925, and Born to Bohr, n.d. [1925] (BSC 9, 2), both quoted in Röseberg, 356–360. Kramers delayed the move to Utrecht until April. Kramers to university Konsistorium, Copenhagen, 10 Mar 1926 (BGC, Konsistorium, 1925–1928, NBA).

55. Pauli to WH, 19 Oct 1926.

56. Trowbridge to Bohr, Apr 1926 (BGC).

57. WH to Bohr, 5 Apr 1926; Bohr to WH, telegram, [Apr 1926] (BSC 11, 2). As a result, Bohr raised Heisenberg's annual salary from 2880 kroner to 3600 kroner. Both sums were relatively much more than the lecturer's annual stipend of 1535 RM he received in Göttingen but less than the 7800 RM offered by Dresden. Gathered from Bohr, notice of job opening, Feb 1926; University Konsistorium, Copenhagen, to WH, 20 May 1926 (BGC, Konsistorium, 1925–1928); University Kurator, Göttingen, to WH, 19 Dec 1925, "Personalakt Dr. Heisenberg" (UA Göttingen, 4/Vc 317); WH to Bohr, 20 Nov 1925 (BSC 11, 2); WH to his parents, 29 Apr [1926].

58. Courant to Bohr, 24 Apr 1926, quoted in MR 4, 280; WH to his parents, 29 Apr [1926].

59. Courant to Bohr, 24 Apr 1926, quoted in English in MR 4, 280.

60. WH to "Herr Professor" [Max von Laue], 19 Apr 1926 (Darmstädter-Sammlung, SPK, F1a(5) 1926).

61. WH to his parents, Berlin, 29 Apr [1926]. Only Planck was missing; he was away on his annual spring vacation.

62. WH, note 61. Courant began immediate efforts to create a chair in Göttingen for Heisenberg. Courant, dean of science faculty, to University Kurator, 8 Jun 1926, "Personalakt Dr. Heisenberg" (UA Göttingen, 4/Vc 317).

63. WH to his parents, 14 May 1926; WH to his mother, 20 Nov 1926.

64. August Heisenberg to Bohr, 17 July 1926 (BSC 11, 2); Bohr to August Heisenberg, 4 Aug 1926 (BSC 11, 2; Röseberg, 360–361).

65. WH to Born, 26 May 1926 (AHQP 18, 2); WH to his parents, 14 May 1926.

66. Ibid.

67. Ibid.

68. WH to his parents, 11 May 1926; lecture notes in HA and in AHQP 45, 5.

69. Pauli to Schrödinger, 24 May [1926].

70. Pauli to Schrödinger, 22 Nov 1926.

71. WH to Pauli, 28 Jul 1926.

72. Register of lectures, "Münchener physikalisches Mittwochscolloquium" (AHQP 20). Schrödinger lectured on July 23 and 24, 1926, a Friday and a Saturday despite the name of the colloquium.

73. Recalled in WH, PB, 73.

74. Wien to Schrödinger, 20 Aug 1926; Schrödinger to Wien, 25 Aug 1926, excerpted in Wien, note 47.

75. WH, ZP, 38 (1926), 411–426, rec. 11 Jun 26 (HCW A1, 456–471); also WH, ZP, 39 (1926), 499–518, rec. 24 Jul 26 (HCW A1, 531–550); WH to Born, 26 May 1926 (AHQP 18, 2).

76. Discussed by van der Waerden, note 15, and MR 4.

77. WH, note 75, ZP 38, 412.

78. WH to Jordan, 28 Jul [1926] (AHQP 18, 2).

79. WH, note 50, 993–994.

80. WH to Jordan et al., 29 Oct 1926 (AHQP 18, 2).

81. WH to Jordan, 24 Nov 1926 (AHQP 18, 2).

82. WH, ZP, 40 (1926), 501–506, rec. 6 Nov 26 (HCW A1, 472–477).

83. WH to Pauli, 4 Nov 26.

84. WH, note 82, p. 504.

85. Schrödinger delivered a lecture to the Fysisk-Forening in Copenhagen on 4 Oct 1926, entitled "Grundlagen der undulatorischen Mechanik." Minute book of Fysisk-Forening, 1908–1946 (AHQP 35, 4).

86. WH, "Quantum theory and its interpretation" (1967), HCW C2, 345–361, on 356.

87. Especially, Schrödinger to Joos, 17 Nov 1926 (AHQP 41, 8); Schrödinger to Kramers, 19 Nov 1926 (AHQP 41, 8); Schrödinger to Bohr, 23 Oct 1926 (BCW 6, 459–461); Bohr to Schrödinger, 2 Dec 1926 (BCW 6, 462–463); Bohr to Darwin, 24 Nov 26 (BSC 9, 4; Röseberg, 361–362); Bohr to Trowbridge, 9 Oct 1926 ("Rockefeller Foundation 1923–25," BGC).

88. Born, ZP, 38 (1926), 803–827, rec. 21 Jul 26. Born's interpretation is examined in greater detail in Beller, note 36; Jammer, Conceptual; Wessels, note 52; and MR 5, 2.

89. Schrödinger, note 31, English translation, p. 60.

90. Quoted in Jammer, Conceptual, p. 288.

91. Born, ZP, 40 (1927), 167–192, rec. 16 Oct 1926. Born's proposal was, of course, more subtle than this description allows.

92. Schrödinger to Joos, 17 Nov 1926 (AHQP 41, 8).

93. Apparently attributed to Bohr, in Schrödinger to Bohr, 23 Oct 1926 (BCW 6, 459–461).

Chapter 12

1. WH, ZP, 43 (1927), 172–198, rec. 23 Mar 1927 (HCW A1, 478–504); English translation in John A. Wheeler and Wojciech Zurek, eds., Quantum theory and measurement (Princeton: Princeton University Press, 1983), pp. 62–84.

2. WH, "Ueber die Grundprinzipien der 'Quantenmechanik,'" Forschungen und Forschritte, 3 (1927), 83 (HCW C1, 21). The following discussion parallels this summary but with embellishments.

3. Ibid.

4. One discussion of this definition may be found in Ernst Cassirer, Determinisumus und Indeterminismus in der modernen Physik (Göteborg: Elanders Boktryckeri, 1937).

5. Heisenberg's route to uncertainty and the rise of the Copenhagen interpretation have been explored by many other authors. Among the most notable are: Jammer, Conceptual; Mara Beller in "The genesis of interpretations of quantum physics, 1925–1927," Ph.D. dissertation, University of Maryland, 1983, "Pascual Jordan's influence on the discovery of Heisenberg's indeterminacy principle," Archive for History of Exact Sciences, 33 (1985), 337–349, and "Experimental accuracy, operationalism, and the limits of knowledge, 1925 to 1935," Science in context, 2 (1988), 147–162; Gerald Holton, "The roots of complementarity" (1970), in Holton, Thematic origins of scientific thought: Kepler to Einstein, rev. ed. (Cambridge, Mass.: Harvard University Press, 1988), pp. 99–145; Arthur I. Miller, "On the origins of the Copenhagen interpretation," Niels Bohr: Physics and the world, H. Feshbach et al., eds. (Chur: Haward Academic Publishers, 1988), 27–43.

6. These are the results of a survey of the literature reported in A. B. Kozhevnikov and O. I. Novik, Analysis of informational ties dynamics in early quantum mechanics (1925–1927) (Moscow: Academy of Sciences, 1987).

7. WH to Pauli, 16 May 1927.
8. Wiener to Sommerfeld, 28 Nov 1926 (SN).
9. Otto Wiener, "Nachruf auf Theodor Des Coudres," SAW, MPK, *Berichte*, 78 (1926), 358–370.
10. Wiener, note 8.
11. Sommerfeld to Wiener, 3 Dec 1926 (SN).
12. Philosophische Fakultät, Protokolle, book VII, p. 337 (UA Leipzig).
13. Born, *ZP*, 40 (1927), 167–192, rec. 16 Oct 1926.
14. Heisenberg acknowledged some of his debts in WH, note 1, pp. 173–174.
15. Pauli to Schrödinger, 12 Dec 1926; Jordan to Schrödinger [May 1927], incorrectly dated 1926 (AHQP 41, 8).
16. Pauli to WH, 19 Oct 1926.
17. WH to Pauli, 28 Oct 1926.
18. WH to Pauli, 23 Feb 1927, and WH, note 1.
19. A and B are noncommuting if $A \times B \neq B \times A$. Heisenberg's mathematical derivation involved transformation of the Gaussian distribution for the uncertainty in one canonical variable to the distribution expressed as a function of the conjugate variable. The uncertainties of the conjugate variables are related by the uncertainty relations.
20. Heisenberg did not indicate in this paper that the uncertainty relations actually represent minimal uncertainties. Note that if h is zero, as it is in classical mechanics, then there is no limitation on the simultaneous uncertainties in p and q, E and t.
21. Heisenberg took pains to emphasize this in his paper, note 1, p. 196.
22. Schrödinger, *Nwn*, 28 (1926), 664–666. Criticized by Born, note 13.
23. Dirac, *PRS*, 113 (1927), 621–641, rec. 2 Dec 1926.
24. Jordan, *ZP*, 40 (1927), 809–838, rec. 18 Dec 1926.
25. WH to Jordan, 24 Nov 1926 (AHQP 18, 2).
26. His faith in matrix mechanics was reinforced by his own latest success in many-body theory. WH, "Mehrkörperprobleme und Resonanz in der Quantenmechanik, II," *ZP*, 41 (1927), 239–267, rec. 22 Dec 1926 (HCW A1, 551–579); and WH to Pauli, 5 Feb 1927.
27. WH to Pauli, 23 Feb 1927 and 23 Nov 1926. See also WH, note 1, p. 173.
28. WH to Pauli, 23 Nov 1926.
29. WH to Pauli, 23 Feb 1927, and WH, note 1, p. 185.
30. Jordan, *Nwn*, 15 (4 Feb 1927), 105–110.
31. Beller, note 5, "Pascual Jordan's influence."
32. WH to his parents, 29 Apr [1926], and WH, *PB*, p. 63.
33. Only Heisenberg's side of the correspondence has been found: WH to Einstein, 16 and 30 Nov 1925, 18 Feb 1926 (EA).
34. Einstein's contributions to quantum theory have been extensively explored in Pais, *Subtle*, and M. J. Klein, "Einstein and the development of quantum physics," *Einstein: A centenary volume*, A. P. French, ed. (Cambridge, Mass.: Harvard University Press, 1979), pp. 133–151, among others.
35. Einstein to Schrödinger, 26 Apr 1926, in K. Przibram, *Briefe zur Wellenmechanik* (Vienna: Springer-Verlag, 1963), p. 26. There is no indication that Einstein already knew of Schrödinger's equivalence proof, which arrived at the publisher several weeks later.
36. WH, *PB*, p. 63, and WH, interview session 2, 30 Nov 1962 (AHQP).
37. Because this paragraph is not integral to the paper, it may be Heisenberg's "capitulation" to the acausal demands of the cultural milieu prevailing in his

homeland. Capitulation by others is argued in Paul Forman, "Weimar Culture, causality and quantum physics," *HSPS*, 3 (1971), 1–115. Since Heisenberg hoped to return soon to his homeland by producing "good papers," perhaps this was further reason to "capitulate," but there were other good reasons, some even stronger ones, to proclaim that the "invalidity of the causal law is definitively proven." WH, note 1, p. 196.

38. Heisenberg and his colleagues were not the first to introduce indeterminism in modern physics. Earlier considerations are explored, for example, in Stephen G. Brush, "Irreversibility and indeterminism: Fourier to Heisenberg," *Journal of the history of ideas*, 37 (1976), 603–630; and in Paul A. Hanle, "Indeterminacy before Heisenberg: The case of Franz Exner and Erwin Schrödinger," *HSPS*, 10 (1979), 225–269.

39. Born, *ZP*, 38 (1926), 803–827, on 804. Thus, for Born, indeterminism on the macroscopic scale cannot be supported, even though, as Forman (note 37) has argued, Weimar culturalists were not silent in their demand for it.

40. WH to Jordan, 7 Mar 1927 (AHQP 18, 2).

41. Dirac, note 23, p. 641.

42. WH, note 1, pp. 179–184.

43. A present-day discussion of Heisenberg's thought experiment is provided in Jammer, *Philosophy*, pp. 61–71.

44. Schrödinger, *AP*, 82 (1927), 257–264, dated 30 Nov 1926.

45. WH, note 1, p. 175.

46. WH to his parents, 17 Feb 1927. Bohr's departure was imminent when Heisenberg wrote this letter.

47. Bohr to Pauli, 25 Mar 1927.

48. WH to Pauli, 4 Apr 1927. Bohr's handling of the microscope and his presentation of complementarity are in Bohr, "The quantum postulate and the recent development of atomic theory," in *Atti del Congresso Internazionale dei Fisici 11–20 Settembre 1927, Como-Pavia-Roma*, vol. 2 (Bologna, 1928), pp. 565–588: lecture delivered 16 Sep 1927 (BCW 6, 113–136). Bohr's derivation is on p. 573.

49. WH to Pauli, 4 Apr 1927.

50. Bohr, note 48, and WH, *Die physikalischen Prinzipien der Quantentheorie* (Leipzig, 1930), pp. 15–16: original version of his summer 1929 Chicago lectures. A later, more complete derivation than that given by either Bohr or Heisenberg may be found in Jammer, *Philosophy*, pp. 64–65.

51. WH to his parents, 16 and 30 May 1927; WH to Pauli, 16 May 1927. Dirac had returned to Cambridge in February.

52. WH, *PB*, p. 79.

53. WH to Pauli, 16 May 1927.

54. WH, note 1, "Nachtrag zur Korrektur," pp. 197–198.

55. WH to Pauli, 16 May 1927, my emphasis.

56. Bohr, note 48. An excellent analysis of Bohr's views and their intellectual origins is provided in Holton, note 5. The discussion here is not intended to be exhaustive.

57. "Discussione sulla comunicazione Bohr," in *Atti del Congresso Internazionale dei Fisici 11–20 Settembre 1927, Como-Pavia-Roma*, vol. 2 (Bologna, 1928), pp. 589–598 (BCW 6, 137–146). WH's remark appears on pp. 593–594.

58. WH to his parents, Copenhagen, 22 Jun 1927. Debye had practically decided on Leipzig by the spring of 1927, as indicated by records in "Prof. Debye Berufg. nach Leipzig" (Archiv, Schweizerischer Schulrat, in ETH Zurich).

59. WH to Bohr, 18 Jun 1927 (BPC).

60. Bayerisches Staatsministerium to University Senate, Munich, 20 Jun 1927 (A. Sommerfeld, Personalakten E II-N, UA Munich); and Sommerfeld to Philosophische Fakultät II, 3 Dec [1925] (Univ. München, Lehraufträge, vol. 2, BHStA, MK 11303).
61. Sommerfeld to WH, 17 Jun 1927 (SN).
62. WH to his parents, 22 Jun 1927.
63. WH to Sommerfeld, 21 Jun 1927 (SN).
64. WH to Bohr, 21 Aug 1927 (BPC).
65. Dr. P. Schindler to the author, 20 Mar 1979, concerning records of Schweizerischer Schulrat, in ETH Zurich. American interest in Heisenberg is indicated in a renewed inquiry from Trowbridge to Bohr, 5 Oct 1927 (BSC 11, 2).
66. WH to Bohr, 5 Dec 1927 (BSC 11, 2).
67. Philosophische Fakultät, note 12, p. 352. Heisenberg was appointed in October 1927, before his 26th birthday, and began lecturing immediately, but he did not officially assume his post until February. The public inaugural lecture was probably "Erkenntnistheoretische Probleme der modernen Physik," manuscript (HA), HCW C1, 22–28.

Chapter 13

1. The missionary zeal is described by J. L. Heilbron, "The earliest missionaries of the Copenhagen spirit," *Revue d'histoire des sciences*, 38 (1985), 195–230, and in *Science in reflection*, Edna Ullmann-Margalit, ed., Boston studies in the philosophy of science, 110 (Dordrecht: Kluwer Academic Publishers, 1988), pp. 201–233. The predominance of this interpretation is explored in James T. Cushing, "Copenhagen hegemony: *Need* it be *so?*" manuscript.
2. WH to Pauli, 16 May 1927.
3. Bohr to Einstein, 13 Apr 1927 (BCW 6, 418–421).
4. Bohr, "The quantum postulate and the recent development of atomic theory," in *Atti del Congresso Internazionale dei Fisici 11–20 Settembre 1927, Como-Pavia-Roma*, vol. 2 (Bologna, 1928), pp. 565–588, on p. 569 (BCW 6, 113–136).
5. Ibid., p. 572. Bohr's Como lecture and complementarity are further discussed in Jammer, *Philosophy*, pp. 86–101; and in Gerald Holton, "The roots of complementarity" (1970), in Holton, *Thematic origins of scientific thought: Kepler to Einstein*, rev. ed. (Cambridge, Mass.: Harvard University Press, 1988), pp. 99–145.
6. "Discussione sulla comunicazione Bohr," note 4, *Atti*, pp. 589–598, on p. 593 (BCW 6, 137–146).
7. WH, "Die Entwicklung der Quantentheorie 1918–1928" (1929), HCW B, 109–115, on 114.
8. WH to Bohr, 23 Jul 1928 (BSC 11, 2).
9. WH to Bohr, 21 Aug 1927 (BPC).
10. Pauli to Bohr, 17 Oct 1927. Pauli's rallying to the cause is described in Hendry, *Creation*.
11. Born, comment in "Discussione," note 6, p. 589.
12. Born and WH, "La mécanique des quanta," in *Electrons et photons. Rapports et discussions du 5. Conseil de Physique . . . Solvay*, ed. Institut International de Physique Solvay (Paris, 1928), pp. 143–181, on p. 143 (HCW B, 58–96).
13. Ibid., p. 178, their italics. The theory of radiation is the theory of the interaction of light with matter.
14. Pauli to Kramers, 27 Jul 25.

15. See J. Mehra, *The Solvay conferences on physics: Aspects of the development of physics since 1911* (Dordrecht: D. Reidel Publishing Company, 1975).

16. Arthur Fine, *The shaky game: Einstein, realism, and the quantum theory* (Chicago: University of Chicago Press, 1986); Jammer, *Philosophy*, pp. 108–158; Pais, *Subtle*, pp. 440–449; Edward MacKinnon, *Scientific explanation and atomic physics* (Chicago: University of Chicago Press, 1982); Bohr, "Discussion with Einstein on epistemological problems in atomic physics," in *Albert Einstein: Philosopher-scientist*, ed. P. A. Schilpp, vol. 1 (La Salle, Ill.: Open Court, 1949), 199–241.

17. Einstein to Born, 4 Dec 1926 (EB, *Briefw.*).

18. Given by Einstein in a discussion remark in *Electrons et photons*, note 12, pp. 253–256 (BCW 6, 101–103).

19. Bohr, note 16, "Discussion with Einstein," pp. 212–216.

20. Einstein to Sommerfeld, 9 Nov 1927 (ES, *Briefw.*): also Einstein to Lorentz, 21 Nov 1927 (EA). Einstein offered another thought experiment at the sixth Solvay Congress in 1930 that concerned the uncertainty relation for energy and time.

21. Pauli to Weyl, 11 Jul 1929. Einstein's and Bohr's differing views are further discussed in the works cited in note 16.

22. WH to his parents, Brussels, 29 Oct 1927.

23. This has been argued in Paul Forman, "*Kausalität, Anschaulichkeit*, and *Individualität*, or How cultural values prescribed the character and the lessons ascribed to quantum mechanics," in *Society and knowledge: Contemporary perspectives in the sociology of knowledge*, ed. Nico Stehr and Volker Meja (New Brunswick, N.J.: Transaction Books, 1984), pp. 333–347.

24. See WH, *Bibl.*

25. These are all reprinted in HCW, C1.

26. Forman, note 23, attributes this to the German cultural milieu.

27. Born, ZP, 37 (1926), 863–867.

28. Born, note 11.

29. WH, "Erkenntnistheoretische Probleme der modernen Physik," manuscript (HA), HCW, C1, 22–28, on 28. This paper and Heisenberg's overall philosophical position have been analyzed by Patrick A. Heelan, *Quantum mechanics and objectivity: A study of the physical philosophy of Werner Heisenberg* (The Hague: Martinus Nijhoff, 1965).

30. WH, "Die Rolle der Unbestimmtheitsrelationen . . ." (1931), HCW C1, 40–47, on 45.

31. WH, "Kausalgesetz und Quantenmechanik" (1931), HCW C1, 29–39, on 29–31.

32. Ibid., p. 39.

33. WH, "Probleme der modernen Physik" (1932), HCW C1, 48–49, on 49.

34. Heisenberg-Schlick correspondence, 1930–1932 (Schlick Papers, Amsterdam). I am grateful to Anne Kox for informing me of this correspondence.

35. WH, "Probleme," note 33, p. 49.

36. Bohr's papers on complementarity are published in BCW 6.

37. Heisenberg recalled these conversations in WH, *PB*, pp. 117–124.

38. Interview with C. F. von Weizsäcker, 30 Apr 1982.

39. K. F. von Weizsäcker, "Ortsbestimmung eines Elektrons durch ein Mikroskop," ZP, 70 (1931), 114–130.

40. Quoted by Fine, note 16, p. 34.

41. Einstein, Podolsky, and Rosen, PR, 47 (1935), 777–780; reprinted in Wheeler and Zurek, eds., *Quantum Theory and Measurement* (Princeton: Princeton University Press, 1983), pp. 138–141. Fine, note 16, argues for Podolosky's authorship.

42. Pauli to WH, 15 Jun 1935.

43. Further historical and philosophical discussions, from which this section has benefited, include Jammer, *Philosophy*, pp. 159–251; MacKinnon, note 16, pp. 338–348; and Fine, note 16.

44. The problem of disturbance is especially emphasized in Fine, note 16.

45. Quoted in Jammer, *Philosophy*, pp. 189–190.

46. Pauli to WH, 15 Jun 1935.

47. WH, "Ist eine deterministische Ergänzung der Quantenmechanik möglich?" manuscript enclosed in WH to Pauli, 2 Jul 1935 (PWB, vol. 2, pp. 409–418; another copy in EA). Heisenberg must have sent the manuscript to Einstein via an indirect route. He could not correspond directly with Einstein from Germany in 1935, and Heisenberg did not travel abroad in those months.

48. Bohr to WH, 2 Jul 1935 (BSC 20, 2), enclosing manuscript of Bohr, "Can quantum-mechanical description of physical reality be considered complete?" *PR*, 48 (1935), 696–702 (Wheeler and Zurek, note 41, 145–151).

49. WH to Bohr, 28 Aug 1935; Bohr to WH, 10 and 15 Sep 1935; and WH to Bohr, 29 Sep 1935 (BSC 20, 2).

50. WH, note 47, p. 417. Heisenberg referred to G. Hermann, *Die naturphilosophischen Grundlagen der Quantenmechanik* (Berlin, 1935). Hermann's views, expounded before the EPR argument, and Heisenberg's later use of them are discussed in Jammer, *Philosophy*, pp. 207–211.

51. WH, "Prinzipielle Fragen der modernen Physik," delivered at University of Vienna, 27 Nov 1935 (HCW C1, 108–119).

52. This is emphasized especially by MacKinnon, note 16, p. 344.

53. Bohr, note 48, "Description," p. 702.

54. WH, note 51, p. 117.

55. WH, note 47, p. 410.

56. Einstein, "Remarks concerning the essays brought together in this co-operative volume," in Schilpp, ed., note 16, vol. 2, pp. 665–688, on pp. 666–667, his parenthetical remarks.

57. WH, *Physics and philosophy: The revolution in modern science* (New York: Harper & Row, 1958), pp. 144 and 129.

58. "Come of age" attributed to physicist J. H. Van Vleck, by K. R. Sopka, *Quantum physics in America 1920–1935* (New York: Arno Press, 1980), p. xiii.

59. This was argued in J. L. Heilbron, "La fisica negli Stati Uniti subito prima della meccanica quantistica," in *Fisici e società negli anni '20* (Milan, 1980), pp. 135–158; and Robert Seidel, "Aspetti istituzionali della transmissione della meccanica quantistica agli Stati Uniti," in *Fisici e società negli anni '20* (Milan, 1980), pp. 189–214. Critiques of these assertions and of alternative and traditional views of the later rise of American quantum theory are offered, among others, in Sopka, note 58; in Albert E. Moyer, "History of Physics," *Osiris*, 1 (1985), 163–182; and in Holton, "On the hesitant rise of quantum physics research in the United States," in Holton, note 5, *Thematic*, pp. 147–187. Philanthropic support of American science is examined in Robert E. Kohler, "Science, foundations, and American universities in the 1920s," *Osiris*, 3 (1987), 135–164. The rise of American theoretical physics is explored in S. S. Schweber, "The empiricist temper regnant: Theoretical physics in the United States 1920–1950," *HSPS*, 17 (1986), 55–98.

60. Schweber, note 59. However, geographic dispersion and disinterest among administrators slowed the development of American quantum physics research.

61. K. T. Compton, *Nature*, 139 (1937), 238–239.

62. List of past and future invited physicists, locations, salaries, and lengths of stay in H. M. Randall to W. F. G. Swann, 19 Feb 1929 (Swann Papers, Am. Philos. Society, Philadelphia).

63. The salaries ranged from $1500 for Hund to $6000 for Heisenberg. By comparison, the annual salary offered Heisenberg a year earlier to teach full time at Columbia University was $10,000. These were all large sums at the time.

64. Unfortunately, such contacts were not as helpful as they might have been.

65. AS, *Atombau und Spektrallinien: Wellenmechanischer Ergänzungsband* (Braunschweig: Springer-Verlag, 1929).

66. WH to Sommerfeld, 6 Feb 1929 (AHQP 83, H).

67. WH to Bohr, 1 Mar 1929 (BSC 11, 2).

68. WH to Bohr, Chicago, 16 Jun 1929 (BSC 11, 2).

69. WH to Pauli, 20 Jul 1929. Heisenberg's and Dirac's travels are described in detail by L. Brown and H. Rechenberg, "Paul Dirac and Werner Heisenberg — a partnership in science," in *Reminiscences about a great physicist: Paul Adrian Maurice Dirac*, ed. B. N. Kursunoglu and E. P. Wigner, (Cambridge: Cambridge University Press, 1987), pp. 117–162.

70. WH, *The physical principles of the quantum theory*, trans. C. Eckart and F. C. Hoyt (Chicago: University of Chicago Press, 1930), p. x (HCW B, 117–166).

71. Sommerfeld, note 65; Born and Jordan, *Elementare Quantenmechanik* (Berlin: Springer-Verlag, 1930; Pauli, "Die allgemeinen Prinzipien der Wellenmechanik," in *Handbuch der Physik*, 2nd ed., vol. 24, part 1 (Berlin, 1933), pp. 83–272. Section A1, pp. 83–90, is entitled "Unbestimmtheitsprinzip und Komplementärität."

72. A brief list of such textbooks is provided in Jammer, *Philosophy*, p. 59.

73. For instance, E. C. Kemble, *The fundamental principles of quantum mechanics* (New York: McGraw-Hill, 1937), p. 5, written after EPR. Kemble's response to quantum theory is discussed in Holton, note 59.

74. I am grateful to Prof. S. S. Schweber for informing me of this recollection.

Chapter 14

1. WH, *PB*, p. 93.

2. L. Weickmann, "Nachruf auf Otto Wiener," SAW, MPK, *Berichte*, 79 (1927), 107–123.

3. Wentzel to Sommerfeld, 26 May 1927 (SN).

4. Note from Ulrich, dated 31 May 1932, on letter of Debye to Saxon Culture Ministry, 26 May 1932 ("Physikalisches Institut der Universität Leipzig," vol. 2, 1929–1939, SStA, Ministerium für Volksbildung, 10230/27). Depictions of the history of Leipzig physics have been assembled in "Sektion Physik der Karl-Marx-Universität 150 Jahre nach der Gründung des Physikalischen Instituts in Leipzig," *Wissenschaftliche Zeitschrift der Karl-Marx-Universität Leipzig, mathematisch-naturwissenschaftliche Reihe*, 34 (1985): 1, 1–87; and in Martin Franke, "Zu den Bemühungen Leipziger Physiker um eine Profilierung der physikalischen Instituts der Universität Leipzig im zweiten Viertel des 20. Jahrhunderts," *NTM-Naturwissenschaft, Technik, Medizin*, 19 (1982), 68–76.

5. Floor plans in Otto Wiener, "Das Physikalische und das Theoretisch-Physikalische Institut," *Festschrift zur Feier des 500jährigen Bestehens der Universität Leipzig*, vol. 4, part 2 (Leipzig: Hirzel, 1909), pp. 24–69.

6. Debye to Ministerialrat von Seydewitz, 19 Sep 1927, and Debye to Notgemeinschaft, 20 Jul 1928 (DP).

7. WH to his parents, 5 and 9 Nov 1927.

8. WH to his parents, 13 Dec 1927, and Akten des Rentamtes (UA Leipzig).

9. WH to his parents, 9 Nov 1927.

10. Sommerfeld to WH, 15 Nov 1927 (SN).

11. WH to his parents, 7 May [1928], and Wentzel to Sommerfeld, 12 May 1928 (SN). The three students were Peierls, Eckart, and Houston, an American visitor.

12. WH to his parents, 27 Jun [1928].

13. In the same period, the total number of Reich university students increased by 20 percent; the number of first-year physics students in Leipzig went from 1 to 14. Data derived from *Vierteljahrshefte zur Statistik des Deutschen Reiches*, 37 (1928), 38 (1929), 39 (1930).

14. Debye to Staatsministerium, 27 Feb 1931 (note 4, Physikalisches Institut).

15. As a result, overcrowding in physics became common. Wien's successor in Munich, Walther Gerlach, reported that about 1000 students attended his lectures in a hall seating 525. This prompted a student delegation to complain "in the most irate form." Gerlach to Rector of Universität München, 16 May 1930 (Acta des K. Akademischen Senats, Das physikalische Cabinet betr., UA München).

16. List compiled by Dr. Michael Eckert, to whom I am grateful for a copy.

17. The establishment and maintenance of this network is indicated by the travels of the assistants and students and by the correspondence of their teachers, especially WH to Pauli, 1 Aug [1929].

18. Interview with Friedrich Hund, Göttingen, April 1981; and interview with Hund by the Project for History of Solid State Physics, Deutsches Museum, Munich, 1982. According to Hund's "Wissenschaftliches Tagebuch" (Deutsches Museum), he accepted the call to Leipzig on February 4, 1929, then left for the United States about the same time Heisenberg did. A photograph of the two in Chicago is extant. Hund returned to Leipzig by the start of the summer semester in May 1929 in order to run the teaching program while Heisenberg stayed abroad until the fall.

19. C. I. Zahn to S. A. Goudsmit, 24 Feb 1933 (PWB, vol. 2, p. 148).

20. WH to Mansel Davies, 16 Jan 1970, published in Mansel Davies, "Peter Joseph Wilhelm Debye," Royal Society of London, *Biographical memoirs*, 16 (1970), 175–232, on 221.

21. Interview with Hund, April 1981. However, Guido Beck later told Peierls that *he* actually picked up the pastries. Rudolf Peierls, *Bird of passage: Recollections of a physicist* (Princeton: Princeton University Press, 1985), p. 32.

22. WH to Bohr, 23 Jul 1928 (BSC 11, 2).

23. WH to his mother, 25 Nov [1928] and 6 Dec 1928.

24. WH to his parents, 27 Jun 1928 and 16 Oct [1928].

25. Felix Bloch, "Reminiscences of Heisenberg and the early days of quantum mechanics," *Physics today*, 29:12 (December 1976), 23–27, on 27.

26. Quoted in Bernhard Schweitzer, *Die Universität Leipzig 1409–1959* (Tübingen: J. C. B. Mohr, 1960), pp. 13–14.

27. Interview with C. F. von Weizsäcker, 30 Apr 1982.

28. WH to his parents, 25 Feb 1930.

29. Interview with Hund, Göttingen, April 1981.

30. Hans Driesch, *Lebenserinnerungen* (Basel: Ernst Reinhardt Verlag, 1951), p. 200.

31. Gerhard Ritter, *Carl Goerdeler und die deutsche Widerstandsbewegung* (Stuttgart: Deutsche Verlagsanstalt, 1954).

32. Draft of nomination, 23 May 1930 (DP), and SAK, MPK, *Berichte*, 82 (1930).

33. WH, "Über Energieschwankungen in einem Strahlungsfeld," delivered 9 January 1931 (HCW A2, 116–131).

34. Hund, note 29, and Weizsäcker, note 27.

35. The early development of quantum field theory and its history to the present are subjects still requiring much research. Summaries of this history, from which this chapter has benefited, include Pais, *Inward*; MR 4; Joan Bromberg, "Dirac's quantum electrodynamics and the wave-particle equivalence," in *History of twentieth century physics*, ed. C. Weiner (New York: Academic Press, 1977), pp. 147–157; Olivier Darrigol, "Les débuts de la théorie quantique des champs (1925–1948)," doctoral dissertation, Université de Paris I (Panthéon-Sorbonne), 1982.

36. Dirac, PRS, 114 (1927), 243–265 and 710–728. Studies of Dirac's life and work, along with more detailed discussions of the physics presented here, have been provided in Helge Kragh, *Dirac: A scientific biography* (New York: Cambridge University Press, 1990), as well as in the works cited in note 35.

37. Commuting variables obey the rule $AB = BA$, or $AB - BA = 0$, for example, $4 \times 3 = 3 \times 4$.

38. Indicated in WH to Pauli, 23 Feb 1927, the same letter in which he first presented to Pauli his notions on uncertainty.

39. WH to Pauli, 4 Apr 1927.

40. Jordan and Pauli, ZP, 47 (1928), 151–173.

41. Indicated in Pauli to Dirac, 17 Feb 1928.

42. WH to Pauli, 23 Feb 1927.

43. Presented on 1 Feb 1928.

44. Dirac, PRS, 117 (1928), 610–624. For details, see Kragh, note 36, and the works in note 35.

45. Jordan and Wigner, ZP, 47 (1928), 631–651.

46. Pauli to Dirac, 17 Feb 1928.

47. The dilemma of negative energies was made worse by Oskar Klein's 1929 paradox, in which negative energy states must be taken into account in the scattering of an electron by a barrier.

48. Dirac delivered his lecture (probably in English) entitled "Über die Quantentheorie des Elektrons" on 22 Jun 1928. Conference program in Debye to Fermi, 9 Jun 1928 (DP). It was published in PZ, 29 (1928), 561–563 and 712. Heisenberg reported a long discussion with Dirac in Leipzig in WH to Jordan, 25 Jun 1928 (Nachlass Jordan, SPK).

49. Pauli to Bohr, 16 Jun 1928.

50. WH to Pauli, 3 May 1928.

51. Sommerfeld, Nwn, 15 (1927), 825–832. A fuller account of the electron theory of metals in this period is provided by L. Hoddeson, G. Baym, and M. Eckert, "The development of the quantum-mechanical electron theory of metals: 1928–1933," *Reviews of modern physics*, 59 (1987), 287–327.

52. Bloch completed his dissertation, entitled "Über die Quantenmechanik der Eletronen in Kristallgittern," on 2 Jul 1928, indicated in "Promotionen," (file MI, UA Leipzig). Heisenberg's and Debye's opinions are in "Promotionsakten" (UA Leipzig). Bloch published a version of the dissertation in ZP, 52 (1928), 555–600.

53. Peierls recalls his work in Leipzig in Peierls, note 21, pp. 32–40.

54. WH to Pauli, 3 May 1928.

55. Heisenberg's earlier efforts and his contributions to the theory of ferromagnetism are discussed in Hoddeson et al., note 51, pp. 304–311.

56. WH, "Zur Theorie des Ferromagnetismus," ZP, 49 (1928), 619–636, rec. 20 May 1928 (HCW A1, 580–597).

57. WH to Pauli, 31 Jul 1928.

58. WH, "Fortschritte in der Theorie des Ferromagnetismus," Metallwirtschaft, 9 (1930), 843–844 (HCW B, 167–168).

59. Papers by WH and Pauli in Probleme der modernen Physik. Arnold Sommerfeld zum 60. Geburtstag gewidmet von seinen Schülern, ed. Peter Debye (Leipzig: Hirzel, 1928); WH's paper reprinted in HCW B, 100–108.

60. Pauli to Bohr, 16 Jan [1929]; Pauli to O. Klein, 18 Feb 1929; WH to Jordan, 22 Jan [1929] (Nachlass Jordan, SPK).

61. WH and Pauli, "Zur Quantendynamik der Wellenfelder" ZP, 56 (1929), 1–61, rec. 19 Mar 1929 (HCW A2, 8–68); WH and Pauli, "Zur Quantentheorie der Wellenfelder, II" ZP, 59 (1930), 168–190, rec. 7 Sep 1929 (HCW A2, 69–91). Their theory is also discussed in the works cited in note 35.

62. J. R. Oppenheimer, ZP, 55 (1929), 725–737, rec. 6 May 1929; and Oppenheimer, PR, 35 (1930), 461–477, rec. 12 Nov 1929.

63. Bohr, J. Chem. Society London, 1932, 349–384, (BCW 6, 371–408).

64. Bohr to Dirac, 29 Aug 1930 (BSC 18, 4).

65. WH to Pauli, 20 Jul 1929.

66. WH to Bohr, 26 Feb [1930], 10 and 23 Mar 1930; Bohr to WH, 18 Mar 1930 (BSC 20, 2).

67. R. E. Peierls, ZP, 53 (1929), 255–266; and Peierls, PZ, 30 (1929), 273–274.

68. Especially, WH to Bohr, 10 Mar 1930 (BSC 20, 2).

69. WH, "Die Selbstenergie des Elektrons" ZP, 65 (1930), 4–13 (HCW A2, 106–115) on p. 4.

70. Ibid., and WH to Bohr, 26 Apr 1930 (BSC 20, 2).

71. WH to his parents, 30 May and 12 Jul 1930.

72. WH to his father, 4 Nov 1930.

73. WH to his parents, 7 Nov 1930.

74. WH to his mother, 15 Dec 1930.

75. WH to Bohr, 18 Mar 1937 (BSC 20, 2).

76. WH, "Zur Theorie der Magnetostriktion und der Magnetisierungskurve" (1931), HCW A1, 598–608; WH, "Über die inkohärente Streuung von Röntgenstrahlen" (1931), HCW A1, 627–630; WH, "Theoretische Überlegungen zur Höhenstrahlung" (1932), HCW A2, 250–272.

77. WH to Bohr, 27 Jul [1931] (BSC 20, 2), containing a discussion of various applications.

78. WH, "Probleme der modernen Physik" (25 Dec 1931), HCW C1, 48–49.

79. Heisenberg, Bohr, and the route to the neutron-proton model of the nucleus have been explored in Joan Bromberg, "The impact of the neutron: Bohr and Heisenberg," HSPS, 3 (1972), 307–341; and discussed in Laurie M. Brown and Helmut Rechenberg, "Nuclear structure and beta-decay (1932–1933)," American journal of physics, 56 (1988), 982–988, and elsewhere.

80. Bromberg, note 79, points out the importance of Heisenberg's style for his nuclear physics. The style portrayed here differs somewhat from Bromberg's portrayal.

81. Guido Beck returned to Leipzig from Copenhagen with a report on Bohr's new theory of superconductivity, as indicated in WH to Bohr, 20 Jun [1932] (BSC 20, 2). But Bohr did apply the neutron to electron-neutron collision phenomena, as pointed out by Bromberg, note 79, p. 330.

82. Niels Bohr, "Atomic stability and conservation laws," in *Convegno di fisica nucleare, Ottobre 1931–IX* (Rome: Reale Accademia d'Italia, 1932), pp. 119–130.

83. WH to Bohr, 3 and 23 Feb [1932]; Bohr to WH, 24 Feb 1932 (BSC 20, 2).

84. Chadwick to Bohr, 24 Feb 1932 (BSC). Heisenberg also learned of the discovery about the same time, that is, after meeting at the ski hut.

85. WH, "Über den Bau der Atomkerne, I" *ZP*, 77 (1932), 1–11, actually rec. 7 Jun 1932 (HCW A2, 197–207). Heisenberg seemed to be protecting his priority by not informing anyone in Copenhagen until June 20, 1932. When earlier announcing his arrival in Copenhagen on April 6, 1932, Heisenberg was more intrigued with a new cosmic-ray phenomenon (showers) than with the neutron. WH to Bohr, 30 Mar [1932]. He later referred only to discussions about the scattering of gamma rays by neutrons during his Easter visit in WH to Bohr, 18 Jul [1932] (BSC 20, 2), and in WH, "Über den Bau der Atomkerne, II" (rec. 30 Jul 1932), HCW A2, 208–216.

86. WH to Bohr, 20 Jun [1932] (BSC 20, 2).

87. WH, note 85, "Über den Bau der Atomkerne, I," 197–198.

88. WH, note 85, "Über den Bau der Atomkerne, II," 216.

89. WH, "Über den Bau der Atomkerne, III" (rec. 22 Dec 1932), HCW A2, 217–226.

90. WH, "Considérations théoretiques générales sur la structure du noyau," report to Solvay Congress, Brussels, October 1933 (HCW B, 179–225); WH, "Die Struktur der leichten Atomkerne" (1935), HCW A2, 227–238; WH, "Bemerkungen zur Theorie des Atomkerns," in *Peter Zeeman, 1865–1935* (The Hague: Nijhoff, 1935), pp. 108–116 (HCW B, 238–246).

91. Interview with C. F. von Weizsäcker, 30 Apr 1982.

Chapter 15

1. "Ein Ermächtigungsgesetz," *LNN*, no. 32 (1 Feb 1933), p. 1, editorial.

2. For example, Ringer; Karl Dietrich Bracher, *The German dictatorship*, trans. Jean Sternberg (New York: Holt, Rinehart and Winston, 1970); Hannah Arendt, *The origins of totalitarianism*, part 3: *Totalitarianism* (San Diego: Harcourt, Brace, Jovanovich, 1951); Hans Mommsen, *Beamtentum im Dritten Reich (Schriftenreihe der Vierteljahrshefte für Zeitgeschichte,* 12) (Stuttgart: DVA, 1966); Kurt Sontheimer, *Antidemokratisches Denken in der Weimarer Republik: Die politischen Ideen des deutschen Nationalismus zwischen 1918 und 1933* (Munich: Nymphenburger Verlag, 1962/1968); Ian Kershaw, *The Nazi dictatorship: Problems and perspectives of interpretation* (London: Edward Arnold, 1985).

3. "Deutsche Hochschullehrer für Adolf Hitler," *Der Führer,* 6 Nov 1932, signed by 55 academics; "Die deutsche Geisteswelt für die Liste I. Erklärung von 300 deutschen Universitätslehrern," *Völkischer Beobachter,* 3 Mar 1933, p. 1, "Appell der deutschen Wissenschaft an die Welt," *Frankfurter Zeitung,* 78 (13 Nov 1933), p. 4, and *Braunschweiger Landeszeitung,* 13 Nov 1933; "Die deutsche Wissenschaft ruft auf: Für Adolf Hitler—Ja!" *Frankfurter Zeitung,* 79 (19 Aug 1934), p. 1, with 68 signatures; Friedrich Schmidt-Ott (former head of Notgemeinschaft), "Für die Erneuerung unseres gesamten Kulturlebens," *LNN,* no. 225 (13 Aug 1934), p. 2.

4. "Ein Hilferuf für die deutsche Wissenschaft," *LNN,* no. 315 (10 Nov 1932), p. 3.

5. Planck's position and compromises are discussed in J. L. Heilbron, *The dilemmas of an upright man: Max Planck as spokesman for German science* (Berkeley: University of California Press, 1986).

6. This is based on election results as reported in *LNN*. For example, in the

election of November 6, 1932, the percent of votes cast in Leipzig (Reich) for the three leading parties was SPD, 33.4 (20.7); KPD, 21.8 (17.0); NSDAP, 27.9 (33). The election of March 5, 1933, yielded the following results: SPD, 31.7 (18.2); KPD, 18.8 (12.2); NSDAP, 36.6 (43.7). Leipzig votes by precinct are given in Karl Dietrich Bracher, *Stufen der Machtergreifung* (part I of *Die nationalsozialistische Machtergreifung*) (Frankfurt am Main: Ullstein, 1983), p. 180.

7. For example, "43 Personen ins Konzentrationslager Dachau gebracht," *LNN,* no. 226 (14 Aug 1933), p. 3. These arrests and the early camps are discussed in Martin Broszat, "Nationalsozialistische Konzentrationslager 1933–1945," in *Anatomie des SS-Staates,* ed. Hans Buchheim et al., vol. 2 (Munich, 1967), pp. 11–133, esp. pp. 15–22.

8. "Gesetz zur Wiederherstellung des Berufsbeamtentums. Vom 7. April 1933," *Reichsgesetzblatt,* no. 34 (1933), pp. 175–177; "Gesetz gegen die Überfüllung deutscher Schulen und Hochschulen. Vom 25. April 1933," *Reichsgesetzblatt,* no. 43, p. 225. The latter specified that, for entering pupils at each level of education, the percent of non-Aryans should not exceed the percent of non-Aryans in the general population. A non-Aryan child was defined as one for whom both parents were non-Aryan, but the definitions of Aryan and non-Aryan were unspecified.

9. Sarah Gordon, *Hitler, Germans and the "Jewish question"* (Princeton: Princeton University Press, 1984). Detlev J. K. Peukert, *Inside Nazi Germany: Conformity, opposition and racism in everyday life,* trans. Richard Deveson (New Haven: Yale University Press, 1982), p. 58, finds inconsistency in popular attitudes toward the persecution of Jews.

10. Gerhard Ritter, *Carl Goerdeler und die deutsche Widerstandsbewegung* (Stuttgart: DVA, 1954), p. 64. However, Goerdeler, a leader of the Deutschnationale Volkspartei, could not or would not prevent the occupation of union halls and the dissolution of the Leipzig unions and the Volkshochschule in the same year.

11. Statement in *New York World-Telegram,* 11 Mar 1933, reprinted in *Einstein on peace,* ed. Otto Nathan and Heinz Norden (New York: Schocken Books, 1960), p. 211.

12. *Einstein on peace,* note 11; and Einstein to George Ellery Hale, 26 Apr 1933 (EA 19-386).

13. "Boykott bei strenger Disziplin," *LNN,* no. 92 (2 Apr 1933), p. 1.

14. Laue to Einstein, 26 Jun 1933 (EA 16-095); Planck to Einstein, 19 Mar 1933 (EA 19-388). Einstein's expulsion from the Berlin Academy and Planck's reactions to the situation are discussed in Heilbron, note 5.

15. Einstein to Laue, 26 May 1933 (EA 16-089).

16. E. Heymann, Academy secretary, press statement of 1 Apr 1933, in Christa Kirsten and Hans-Jürgen Treder, eds., *Albert Einstein in Berlin, 1913–1933,* vol. 1 (Berlin: Akademie-Verlag, 1979), p. 248.

17. This was Arnold Berliner's assessment in Berliner to Einstein, 1 Jan 1934 (EA 7-038). Berliner was editor of *Die Naturwissenschaften.*

18. "Gesetz zur Wiederherstellung," note 8, pars. 3 and 4. Further discussed in Beyerchen and elsewhere.

19. Claimed in Kopfermann to Bohr, 23 May 1933 (BSC 22, 2). Bohr was at that time in the United States seeking support for refugee organizations. The definition of non-Aryan in the Nürnberg laws of 1935 was much harsher but never completely fixed. The changing definitions are discussed in Raul Hilberg, *The destruction of the European Jews* (Chicago: University of Chicago Press, 1967), esp. ch. 4.

20. WH to his mother, 6 Oct 1933.

21. Born to Einstein, 2 Jun 1933 (EB, *Briefw.,* 162–164); Born to Heisenberg, 11

Jun 1933 (HA); Born to Mrs. Schrödinger, 15 May 1933 (PWB, vol. 2, p. 207). Recalled by Born, *My life: Recollections of a Nobel laureate* (New York: Charles Scribner's Sons, 1975).

22. The decimation of Göttingen physics is described in Beyerchen, and in Heinrich Becker et al., eds., *Die Universität Göttingen unter dem Nationalsozialismus* (Munich: Saur-Verlag, 1987). Peter Kröner, *Vor fünfzig Jahren: Die Emigration deutschsprachiger Wissenschaftler 1933–1939* (Münster: Gesellschaft für Wissenschaftsgeschichte, 1983), pp. 53–55, lists a total of 50 academics who had emigrated from Göttingen by 1939; of these 32 were in the natural sciences. He does not indicate rank or circumstances of the emigration.

23. WH, reviews of Planck's *Postivismus und reale Aussenwelt* (1931), HCW C4, 238; *Wege zur physikalischen Erkenntnis* (1933), HCW C4, 239; *Die Physik im Kampf um die Weltanschauung* (1935), HCW C4, 240.

24. As indicated in Planck to WH, 26 Apr 1933 (HA), Heisenberg had requested a meeting on 20 April.

25. "Neue Unruhen an der Universität Leipzig," *LNN*, no. 335 (30 Nov 1932), p. 3; recollections in W. Goetz, *Historiker in meiner Zeit* (Cologne, 1958).

26. For example, the authorities used paragraph 4 of the civil service law to force a music professor into retirement for signing a pacifist petition and supporting dismissed Jewish colleagues. Hans Driesch, *Lebenserinnerungen* (Leipzig, 1951), pp. 270–272. Other incidents are reported in Herbert Helbig, *Universität Leipzig* (Frankfurt am Main: Wolfgang Weidlich, 1961); and Manfred Unger, "Georg Sacke — Ein Kämpfer gegen den Faschismus," in *Karl-Marx-Universität Leipzig 1409–1959*, vol. 2 (Leipzig: Verlag Enzyklopädie, 1959), pp. 307–330.

27. Heilbron, note 5, p. 150. Also recalled by WH, *PB*, pp. 150–151.

28. Warren Weaver (an official of the Rockefeller Foundation), diary entry for 6 Dec 1934, after a visit with Leipzig Pro-Rector Ludwig Weickmann (RAC Record Group 1.1, Series 717 Germany, Box 14, folders 127–137). Other professionals faced a similar dilemma. The case of newspaper publishers is discussed in Oron J. Hale, *The captive press in the Third Reich* (Princeton: Princeton University Press, 1964), p. 78.

29. Notation by Staatssekretär Lammers on thank-you note from Planck to Hitler of 9 May 1933 (BA Koblenz, Reichskanzlei, Bd. R43 II/1227a). I thank Dr. Werner of the archive for this information. Dr. Werner could find no evidence of a possible second visit nor any record or confirmation of the 16 May visit.

30. Planck, "Mein Besuch bei Adolf Hitler," *PBl*, 3 (1947), 143, a one-paragraph statement; Einstein to Ludwik Silberstein, 20 Sep 1934 (carbon copy, EA 19-404). Later developments may have influenced Planck's recollections. In Planck's hindsight, Hitler's intransigence only confirmed Planck's fatalism.

31. Heisenberg's 1969 recollection of a meeting with Planck, during which he was informed of Planck's meeting with Hitler (WH, *PB*, pp. 149–154), is placed in 1933, but the recollected events in Leipzig surrounding Heisenberg's visit actually occurred in 1935. It is possible, though unlikely, that Planck visited Hitler again in 1935. The account here relies as far as possible on documentation from that period.

32. WH to Born, 2 Jun 1933, typed excerpt in a letter from Born to Ehrenfest, 11 Jun 1933 (EHR 18, 5); published in PWB, vol. 2, p. 168.

33. "I have spoken or corresponded often with Planck and Laue; we have attempted to hold Franck and Born." WH to Bohr, 30 Jun [1933] (BSC 20, 2).

34. Kopfermann to Bohr, 23 May 1933 (BSC 22, 2). Kopfermann made no mention, however, of the Planck-Hitler meeting. According to the editorial introduction to Planck's recollection (note 30), only a few people were informed then of the meeting.

35. Planck to WH, 18 Jun 1933 (HA). The petition and Courant's reaction are discussed, along with the differing reactions of Franck and Born, in Beyerchen and (save for the petition) in Constance Reid, *Courant in Göttingen and New York: The story of an improbable mathematician* (New York: Springer-Verlag, 1976).
36. WH, note 32.
37. Born to Ehrenfest, 11 Jun 1933 (EHR 18, 5).
38. Born to WH, 11 Jun 1933 (HA).
39. Planck to WH, 18 Jun 1933 (HA).
40. Ibid., and Arnold Eucken to WH, 5 Jul 1933 (HA).
41. WH, affidavit for Professor Achelis, 26 Jun 1946 (HA), and WH to his mother, 22 Jun and 31 Jul 1933. In both letters Heisenberg referred to visits to the Preussisches Kultusministerium, which at that time had authority over Göttingen but not over Leipzig. He did not divulge the purpose or the content of his visits.
42. Robin E. Rider, "Alarm and opportunity: Emigration of mathematicians and physicists to Britain and the United States, 1933–1945," *HSPS*, 15 (1984), 107–176; Donald Fleming and Bernard Bailyn, eds., *The intellectual migration: Europe and America, 1930–1960* (Cambridge, Mass.: Harvard University Press, 1969); Norman Bentwich, *The rescue and achievement of refugee scholars: The story of displaced scholars and scientists, 1933–1952* (The Hague: Nijhoff, 1953); Leonard Dinnerstein, *America and the survivors of the Holocaust* (New York: Columbia University Press, 1982).
43. Born to WH, 4 Jul 1933 (HA).
44. Born to Sommerfeld, 1 Sep 1933 (SN).
45. Bohr to WH, 17 Aug 1933 (BSC 20, 2); Bohr to Pauli, 25 Aug 1933. Karl von Meyenn, ed., PWB, vol. 2, p. 215, suggests that Heisenberg may have wanted to avoid the confrontation brewing between Stark and Laue.
46. WH to his mother, 8 Sep and 6 Oct 1933.
47. Laue's activities are indicated in Laue to WH, 10 May 1933 (HA); Sommerfeld to Laue, 19 May 1933 (SN); and Ladenburg to Bohr in Pasadena, 24 May 1933 (BSC 17, 3). Bohr's work in behalf of refugees in 1933 and 1934 is discussed by Finn Aaserud, *Redirecting science: Niels Bohr, philanthropy and the rise of nuclear physics* (Cambridge: Cambridge University Press, 1990), ch. 5.
48. Heisenberg's possible role as conduit is indicated in WH to Bohr, 14 Oct 1933 (BSC 20, 2; quoted in part in Hermann, *Jahrhundert*, p. 116).
49. Beyerchen; Hermann, *Jahrhundert;* Einstein to Laue, 23 Mar 1934 (EA 16-102); Planck to Laue, 22 Mar 1934 (Max von Laue papers, 1964–6/136a, b, Sondersammlungen, Deutsches Museum, Munich).
50. Planck to Ministerialdirektor Gerulis, 30 Aug 1933, forwarded to Ministerialrat Achelis (Archiv der Akademie der Wissenschaften, Berlin; partial facsimile in Hans Wussing, ed., *Geschichte der Naturwissenschaften* [Leipzig: Edition Leipzig, 1983], p. 434); and Planck to Laue, 11 Sep 1933 (Laue Papers, note 49). The circumstances of the funeral service are described in Beyerchen and in Heilbron, note 5.
51. Franz Senger, ed., *Reichs-Habilitations-Ordnung: Amtliche Bestimmungen über den Erwerb des Dr. Habil. und der Lehrbefugnis an den deutschen wissenschaftlichen Hochschulen* (Berlin: Wiedemannsche Verlagsbuchhandlung, 1939), first promulgated 13 Dec 1934, revised 1 Oct 1938. Meitner's career is described by Patricia Rife, *Lise Meitner: Ein Leben für die Wissenschaft* (Düsseldorf: Classen, 1990).
52. WH to his mother, 17 Sep 1933. See also Planck to Laue, note 50. Weyl's wife was Jewish. Schrödinger's resignation is discussed in Walter Moore, *Schrödinger: Life and Thought* (Cambridge: Cambridge University Press, 1989).

53. Planck to Laue, 11 Sep 1933 (Laue Papers, note 49).
54. WH to his mother, 6 Oct 1933.
55. Planck to WH, 8 and 15 Nov 1933 (HA). The Schrödinger-Heisenberg correspondence on this subject has not been found.
56. Casimir to WH, 23 Oct 1933 (HA); and Kramers to WH, 15 Dec 1933 (HA).
57. WH to his mother, 6 Oct 1933 (HA).
58. Courant, writing from Göttingen, notified Heisenberg that he was following Born into temporary exile. Courant to WH, 26 Dec 1933 (HA).
59. WH to Franck, 10 Jan 1934 (Franck Papers, Library, University of Chicago).
60. Planck reiterated his ethics to Laue, 22 Mar 1934 (Laue Papers, note 49).
61. Brandi to WH, 12 Dec 1933 and 1 Feb 1934 (HP).
62. Ibid., and M. Reich, Dean of the mathematisch-naturwissenschaftliche Fakultät, to WH, 9 Feb 1934 (HA).
63. Debye to Clemens Schäfer, 14 Feb 1934 (carbon copy, DP).
64. Fritz Sauter to WH, 13 and 18 Apr 1934 (HA).
65. Sauter to WH, 5 May 134 (HA).
66. Pohl to WH, 13 Jun 1934 (HA).
67. Discussed in the following chapters.
68. Indicative of the situation, Sauter was named acting head of Born's institute but retained the rank of Assistent. He remained in the position until relieved in 1937 by the theorist Richard Becker, who was ordered to Göttingen on the dissolution of his chair for theoretical physics at the TH Berlin. Details in DC, "TH Berlin."
69. Planck's admiration for and gratitude to Heisenberg radiate from his letters to Heisenberg on 8 and 15 November 1933 (HA).

Chapter 16

1. WH to his mother, 6 Oct 1933.
2. WH to his mother, 6 Feb 1938.
3. WH to his mother, 25 Jun [1933].
4. WH to Margrethe Bohr, 31 Jan 1935 (BSC 20, 2).
5. WH to his mother, 17 Mar [1935]. Writing in his diary, Carl Friedrich von Weizsäcker's father also expressed little objection to the new law: "For our sons this law also does not mean little, and I certainly see Carl Friedrich as a soldier." *Die Weizsäcker-Papiere 1933–1950,* ed. Leonidas E. Hill (Frankfurt am Main: Propyläen Verlag, 1974), p. 88.
6. WH to his mother, 12 Jul [1934]; also indicated in WH to Bohr, 17 Jun [1934] (BSC 20, 2).
7. This is emphasized, for example, in Hans Buchheim, "Die SS—das Herrschaftsinstrument," in Buchheim et al., *Anatomie des SS-Staates,* vol. 1 (Munich: Deutscher Taschenbuch Verlag, 1967), pp. 15–212, esp. pp. 15–29.
8. Certification by Leipzig Universitätsrat Sperling to Kaiser Wilhelm-Institut für Physik in Berlin, 13 Apr 1943 (HA).
9. The law and the wording of each oath are given in "Gesetz über die Vereidigung der Beamten und der Soldaten der Wehrmacht, vom 20. August 1934," *Reichsgesetzblatt,* no. 98 (22 Aug 1934), p. 783.
10. EH, *Recoll.,* p. 38.
11. Karl Seidelmann, lecture delivered Easter Sunday, 1931 (Nachlass Seidelmann, ADJ).
12. Oberstudiendirektor Wimmer (1. Vorsitz der Neupfadfinder) to Registergericht

München, 15 Jul 1934, in "Registerakten des Amtsgerichtes München in Sachen: Bayerischer Pfadfinderbund e.V. Landesverband zur Förderung der Pfadfinderbewegung" (StAM AG 33157), and in "Sachen: Bayerischer Pfadfinderbund München" (StAM AG 33158). The last year of the Neupfadfinder is recounted by Franz Ludwig Habbel, "The story of the German youth movement 1896–1933" (1947; IZ, Ms 73). The last official meeting of Heisenberg's group is recalled by Rolf Wägele, manuscript of 15 May 1962 (HA). Some of the "boys" of course met privately after 1934. Heisenberg remained especially close to Robert Honsell. In 1947 Karl Sonntag revived the organization, which met regularly, with widows, wives, and children, until the 1980s.

13. EH, note 10.

14. Records and correspondence in BDC and HA.

15. WH, "Die aktive und die passive Opposition im Dritten Reich," 12 Nov 1947 (HA). This differs considerably from the affidavit Heisenberg subsequently submitted on Ernst von Weizsäcker's behalf: WH, affidavit, 3 Apr 1948, Weizsäcker defense exhibit no. 303 (NARA, microfilms M897, roll 119). Other statements, apparently derivative of the manuscript, include WH, PB, pp. 152–154, and a letter from Heisenberg to Robert Jungk, excerpted in Jungk, Brighter than a thousand suns: A personal history of the atomic scientists, trans. James Cleugh (New York: Harcourt Brace Jovanovich, 1958), p. 91, note.

16. This is noted and discussed in Walker.

17. Heisenberg to S. A. Goudsmit, 5 Jan 1948 (GP; and Hermann, Jahrhundert, p. 119). Heisenberg's colleague, Friedrich Hund, made a similar argument years later: one doesn't leave a country just because one doesn't like its politics (interview with Hund, Göttingen, 28 Nov 1979).

18. Hund, note 17.

19. Lion Feuchtwanger, Exil: Roman (reprint, Frankfurt am Main: Fischer Taschenbuch Verlag, 1979), p. 363. I am grateful to Dr. Michael Eckert for bringing this passage to my attention.

20. LNN, no. 80 (21 Mar 1933), p. 3; and "Hochschule im neuen Staat. Die Bewegung der Besucherzahl an der Universität Leipzig," LNN, no. 192 (11 Jul 1934) (newspaper clipping collection, BA Koblenz, Zsg. 129/643).

21. Hans Peter Bleuel and E. Klinnert, Deutsche Studenten auf dem Weg ins Dritte Reich: Ideologien, Programme, Aktionen, 1918–1935 (Gütersloh: Sigbert Mohn-Verlag, 1967).

22. The victim was Leon Lichtenstein, professor of mathematics. Herbert Helbig, Universität Leipzig (Frankfurt am Main: Weidlich, 1961), p. 105; "Aufruf der Deutschen Studentenschaft gegen den undeutschen Geist," LNN, no. 96 (12 Apr 1933), p. 3.

23. Ettore Majorana to his mother, Leipzig [early 1933], quoted in English in Leonardo Sciascia, The Moro affair and the mystery of Majorana, trans. Sacha Rabinovitch (New York: Carcanet Press, 1987), pp. 151–152.

24. L. W. Jones and Warren Weaver, diary entry of 29 May 1933, Leipzig (RAC, Record Group 1.1, Series 717 Germany, Box 14, folders 127–137).

25. Staatsminister des Inneren Fritsche to SS-Sturmführer Wolf Friedrich, 31 Jul 1933, and Friedrich to Ministerpräsident von Killinger, 12 Aug 1933 (IZ, records of the Nazi Students League, microfilm MA-228, frames 5024396 and 5024403).

26. Helbig, note 22. By the end of May 1933, 8 professors and 11 assistants had been dismissed. Report of Dean Weickmann to Jones and Weaver, note 24. Peter Kröner, Vor fünfzig Jahren: Die Emigration deutschsprachiger Wissenschaftler

1933–1939 (Münster: Gesellschaft für Wissenschaftsgeschichte, 1983), pp. 66–67, lists a total loss of 35 academics (without indication of rank).

27. For Dornfeld: Heisenberg to Universitäts-Rentamt, 14 Oct 1933, "Akten des Universitätes-Rentamtes zu Leipzig betreffend: das theoretisch-physikalische Institut" (UA Leipzig, RA 1407).

28. "Philosophische Fakultät, Protokolle, Buch VIII, 1928–1947" (UA Leipzig).

29. Ibid., and "Akten der philosophischen Fakultät zu Leipzig betr. Berufungsverfahren" (UA Leipzig, A2/21).

30. Hahn to Seydewitz, 30 May 1933, in "Physikalisches Institut der Universität Leipzig, Band II: 1929–1939" (SStA Ministerium für Volksbildung, 10230/27).

31. Debye to Rector Achelis, 26 Jun 1933; Seydewitz, press release, 4 Jul 1933; Debye to Ministerium für Volksbildung, asking that Bewilogua be named as Sack's replacement, 30 Jun 1933. All in "Physikalisches Institut," note 30.

32. Swiss Consul to Sächsisches Aussenministerium, Dresden, 28 Jul 1933, in "Physikalisches Institut," note 30.

33. Bloch to Bohr, 6 Apr 1933 (BSC 17, 3); interview with Bloch by Charles Weiner, 15 Aug 1968 (AIP).

34. Bloch, note 33.

35. WH to Bohr, 30 Jun [1933], and Bohr to WH, Pasadena, 19 May 1933 (BSC 20, 2).

36. Bloch to Bohr, Rome, 10 Feb 1934 (BSC 17, 3).

37. Polanyi to WH, 19 Jun 1934 (HA), and WH to Polanyi, 9 Jul 1934 (Archive, University of Chicago). I am grateful to William Lanouette for bringing the latter to my attention.

38. Kopfermann to Bohr, 23 May 1933 (BSC 22, 2).

39. Sponer to WH, 25 Oct 1933 (HA).

40. WH to Ehrenfest, 2 Feb 1932 (EHR 21, 4); Reinhold Fürth to WH, 5 Oct 1933 (HA); Kramers to WH, 15 Dec 1933 (HA). Bohr's efforts on Beck's behalf are discussed by Finn Aaserud, *Redirecting science: Niels Bohr, philanthropy and the rise of nuclear physics* (Cambridge: Cambridge University Press, 1990), pp. 115–117. Beck's odyssey is recalled in Rudolf Peierls, *Bird of passage: Recollections of a physicist* (Princeton: Princeton University Press, 1985), p. 39, and in the obituary by H. M. Nussenzveig, "Guido Beck," *Physics today*, 43: 12 (1990), 89–90.

41. Philosophische Fakultät," note 28.

42. Discussed in Martin Franke, "Zu den Bemühungen Leipziger Physiker um eine Profilierung der Physikalischen Institute der Universität Leipzig im zweiten Viertel des 20. Jahrhunderts," *NTM-Naturwissenschaft, Technik, Medizin,* 19 (1982), 68–76, based on documents in "Physikalisches Institut," note 30; and in Wolfgang Windsch and Martin Franke, "1927–1945: Blütezeit und Niedergang der Leipziger Physikalischen Institute," *Wissenschaftliche Zeitschrift der Karl-Marx-Universität Leipzig, mathematisch-naturwissenschaftliche Reihe,* 34: 1 (1985), 30–42.

43. Kirchner to Hans Bethe, 20 May 1935 (Bethe Papers, Library, Cornell University, Ithaca, New York, Box 3, 5-57-H-5). Robert Döpel replaced Kirchner in 1938.

44. One of Debye's assistants was listed as a representative of the Teachers League at meetings of the Philosophical Faculty starting in 1937 ("Philosophische Fakultät," note 28). Heisenberg's efforts in the situation are indicated in Heisenberg to Debye, 5 Nov 1934, and Debye to WH, carbon copy, 10 Nov 1934 (DP). Hund always suspected an informant in the institute (Hund, note 17). Hahn's letter to Seydewitz (note 30) recounts secondhand institute gossip.

45. Gerhard Geissler, "Die Universität. Ein neues Semester beginnt mit neuen Aufgaben," *Neue Leipziger Zeitung,* 29 Oct 1933 (newspaper clipping collection, BA Koblenz, Zsg. 129/643).

46. "Jahresbericht des Rektors Dr. Arthur Golf," in *Akademische Reden: Gehalten am 31. Oktober 1934 in der Aula der Universität Leipzig aus Anlass des 525. Jahrestages ihrer Gründung* (Leipzig: A. Edelmann, 1934), pp. 3–17; "Dienstplan der Studentenschaft," 18 Oct 1934 (records of Reichsstudentenführung, IZ, microfilm MA-528, frames 5021537 and 5021538). The students were assigned to vacation work camps, SA service, and "Kameradschaft" living quarters.

47. "Feierliche Gründung der Leipziger Dozentenschaft: Erziehung statt Bildung," *LNN*, no. 198 (17 Jul 1934), p. 2; "Entwurf der Dozentenschaft der Universität Leipzig," Dec 1933 (records of Nazi Students League, IZ, microfilm MA-228, frames 5024324 and 5024325).

48. According to Rector Golf (note 46, p. 14), the university received from the Saxon Education Ministry a new constitution on 1 January 1934, introducing the "Führerprinzip."

49. Dr. Heinz Woltereck, "Lutherfeier an der Universität," *LNN*, no. 315 (11 Nov 1933), p. 3. Golf reiterated such a relationship a year later (note 46, p. 7).

50. WH, *PB*, pp. 141–149.

51. WH to Bohr, 30 Jun 1933 (BSC 20, 2). But the conversation, as Heisenberg recalls it, contains a quotation by Heisenberg from Jakob Burckhardt, whom Heisenberg may not have read before 1934. He listed Burkhardt among his readings in WH to Bohr, 17 Jun 1934 (BSC 20, 2).

52. For instance, Heisenberg complained of the "proletarianization of scholarship" shown by the sloppiness of the faculty march on May Day 1934. Two deans had the audacity to amble together while slurping ice cream cones (WH to his mother, 4 May 1934).

53. Official communication from H. Pleijel (Stockholm Academy) to WH, 9 Nov 1933 (facsimile in Hermann, *WH*, p. 53).

54. The rally is reported in "Appell der deutschen Wissenschaft an die Welt," *Frankfurter Zeitung*, 78 (13 Nov 1933), p. 4; "Die deutsche Wissenschaft spricht ihr 'Ja'! Der N.S.-Lehrerbund und der 12. November," *Völkischer Beobachter*, 1 Nov 1933, Beiblatt: Aus der Bewegung, p. 1; and *Bekenntnis der Professoren an den deutschen Universitäten und Hochschulen zu Adolf Hitler und dem nationalsozialistischen Staat, überreicht vom NS Lehrerbund Deutschlands/Sachsen* (Dresden, 1934). The echo could be heard even abroad: "Leipzig: Professors Plead for Better Understanding of Germany," *New York Times*, 12 Nov 1933, sec. 3, p. 2.

55. Hedwig Goerlich to WH, 9 Nov 1946 (HA).

56. EH, *Recoll.*, pp. 36–39. In addition, Friedrich Hund recalls that Heisenberg had to throw a student out of his office (Hund, note 17). Heisenberg apparently lodged a complaint with a Professor Knick (later rector of the university) against the Führer of the mathematics Studentenschaft. SS-Obersturmführer Friedrich to WH, 19 Nov 1933 (HA).

57. WH to his mother, 11 Nov 1933. This was after the annual Luther celebration described in Woltereck, note 49.

58. Kreisführer of Kreis IV, Nazi Students League, to Saxon Minister of Education, 19 Nov 33 (records of Nazi Students League, IZ, microfilm MA-228, frame 5024355).

59. Niels H. deV. Heathcote, *Nobel prize winners in physics 1901–1950* (Freeport, NY: Books for Libraries, 1953), pp. 306 and 313. The award of the physics prizes is explored in Elisabeth Crawford, *The beginnings of the Nobel institution: The science prizes* (Cambridge: Cambridge University Press, 1984). The award of the prizes in 1933 is further discussed in Walter Moore, *Schrödinger-Life and Thought* (Cambridge: Cambridge University Press, 1989), pp. 281–292.

60. Born to Schrödinger, 6 Nov 1960 (PWB, vol. 2, p. 228).

61. For instance, Arnold Kramish, *The Griffin* (Boston: Houghton Mifflin, 1986).
62. WH to Bohr, 27 Nov 1933 (BSC 20, 2); WH to Born, Zurich, 25 Nov 1933, quoted in English in Born, *My life: Recollections of a Nobel laureate* (New York: Charles Scribner's Sons, 1975).
63. Martin Wein, *Die Weizsäckers: Geschichte einer deutschen Familie* (Stuttgart: DVA, 1988).
64. WH to his mother, 8 Oct 1934.
65. Minutes of the faculty meeting on 8 May 1935 ("Philosophische Falcultät," note 28). The orders, at least for the forced retirements, were based on the arbitrariness allowed by paragraph 6 of the Law for the Restitution of the Civil Service: "In order to simplify administration, civil servants can be placed in retirement even though they are not yet incapable of service." *Reichsgesetzblatt,* no. 34 (1933), p. 176.
66. Minutes of faculty meeting on 8 May 1935, note 65.
67. Van der Waerden to Armin Hermann [1970s] (Hermann, *Jahrhundert,* p. 117). Van der Waerden, a Dutch citizen, was told to stay out of German affairs (interview with van der Waerden, Zurich, 25 Apr 1985). There is, however, no indication of a reprimand in Heisenberg's personnel file (UA Leipzig, PA 560).
68. WH to his mother, 9 Apr 1935. Heisenberg was, however, listed as a member of the university senate from 1935 until he left Leipzig in 1942, in *Rektorenwechsel an der Universität Leipzig* (Leipzig: A. Edelmann, annual).
69. Berve to WH, 13 Mar 1946 (HA). Nevertheless, the Education Ministry named Berve prorector of the university in 1937 and rector in 1940. *Rektorenwechsel,* note 68.
70. Van der Waerden, note 67.
71. WH, *PB,* p. 149.
72. Ibid., pp. 149–154.
73. Since Heisenberg wrote this recollection long after the end of the war, it is uncertain what sort of future world Planck and Heisenberg had in mind; perhaps Heisenberg had read this into his recollection of their meeting.
74. WH to Born, 2 Jun 1933, typed excerpt in letter from Born to Ehrenfest, 11 Jun 1933 (EHR 18, 5, published PWB 2, p. 168).
75. WH to his mother, 5 Oct 1935.

Chapter 17

1. P. M. S. Blackett and G. Occhialini, *PRS,* 139 (1933), 699–727.
2. Carl D. Anderson, *Science,* 76 (1932), 238–239. See Michelangelo de Maria and Arturo Russo, "The discovery of the positron," *Rivista di storia della scienza,* 2 (1985), 237–286.
3. P. A. M. Dirac, *PRS,* 117 (1928), 610–624, and 118 (1928), 351–361.
4. P. A. M. Dirac, *PRS,* 126 (1930), 360–365, and 133 (1931), 60–72. Historical accounts include Joan Bromberg, "The concept of particle creation before and after quantum mechanics," *HSPS,* 7 (1976), 161–191; Pais, *Inward,* pp. 346–352; and Helge Kragh, *Dirac: A scientific biography* (Cambridge: Cambridge University Press, 1990).
5. Blackett and Occhialini, note 1, pp. 713 ff.
6. According to the "Copenhagen registers and minute books" (AHQP 35, 2), Heisenberg visited Bohr's institute from September 10 to October 3, 1933. The conference was held September 14–20. Bohr to WH, 17 Aug 1933 (BSC 20, 2).

7. Furry and Oppenheimer, *PR*, 45 (1934), 245–262; Bethe and Heitler, *PRS*, 146 (1934), 83–112.

8. B. Rossi, *Nwn*, 20 (1932), 65.

9. An attempt to reconstruct these arguments and their impact is offered in DC, "Cosmic rays," 6–7.

10. WH to Sommerfeld, 9 Oct 1933 (AHQP 83, H). The conference and its memory were marred by two tragic events. Mrs. Maar, who had taught Heisenberg Danish and had been housekeeper to numerous institute guests, died during the meeting after calling everyone at the meeting to her bedside the night before. WH to his mother, 26 Sep 1933. Paul Ehrenfest, a leading figure in quantum theory since 1905, committed suicide after his return to Leyden. Pauli wrote a moving obituary in *Nwn*, 21 (1933), 841–843.

11. WH to Bohr, 17 Jun 1934 (BSC 20, 2).

12. WH to Sommerfeld, 9 Oct 1933 (AHQP 83, H).

13. Dirac, "Théorie du positron," in *Structure et propriétés des noyaux atomiques (Conseil International de Physique Solvay, VII, rapports et discussions)* (Paris, 1934), pp. 203–212. See Pais, *Inward*, pp. 377–381, and his introduction to Heisenberg's papers on positron theory (HCW A2, 95–105).

14. While 20 Heisenberg-Pauli letters are extant in 1933, 50 letters are extant for 1934 (PWB), not counting the Weisskopf-Heisenberg exchanges.

15. Pauli to WH, 14 Jun 1934, and remarks in response to Dirac, note 13, pp. 213–214.

16. WH to Paul Goerns, 18 Apr 1934 (HA). Lecture delivered 2 Jun 1934, published in *Stahl und Eisen*, 54 (1934), 749–752 (HCW C1, 92–95).

17. WH, "Bemerkungen zur Diracschen Theorie des Positrons," *ZP*, 90 (1934), 209–231, rec. 21 Jun 1934 (HCW A2, 132–154).

18. The physics of Heisenberg's papers on positrons in this period is summarized by Pais in HCW A2, 95–105; in Pais, *Inward*, pp. 374–388; and from a different perspective in DC, "Cosmic rays." Weisskopf published his results in *Nwn*, 23 (1935), 631–637, 647–653, 669–674; and in *Kgl. danske vid. selskab, math.-fys. medd*, 14 (1936): 6, 3–39.

19. Pauli to WH, 14 Jun 1934.

20. WH to Pauli, 16 Jun 1934.

21. Pauli and Weisskopf published their formalism for scalar (spin zero) fields in *Helvetia physica acta*, 7 (1934), 709–731, rec. 27 Dec 1934.

22. WH to Pauli, 11 Jul 1934.

23. Stark thus refused to attend Heisenberg's lecture to the mining engineers. Laue to Dr. W. F. Berg, 21 Aug 1934, transcript by Berg forwarded to Einstein (EA 16-107); WH to Bohr, 17 Jun 1934 (BSC 20, 2).

24. Based on Laue to Einstein, 22 Aug 1934 (EA 16-105); Stark to Laue, 21 Aug 1934, transcripts (Nachlass Laue, MPG, and Deutsches Museum, Munich); and Beyerchen, pp. 118 and 243. Beyerchen cites Heisenberg-Stark correspondence in the Heisenberg Papers from August 1934, which could not be found during a later search. The author of a 1937 assault on Heisenberg and physics in *Das Schwarze Korps* claims to quote from Heisenberg's response to Stark.

25. Stark, note 24.

26. In *Nwn*, 22 (1934), 669–675 (HCW C1, 96–101).

27. WH to his mother, 21 Sep 1934.

28. WH, "Grundfragen der modernen Physik," lectures delivered in winter semester 1934–1935 (HA). The main section headings are "Ausführliche historische

Einleitung," "Die Relativitätstheorie" (both special and general theories), and "Die Quantentheorie."

29. Dr. Dalfarts(?) to WH, 4 Dec 1934 (HA).

30. Dr. Rosskothen to Rosenberg, and Rosenberg to Rosskothen both 1934, quoted without exact dates or source in Werner Haberditzl, "Der Widerstand deutscher Naturwissenschaftler gegen die 'Deutsche Physik' und andere faschistische Zerrbilder der Wissenschaft," *NTM — Naturwissenschaft, Technik, Medizin,* Beiheft (1963), 320–326, on 323.

31. Euler and Kockel, *Nwn,* 23 (1935), 246–247.

32. From Heisenberg's evaluation of Euler's dissertation, 1 Nov 1935 (Promotions-akten, Philosophische Fakultät, UA Leipzig).

33. WH to Pauli, 25 Apr 1935.

34. Euler's dissertation was entitled "Über die Streuung von Licht an Licht nach der Diracschen Theorie" (Anmeldungsschein, Promotionen MI and MII, UA Leipzig). Heisenberg and Euler, "Folgerungen aus der Diracschen Theorie des Positrons," *ZP,* 98 (1936), 714–732, rec. 22 Dec 1935 (HCW A2, 162–180). Their work is also summarized in Pais, *Inward,* pp. 385–387.

35. WH to Pauli, 29 Apr 1935; Pauli to WH, 15 Jun 1935; and Bohr to WH, 22 May 1935 (BSC 20, 2).

36. Born and Infeld, *PRS,* 144 (1934), 425–451, and many sequels.

37. Heisenberg and Euler, note 34, 732. They used the same critical field as that introduced earlier by E. A. Uehling, *PR,* 48 (1935), 55–63.

38. WH to Born, 18 Mar 1935 (Nachlass Born, SPK), Bohr to E. J. Williams, 16 Mar 1935 (BSC 26, 4).

39. Pauli to Kronig, 20 Nov 1935. The comparison is with Pauli to Kronig, 21 May 1925.

40. Heisenberg and Euler, note 34, 732.

41. WH, affidavit on Kockel's political past, 7 Jan 1947 (HA).

42. WH to Debye, 14 Aug 1936 (DP). Euler received his doctorate on 25 June 1936 (UA Leipzig, Promotionen MI and MII). The falling-out with Weizsäcker is discussed in chapter 19.

43. D. Hoffmann, "Hans Euler (1909–1941)," *PBl,* 45 (1989), 382–384, finds no indication that Euler was a "staunch communist," as Heisenberg put it, WH, *PB,* p. 176.

44. Claimed in WH, affidavit for Berve, 19 Nov 1946 (HA).

45. Euler, declarations in his doctorate file (Promotionsakten, Philosophische Fakultät, UA Leipzig).

46. Beyerchen, pp. 51–57.

47. Rust, lecture at University of Berlin, 6 May 1933, in *Die nationalsozialistische Revolution 1933,* ed. Friedrichs (Berlin: Junker, 1937), pp. 278–285.

48. Herbert Mehrtens and Steffan Richter, eds, *Naturwissenschaft, Technik und NS-Ideologie* (Frankfurt am Main: Suhrkamp, 1980).

49. Perhaps the most insightful summary is offered in Beyerchen, pp. 79–122.

50. Statements for Stark's denazification trial: Einstein to Senatsvorsitzender Schliefer, 14 May 1949 (EA 22-366); WH to Schliefer, 24 May 1949 (HA). Stark's trial has been discussed in Andreas Kleinert, "Das Spruchkammerverfahren gegen Johannes Stark," *Sudhoffs Archiv,* 67 (1983), 13–24. Lenard was not brought to trial because of his age.

51. Sommerfeld to Rector, University of Munich, to be forwarded to Bavarian Culture Ministry, 26 Jul 1937 (UA Munich, Sommerfeld, Personalakten, E II-N).

52. "Deutsche" physics has been described in Beyerchen, pp. 123–140, and especially

in Steffen Richter, "Die 'Deutsche Physik,'" in Mehrtens and Richter, note 48, pp. 116-141.

53. Despite this difference, the relativity and quantum theories suffered proscription under both systems—perhaps an indication that, when seeking to further one's aims, often any dogma will do.

54. Philipp Lenard, *Deutsche Physik*, vol. 1: *Einleitung und Mechanik* (Munich: J. F. Lehmanns Verlag, 1936), p. ix.

55. Stark to Lenard, 3 Feb 1933, and Lenard to Hitler, 21 Mar 1933, both quoted in Andreas Kleinert, "Lenard, Stark und die Kaiser-Wilhelm-Gesellschaft," *PBl, 36* (1980), 35-43, on 35.

56. Stark, *Nationalsozialismus und Wissenschaft* (Munich: Zentralverlag der Nationalsozialistischen Deutsche Arbeiterpartei, 1934), quotations on pp. 13-14. Beyerchen, pp. 115-122, provides a further account of Stark's activities in this period.

57. Published as J. Stark, "Organisation der physikalischen Forschung," *Zs. für technische Physik, 14* (1933), 433-435.

58. For example, Beyerchen, pp. 115-122; Ludwig, pp. 210-216; and the discussion of the "Fall Wildhagen," the attack on Stark's manager at the German Research Association, in Helmut Heiber, *Walter Frank und sein Reichsinstitut für Geschichte des neuen Deutschlands* (Stuttgart: DVA, 1966), pp. 821-847. Stark claimed after the war that he had indeed attempted to "free" German physics from domination—not of the "Jewish-Marxist" variety—but by the REM. J. Stark, "Zu den Kämpfen in der Physik während der Hitler-Zeit," *PBl, 3* (1947), 271-272; also, Stark, *Erinnerungen eines deutschen Naturforschers*, ed. A. Kleinert (Mannheim: Bionomica-Verlag, 1987).

59. Heiber, note 58, p. 796.

Chapter 18

1. "Gesetz über die Entpflichtung und Versetzung von Hochschullehrern aus Anlass des Neuaufbaus des deutschen Hochschulwesens, vom 21. Januar 1935," *Reichsgesetzblatt, 1935*, part I, pp. 23-24, signed by Hitler and Rust. Previously, the retirement age had been 68.

2. WH to Sommerfeld, 18 Jan [1935] (SN); WH to his mother, 9 Apr 1935.

3. Oberregierungsrat Studentkowski, Akten-Notiz, after 7 Oct 1935, in "Besetzung des Ord. Lehrstuhls für Experimentalphysik, Nachf. Debye," Band II, Heft 2 1935-1937 (SStA, Akten des Ministeriums für Volksbildung, 10230); Dean of Philosophische Fakultät, II. Sektion, to Rector, Universität München, 24 Mar 1935, in personnel file "Dr. Arnold Sommerfeld 1905-1951" (UA München, Akten des Rektorats, Personalakte EII-N).

4. Correspondence January to June 1935, in Sommerfeld's personnel file, note 3; minutes of faculty meeting, 15 Feb 1935 (UA München, Sitzungsprotokolle, Phil. Fak. II. Sektion, OC-N1d). Sommerfeld was a civil servant in the Bavarian Academy, not the university, since his chair had been established and administered by it.

5. Dean of Philosophische Fakultät, note 3.

6. Debye to Sommerfeld, 16 Apr 1935 (ETH) and 19 Jul 1935 (DP); Sommerfeld to Debye, 7 Jul 1935 (DP).

7. There are differing reports of the content and dating of the original list, which has not been found. The minutes of the faculty meeting of April 24, 1935, list Heisenberg, Debye, and Richard Becker (Sitzungsprotokolle, note 4). In a letter from Carathéodory, Sommerfeld, Gerlach, and Wieland to the dean of the faculty, 18 Jan

1940 (UA München, Nachfolgeakt Sommerfeld, OC-N 10a), the authors note that the first list, submitted on June 13, 1935, had Heisenberg and Debye in first place and Becker in second. The second list, submitted on November 4, 1935 (SN), refers to its predecessor only as "die Kandidatur Heisenberg" and quotes from the faculty opinion, submitted on July 13, 1935. The quote is given in the text.

8. Studentkowski headed the new seminar for political education in Leipzig until called in 1934 to the Saxon ministry in Dresden. "Politische Erziehung der Leipziger Studentenschaft," *LNN,* no. 302 (29 Oct 1933), p. 2.

9. Studentkowski, note 3; Debye to Sommerfeld, Berlin, 20 Sep 1935 (ETH).

10. Ibid.

11. Submission of 4 Nov 1935 (transcription in SN).

12. Martin Kersten, "Richard Becker, 1887–1955," *PBl,* 34 (1978), 379–382. Such forcible transfers were made legal in the law of January 21, 1935 (note 1).

13. Most notably, Beyerchen, ch. 8; Wolfgang Schlicker, "Physiker im faschistischen Deutschland: Zum Geschehen um eine naturwissenschaftlich-technische Grundlagendisziplin seit 1933," *Jahrbuch für Geschichte,* 27 (1983), 109–142; Mark Walker, "National Socialism and German physics," *Journal of contemporary history,* 24 (1989), 63–89; and Hermann, *Jahrhundert,* pp. 126–147; from which the discussion in this section has benefited.

14. Quoted in Hans Buchheim, "Die SS—das Herrschaftsinstrument," in Buchheim et al., *Anatomie des SS-Staates,* vol. 1 (Munich: Deutscher Taschenbuch Verlag, 1967), pp. 98–99.

15. Also at that time a number of Nazi science and engineering journals began to appear. For instance, Theodor Vahlen, head of REM office W-I, founded the Nazi journal *Deutsche Mathematik.*

16. Stark, "Philipp Lenard als deutscher Naturforscher, Rede zur Einweihung des Philipp-Lenard-Instituts in Heidelberg am 13. Dezember 1938," *Nationalsozialistische Monatshefte,* 7 (Feb 1936), 106–112. The speech is signed, however, by both Lenard and Stark.

17. Oron J. Hale, *The captive press in the Third Reich* (Princeton: Princeton University Press, 1964), pp. 31–32.

18. Stark, note 16, and Willi Menzel, "Deutsche Physik und jüdische Physik," *Völkischer Beobachter,* 49: 29 (29 Jan 1936), 7. The latter was reprinted in the Students League newspaper for the TH Berlin as "Deutsche Physik," *Die Technische Hochschule,* 14: 4 (15 Feb 1936), 69–71.

19. Willi Menzel is probably the "W. Menzel" listed as "student leader for physics and mathematics" in room 101 of the TH Berlin Physics Institute, in *Die Technische Hochschule,* 13 (1934–1935), 13. If so, he was probably coached by Stark at the nearby Physical-Technical Institute.

20. Heisenberg's most recent physics was indeed more formalistic than conceptual, but it is unlikely that Stark and Menzel knew much about it.

21. WH to his mother 15 Feb [1936], reporting the meeting with Kölbl.

22. WH to Sommerfeld, 14 Feb 1936 (SN).

23. "German science goose-steps," editorial, *The New York Times,* 12 Mar 1936, p. 20; "Zum Artikel: Deutsche und jüdische Physik. Entgegnung von Prof. Dr. Heisenberg," *Völkischer Beobachter,* 49: 59 (28 Feb 1936), 6 (HCW C5, 10–11). According to Hale, note 17, Rosenberg actually had little to do with the daily editorial affairs of the *VB.*

24. Indicated by Studentkowski to Rudorf, Dekan der Philosophischen Fakultät,

mathematisch-naturwissenschaftliche Abteilung, Universität Leipzig, 24 Mar 1936 (UA Leipzig, Heisenberg, Personalakte, PA 560).

25. WH to his mother, 28 Feb [1936].

26. Note 23.

27. Rosenberg, editorial comment preceding Heisenberg, note 23, followed by "Stellungnahme von Prof. Dr. J. Stark," *Völkischer Beobachter,* 49: 59 (28 Feb 1936), 6. Rosenberg's comment is also published in HCW C5, editorial introduction, p. 4.

28. Studentkowski, "Aktenvermerk," in "Besetzung," note 3, referring to a meeting with Mentzel on March 2, 1936. Studentkowski met with Mentzel, rather than Bachér, probably because of the latter's waning influence in the REM. He was then embroiled in a controversy with his boss, Vahlen, whose interests had already run to "deutsche" mathematics. Vahlen and Bachér were replaced within a year. Helmut Heiber, *Walter Frank und sein Reichsinstitut für Geschichte des neuen Deutschlands* (Stuttgart: DVA, 1966), pp. 644–645.

29. Studentkowski, note 24.

30. At least two other articles of a related nature did appear: Christian J. Hansen, "Kant und die deutsche Naturwissenschaft," *Volkishcer Beobachter,* 49: 70 (11 Mar 1936) 5; and "Intellectualistische Wissenschaft," *Volkischer Beobachter,* 49: 74 (14 March 1936) 5.

31. Beyerchen, p. 154.

32. Rudorf to Heisenberg, 30 Mar 1936, and to Rudorf to Prorector Golf, 30 Mar 1936 (UA Leipzig, Heisenberg, Personalakte, PA 560).

33. Form letter with Sommerfeld's name in the blank for the addressee, with typed signatures of M. Wien, H. Geiger, W. Heisenberg (SN). Quoted in part and dated 11 May 1936 in Hermann, *Jahrhundert,* p. 141.

34. Copy of the undated petition "An den Herrn Reichsminister für Erziehung, Wissenschaft und Volksbildung," with a typed list of signatories (Friedrich Hund Papers, Göttingen); reprinted without the names of the signatories in HCW C5, 12–13. It was sent to Heisenberg's private address so that he could not be accused of using state facilities for antistate activities.

35. For example, Lenard to Dr. Wacker, 10 Jun 1936 (transcription in Lenard Nachlass, private hands).

36. This is shown also in Abraham Esau, "Nachruf auf Max Wien," *Zeitschrift für technische Physik,* 20 (1939), 1–3.

37. Studentkowski, note 3. Geiger replaced Hertz at the TH Berlin. DC, "TH Berlin." Gerhard Hoffmann eventually replaced Debye, who moved to Berlin in April 1936.

38. The memo is probably "Die Physik an den deutschen Hochschulen," no date or author (SN). Its origins and impact are explored by Dieter Hoffmann, "Die Physikdenkschriften von 1934/36 und zur Situation der Physik im faschistischen Deutschland," in *Wissenschaft und Staat: Denkschriften und Stellungnahmen von Wissenschaftlern als Mittel wissenschaftspolitischer Artikulation* (Berlin: Akademie der Wissenschaften der DDR, 1989), pp. 185–211. Further background to the 1934 memo may be gleaned from the Debye-Wien correspondence, 1934–1935 (DP).

39. Private communication from Dr. Helmut Fischer.

40. Petition, note 34. Among the signatories were even Professors Karolus and Schiller from the technical divisions of the Leipzig physics institute.

41. Zschintzsch, file copy of letter of 2 Oct 1936, "Dem Herrn Reichsminister weitergeleitet," in "Korrespondenz des Staatssekretärs Zschintzsch," vol. 3:

"Schreiben und Vorlagen an Reichsminister Rust 1936–1945 (BA Koblenz, R21/203); original in "Research: Wi Heisenberg" (BDC).
42. Beyerchen, pp. 121–122; WH to Pauli, 26 Nov 1936.
43. Sommerfeld to WH, 9 Nov 1936 (HA).
44. The expectation is examined in DC, "Cosmic Rays," from which the following derives.
45. C. F. von Weizsäcker, ZP, 88 (1934), 612–625; E. J. Williams, PR, 45 (1934), 729–730; L. D. Landau, Physikalische Zeitschrift der Sowjetunion, 5 (1934), 761–764.
46. J. R. Oppenheimer, PR, 47 (1935), 44–52; L. W. Nordheim, PR, 49 (1936), 189–191.
47. WH to Pauli, 26 May [1936].
48. E. Fermi, ZP, 88 (1934), 161–177.
49. Heisenberg's studies are indicated in Fermi to WH, 30 Jan and 6 Feb 1934 (AHQP 45, 1); WH-Meitner correspondence in November 1935 (Meitner Collection, Archives, Churchill College, Cambridge); and WH, "Bemerkungen zur Theorie des Atomkerns," in Pieter Zeeman, 1865–25 Mei 1935. Verhandelingen op 25 Mei 1935 Aangeboden aan Prof. Dr. P. Zeeman (The Hague: Martinus Nijhoff, 1935), pp 108–116 (HCW B, 238–246). Heisenberg's work on Fermi-force nuclear physics is discussed in Laurie Brown and Helmut Rechenberg, "The origin of the concept of nuclear forces II: The Fermi-field theory of nuclear forces (1933–1937)," manuscript; and Olivier Darrigol, "The quantum electrodynamical analogy in early nuclear theory or the roots of Yukawa's theory," Reviews of history of Science, 41 (1988), 225–297.
50. Brown and Rechenberg, note 49, and WH, summary of his Cambridge lectures (HA).
51. Summarized in WH, note 50, and "Bemerkungen," note 49. The alternative Yukawa force was apparently not then known in Europe.
52. Heisenberg's theory and its cosmic-ray context are summarized from different perspectives in Darrigol, note 49; DC, "Cosmic rays"; and Brown and Rechenberg, "The origin of the concept of nuclear forces III: Quantum field theories, nuclear forces, and the cosmic rays (1934–1938)," manuscript.
53. Indicated in WH to Pauli, 30 May [1936].
54. WH, "Zur Theorie der 'Schauer' in der Höhenstrahlung," ZP, 101 (1936), 533–540, rec. 8 Jun 1936 (HCW A2, 275–282); WH, "Theorie der Schauer," abstract of paper delivered to Copenhagen conference, 17–20 Jun 1936 (HCW B, 247).
55. WH, note 54, "Höhenstrahlung," 540.
56. Heitler, Angewandte Chemie, 49 (1936), 690.
57. Anderson to Heitler, 21 May 1936 (BP 3).
58. Anderson and Neddermeyer, PR, 50 (1936), 263–271, rec. 7 Jun 1936.
59. J. R. Oppenheimer, abstract of paper delivered to American Physical Society, June 1936, PR, 50 (1936), 389; H. J. Bhabha and W. Heitler, letter to editor, Nature, 138 (Jul 1936), 401. They presented their complete theories in Carlson and Oppenheimer, PR, 51 (1937), 220–231; and Bhabha and Heitler, PRS, 159 (1937), 432–458.
60. WH, note 54, "Höhenstrahlung."
61. WH to Pauli, 30 May [1936].
62. Gerlach to Goudsmit, 7 Nov 1936 (GP).
63. WH, "Über die 'Schauer' in der kosmischen Strahlung," Forschungen und Fortschritte, 12 (1936), 341–342 (not in HCW).

64. G. Herzog and W. Scherrer, "Tagung über Kernphysik an der Technischen Hochschule Zürich, 30. Juni bis 4. Juli 1936," *Nwn,* 24 (1936), 718–720.

65. H. Geiger, *Preussische Akademie der Wissenschaften, Vorträge und Schriften,* 3 (1940), 1–33, esp. 16.

66. Pauli to WH, 9 Jun 1936.

67. The eight weeks of active duty took place from August 25 to October 19, 1936, according to his military questionnaire, dated 9 February 1940, note 24.

68. "Harvard visit off for Reich physicist. Dr. Heisenberg, scheduled to give a paper at exercises, is serving 8 weeks in army," *The New York Times,* Sunday, 30 Aug 1936, Section II, p. 8.

69. WH to Goudsmit, 9 Apr [1936] (HA).

70. WH to his mother, 28 Aug and 4 Sep 1936, also WH to Born, 3 Nov 1936 (Nachlass Born, SPK).

71. Pauli to WH, 26 Oct 1936.

72. Ibid.

73. WH to Born, 3 Nov 1936 (Nachlass Born, SPK).

Chapter 19

1. WH, *PB,* p. 165.

2. Ibid.

3. WH to his mother, 19 Mar 1936.

4. WH to his mother, 30 Mar 1936.

5. *Die Weizsäcker-Papiere 1933–1950,* ed. Leonidas E. Hill (Frankfurt am Main; Verlag Ullstein, 1974), pp. 131, 631 (n. 13), 503 (n. 79); Martin Wein, *Die Weizsäckers: Geschichte einer deutschen Familie* (Stuttgart: DVA, 1988). In a sad letter to Werner a year later (27 Aug 1937 [HP]), Adelheid seemed to regret her decision not to marry him and asked his forgiveness.

6. WH, note 4.

7. WH to his mother, 9 Apr 1936.

8. WH to his mother, 12 Nov 1936.

9. WH to his mother, 12 Feb 1937.

10. "I am now playing rather a lot of music and experience through it many enjoyable moments." WH to his mother, 25 Jan 1937.

11. WH, *PB,* p. 166.

12. EH, *Recoll.,* p. 47. She remembers this as having occurred in the Mittelstädt home.

13. Dates based on WH to his mother, 6 and 12 Feb 1937.

14. Klaus Schwabe, "Ursprung und Verbreitung des alldeutschen Annexionismus in der deutschen Professorenschaft im Ersten Weltkrieg (zur Entstehung der Intellektuelleneingaben vom Sommer 1915)," *Vierteljahrshefte für Zeitgeschichte,* 14 (1966), 105–188.

15. Fritz Schumacher, *Small is beautiful: Economics as if people mattered* (New York: Harper & Row, 1973).

16. Erich Kuby's war experiences, along with references to his wife and brother-in-law, are given in *Mein Krieg: Aufzeichnungen aus 2129 Tagen* (Munich: Nymphenburger Verlagshandlung, 1975).

17. Barbara Wood, *E. F. Schumacher: His life and thought* (New York: Harper & Row, 1984), pp. 2 and 6.

18. Interview with EH, Göttingen, February 8–9, 1982.
19. Ibid.
20. EH, note 12.
21. WH to his mother, 12 Feb 1937.
22. WH to his mother, 6 Dec 1935.
23. "Hilfsaktion des WHW zum 30. Januar," *LNN*, no. 23 (23 Jan 1937), p. 1.
Goebbels apparently got his wish. Official data indicate that street collections nearly doubled to a claimed 30,532 million RM for 1936–1937. *Statistisches Jahrbuch für das Deutsche Reich*, 56 (1937), 570.
24. WH, *PB*, p. 166.
25. Ibid.
26. WH to Elisabeth Schumacher, 23 Mar 1937 (HP).
27. EH, note 18.
28. Euler to WH, 13 Mar 1937 (AHQP 45, 1); and Pauli to WH, 10 Mar 1937.
29. WH to his mother, 23 Mar 1937.
30. WH to his mother, 23 and 27 Mar 1937.
31. That is, as of 1 April. WH to his mother, 27 Mar 1937; Wilmanns (Dean of math.-naturwiss. Abteilung der Phil. Fak., Leipzig), summary of conversation with Hund, 8 Apr 1937 (Heisenberg personnel file, UA Leipzig, PA 560); WH to Sommerfeld, 27 Oct 1940 (SN).
32. WH, note 30.
33. WH to Bohr, 18 Mar 1937 (BSC 20, 2).
34. Ibid. He wrote nearly the same words to his mother, 23 Mar [1937].
35. WH to Pauli, 26 Apr [1937].
36. Invitation and menu (HP).
37. Pauli to WH, 22 Feb 1938.
38. EH, note 18.
39. Pauli to WH, 19 Jan 1937; WH to Pauli, 21 Jan 1937; C. D. Anderson and S. H. Neddermeyer, *PR*, 50 (1936), 263–271; H. J. Bhabha and W. Heitler, *PRS*, 159 (1937), 432–458, rec. 11 Dec. 1936; J. F. Carlson and J. R. Oppenheimer, *PR*, 51 (1937), 220–231, rec. 8 Dec 1936. The following discussion follows DC, "Cosmic Rays." Discussed by Brown and Rechenberg, "The origin of the concept of nuclear forces III: Quantum field theories, nuclear forces, and the cosmic rays (1934–1938)," manuscript.
40. Anderson and Neddermeyer, note 39, p. 268.
41. Bhabha and Heitler, note 39, p. 435.
42. Carlson and Oppenheimer, note 39, p. 221.
43. L. W. Nordheim, G. Nordheim, J. R. Oppenheimer, R. Serber, *PR*, 51 (1937), 1037–1045, on 1038.
44. L. W. Nordheim to Sommerfeld, 24 Oct 1936 (AHQP 32, 9); Nordheim and Nordheim, *PR*, 51 (1937), 379.
45. Nordheim et al, note 43, p. 1038.
46. WH, "Über die in der Theorie der Elementarteilchen auftretende universelle Länge," *AP*, 32 (1938), 20–33, on 26–27 (HCW A2, 301–314).
47. WH to Pauli, 16 Jan 1937.
48. WH to Pauli, 12 Jun 1937. Brown and Rechenberg, note 39, perceive no difference at all between the approaches and programs pursued by Heisenberg and Pauli on the one hand, and Oppenheimer, Carlson, Nordheim, Heitler, and Bhabha on the other. I still disagree.
49. The influence of Weimar culture on quantum physics during the 1920s has been

argued by Paul Forman, "Weimar Culture, causality, and quantum physics," *HSPS,* 3 (1971), 1–115.

50. WH to Bhabha, 21 Jan 1937.
51. Ibid., and WH to Pauli, 21 Jan 1937.
52. Pauli to WH, 20 Nov [1936].
53. WH to Pauli, 2 and 14 Feb 1937. The procedure was equivalent to the later S-matrix approach.
54. WH to Pauli, 26 Apr 1937.
55. Pauli to WH, 2 May 1937.
56. WH to Pauli, 12 Jun 1937.
57. WH to Bohr, 5 Jul 1937 (BSC 20, 2); S. Neddermeyer and C. D. Anderson, *PR,* 51 (1937), 884–886. Independently found by J. C. Street and E. C. Stevenson, *PR,* 51 (1937), 1005.
58. WH to Kramers, 22 May 1937 (AHQP 10, 3).
59. WH to his mother, 1 Apr and 17 Jun 1937; EH, *Recoll.,* p. 47.
60. WH to his mother, 10 Jul 1937.
61. WH to his mother, 9 Jun 1937.
62. Quoted in EH, *Recoll.,* p. 47.

Chapter 20

1. "'Weisse Juden' in der Wissenschaft," *Das Schwarze Korps* (15 Jul 1937), p. 6. Parallel discussions of this episode and of some of the materials cited here—but, again, with widely varying aims and methods—may be found in Beyerchen, pp. 156–167; Hermann, *Jahrhundert,* pp. 143–147; Walker, pp. 61–66; EH, *Recoll.,* pp. 47–70; and Wolfgang Schlicker, "Physiker im faschistischen Deutschland: Zum Geschehen um eine naturwissenschaftlich-technische Grundlagendisziplin seit 1933," *Jahrbuch für Geschichte,* 27 (1983), 109–142. Heisenberg himself chose not to discuss it in his memoirs.
2. Ludwig Wesch file, in "Hauptamt Wissenschaft," (IZ, microfilm MA-116/117). Beuthe performed mainly spectroscopic analyses at Stark's institute. His admiration for Stark is indicated in Beuthe, "Johannes Stark," *PZ,* 40 (1939), 321–322.
3. Hans Buchheim, "Die SS—das Herrschaftsinstrument," in *Anatomie des SS-Staates,* vol. 1, ed. Hans Buchheim et al. (Munich: Deutscher Taschenbuch Verlag, 1967), pp. 15–212. An insight into the suffocating world of the SS is provided in Gunter d'Alquen's manual for members, *Die SS: Geschichte, Aufgabe und Organisation der Schutzstaffeln der NSDAP* (Berlin, 1939).
4. Allied interrogation of d'Alquen, 16 Feb 1948 (NARA, microfilm M-1019, roll 2).
5. This third section is reprinted in Léon Poliakov and Josef Wulf, eds. *Das Dritte Reich und seine Denker: Dokumente* (Berlin-Grunewald: Arami Verlags-GMBH, 1959), pp. 299–300. "Graue Theorie" (grey theory) may also be translated "abstract theory."
6. Beuthe to Wesch, 1 Nov 1937 (Ludwig Wesch papers, UA Heidelberg). Beyerchen, p. 158, attributes the second and third sections to Stark and the first to d'Alquen or a staff member.
7. Kurt R. Grossmann, *Ossietsky: Ein deutscher Patriot* (Munich: Kindler-Verlag, 1963).
8. Richard Becker et al. to Gesellschaft der Wissenschaften zu Göttingen, 19 Nov 1937; WH to secretary of same, 9 Dec 1937 (Göttingen Academy, Archiv);

Sächsische Akademie der Wissenschaften, mathematische-physikalische Klasse, Protokollbuch, 1926–1956 (SAW, archive, Leipzig); annual *Berichte über die Verhandlugen der sächsischen Akademie der Wissenschaften zu Leipzig, mathematisch-physikalische Klasse.*

9. Sommerfeld to Kölbl, 26 Jul 1937 (Sommerfeld personnel file, UA München, E II-N).

10. Kölbl to Bavarian State Minister for Instruction and Culture, 22 Oct 1937 (Sommerfeld personnel file, note 9).

11. Hund to Reichsminister Rust, 20 Jul 1937, via Koebe; Hund to Koebe, 16 Jul 1937 (both in Heisenberg personnel file, UA Leipzig, PA 560).

12. Debye to Hund, 22 Jul 1937; Hund to Debye, 21 Jul 1937 (both in DP).

13. Others indicated by Ernst von Weizsäcker to Herr Professor [Sommerfeld?], 30 Sep 1937 (SN); Karl Friedrich Bonhoeffer to WH, 6 Aug 1937 (HA).

14. WH to Sommerfeld, 14 Apr 1938 (AHQP 31, 5; original in SN).

15. WH to Berve, 17 Jul 1937 (Heisenberg file, note 11).

16. Berve to Koebe, 19 Jul 1937 (Heisenberg file, note 11).

17. Studentkowski, memo, "Wegberufung von Prof. Heisenberg," 3 Jun 1937 (Heisenberg file, note 11), after discussion with Dames.

18. "Akten des Ministeriums für Volksbildung. Besetzung des ord. Lehrstuhls für Experimentalphysik. Nachf. Debye," vol. 2, book 2, 1935–1937 (SStA 10230).

19. WH to Wacker, 28 Jul 1937, cited in Beyerchen, p. 254, n. 65, as being in HA. It has not been found.

20. Wacker to Bavarian Ministry for Instruction and Culture, and Wacker to Sommerfeld, both 16 Nov 1937 (both in Sommerfeld file, note 9).

21. WH to Himmler, 21 Jul 1937, excerpted without location in Hermann, *Jahrhundert,* p. 144; cited in Beyerchen, p. 254, n. 64, as being in HA. It has not been found. Elisabeth Heisenberg recalls her reaction in EH, *Recoll.,* pp. 51–52.

22. Quoted (probably in translation) in Beyerchen, p. 160, from his interview with WH, 13 Jul 1971. See also EH, *Recoll.,* p. 54. Before Mrs. Heisenberg left, Mrs. Himmler turned to her to ask: "Or do you think, Mrs. Heisenberg, that my little Heinrich is not on the right path after all?"

23. Himmler, Terminbücher 1938–1939 (NARA, microfilm 37A; original in BA Koblenz, NS19/1437).

24. In his correspondence with his mother (HP), Heisenberg refers to efforts by Annie's sister Marie to influence a Schmidt in Himmler's office through his aged mother in Munich. Perhaps she was the widow of another schoolmaster. The letter from Himmler to WH, 4 Nov 1937, cited in Beyerchen, p. 254, n. 66, as being in HA, has not been found.

25. Himmler's private papers (IZ, microfilm MA-320; originals in BA Koblenz).

26. WH to Himmler, 7 Nov 1937, cited in Beyerchen, p. 254, n. 66, has not been found in HA. The paraphrase below follows Beyerchen, p. 161.

27. Stark had conflicted in 1936 with the SS over one of Himmler's favorite projects, "Ahnenerbe" research. This had led to a messy battle between Stark and the vice president of the DFG, Wildhagen, in which Heisenberg had played a backstage role. Wildhagen to WH, 22 Nov 1966 (HA). Low opinions of Stark were communicated to Himmler in late 1936. "Ahnenerbe" files "Stark, Johannes" and "Weigel, Karl-Theodor" (BDC).

28. The visitor was a Professor Betz.

29. WH to his mother, 27 Sep 1938.

30. WH to Sommerfeld, 31 Aug 1938 (HA); George B. Pegram to WH, 3 Jan and

31 Mar 1938 (HA).

31. On December 8, 1937, Ladenburg, having returned to the United States from a visit to Europe, reported Heisenberg's mood to Goudsmit (in German): "Heisenberg was very optimistic and was of the opinion that he would get Stark" (GP).

32. Elisabeth Heisenberg to Annie Heisenberg, 1 Nov 1937 (HP).

33. Heisenberg had even learned that the positive report lay on Rust's desk but that Rust refused to sign it until Himmler acted. WH to Sommerfeld, 16 Jan, 12 and 23 Feb 1938 (SN; AHQP 31, 5).

34. Heisenberg's SS files have not been found. The SS destroyed many files at war's end. However, enough bureaucratic and private correspondence, as well as official reports, remain to reconstruct events.

35. Note 2. Wesch belonged to SS-Standarte 32 (for the state of Baden) and had organized and headed his own intelligence unit (what he called a Nachrichten-Sturm). Private communication from his colleague, Dr. Helmut J. Fischer, 2 Apr 1983.

36. For example, Hugo Dingler, "Die 'Physik des 20. Jahrhunderts,'" *Zs. f. ges. Naturwiss.*, 3 (1937), 321–335; Dingler, "Pascual Jordan, 'Die Physik des 20. Jahrhunderts' [2nd. ed.]," *Zs. f. ges. Naturwiss.*, 4 (1938–1939), 389–393; Ludwig Bieberbach, "Ein Physiker über deutsche Art in der Mathematik," *Deut. Math.*, 1 (1936), 109. Dingler's reviews were privately countered by a Professor A. von Antropoff, physicist and SS officer, as indicated in correspondence of Dr. Fritz Kubach (Reichsstudentenführung, Amt Wissenschaft und Facherziehung, IZ, microfilm MA-1160).

37. Note 36. Also, Dingler, "Zur Entstehung der sogen. modernen theoretischen Physik," *Zs. f. ges. Naturwiss.*, 4 (1938–1939), 329–341. He was rewarded for his efforts by the Munich Teachers League which, in the person of Thüring, tried to have him named an adjunct professor of physics. Sommerfeld and colleagues successfully opposed the plan. Sitzungsprotokolle, Phil. Fak. II, for 11 Nov 1937, UA München, OC-N1d.

38. It was called the "Amt Wissenschaft und Facherziehung," under Dr. Fritz Kubach.

39. This is indicated by the list of advisors and contributors published in both journals.

40. Beyerchen, p. 164.

41. All except Führer were on the masthead of the *Zs. f. ges. Naturwiss.*

42. Indicated in Friedrich von Faber, Dean of Phil. Fak. II, to Rektorat, Universität München, 29 Dec 1938 (Lehrstuhl für theoretische Physik, Nachfolge Sommerfeld, UA München, OC-N 10a).

43. Asserted in Helmut Heiber, *Walter Frank und sein Reichsinstitut für Geschichte des neuen Deutschlands* (Stuttgart: DVA, 1966).

44. Correspondence (UA Heidelberg).

45. "§ 175 (Unzucht zwischen Männern)," cited with commentary in, for instance, Schönke-Schröder, ed., *Strafgesetzbuch* (Munich: C. H. Beck, 1967), pp. 883–887.

46. WH to his mother, 14 Nov 1937.

47. WH to his mother, 21 Nov 1937.

48. Interview with EH, Göttingen, April 1983.

49. EH, *Recoll.*, p. 56.

50. This information is derived from postwar affidavits submitted by Heisenberg in support of their cases during denazification hearings.

51. *Catalogus professorum: Der Lehrkörper der TH Hannover* (Hannover, 1956); and private communication, 6 Aug 1983.

52. "Das erste deutsche Mathematikerlager," *Deut. Math.*, 4 (1939), 109; "Eine Exkursion der Fachgruppe Naturwissenschaften der Studentenführung der

Universität Berlin," *Zs. f. ges. Naturwiss.*, 6 (1940), 317–318. Jules was group leader of the latter.

53. Private communication, 6 Aug 1983.

54. Ibid.

55. It was common practice to write textbooks after losing one's job to denazification.

56. Report on Heisenberg, enclosed in letter from office of Reichsführer-SS to REM, 26 May 1939 (ZStA Potsdam, REM 2943, 370-1, transcription by Mark Walker). I am grateful to Dr. Walker for providing this transcription.

57. Heisenberg was always watched during his trips abroad, especially since he was under investigation by the SD. Beuthe was apparently in possession of travel reports on Heisenberg, which he forwarded to Wesch. Beuthe to Wesch, 25 Jul 1938; Wesch to Kreisaussenstelle Heidelberg, 1 Aug 1938 (both in Wesch papers, UA Heidelberg).

58. Quotation from letter to Heisenberg from an unnamed "lower-level SS-Führer," quoted in WH to Sommerfeld, 14 Apr 1938 (SN; AHQP 31, 5).

59. Ibid. A possible reference to the Heisenberg case occurs in one of Himmler's "Tischgespräche" ("table talks") on February 19, 1938, recorded by Dr. Brandt: "It is a tragic fate for a scientist when he researches his life long and then finds that he has gone in the wrong direction." If such a scientist has "the greatness" to admit this, then he deserves a "laurel wreath" because that, too, is a great service to science. (BA Koblenz, NS 19/neu 1446). He did not say what a scientist deserved if the contrary were true.

60. WH to his mother, 12 Jun 1938.

61. C. Wieselsberger, "Ludwig Prandtl," *Zs. f. tech. Physik*, 17 (1935), 25–27.

62. From the opening paragraph of Prandtl to Himmler, 12 Jul 1938 (initialed transcript copy, Nachlass Prandtl, UA Göttingen).

63. Ibid.

64. Himmler to WH, 21 Jul 1938 (facsimile in *Alsos*, p. 119; transcript in SN). Heisenberg did teach his course on the special theory of relativity once again in the spring of 1940, using his lecture notes from 1931–1932 (HA), but he changed the course name to the title of Einstein's original paper: "The electrodynamics of moving bodies." Lecture listings in *Physikalische Zeitschrift* for that semester. Einstein's name appears in Heisenberg's earlier notes. It cannot be ascertained how he treated Einstein in the later course.

65. Himmler to Heydrich, 21 Jul 1938 (facsimile of certified copy in *Alsos*, p. 116; transcript in "Research: Ahnenerbe, Heisenberg, Werner," BDC Berlin). Himmler's phrase "tot zu machen" is translated as "to silence." It may also mean "to kill" (literally: "to make dead").

66. WH to Himmler, Fischen (Allgäu), 23 Jul 1938 (GP).

67. Heisenberg immediately notified Sommerfeld, WH to Sommerfeld, 23 Jul 1938 (SN). When a copy of Himmler's letter to WH, forwarded to the REM via Leipzig, did not arrive by the middle of October 1938, Sommerfeld sent a copy directly to Berlin in the hope that the REM would act immediately. Sommerfeld to REM, 17 Oct 1938 (AHQP 33, 6).

68. Heisenberg served from August 2 to October 15, 1938, according to his military questionnaire, dated 9 Feb 1940 (Heisenberg file, note 11).

69. WH to his mother, barracks near Sonthofen, 18 Sep 1938.

70. Ibid.

Chapter 21

1. Sommerfeld to WH, 28 Feb 1939 (carbon copy in SN).
2. Dean von Faber to rectorate, 29 Dec 1938 (UA Munich, Lehrstuhl für theoretische Physik, Nachfolge Sommerfeld, OC-N 10a).
3. WH to Sommerfeld, 5 Nov 1938 (SN); notice in *PZ*, 39 (1938), 136.
4. WH to his mother, 4 Nov 1938.
5. Hess to REM, 12 Jan 1939, referring to an earlier complaint (ZStA Potsdam, REM 2943, courtesy of Mark Walker).
6. Himmler to REM, 26 May 1939 (ZStA Potsdam, REM 2943, courtesy of Mark Walker).
7. WH to Sommerfeld, 15 Feb and 13 May 1939 (SN).
8. Minutes of meetings, Philosophische Fakultät, II. Sektion (UA Munich, OC-N 1d).
9. Dean von Faber, note 2.
10. WH to Sommerfeld, 13 May 1939 (SN).
11. WH to Sommerfeld, 27 Oct 1940 (SN); complaint by Müller to Dean von Faber, 11 Sep 1940 (SN).
12. Hugo Rössner, District School Leader, Vienna, to Dr. Haertle, NSDAP Amt Wissenschaft, 12 Jun 1942 (Rosenberg-Akten, Hauptamt Wissenschaft, Heisenberg file, IZ, microfilm MA 116/5).
13. WH, note 10.
14. Dames to Geiger, Pohl, and Gerlach, 1 Feb 1938; Führer to series of physicists, 29 Jul 1941 (both BA Koblenz, R21/500).
15. UA Leipzig, Promotionen, MI and MII.
16. Wacker to REM functionaries, 27 May 1938; Dames, report to Wacker, [Oct] 1938 (BA Koblenz, R21/349).
17. At that time, forced labor came with the usually unfulfilled promise of buying one's freedom in forced emigration.
18. Lucy S. Dawidowicz, *The war against the Jews 1933–1945* (Toronto: Bantam Books, 1975, 1986), p. 99.
19. This account follows Dawidowicz, op. cit., pp. 99–106.
20. David H. Buffum, U.S. Counsul in Leipzig, diplomatic report of 21 Nov 1938, published in *Nazi conspiracy and aggression*, vol. 7, ed. Office of U.S. Chief of Counsel for Prosecution of Axis Criminality (Washington: GPO, 1946), trial Doc. L-202, pp. 1037–1041.
21. Ibid.
22. WH to his mother, 12 Nov 1938.
23. WH to his mother, 6 Apr 1939.
24. WH to Sommerfeld, Badenweiler, Easter Sunday [1939] (SN). Badenweiler had its own dark association. Hitler's "signature tune" was the Badenweiler March, played whenever he entered a hall or stadium.
25. Interview with EH, Göttingen, 1983.
26. WH, note 23.
27. WH to his mother, 23 Jun 1939. The house was not sold until after the war.
28. WH to his mother, 21 Nov 1937.
29. Ibid. "Die gegenwärtigen Aufgaben der theoretischen Physik," published in excerpt in *Frankfurter Zeitung*, 26 Aug 1937, and in *Scientia*, 1938 (HCW C1, 133–141).
30. "Der Durchgang sehr energiereicher Korpuskeln durch den Atomkern," presented

to Bologna conference, 18–21 Oct 1937 (HCW B, 256–259); *Nwn,* 25 (1937), 749–750 (HCW A2, 283–284); Saxon Academy of Sciences, 8 Nov 1937 (HCW A2, 285–300); Physikalischer Verein zu Frankfurt, 15 Nov 1937 (abstract in HCW B, 260); Saxon district meeting of German Physical Society, Dresden, 8–9 Jan 38 (abstract in HCW B, 261).

31. "Über die in der Theorie der Elementarteilchen auftretende universelle Länge," *AP,* 32 (1938), 20–33, rec. 13 Jan 1938 (HCW A2, 301–314).

32. "Wahrscheinlichkeitsaussagen in der Quantentheorie der Wellenfelder," presented at University of Geneva, 11–16 Oct 1937 (HCW B, 249–255).

33. Abraham Esau to REM, 13 Jul 1939, indicating earlier observations; Hess to REM, 22 June 1939 (ZStA Potsdam, REM 2943).

34. WH to Bohr, 4 May and 14 Jun 1938; Bohr to WH, 13 Jun 1938 (BSC 20, 2).

35. Conversation with Res Jost, Zurich, 1986.

36. Yukawa, *Proceedings of the Physico-Mathematical Society of Japan,* 17 (1935), 48–57. This discussion follows DC, "Cosmic rays." Other historical works consulted, to which readers are referred for details, include Visvapriya Mukherji, *Archive for history of exact sciences,* 13 (1974), 27–102, Laurie M. Brown, *Centaurus,* 25 (1981), 71–132; Peter Galison, *Centaurus,* 26 (1983), 262–316; Olivier Darrigol, *Reviews of history of science,* 41 (1988), 225–297.

37. The particles are "virtual" because their existence in a bound system for longer than allowed by the uncertainty principle would violate energy conservation.

38. Yukawa, note 36, p. 57.

39. J. R. Oppenheimer and R. Serber, *PR,* 51 (1937), 1113; Serber, *PR,* 53 (1938), 221.

40. Kemmer, *PRS,* 166 (1938), 127–153.

41. Fröhlich, Heitler, and Kemmer, *PRS,* 166 (1938), 154–177, on 155; Kemmer, *Proceedings of the Cambridge Philosophical Society,* 34 (1938), 354–364; Bhabha, *PRS,* 166 (1938), 501–528.

42. Kemmer, note 40, p. 148.

43. WH to his wife, Manchester [Mar 1938] (HP).

44. This scheme is based on Heitler, *PRS,* 166 (1938), 529–543. It was later completely reinterpreted in the light of two different mesons.

45. WH, "Die Grenzen der Anwendbarkeit der bisherigen Quantentheorie," *ZP,* 110 (1938), 251–266, rec. 24 Jun 1938 (HCW A2, 315–330).

46. WH to Bohr, 4 May 1938 (BSC 20, 2).

47. Euler, *ZP,* 110 (1938), 692–716.

48. Euler and WH, "Theoretische Gesichtspunkte zur Deutung der kosmischen Strahlung" (1938), HCW B, 262–330.

49. WH, "Die Absorption der durchdringenden Komponente der Höhenstrahlung," *AP,* 33 (1938), 594–599, dated 6 Oct 1938 (HCW A2, 331–336).

50. WH to Pauli, 15 Jul 1938.

51. Pauli to Peierls, 18 Jul, to Rubinowicz, 22 Jul, and to WH, 15 Aug 1938.

52. WH to Bohr, 19 Jun and 30 Jul 1938, and WH to Blackett, 20 Jul 1938 (all BSC 20, 2); WH to Pauli, 15 Jul 1938.

53. Lectures delivered to Physikalisches Colloquium, Hamburg, 1 Dec 1938 (HCW B, 331–332); Bavarian district meeting of the German Physical Society, 4 Dec 1938 (abstract in HCW B, 333); Saxon district meeting of the German Physical Society, Leipzig, 7–8 Jan 39 (abstract in HCW B, 334).

54. For example, Hans Bethe and Lothar Nordheim, *PR,* 57 (1940), 998–1006.

55. J. G. Wilson, *PRS,* 172 (1939), 517–529, and *PRS,* 174 (1940), 73–85.

56. Blackett to WH, 10 Sep 1938 (AHQP 45, 1); Paul Ehrenfest, Jr., to WH, 30 Nov 1938 (AHQP 45, 1); Lovell and Wilson, *Nature*, 144 (1939), 863–864.

57. C. G. and D. D. Montgomery, *Reviews of modern physics*, 11 (1939), 255–265, on 257.

58. These views and the "failed revolution" against QED are discussed in Galison, note 36.

59. Bhabha, *Nature*, 143 (1939), 276–277; Dirac, *PRS*, 167 (1938), 148–169, on 149.

60. Bhabha, *PRS*, 172 (1939), 384–409, on 385; and Bhabha to Bohr, 17 Dec 1938 (BSC 17, 2).

61. Bhabha, note 60, 386.

62. J. R. Oppenheimer, H. Snyder, and R. Serber, *PR*, 57 (1940), 75–81, on 75.

63. WH to Pauli, 19 May 1939.

64. WH, *ZP*, 113 (1939), 61–86, rec. 5 May 1939 (HCW A2, 337–362). He repeated the argument in Chicago in June 1939 (summary in HCW B, 345) and for the Solvay Congress planned for October 1939 and canceled (HCW B, 346–358).

65. WH, note 64, *ZP*, p. 82, Bhabha's response and the eventual demonstration that their two opposite approaches are in fact equivalent are indicated in DC, "Cosmic rays."

66. Heisenberg's travels are discussed in Mark Walker, "Propaganda in science," manuscript for *HSPS*. I thank Dr. Walker for a copy of this manuscript.

67. WH, interview session 10, 28 Feb 1963 (AHQP); Oppenheimer, discussion remark, *Reviews of modern physics*, 11 (1939), 264–266; and C. G. and D. D. Montgomery, note 57.

68. WH to Sommerfeld, Fischen, 31 Aug 1938 (SN).

69. George Pegram was the treasurer, W. L. Severinghaus, also of Columbia, the secretary.

70. A. H. Compton to his son Arthur, 16 Jul 1937, and Compton to WH, 16 Jul 1937 (UA Chicago, Arthur H. Compton Papers, series 2). The earlier offer is indicated in K. T. Compton to WH, 31 Jul 1929 (carbon copy, UA Princeton, K. T. Compton papers). I thank Ron Doel for informing me of this letter.

71. WH, *PB*, pp. 170–171.

72. Sir Nevill Mott and Sir Rudolf Peierls, "Werner Heisenberg, 5 December 1901–1 February 1976," *Biographical memoirs of Fellows of the Royal Society*, 23 (1977), 213–251, on 232.

73. WH to his mother, 23 Jan 1939.

74. WH to his wife, 18 and 24 Jul 1939 (HP).

75. WH, *PB*, pp. 169–172. Max Dresden recalls the following discussion occurring at a party given by Otto Laporte. Dresden, letter to the editor, *Physics Today*, 44:5 (1991), 92 and 94.

76. Goudsmit, "Werner Heisenberg (1901–1976)," *Yearbook of the American Philosophical Society*, 1976, pp. 74–80, on p. 76. He noted that by "damage" Heisenberg meant damage to German physics.

77. Discussion with Max Dresden, Stony Brook, New York, 1989.

78. WH, *PB*, p. 172; and WH to his wife, 28 Jul 1939 (HP).

Chapter 22

1. Ernst von Weizsäcker, *Memoirs*, trans. John Williams (London: Victor Gollancz, 1951), p. 212.

2. WH to his mother, 4 Dec 1939.

3. Erich Bagge, Kurt Deibner, Kenneth Jay, *Von der Uranspaltung bis Calder Hall* (Hamburg: Rowohlt, 1957), p. 22.

4. Ibid., p. 23. Heisenberg's role as Bagge's advisor and provider of research funds is indicated in Bagge file, Deutsche Forschungs-Gemeinschaft. (BA Koblenz, R73/10147).

5. WH to Sommerfeld, 4 Sep 1939 (AHQP 31, 5).

6. Bagge's diary, entry for 25 Sep 1939 (DJ 29, 106–143), and WH, "Militaire Fragebogen," 9 Feb 1940 (UA Leipzig, Heisenberg file PA 560).

7. The history of fission research has been told many times. An excellent account is offered in Richard Rhodes, *The making of the atomic bomb* (New York: Simon and Schuster, 1986). A significant scholarly history of the German effort, which parallels much of the discussion here, is offered in Walker. See also Weart and the nontechnical and not always accurate Irving.

8. Quoted in *Einstein on peace*, ed. Otto Nathan and Heinz Norden (New York: Schocken Books, 1960), p. 289. Surveys of the Allied fission projects are provided in Rhodes, note 7; Weart; Margaret Gowing, *Britain and atomic energy 1939–1945* (New York: St. Martin's Press, 1964); and Henry DeWolf Smyth, *Atomic energy for military purposes* (Princeton: Princeton University Press, 1946).

9. The letter is quoted and the delay discussed in *Einstein on peace*, note 8; Rhodes, note 7; among others.

10. Siegfried Flügge, *Nwn*, 27 (1939), 402–410.

11. WH, *PB*, p. 170.

12. Ibid.

13. HCW A2, 378–396, dated 6 Dec 1939. All of the available declassified secret German reports written and cowritten by Heisenberg are published in HCW A2.

14. Bohr and Wheeler, *PR*, 56 (1939), 426–450.

15. WH, note 13, p. 394.

16. Radiation poisoning was also little appreciated. Although some measures were taken to counter the hazard, Fischer and Bopp both suffered effects. Dr. Auer, University Clinic, to WH, 20 May 1942 (HA). Elisabeth Heisenberg (*Recoll.*, p. 89) writes that Heisenberg once broke out in "herpeslike inflammations" all over his body. She attributes this to stress, but it may have been caused by radiation.

17. WH, note 13, p. 396. It is unclear at this point whether or not Heisenberg knew that U-235 is also fissionable by fast neutrons.

18. WH, "Bericht über die Möglichkeit technischer Energiegewinnung aus der Uranspaltung, II," 29 Feb 1940 (HCW A2, 397–418).

19. DC, "TH Berlin."

20. WH, note 18, 397; K. H. Höcker, "Berechnung der Energiegewinnung in der Uranmaschine II: Kohle als Bremssubstanz," 20 Apr 1940 (DJ 29, 427–433); P. O. Müller, "Berechnung der Energieerzeugung in der Uranmaschine III: D_2O als Bremssubstanz," 29 Apr 1940 (DJ 29, 434–436).

21. Bothe, "Einige Eigenschaften des U und der Bremsstoffe. Zusammenfassender Bericht über die Arbeiten im Institut für Physik im KWI für med. Forschung, Heidelberg," 8 Mar 1941 (DJ 31, 117–127); Bothe and P. Jensen, "Die Absorption thermischer Neutronen in Elektrographit" (1941), *ZP*, 122 (1944), 749–752.

22. C. F. v. Weizsäcker, "Eine Möglichkeit der Energiegewinnung aus U-238," 17 Jul 1940 (DJ 29, 451–455).

23. McMillan and Abelson, *PR*, 57 (1940), 1185–1186.

24. Sir Nevill Mott and Sir Rudolf Peierls, "Werner Heisenberg, 5 December 1901–1 February 1976," *Biographical memoirs of Fellows of the Royal Society*, 23 (1977), 213–251, on 232.

25. Letter to DC, 1 Jul 1990.

26. Debye to W. E. Tisdale, Rockefeller Foundation, 7 Oct 1939 (DP). Debye could not take his salary out of Germany.

27. Warren Weaver, officer's diary, entry for 6 Feb 1940 (Rockefeller Foundation Archives, RG 12.1, RAC).

28. WH, *PB*, p. 172, my emphasis; interview of WH by J. J. Ermenc Urfeld, 29 Aug 1967 (transcript in Papers of Gen. Leslie Groves, NARA Gift Collection).

29. German work in each of these areas is thoroughly discussed in Walker.

30. WH to Sommerfeld, 29 Oct 1940 (SN); WH to R. Ortvay, 8 Dec 1941 (HA). Walker, pp. 52–53, lists nine main groups by 1942. "Energiegewinnung aus Uran," a report to the HWA, dated February 1942 (Bagge Papers), p. 135, lists a total of 22 participating institutes at 12 different locations. I thank Dr. Walker for a copy of this report.

31. Oral communication from Wirtz to H. Rechenberg, 1983. I thank Dr. Rechenberg for this information.

32. WH to Harteck, 29 Apr 1940 (DJ 29). Auer later engaged the Degussa Company to turn the oxide into metal powder and plates.

33. Harteck to WH, 15 Jan 1940 (DJ 29).

34. WH to Harteck, 18 Jan 1940 (DJ 29). Robert and Klara Döpel and WH, "Bestimmung der Diffusionslänge thermischer Neutronen in schwerem Wasser" 7 Aug 1940 (HCW A2, 419–426).

35. Irving, ch. 7.

36. WH to Bagge in Paris, 6 Jun 1941 (HA); Bagge's diary, note 6, entries for 1941. Joliot and French fission research during the war are well discussed in Weart.

37. This is especially emphasized in Walker.

38. WH to his mother, letters in early 1940.

39. WH to his mother, 11 Nov 1940.

40. WH to his mother, 10 Feb 1940.

41. A list of courses offered by the Leipzig Physics Institute appeared annually in PZ. According to university records (UA Leipzig, Promotionen, MI and MII), after Bagge passed his exams in 1938 Heisenberg examined only three other students in Leipzig.

42. "Aktennotiz: Betrifft Prof. Dr. Hans V. Universität Leipzig, 15.6.42," in Hauptamt Wissenschaft, Amt Wissenschaftsbeobachtung u. -wertung (Erxleben) (IZ, MA 129/9). Also recalled by former Leipzig professor Hans-Georg Gadamer, *Philosophical apprenticeships*, trans. Robert R. Sullivan (Cambridge, Mass.: Massachusetts Institute of Technology Press, 1985), 93–102.

43. "Goerdeler und die Deportation der Leipziger Juden," *Vierteljahrschefte für Zeitgeschichte*, 13 (1965), 338–339.

44. Heisenberg refereed Gora's dissertation, but since Heisenberg had already left Leipzig, Hund served as his examiner in physics (UA Leipzig, Promotionen, MII); WH to Hund, 13 Nov 1942 (HA). Gora's story is based on WH to Edwin Gora, 3 Jan 1941 (Gora Papers, Providence, Rhode Island); Gora to Ch. Kleint, 20 Feb 1985 (Gora Papers); interview with Professor Gora, Boston, Mar 1985; Edwin K. Gora, "One Heisenberg did save," *Science news,* 109 (1976), 179.

45. The RFR supported Gora with a stipend to work on high-pressure physics with Gerlach, then head of the physics section of the RFR. Gerlach to Fischer, 13 May 1944 (LC MF 124, File 18750).

46. Heisenberg recalled Euler and Grönblom in WH, *PB,* pp. 176–179.

47. This and other letters from Euler to WH are in HA, excerpts published in WH, note 46.

48. E. Zeigner, friend of the Euler family, to WH, 1 Oct 1941 (HA). Mrs. Martha Euler to WH, 11 Apr 1942; Deutsches Rotes Kreuz to Marita Euler, sister, 31 Jul 1941; Oberleutnant Heppner to Martha Euler, 1 Dec 1941 (all HA).

49. Marita Euler to WH, 9 Nov 1946; WH to Marita Euler, 26 Nov 1946 (HP). Heisenberg tried again through Blackett while in postwar British captivity.

50. WH, B. O. *Grönbloms wissenschaftliche Arbeiten* (1943), HCW C4, 45–61.

Chapter 23

1. The technical features of the Leipzig and Berlin experiments are briefly summarized in WH-Wirtz. For Leipzig research, see also Ch. Kleint, "Aus der Geschichte der Leipziger Uranmaschinenversuche — Zum 90. Geburtstag von Robert Döpel," *Kernenergie*, 29 (1986), 245–251; as well as Walker and Irving.

2. A detailed description of the Aussenlabor is included in WH, "Bericht über die Versuche mit Schichtenanordnungen von Präparat 38 und Paraffin am KWI für Physik in Bln-Dahlem" (March 1941), HCW A2, 432–462, on 435.

3. Reports on experiments *B*-1 to *B*-4 through January 1942, co-written by WH, are published in HCW A2. Heisenberg's view of Wirtz's role is described in WH to Dean of science faculty, University of Tübingen, 6 Oct 1947 (HA).

4. R. and K. Döpel and WH, "Der experimentelle Nachweis der effektiven Neutronenvermehrung in einem Kugel-Schichten-System aus D_2O und Uran-Metall," July 1942 (HCW A2, 536–543). The first multiplication ever had occurred already in Cambridge, England, in December 1940.

5. R. and K. Döpel and WH, "Versuch mit einer Schichtenanordnung von D_2O und Präp[arat] 38," 28 Oct 41 (HCW A2, 481–498).

6. R. and K. Döpel and WH, "Die Neutronenvermehrung in einem D_2O-38-Metallschichtensystem" 26–28 Feb 1942 (HCW A2, 526–528); W. Bothe, "Bemerkungen zum Leipziger D_2O-Versuch," 7 Jan 1942 (DJ 30, 023–024).

7. This was indicated in R. and K. Döpel and WH, note 4.

8. R. Döpel, "Bericht über zwei Unfälle beim Umgang mit Uranmetall," 9 Jul 1942 (DJ 29, 539–548; DJ 30, 298–303).

9. Döpels and WH, note 4, p. 543.

10. WH-Wirtz, p. 149. Heisenberg and Wirtz remembered "the first certain proof" as occurring in February 1942, which seems a little too early.

11. Interview of WH by D. Irving, 23 Oct 1965 (DJ 31, 526–567), quoted in Irving, p. 102.

12. Fritz G. Houtermans, "Zur Frage der Auslösung von Kern-Kettenreaktionen," Aug 1941 (DJ 30, 704–719).

13. Quoted by H. Rechenberg, "Einleitung," in WH, *Ordnung der Wirklichkeit*, ed. H. Rechenberg (Munich: Piper, 1989), p. 17 (original in HA).

14. Surmised from WH, "Bericht über die Teilnahme an einer astrophysikalischen Arbeitstagung im Deutschen Wissenschaftlichen Institut in Kopenhagen," 23 Sep 1941 (ZStA Potsdam, REM 2943, Bl. 547). I thank Mark Walker for a transcription of this document.

15. WH, affidavit on 1941 visit to Copenhagen (manuscript and typescript [1948] in HA). A similar, briefer account, perhaps derivative of this draft, is WH to B. L. van der Waerden, 28 Apr 1948, transcription by van der Waerden forwarded to S. A. Goudsmit (DJ 29, 1190–1191). The official trial affidavit is Weizsäcker defense exhibit no. 303 (NARA, microfilms M897, roll 119). He may have submitted it in connection with Carl Friedrich's interrogation (NARA, microfilms M1019, roll 78)

and later testimony (NARA, microfilms M897, roll 10, 24 June 1948, vol. 25, pp. 10007–10040).

Use of the word *bomb* is somewhat anachronistic. It was not yet determined that the explosive could be delivered as a bomb. If it were too unwieldy for flight, it would have to be delivered by ship.

16. Walker, pp. 223–228, as well as Walker, "Physics and propaganda," manuscript submitted to *HSPS*, provide the most extensive accounts.

17. WH to B. L. van der Waerden, typed transcription by van der Waerden, 28 Apr 1948 (DJ 29, 1190–1191).

18. WH, *PB*, p. 170. Interview of WH by David Irving, 23 Oct 1965 (transcript DJ 31, 526–567).

19. This according to WH, note 15, affidavit.

20. EH, *Recoll.*, p. 79.

21. "Energiegewinnung aus Uran," report to Army Ordnance, Feb 1942 (Bagge Papers).

22. WH, note 15, affidavit.

23. William Dan Andersen, "The German Armed Forces in Denmark 1940–1943: A study in occupation policy," Ph.D. thesis, University of Kansas, 1972; Erich Thomsen, *Deutsche Besatzungspolitik in Dänemark 1940–1945* (Düsseldorf: Bertelsmann Verlag, 1971).

24. Niels Bohr, affidavit, 20 Dec 1947, Weizsäcker defense exhibit no. 301; also, Werner Best (occupation head after 6 Nov 1942), affidavit, 27 Aug 1947, Weizsäcker defense exhibit no. 302; WH, affidavit, 3 April 1948, Weizsäcker defense exhibit no. 303 (all NARA, microfilms M897, roll 119).

25. The U.S. State Department allowed the foundation to continue its funding, despite the occupation, until 1943. Grant documents and report by Thomas Lauritsen to Rockefeller Foundation, after return from Denmark to United States in November 1940, reported in "Excerpt from Trustees Confidential Bulletin, Feb 1941, 'Recent News from Occupied Denmark,'" (RAC, RF 1.1, 713D, folder "Biophysics, Univ. Copenhagen, Bohr").

26. Dr. Brauwiler, "Bericht betreffend Reise nach Kopenhagen, 24./25. Mai 1940," report to Reichsministerium für Volksaufklärung und Propaganda (BA Koblenz, R58/1091, Bl. 32–40).

27. Recollection of Stefan Rozental recounted in Walker, p. 224.

28. C. F. von Weizsäcker recognized the propaganda function of this institute in Weizsäcker to Lamberts (DAAD), 22 Jul 1941 (ZStA Potsdam, REM 2943). The exclusive propaganda function of these institutes was admitted by Fritz von Twardowski, head of the Culture Department in the Foreign Office from 1939 to 1943, affidavit of 17 Apr 1948, Weizsäcker defense exhibit no. 208 (NARA, microfilms M897, roll 119). This function is discussed at length in Walker. I am grateful to Dr. Walker for transcripts of REM documents cited here from ZStA Potsdam.

29. Plenipotentiary of German Reich in Denmark to Berlin Foreign Office, 27 Mar 1941 (ZStA Potsdam, REM 2943).

30. C. F. von Weizsäcker to WH, 26 Mar 1941 (HA).

31. C. F. von Weizsäcker to Lamberts (DAAD), 22 Jul 1941, and Foreign Office to REM, 2 Aug 1941 (ZStA Potsdam, REM 2943).

32. C. F. von Weizsäcker to Niels Bohr, 15 Aug 1941 (BSC 26, 2).

33. WH, note 15, affidavit.

34. REM documents in August and September 1941 (ZStA Potsdam, REM 2943).

35. I thank Erik Rüdinger for Mrs. Bohr's remark.

36. Weizsäcker, note 30. The Copenhagen view of that visit is repeated in Arnold Kramish, *The griffin* (Boston: Houghton Mifflin, 1986), p. 120.

37. Quoted in Margaret Gowing, *Britain and atomic energy 1939–1945* (New York: St. Martin's Press, 1964), p. 246.

38. "Notes on meeting of sub-committee September 10, 1943 (R.C.T.)" (NARA, RG 77, file 334, British Interchange Sub-Committee).

39. Schumann to directors and researchers, 5 Dec 1941, quoted in Bagge et al., *Von der Uranspaltung bis Calder Hall* (Hamburg: Rowohlt, 1957), p. 28.

40. Report, note 21, p. 133. The title is taped onto the first available page. Two of the authors were probably Bagge and Diebner.

41. Ibid., pp. 5 and 132.

42. Surmised in Irving, p. 106.

43. List of lectures, 26–28 Feb 1942 (DJ 29, 998–1005).

44. List of lectures, 26 Feb 1942 (DJ 29, 705; facsimile in Irving, p. 109, and elsewhere).

45. WH, "Die theoretischen Grundlagen für die Energiegewinnung aus der Uranspaltung," 26 Feb 1942 (HCW A2, 517–521, quote on 518–519).

46. Walker, p. 58.

47. This campaign is described in Beyerchen, pp. 190–191; Ludwig, pp. 241–242; and Walker, pp. 119–122.

48. Finkelnburg to WH, 6 May 1942 (HA).

49. WH to Finkelnburg, 22 May 1942 (HA). Even Goebbels noted uranium research with pleasure. Joseph Goebbels, *The Goebbels diaries 1942–1943*, trans. Louis Lochner (New York: Doubleday, 1948), entry for 21 March 1942, p. 140.

50. "Niederschrift über die Sitzung des Senats der KWG," 24 Apr 1942 (MPG).

Chapter 24

1. WH, untitled typescript (HA); published as "Ordnung der Wirklichkeit," HCW C1, 218–306; and as *Ordnung der Wirklichkeit* (Munich: Piper, 1989).

2. WH, "Die Goethe'sche und die Newton'sche Farbenlehre im Lichte der modernen Physik," delivered to Gesellschaft für kulturelle Zusammenarbeit, Budapest, 28 Apr 1941 (HCW C1, 146–160); "Die Einheit des naturwissenschaftlichen Weltbildes," delivered to Zurich student body, 27 Nov 1942, published as "Über das Weltbild der Naturwissenschaft" (HCW C1, 201–215); "Die Einheit des naturwissenschaftlichen Weltbildes," delivered to University of Leipzig, 26 Nov 1941 (HCW C1, 161–192); "100 Jahre Energiegesetz," radio lecture, Aug 1942, widely published (HCW C1, 202–206).

3. Helmut Rechenberg et al., "Vorbemerkung der Herausgeber," HCW C1, 217.

4. WH, note 1, 218. All references are to the HCW edition.

5. Ibid., 226.

6. Ibid., 232. Goethe's color theory was, of course, nonsense as science.

7. WH, note 2, "Farbenlehre," 160.

8. WH, note 1, 304–306.

9. Ibid., 298.

10. Ibid., 305.

11. Ibid., 304.

12. The details of these changes are discussed in Walker; recollections in Speer, *Inside the Third Reich: Memoirs,* trans. R. and C. Winston (New York: Macmillan, 1970), p. 276; documents on DJ and LC Microfilm 107, File 12847.

13. Göring, order of 8 Dec 1942 (DJ 29, 1031).

14. WH to Telschow, 11 June 1942 (HA); WH to KWG, 2 Jul 1942 (HA), on return of the institute to the KWG.

15. WH to dean of Philosophische Fakultät, Leipzig, 9 Sep 1942 (UA Leipzig, Heisenberg, personnel file, PA 560); Head of Saxon education ministry to Rector, Universität Leipzig, 22 Sep 1942 (HA), approval; WH to H. Falkenhagen, 19 Jun 1942 (HA).

16. Telschow, Aktennotiz, 8 Feb 1943 (HA and DJ 29, 1049–1050), on meeting between Vögler, Mentzel, and Telschow; Esau to Mentzel, 5 Apr 43 (DJ 29, 1060–1061), on financing. The ordnance office also contributed to the support of Harteck's and Clusius's institutes.

17. Return addresses on correspondence (HA). The institute library still consisted largely of Debye's private collection. Eberhard Scholz, attorney, to Telschow, 16 Feb 1943 (HA).

18. Opposition to Heisenberg's call is indicated by REM correspondence with the Party Chancellery in that period (BDC, REM files 5512–5518). The situation regarding political approval at that time is summarized in Beyerchen, ch. 9.

19. WH to Fues, 10 Nov 1942 (HA).

20. EH, *Recoll.*, p. 91.

21. Published in part as "Eingabe an Rust," *PB*, 3 (1947), 43–47. Also, C. Ramsauer, "Zur Geschichte der Deutschen Physikalischen Gesellschaft in der Hitlerzeit," *PB*, 3 (1947), 110–114.

22. Erxleben to Bechtold, Partei-Kanzlei, 9 Sep 1942 (IZ, Hauptamt Wissenschaft, Heisenberg file, MA 116/5).

23. WH to Pascual Jordan, 31 Jul 1942 (HP). The date of the meeting is given in Dr. Borger to Partei-Kanzlei, 9 Sep 1942 (IZ, MA 116/5).

24. Dr. Borger to Partei-Kanzlei, 9 Sep 1942 (IZ, MA 116/5).

25. Erxleben to Bechtold, 9 Sep 1942 (IZ, MA 116/5).

26. The Seefeld meeting and the compromises reached are discussed in Beyerchen, ch. 9, and Walker. Heisenberg's attendance is indicated in WH to Hotel Tiroler Weinstube, Seefeld, 29 Oct 1942 (HA); Bruno Thüring to WH, 17 Nov 1942 (HA); and WH, affidavit for Borger, 6 Jun 1947 (HA).

27. REM to WH, 26 Feb 1943 (HA); Telschow, Aktennotiz, 8 Feb 1943 (DJ 29, 1049–1050, and HA); Dean Schwender to Rector, Universität Leipzig, 10 Mar 1943 (UA Leipzig, Heisenberg file, PA 560). Schwender, referring to the REM's written appointment of Heisenberg, describes Heisenberg as "Direktor *des* Kaiser Wilhelm-Instituts für Physik in Berlin-Dahlem."

28. Walker, p. 79.

29. Hitler, decree of 9 Jun 1942, note 12, RFR documents.

30. WH, Farm Hall statement, August 1945 (HCW C5, 26–27), and elsewhere.

31. Quoted in Speer, note 12, p. 677, n. 24.

32. Speer, note 12, p. 301.

33. Ibid.

34. Ibid. Recollections of Milch's question and of Heisenberg's reply are recounted in Irving, p. 120. Heisenberg's "pineapple" remark was off by about a factor of 10.

35. Speer, note 12, p. 302.

36. Quoted in Speer, note 12, p. 677, n. 26.

37. Based on R. Döpel, "Bericht über zwei Unfälle beim Umgang mit Uranmetall," rec. 9 Jul 1942 (DI 29, 539–547).

38. WH, "Bemerkungen zu den geplanten halbtechnischen Versuch mit 1,5 to D_2O

und 3 to 38-Metall," 31 Jul 1942 (DJ 30, 228–232; HCW A2, 545–552); Bothe to WH, 18 Aug 1942 (HA), encouraging the use of plates.

39. WH to Bothe, 23 Oct 1942 (HA). The bunker was completed in November 1943 and dedicated during a well-provisioned institute party. WH to his mother, 7 Nov 1943 (HP). The classification changed up and down over the years.

40. Esau to Mentzel, 5 Apr 1943 (DJ 29, 1060–1061).

41. Bopp, Fischer, WH, Weizsäcker, Wirtz, "Messungen an Schichtenanordnungen aus 38-Metall und Paraffin," 30 Oct 1942 (DJ 30, 342–356), report on experiments B-3 to B-5 (HCW A2, 553–561). The last plates for B-6 and B-7 may not have been delivered until as late as 15 January 1944. WH to Bothe, 13 Dec 1943 (HA).

42. WH to his mother, 20 May 1943 (HP); WH, ed., Kosmische Strahlung: Vorträge gehalten im Max Planck-Institut, Berlin-Dahlem (Berlin: Springer-Verlag, 1943).

43. Bäumker, Aeronautical Academy, to WH, 28 Sep 1942 (HA); Academy notice, 20 Oct 1942 (LC microfilm 128, file 19059).

44. Professor Dr. Scheel, director of Prussian Academy, to WH, 21 Apr 1943 (HA).

45. Klaus Scholder, Die Mittwochs-Gesellschaft: Protokolle aus dem geistigen Deutschland 1932 bis 1944 (n.p. [1982]), p. 305.

46. Geordeler's role is described in Gerhard Ritter, Carl Goerdeler und die deutsche Widerstandsbewegung (Stuttgart: DVA, 1954). The conspiracy is treated in Peter Hoffmann, The history of the German resistance 1933–1945, trans. Richard Barry (Cambridge, Mass.: Massachusetts Institute of Technology Press, 1977).

47. Quoted in Scholder, note 45, p. 326. In German, Schimpanski sounds similar to Schimpanse (chimpanzee).

48. Works cited in note 46; also, Eduard Spranger, Berliner Geist (Tübingen: Wunderlich Verlag, 1966). Assassination would, among other things, release army officers from their oath of allegiance to Hitler.

49. EH, Recoll., pp. 98–99.

50. Scholder, note 45.

51. Heisenberg, summary of his lecture for the society's minutes, in Scholder, note 45, pp. 351–353.

52. Quoted by Scholder, note 45, p. 351.

53. Heisenberg only briefly recalled these events in WH, PB, pp. 189–190.

54. WH, affidavit for Dr. J. H., 23 Nov 1947 (HA).

55. Dr. Graun, Dozentenführer der Freien Forschungsinstitute, to WH, 20 Aug 1942 (HA).

56. Dr. Helmut Joachim Fischer, "Feuerwehr für die Forschung," recollections, 1970 (IZ). A shortened version of the story of his visit appears in Fischer, Erinnerungen, Part 1: Von der Wissenschaft zum Sicherheitsdienst, Zeitgeschichtliche Quellenstudien, vol. 3 (Ingolstadt, 1984).

57. WH to Sommerfeld, 8 Oct 1942 (SN); Sommerfeld to WH, 14 Oct 1942 (HA); Sommerfeld to W. Becker, Akademische Verlagsgesellschaft, 15 Oct 1942 (SN).

58. Mentzel to Laue, 22 May 1943 (HA); Weizsäcker to Laue, 2 Jun 1943 (DJ 29, 1066).

59. Based on WH, PB, pp. 188–189; WH to Karl and Helen Heisenberg, 1 Nov 1945 (HP); WH to his mother, 11 Mar 1943.

60. WH, affidavit for E. S., 30 Jul 1946 (HA). Heisenberg's brother had already left Berlin for a position with Agfa near Stuttgart.

61. WH to NSDAP Ortsgruppe München, 11 Oct and 1 Nov 1943 (HA), for permission to use a truck for the move. His mother had moved to a second-floor

apartment at Ainmillerstrasse 34 in 1938. *Adressbuch München* for the years 1938 to 1944.

62. WH to R. Döpel, 18 Dec 1943 (HA).

63. WH to Dr. Boseck, Sachbearbeiter in Turowsky's office, 26 Jun 1942 (HA); WH to Himmler, 4 Feb 1943 (HA).

64. Thüring to WH, 17 Nov 1942 (HA); also, Thüring to WH, 20 Oct 1942 (HA), and WH to Thüring, 26 Oct 1942 (HA).

65. WH, *Zeitschrift für die gesamte Naturwissenschaft,* 9 (1943), 201–212, rec. 20 May 1943 (HCW C5, 14–25); WH to Himmler, 4 Feb 1943 (HA); R. Brandt, member of Himmler's personal staff, to WH, 15 Feb 1943 (HA).

66. WH to Vahlen, 10 Sep 1942 (HA); Vahlen to WH, 8 Sep 1942 (HA).

67. WH to Dr. Gustav Borger, 11 Jun 1943 (HA). By then the Copernicus Prize was awarded by Frank's Institut für deutsche Ostarbeit. Michael Burleigh, *Germany turns eastwards: A study of Ostforschung in the Third Reich* (New York: Cambridge University Press, 1989).

68. Staff member of *Völkischer Beobachter* to WH, 27 Mar 1943 (HA).

69. Telschow to WH, 26 Nov 1943 (HA); Esau to WH, 19 Nov 1943 (HA).

70. "Max Planck," *Das Reich,* 1943, 11 Apr, p. 1; "Werner Heisenberg," *Das Reich,* 1944, 14 May, p. 1.

71. Dr. Heinrich Dahnke, former Referent in Auslandsabteilung of REM, to WH, 21 Apr 1947 (HA).

72. Hans-Georg Gadamer, *Philosophical apprenticeships,* trans. Robert R. Sullivan (Cambridge, Mass.: Massachusetts Institute of Technology Press, 1985), p. 99.

73. Foreign Office to REM, 27 Nov 1941 (ZStA Potsdam, REM 2943, Bl. 557).

74. Fritz von Twardowski, affidavit, 17 Apr 1948, Weizsäcker defense exhibit no. 208, doc. 352 (NARA, Weizsäcker case, M897, roll 119); also German Embassy, Prague, to Foreign Office, 6 Jan 1943 (ZStA Potsdam, REM 2943) on Budapest lectures by Heisenberg, Planck, and Weizsäcker.

75. Hendrik Casimir, *Haphazard reality: Half a century of science* (New York: Harper & Row, 1983), p. 209; also remarks by Casimir in "Heisenberg im Urteil seiner Schüler," *Bild der Wissenschaft,* 1985, pp. 138–147.

76. WH, note 1, p. 305.

77. WH, affidavit for Dr. Schumburg, formerly of Foreign Office, 29 Jul 1947 (HA). REM documents on party objections to Heisenberg's travels are in ZStA Potsdam, REM 2943.

78. Heisenberg's travels are also discussed in Walker, pp. 105–118 and in Walker, "Physics and propaganda," manuscript submitted to *HSPS.*

79. Dr. Coblitz, director of Institut für deutsche Ostarbeit, Cracow, to WH, 25 May and 29 Sep 1943 (HA); WH to Coblitz, 1943 (HA); WH to Harteck, 8 Dec 1943 (IIA).

80. Lucy S. Dawidowicz, *The war against the Jews 1933–1945* (New York: Holt, Rinehart and Winston, 1975).

81. EH, *Recoll.,* p. 49, recalls an incident in which Heisenberg informed her father of war crimes of which he had learned.

82. Based on records in HA.

83. REM, Merkblatt on travel abroad, 1939 and revised annually; version of 1 Mar 1943 (HA). Also, WH to Abwehrbeauftragter of KWI für Physik, 17 Apr 1944, on items to be taken to Copenhagen.

84. William Dan Andersen, "The German armed forces in Denmark 1940–1943: A

study in occupation policy," Ph.D. thesis, University of Kansas, 1972; and Erich
Thomsen, *Deutsche Besatzungspolitik in Dänemark 1940–1945* (Düsseldorf:
Bertelsmann Verlag, 1971).

85. Telegrams from Best's office to Wehrmacht Headquarters, Sep to Oct 1943 (DJ
31, 1000–1002).

86. Stefan Rozental, "The 40s and 50s," in *Niels Bohr* ed. Rozental (Amsterdam:
North-Holland, 1968), pp. 149–190. According to one count, all but 400 escaped to
Sweden. Those who didn't escape were sent to Theresienstadt, where 51 died.
Dawidowicz, note 80, pp. 374 and 403.

87. J. G. Crowther, *Science in liberated Europe* (London: Pilot Press, 1949), pp.
105–109.

88. Based on "Rapport over Begivenhederne under Besættelsen af Universitetets
Institutet for teoretisk Fysik fra d. 6. December 1943 til d. 3. Februar 1944" (BGC).
I thank Dr. Frederick Nebeker for translating this report.

89. Werner Osenberg of the RFR made the suggestion in "Die Kriegslage als Folge
unzureichender Auslastung verfügbarer Kapazitäten der deutschen Forschung," 28
Dec 1943 (BA Koblenz, RFR docs., R 26III/120). Heisenberg may have learned of
this through his SD contacts, in addition to Suess.

90. Interview of Walther Gerlach by David Irving, 5 Dec 1965 (DJ 29, 1230).

91. WH to his mother, 23 Jan 1944 (HP).

92. "Rapport," note 88.

93. Hans von Euler to WH, 8 Jan 1944 (HA).

94. "Rapport," note 88. Bohr had already burned any compromising
correspondence before he left Copenhagen. Communication from Erik Rüdinger.

95. The above account is based on "Rapport," note 88.

96. "Rapport," note 88, and facsimile of the institute's certificate of release in
Rozental, note 86, p. 161.

97. REM to WH, via Bieberbach, 1 Mar 1944 (HA).

98. This view was expressed in EH, *Recoll.*, p. 82.

99. Crowther, note 87, p. 108.

100. Seyss-Inquart, "Report of the situation and developments in occupied territories
of the Netherlands, 29 May–19 July 1940," translation of doc. 997-PS, in *Nazi
conspiracy and aggression*, vol. 3, pp. 641–656, on p. 653.

101. Werner Warmbrunn, *The Dutch under German occupation 1940–1945*
(Stanford, Calif.: Stanford University Press, 1963), pp. 146–147.

102. Warmbrunn, note 101, p. 149; Casimir, note 75, *Haphazard reality*, p. 202.

103. Van Dellen, General Secretary, Dutch Education Ministry, to WH, 28 May
1943 (HA); REM to WH, 15 Jun 1943 (HA); WH to van Dellen, 21 Jun 1943 (HA).

104. Kramers to WH, 29 Jul 1943, rec. 19 Aug 1943; WH to Kramers, 20 Aug 1943
(HA).

105. Scherrer to Kramers, 24 Jul 1943 (AHQP 13, 3).

106. German designs on the well-equipped Dutch laboratories were strongly
encouraged by Osenberg, note 89.

107. Dr. Plutzar, Hauptabteilung Wissenschaft, Volksbildung und Kulturpflege, to
WH, 15 Sep 1943 (HA).

108. Based on WH, "Bericht über eine Reise nach Holland vom 18.–26.10.43," 10
Nov 1943 (HA); testimony by Seyss-Inquart during his Nürnberg trial, 11 Jun 1946,
in *Der Prozess gegen die Hauptkriegsverbrecher vor dem internationalen
Militärgerichtshof Nürnberg 14. November 1945–1. Oktober 1946*, vol. 16 (Nürnberg,
1948), p. 14.

109. WH, note 108; WH to Kramers, 23 May 1944 (AHQP 12, 5); Rosenfeld to WH, 10 Dec 1943 and 14 Apr 1944 (HA).

110. Rosenfeld, note 109; Kramers to WH, 1 Dec 1943 (HA).

111. G. P. Kuiper to Major Fischer, 30 Jun 1945 (University of Arizona Library, Kuiper Papers, Box 28). I thank Dr. Ronald Doel for informing me of this letter.

112. Walker, p. 113.

113. WH to Gregory Breit, 9 Jan 1951, carbon copy (HA).

Chapter 25

1. WH to Geiger, 23 Oct 1942 (HA).

2. WH, "Die 'beobachtbaren Grössen' in der Theorie der Elementarteilchen," *ZP*, 120 (1943), 513–538, rec. 8 Sept 1942; and *ZP*, 120 (1943), 673–702, rec. 30 Oct 1942 (HCW A2, 611–636, 637–666).

3. WH, *ZP*, 123 (1944), 93–112, rec. 12 May 1944 (HCW A2, 667–686). The fourth paper, probably "Die Behandlung von Mehrkörperproblemen mit Hilfe der *S*-Matrix" (HA), was published in HCW A2, 687–698.

4. WH, "Die Grenzen der Anwendbarkeit der bisherigen Quantentheorie" (1938), HCW A2, 315–330; WH and Pauli, "Bericht über die allgemeinen Eigenschaften der Elementarteilchen," report for Solvay Congress of 1939 (canceled), parts 2 and 3 by WH published in HCW B, 346–358. Motivations for this view are also discussed in James T. Cushing, "Models and methodologies in current theoretical high-energy physics," *Synthese*, 50 (1982), 5–101.

5. This length corresponded to a minimum time of interaction of $t_0 = r_0/c = 3 \times 10^{-24}$ sec.

6. Pauli to WH, 26 Oct 1936; WH to Pauli, 16 Jan 1937.

7. Sommerfeld to WH, 17 Jun 1942 (HA); WH to Sommerfeld, 19 Jun 1942 (SN); WH to Wick, 19 Jun 1942 (HA).

8. WH, note 2, first paper, 533. J. A. Wheeler had proposed the *S*-matrix already in 1937; Heisenberg may have known about it. Aspects of Heisenberg's *S*-matrix theory have been discussed in James T. Cushing, *Theory construction and selection in modern physics: The S matrix* (Cambridge: Cambridge University Press, 1990); Inge Grythe, "Some remarks on the early *S*-matrix," *Centaurus*, 26 (1982–1983), 198–203; Reinhard Dehme, "Theory of the scattering matrix (1942–1946)," HCW A2, 605–610; and H. Rechenberg, "The eary *S*-matrix theory and its propagation (1942–1952)," in *Pions and quarks: Particle physics in the 1950s*, ed. L. M. Brown et al. (Cambridge: Cambridge Univesity Press, 1989), pp. 551–578. The following discussion has benefited from these works.

9. Adopting the cutoff of divergent terms in QED in 1938, Heisenberg had expressed the limit in terms of changes in the relativistic energy-momentum four-vector **k**. In the scattering of a particle from initial relativistic momentum and energy **k**(1), $k_0(1) = E(1)/c$ to final momentum and energy **k**(2), $k_0(2) = E(2)/c$ the net change in momentum and energy must be less than $(h/2\pi r_0)^2$ for quantum theory to apply. This he had expressed by the condition $|[\mathbf{k}(1) - \mathbf{k}(2)]^2 - [k_0(1) - k_0(2)]^2| \leq (h/2\pi r_0)^2$.

10. Perhaps recalling his 1926 discussion with Einstein, Heisenberg conceded Einstein's purported point when he declared that in the end it is the theory that decides what is observable. But Heisenberg argued that even without a theory, one

could utilize presently observable quantities as "clues" for the future theory, in WH, note 2, first paper, 514.

11. The energy values of the stationary states of bound particles were also observable.

12. WH, note 2, second paper. He also showed that the S-matrix can be written in the form $S = 1 + R$, where 1 is the unit matrix and R is a Hermitian matrix.

13. Pauli to Dirac, 21 Dec 1943 (Pauli Letter Collection, CERN, Geneva).

14. WH, note 2, first paper, p. 538.

15. The postwar uses of the S-matrix, which did not usually follow Heisenberg's original program, are described in Cushing, note 8, and Pais, *Inward*, pp. 497–505.

16. Works in notes 8 and 15. Heisenberg's presentation of the theory in Leiden, as recalled by some of the attendees and Kramers family members, is recounted in Dresden, pp. 453–458, which also includes an evaluation of some of the subsequent Kramers-Heisenberg correspondence.

17. WH to Kramers, 31 Oct 1943 (AHQP 12, 5).

18. WH, note 3, third paper. The zeros correspond to poles of the η matrix in the complex k-plane.

19. Kramers to WH, 1 Dec 1943 (HA)

20. Dresden, p. 457.

21. WH to Kramers, 31 Oct 1943 and 16 Mar 1944 (AHQP 12, 5) and 10 Jan 1944 (HA).

22. Kramers to WH, 12 Apr 1944 (AHQP 12, 5 and HA).

23. WH, note 3, third paper. Correspondence followed the visit. WH-Møller correspondence (HA), also discussed in Grythe, note 8.

24. H. A. Kramers, "Fundamental difficulties of a theory of particles," address to Symposium on Elementary Particles at Utrecht, April 14, 1944, *Nederlands Tijdschrift voor Natuurkunde*, 11 (1944), 134–140; reprinted in H. A. Kramers, *Collected Scientific Papers,* ed. H. B. G. Casimir (Amsterdam: North-Holland, 1956), pp. 838–844.

25. WH, "Der mathematische Rahmen der Quantentheorie der Wellenfelder" (1946), HCW A2, 699–713; "The present situation in the theory of elementary particles," delivered at the Cavendish Laboratory, Cambridge, Dec 1947 (HCW B, 444–449).

26. These were apparently also found by Res Jost. See Oehme, note 8, p. 607.

27. Pais, *Inward,* pp. 497–505; Grythe and Rechenberg, note 8.

28. Ludwig, p. 241. For instance, W. Süss, "Die gegenwärtige Lage der deutschen Wissenchaft und der deutschen Hochschulen," manuscript, 26 Aug 1943 (HA); Carl Ramsauer, "Programm der Deutschen Physikalischen Gesellschaft für den Ausbau der Physik in Grossdeutschland," *Verhandlungen der Deutschen Physikalischen Gesellschaft*, 25 (1944), 1–6; lectures by Ramsauer listed in Ramsauer to WH, 21 Jul 1943 (HA). Parallel discussions of the nuclear research topics in this chapter are offered in Walker and Irving.

29. Ramsauer to WH, 21 July 1943 (HA).

30. Ramsauer, "Über die Leistung und Organisation der angelsächsischen Physik," delivered to the Aeronautical Academy, 3 Apr 1943, English translation by Allied intelligence (DJ 31, 157–170).

31. Telschow, Aktennotiz, 8 Feb 1943 (DJ 29, 1049–1050); Esau to Mentzel, 5 Apr 1943 (DJ 29, 1060–1061).

32. WH to Telschow, 25 Mar 1943 (HA).

33. Bothe to WH, 10 Feb 1943 (HA); WH to Bothe, 12 Feb 1943 (HA).

34. Esau to Mentzel, 28 Oct 1943 (DJ 29, 1082); Mentzel to Görnnert, Göring's

office, 6 Nov 1943 (DJ 29, 1084); Speer to Görnnert, 17 Nov 1943 (DJ 29, 1088); Göring to Gerlach, 2 Dec 1943 (DJ 29, 1091).

35. WH, "Die Energiegewinnung aus der Atomkernspaltung," published for limited circulation in *Probleme der Kernphysik,* Deutsche Akademie der Luftfahrtforschung, Schriften 1943/44 (Berlin, 1943), pp. 29–36 (microfilm DJ 31, 197–204; HCW A2, 570–575).

36. Bäumker, Secretary of Akademie der Luftfahrtforschung, to WH, 17 Feb 1943 (HA).

37. WH, note 35, p. 30.

38. Assessment based on data derived from Diebner to Göring, 18 Apr 1944 (DJ 29, 1110–1112); Esau to Mentzel, 5 Apr 1943 (DJ 29, 1060–1061); Gerlach to Mentzel, 26 May 1944 (DJ 29, 1113–1115). Nuclear pile experiments in four institutes, including Heisenberg's, received the second of three ratings and the lowest for continuation of work during wartime. The RFR delegated the smallest portion of its nuclear research budget to pile research (which received its main funding from the regular budget of the KWG).

39. Graue, Leiter des Geschäftsführenden Beirates des RFR, notes on a meeting with Mentzel and other policy officials on 7 Sep 1943, dated 13 Sep 1943 (LC 107, file 12853); OKW, Wehrersatzamt, Rundschreiben, 18 Dec 1943 (BA Koblenz, R26 III/108); Göring to Mentzel, 24 Aug 1944 (DJ 29, 1141–1142).

40. Max von Laue, "Die Kriegstätigkeit der deutschen Physiker," *PBl,* 3 (1947), 424–425. The exemptions were reviewed periodically.

41. EH, *Recoll.,* p. 90.

42. Report of Samuel A. Goudsmit's remarks on Heisenberg by Dietrich E. Thomsen, *Science news,* 109 (1976), 157.

43. Dresden, p. 458.

44. WH to Himmler's personal staff, 9 Mar 1943 (HA).

45. Records in HA.

46. Kramers to WH, 12 Apr 1944 (HA); WH to Kramers, 4 May 1944 (AHQP 12, 5).

47. WH to Coster, 16 Feb 1943 (original: GP; copy: DJ 29, 1051; carbon: HA). Coster's letter to WH could not be found.

48. David Irving, note written on his copy of WH to Coster, 16 Feb 1943 (DJ 29, 1051), indicating that Goudsmit's parents died on 11 Feb 1943.

49. *Alsos,* p. 48.

50. Goudsmit, "Werner Heisenberg (1901–1976)," *Yearbook of the American Philosophical Society,* 1976, pp. 74–80, on p. 78.

51. Mentzel to Görnnert in Göring's office, 8 Jul 1943 (DJ 29, 1077), transferral of report by Esau.

52. WH, "Bemerkungen zu dem geplanten halbtechnischen Versuch mit 1,5t D_2O und 3t 38-Metall," 31 Jul 1942 (DJ 30, 228–232; HCW A2, 545–552).

53. Bothe to WH, 30 Jul 1943 (HA).

54. Esau, "Bericht über den Stand der Arbeiten auf dem Gebiete der Kernphysik am 31.3.44," 21 Jul 1944 (DJ 29, 1102–1108). Discussions of Diebner and his research may be found in Walker and Irving.

55. WH, note 35, p. 34.

56. Albert Vögler to WH, 18 Jan 1944 (HA).

57. Beuthe to Mentzel, 22 Dec 1943 (DJ 29, 1095–1096).

58. Described after the war in WH-Wirtz, p. 157.

59. Derived from WH-Wirtz, pp. 155–156.

60. Ibid., pp. 157–158, and WH et al., "Bericht über einen Versuch mit 1,5 to

D$_2$O und U und 40 cm Kohlerückstreumantel (B-7)," 3 Jan 1945 (DJ 29, 565–572; HCW A2, 595–601).

61. WH-Wirtz, p. 156.

62. Laue to WH, 20 Aug 1943 (HA).

63. WH to Stetter, head of reactor research at the University of Vienna, 2 Jul 1943 (HA); Stetter to WH, 11 Aug 1943 (HA).

64. WH to Adolf Hornung, institute technician, 20 Jul 1943 (HA); Erich Bagge, diary, entries for Aug–Sep 1943 (DJ 29, 106–133); list of 55 members of the institute, with titles and locations, as of 24 Jan 1944 (HA).

65. They also occupied several rooms in a brewery. *Alsos*, p. 106.

66. Recalled by K. F. Bonhoeffer to WH, 7 Mar 1946 (HA).

67. WH to the Mayor of Kochel, 3 Apr 1944 (HA); WH to Kurt Staun, Leipzig official, 24 Apr 1944 (HA).

68. EH, *Recoll.*, pp. 93–94; WH to his mother, 5 Sep 1943, and WH to his uncle Karl, 1 Nov 1945 (HA).

69. Ibid., and discussion with Frau Roy, Stuttgart, formerly of Hechingen, who assisted Heisenberg in his piano playing.

70. WH to his mother, 19 May 1944.

71. WH to his mother, 15 Jun 1944.

72. WH to his mother, 18 Aug 1944.

73. The activities of one purported informant are described in Arnold Kramish, *The griffin* (Boston: Houghton Mifflin, 1986), but this work is open to question.

74. Victor Weisskopf to J. Robert Oppenheimer, 1942, published in S. S. Schweber, "Some chapters for a history of quantum field theory: 1938–1952," in *Les houches*, ed. B. S. DeWitt and R. Stora, session XL (Amsterdam: North-Holland, 1984), pp. 37–220, on pp. 126–128.

75. This account of Berg's activities is based on unpublished War Department reports of 5 December 1945, 25 September 1946, and 30 September 1946 supporting a hero's citation for Berg, and Louis Kaufmann et al., *Moe Berg: Athlete, scholar, spy* (Boston: Little, Brown, 1974). Although these sources are questionable, I am grateful to Neil Goldstein for calling my attention to them.

76. This and following accounts of the activities of the Alsos mission are based on mission reports and correspondence in NARA, microfilms M1109, 5 rolls; NARA RG 77 (Chief of Engineers), 371.2 (Goudsmit Mission); as well as the following works: *Alsos*; Boris T. Pash, *The Alsos Mission* (New York: Award House, 1969); Leslie R. Groves, *Now it can be told: The story of the Manhattan Project* (New York: Harper & Row, 1962); Leo James Mahoney, "A history of the War Department scientific intelligence mission (Alsos), 1943–1945," Ph.D. dissertation, Kent State University, 1981.

77. *Alsos*, pp. 46–49. The size, membership, and organization of the mission in January 1945 may be gleaned from mission papers, note 76.

78. Jozef Garlinski, *The Swiss Corridor: Espionage networks in Switzerland during World War II* (London: Dent, 1981), p. 17.

79. References in note 75 and discussion with Professor Res Jost, Zurich, 1985.

80. Kaufmann et al., note 75, p. 195. However, this story is not in the War Department reports cited in note 75.

81. *Alsos*, p. 114. Heisenberg denied saying this, in WH to van der Waerden, 22 Apr 1948 (GP).

82. Recalled in Gerlach to WH, 16 Apr 1946 (HA); and in interview of Walther Gerlach by David Irving, 5 Dec 1965 (transcript DJ 29, 1230–1240).

83. Gerlach, "Bericht: Über die Arbeiten auf kernphysikalischem Gebiet vom 1. Febr. bis 31. Mai 1944," n.d. (DJ 29, 1118–1122).

84. The official report of the B-8 experiment could not be found. The following account is based on independently confirmed technical data in WH-Wirtz, pp. 158–165, and in part on the somewhat embellished account apparently derived from the recollections of the participants in Irving, pp. 266–271.

85. Interview with the innkeeper's son, Haigerloch, 1977.

86. "Erlass des Führers über die Konzentration der Rüstung und Kriegsproduktion vom 19.6.1944" and Speer, "Anordnung zum Erlass des Führers," 12 Jul 1944 (both LC M107, file 12853).

87. "Sitzungsbericht vom 21. August 1944 des Reichsmin. f. Rüstung u. Kriegsproduktion" (BA Koblenz, R26 III/92); Bormann, Rundschreiben, 3 Sep 1944, and Osenberg, Rundschreiben, 7 Sep 1944 (both BA Koblenz, R26 III/108); Osenberg, Rundschreiben, 12 Feb 1945 (BA Koblenz, R26 III/42); Himmler to SS-General Jüttner, Jul 1944 (English translation in Irving, p. 256). The mobilization required regular physical fitness exercises and weapons training, regardless of age or physical condition.

88. Not only did the Allies have no need of a German reactor, they never wavered from their 1943 Casablanca demand of unconditional surrender.

89. WH to Helmut Volz, 24 Apr 1944 (HA).

90. Based in part on Irving, pp. 271–272, and Walker.

91. WH-Wirtz, p. 164.

92. Indicated in Colonel Lansdale, report to General Groves, 5 May 1945 (NARA, M1109, roll 2).

93. Recounted in the references in note 76.

94. "Organizational disposition," 15 Apr 1945 (NARA, M1109, roll 4). Paris served as Alsos headquarters, Aachen as Advance Base, North.

95. Groves to Chief of Staff, 23 Apr 1945 (NARA, M1109, roll 2).

96. Lansdale, note 92.

97. Ibid.

98. WH to Wolfgang Schadewaldt, 19 Jan 1946 (HA). This account of Heisenberg's activities draws on this letter and on WH to Karl and Helen Heisenberg, Farm Hall, 1 Nov 1945 (HA).

99. WH to Fritz Schumacher, 6 May 1945 (GP).

100. Colonel Boris T. Pash to Chief, Military Intelligence Service, War Department, "Subject: Alpine Operation," 18 May 1945 (NARA, M1109, roll 4); and Pash, note 76, pp. 219–241. The Munich operation is described in Major R. C. Ham, report to Colonel Boris T. Pash, 12 May 1945 (NARA, M1109, roll 4).

Chapter 26

1. *Alsos*, p. 113.

2. Ibid., p. 112. The invitation was repeated and declined again while Heisenberg was in British captivity. WH to his wife, 25 Jan 1946 (HP).

3. Max von Laue to his son, Theodor, 29 May 1945 (Nachlass von Laue, Deutsches Museum, 1976–20). Dustbin was the main holding camp for German scientists and technicians. It's operation is described in John Gimbel, "U.S. policy and German scientists: The early cold war," *Political Science Quarterly*, 101 (1986), 433–451.

4. Max von Laue to his son, 26 May 1945 (Nachlass von Laue).

5. R. V. Jones, *The wizard war: British scientific intelligence 1939-1945* (New York: Coward, McCann and Geoghegan, 1978), p. 481. The story also appears in Jones's introduction to *Alsos*, reprint edition.

6. Max von Laue to his son, 7 Aug 1945 (Nachlass von Laue).

7. Ibid.; Jones, note 5, p. 483; Leslie R. Groves, *Now it can be told: The story of the Manhattan project* (New York: Harper & Row, 1962; Plenum, 1983); Laue to Paul Rosbaud, 4 Apr 1959, English translation in Arnold Kramish, *The griffin* (Boston: Houghton Mifflin, 1986), pp. 245-247.

8. Jones, note 5, p. 483.

9. Groves, note 7, p. 333.

10. Laue, note 6. It is not certain that these letters were actually sent at the time.

11. The question was asked again by Sir Nevill Mott and Sir Rudolf Peierls, "Werner Hesienberg 5 December 1901-1 February 1976," *Biographical memoirs of fellows of the Royal Society*, 23 (1977), 213-251.

12. HA; HCW C5, 26-27.

13. Erich Bagge, diary entry for 10 Aug 1945, which also refers to "various difficulties," in Bagge, Diebner, and Kenneth Jay, *Von der Uranspaltung bis Calder Hall* (Hamburg: Rowohlt, 1957), p. 58.

14. Max von Laue to his son, 7 Aug and 22 Sep 1945, and typed draft of press release, with small emendations in Laue's hand, 8 Aug 1945 (Nachlass von Laue), excerpted in English translation in Groves, note 7, pp. 336-337.

15. Walker, ch. 5-7; Walker, "Legenden um die deutsche Atombombe," *Vierteljahrshefte für Zeitgeschichte*, 1990, 45-74.

16. Quoted in Groves, note 7, p. 335.

17. Laue, note 6.

18. Statement, notes 12 and 14.

19. Groves, note 7, p. 334.

20. For instance, Paul Forman, "Behind quantum electronics: National security as a basis for physical research in the United States, 1940-1960," *HSPS*, 18 (1987), 149-229; Dan Kevles, "Cold war and hot physics: Science, security, and the American state, 1945-1956," *HSPS*, 20 (1990), 239-264.

21. For instance, B. J. Bernstein, "Four physicists and the bomb: The early years," *HSPS*, 18 (1988), 231-263.

22. Henry DeWolf Smyth, *Atomic energy for military purposes: The official report on the development of the atomic bomb under the auspices of the United States government, 1940-1945* (Princeton: Princeton University Press, 1946), p. 223.

23. This was not unique to the scientists. German citizens in general fell suddenly silent on their past.

24. Philip Morrison, "Alsos: The story of German scientists," *Bulletin of the atomic scientists*, 3 (1947), 354, 365.

25. WH, "Über die Arbeiten zur technischen Ausnutzung der Atomkernenergie in Deutschland," *Nwn*, 33 (15 Dec 1946), 325-329 (HCW C5, 28-32), abridged English translation in *Nature*, 160 (16 Aug 1947), 211-215; "Es gab niemals eine deutsche Atombombe," interview of WH, *Schwäbische Donau-Zeitung*, 2 Aug 1948, p. 3 (HCW C5, 33-34): "Für Bombe unzuständig," *Die Welt*, 5: 28 (2 Feb 1950), p. 2; Waldemar Kaempffert, "Why the Germans failed to develop an atomic bomb is now revealed in two reports," *New York Times*, 26 Oct 1947, p. E9; Kaempffert, "Nazis spurned idea of an atomic bomb," interview of WH, *New York Times*, 28 Dec 1948, p. 10; WH, "German atomic research," letter to the editor, *New York Times*, 30 Jan 1949, sec. 4, p. 8 (HCW C5, 41-42).

26. Kaempffert, "Nazis spurned idea," note 25, quoting WH.

27. WH, note 25, *Nature*.

28. Ibid.

29. Ibid. The same formulation appears as early as WH to the Schumachers, 11 Feb 1946 (HP).

30. This is also stated in Irving, pp. 296–297.

31. Jungk, *Brighter than a thousand suns: A personal history of the atomic scientists*, trans. James Cleugh (New York: Harcourt Brace Jovanovich, 1958), p. 91. Also, WH, draft of *Nwn* report, note 25, enclosed in WH to Bothe et al., 29 Nov 1946; Bothe to WH, 7 Dec 1946 (all Bothe papers, MPG).

32. These events took place before Heisenberg's June meeting with Speer.

33. WH, *PB*, p. 218.

34. *Alsos*; Goudsmit "Heisenberg on the German nuclear power project," *Bulletin of the atomic scientists*, 3 (1947), 64 and 67; Goudsmit, "German atom research," *New York Times*, 9 Nov 1947, p. E8, letter to the editor of 29 Oct 1947; Goudsmit, "German war research," *New York Times*, 9 Jan 1949, sec. 4, p. 8, letter to editor of 4 Jan 1949. Heisenberg's side of the exchange is cited in note 25; correspondence in GP and BSC 29, 4.

35. Goudsmit, note 34, "German war research."

36. *Alsos*, p. 121.

37. *New York Times* exchanges, notes 25 and 34; WH-van der Waerden correspondence (HA); WH-Weizsäcker correspondence (HP). Weizsäcker's efforts on behalf of the German position are evident in his long memos to Einstein biographer Carl Seelig (Seelig Collection, ETH).

38. Kaempffert, note 25, "Why the Germans failed."

39. Goudsmit, note 34, "German atom research," and reply by Kaempffert. Henry Schuman to Einstein, 11 Nov 1947 (EA 12-176). Einstein responded that they may lie under the pressure of circumstances. Einstein to Schuman, 17 Nov 1947 (EA 12-177).

40. This was asserted in a response to Morrison's review (note 24) by Max von Laue, "Die Kriegstätigkeit der deutschen Physiker," *PBl*, 3 (1947), 424–425, and repeated by Heisenberg.

41. WH to Bohr, 25 Apr 1946 (BSC 29, 4); Karl Heisenberg to WH, 16 May 1946 (HP); R. Fraser, London, to Bohr, 18 Jul 1947 (BSC 29, 4); WH to Goudsmit, 23 Sep 1947 (GP and BSC 29, 4). Goudsmit presented his views to Bohr in Goudsmit to Bohr, 2 Dec 1947 (BSC 29, 4).

42. WH to B. L. van der Waerden, 28 Apr 1948 (HA).

43. B. L. van der Waerden, "Aide-mémoire," 12 Dec 1948 (BSC 33, 1); Waerden to WH, 28 Feb 1949 (HA).

44. WH to Goudsmit, 23 Sep 1947 (GP and BSC 29, 4).

45. WH, "Die aktive und die passive Opposition im Dritten Reich," 12 Nov 1947 (HA); WH, "Ordnung des Wirklichkeit," in HCW C1, 218–306.

46. Goudsmit to WH, 1 Dec 1947 (GP and BSC 29, 4).

47. WH-Wirtz.

48. Goudsmit, note 34, "German war research"; Kaempffert, note 25, "Nazis spurned idea"; Goudsmit to WH, 20 Sep 1948 (GP).

49. For instance, Goudsmit to WH, 1 Dec 1947 (GP and BSC 29, 4).

50. *Alsos*, p. 115.

51. C. F. von Weizsäcker to WH, Chicago, 14 Oct 1949 (HP).

52. Goudsmit to WH, 19 Mar 1973 (HA); WH to Goudsmit, 11 Apr 1973 (GP).

53. Goudsmit, "Werner Heisenberg (1901–1976)," *Yearbook of the American*

Philosophical Society, 1976, pp. 74–80, on p. 79. Goudsmit's assessment is supported by subsequent research; e.g., S. S. Schweber, "The empiricist temper regnant: Theoretical physics in the United States 1920–1950," *HSPS*, 17 (1986), 55–98.

54. Weizsäcker, note 51, wrote Heisenberg of the depressed feelings of many of the American physicists at the prevailing state of affairs.

55. Jungk, note 31; German version: *Heller als tausend Sonnen* (Bern: Alfred Scherz Verlag, 1956).

56. Jungk, note 31, p. 88. The Americans were not so personalized.

57. Ibid., p. 91.

58. Ibid., p. 88.

59. Jungk and Walker later attributed this to Jungk's aversion to McCarthyism and to cold-war posturing in the United States during those years. Walker, note 15, "Legenden um die deutsche Atombombe."

60. WH to Oskar Huemer, 22 Sep 1965 (HA).

61. Quoted in Kramish, note 7, p. 247. Rosbaud's wartime role is maintained by Kramish and by R. V. Jones, foreword to *Alsos*.

62. Laue to Rosbaud, note 7.

63. Quoted in WH to Hans Bethe, 27 Apr 1964 (HA).

64. Ibid. Bethe seemed to agree with Heisenberg that conscience must have played a role. Bethe to WH, 15 May 1964 (HA).

65. Irving.

66. WH, "The Third Reich and the atomic bomb," *Bulletin of the atomic scientists*, 24 (1968): 6, 34–35, translation of "Das Dritte Reich versuchte nicht, die Atombombe zu bauen," *Frankfurter Allgemeine Zeitung*, 6 Dec 1967, magazine "Bilder und Zeiten," p. 4 (HCW C5, 50–52); also interview of WH by editors of *Der Spiegel*, 1967 (HCW C5, 45–48).

67. WH, note 66, *Bulletin*, p. 35.

68. Eugene Rabinowitz, "The Virus House: The German atomic bomb project," *Bulletin of the atomic scientists*, 24 (1968): 6, 32–35.

69. For instance, Mark Walker, "Heisenberg, Goudsmit and the German atomic bomb," *Physics Today*, 43 (1990):1, 52–60, and the widely divergent responses to it: "Heisenberg, Goudsmit and the German 'A-bomb,'" letters to the editor, *Physics Today*, 44(1991):5, pp. 13, 15, 90–96.

70. Goudsmit, note 53, p. 80.

Chapter 27

1. Max von Laue to Theodor von Laue, Göttingen, 19 May 1946 (Nachlass von Laue, Deutsches Museum, 1976–20).

2. WH to the Schumachers, 11 Feb 1946 (HP).

3. WH to Karl Heisenberg, 8 May 1946 (HP), and WH to Bohr, 25 Apr 1946 (BSC 29, 4).

4. WH to Karl Heisenberg, 8 May 1946 (HP), and Johann Dieckmann to WH, 10 Nov 1946 (HA).

5. Hund to WH, 1 May 1946 (HA); Hund, "Wissenschaftliches Tagebuch," entry for 6 Apr 1945 (Hund papers, Göttingen).

6. WH, note 3, and WH to Annaliese Clar, 18 Apr 1947 (HA).

7. WH to his wife, 25 Jan 1946 (HP); WH to Fritz Schumacher, 6 May 1945 (GP) and 11 Nov 1946 (HA).

8. Johann Dieckmann to WH, Osnabrück, 10 Nov 1946 (HA); Firma Hartmann to WH, Osnabrück, 26 Jun 1947; WH to Hartmann, 2 Jul 1947 (HA).

9. Erich Bagge, diary, entry for 7 Jan 1946, in Bagge, Diebner, and Jay, *Von der Uranspaltung bis Calder Hall* (Hamburg: Rowohlt, 1957), p. 71.

10. Interview with EH, Göttingen, 1982; WH to the Schumachers, Alswede, 11 Feb 1946 (HP).

11. WH to his wife, Aug 1945 (HP); WH to Karl and Helen Heisenberg, 1 Nov 1945 (HP); WH to Sommerfeld, 5 Feb 1946 (SN).

12. EH, note 10, WH to Karl and Helen Heisenberg, 1 Nov 1945 (HP), and WH to Wolfgang Schadewaldt, 19 Jan 1946 (HA).

13. WH to his wife, 20 Jan 1946 (HP); WH to the Schumachers, 11 Feb 1946 (HP); Max von Laue, letters to his son, note 1.

14. WH to P. M. S. Blackett, 5 Oct 1945 (HA).

15. Bagge, note 9, entry for 3 Feb 1946; WH to Sommerfeld, 5 Feb 1946 (SN). Max von Laue later moved to West Berlin.

16. WH to Sommerfeld, 5 Feb and 29 Jun 1946 (SN); WH to Gerlach, 16 Jul 1946 (HA).

17. Sommerfeld to WH, 17 Feb 1946 (HA).

18. WH to B. Schweitzer, 17 Apr 1947; WH to E. Sethe, 30 Jul 1946 (HA).

19. M. Pahl to WH, Hechingen, 3 Mar 1946; WH to Otto Hahn, Urfeld, 16 Aug 1946; WH to F. von Hermann, 2 Dec 1946 (HA).

20. WH to Sommerfeld, 7 Feb 1947 (SN); Sommerfeld to WH, 24 Sep 1947 (HA).

21. Discussions with colleagues.

22. Armin Hermann, "Germany's part in the setting-up of CERN;" in A. Hermann et al., *History of CERN*, vol. 1 (Amsterdam: North-Holland, 1987), pp. 383–429.

23. Heisenberg's enthusiasm for reestablishing international relations is evident from his talk to Göttingen students, "Wissenschaft als Mittel zur Verständigung unter den Völkern" (1946), HCW C5, 384–394. His pleasure at serving as president of the Humboldt Foundation is evident from his publications in that capacity, HCW C5, group 6.

24. Allied Control Law 25, 29 Apr 1946, in Felix Brandl, ed., *Das Recht der Besatzungsmacht* (Munich: Oldenbourg, 1947), pp. 674–687.

25. Laue, note 1.

26. American and British denazification policies are discussed in Tom Brower, *The pledge betrayed: America, Britain and the denazification of postwar Germany* (New York: Doubleday, 1981); and Lutz Niethammer, *Entnazifizierung in Bayern* (Frankfurt, 1972).

27. Heisenberg's cetificates are discussed in Walker, pp. 195–201, based on documents in HA. Refusals and inquiries about forgotten individuals are also in HA.

28. For instance, British Research Branch, certificate for WH, 27 Feb 1947 (HA).

29. Heisenberg's many papers, addresses, and interviews on German atomic energy policy are reprinted in HCW C5.

30. Heisenberg, the history of postwar German nuclear technology, and the scientists' opposition to nuclear weapons have been described in numerous works. They include Joachim Radkau, *Aufstieg und Krise der deutschen Atomwirtschaft 1945–1975* (Reinbek: Rowohlt, 1983); Hans Karl Rupp, *Ausserparlamentarische Opposition in der Ära Adenauer: Der Kampf gegen die Atombewaffnung in den fünfziger Jahren* (Cologne: Pahl-Rugenstein, 1970); Mark Cioc, *Pax atomica: The nuclear defense debate in West Germany during the Adenauer era* (New York: Columbia University Press, 1988); Hermann, *Jahrhundert*; Michael Eckert, "Neutrons and politics:

Maier-Leibnitz and the emergence of pile neutron research in the FRG," *HSPS*, 19 (1988), 81–113; and Eckert, "Primacy doomed to failure: Heisenberg's role as scientific adviser for nuclear policy in the FRG," *HSPS*, 21 (1990), 29–58.

31. Some of the later science policy history is summarized in Otto Keck, "West German science policy since the early 1960s: Trends and objectives," *Research policy*, 5 (1976), 116–157; and Keck, "Government policy and technical choice in the West German reactor programme," *Research policy*, 9 (1980), 302–356. Works devoted to the early history include Thomas Stamm, *Zwischen Staat und Selbstverwaltung: Die deutsche Forschung im Wiederaufbau 1945–1965* (Cologne, 1981); and Maria Osietzki, *Wissenschaftsorganisation und Restauration: Der Aufbau ausseruniversitärer Forschungseinrichtungen und die Gründung des westdeutschen Staates 1945–1952* (Cologne: Böhlau Verlag, 1984).

32. The refounding of the Notgemeinschaft is discussed in Stamm and Osietzki, note 31.

33. Indicated, for example, in the "Düsseldorfer Leitsätze vom 15. Juli 1949," of the CDU, in *Einigkeit und Recht und Freiheit: Westdeutsche Innenpolitik 1945–1955*, ed. Theo Stammen (Munich: DTV, 1965), pp. 94–97.

34. "Military governors' aide mémoire for the Parliamentary Council," 22 Nov 1948, in John Ford Golay, *The founding of the Federal Republic of Germany* (Chicago: University of Chicago Press, 1958), pp. 263–264, on pp. 263.

35. Zierold described his activities in *Forschungsförderung in drei Epochen: Deutsche Forschungsgemeinschaft—Geschichte, Arbeitsweise, Kommentar* (Wiesbaden: Franz Steiner Verlag, 1968).

36. WH to "Herr Minister" [1948] (HA), concerning agreement with Dr. Fraser, Göttingen representative of the British Research Branch, on the establishment of the DFR.

38. WH, Regener, Rein, and Zenneck to Carlo Schmid, 15 Dec 1948, letter and stenographic report of debate in *Parlamentarischer Rat: Verhandlungen des Hauptausschusses* (Bonn, 1949), report of 30th session on 6 Jan 1949.

38. Lists of members and other useful information are available in *Abschlussbericht des Deutschen Forschungsrats (DFR) über seine Tätigkeit*, ed. H. Eickemeyer (Munich: R. Oldenbourg, 1953).

39. WH, manuscript of lecture, 9 Mar 1949 (HA).

40. Art. 2, sec. 4 of the bylaws of the DFR, reprinted in *Abschlussbericht*, note 38, pp. 82 and 85.

41. WH to Gerhard Hess, President of the University of Heidelberg, 8 May 1951 (HA); Col. Bertie Blount, head of Research Branch of the British Control Commission, to WH, 3 May 1949 (HA).

42. WH to Karl Geiler, President of the Notgemeinschaft, draft, 11 Jan 1951 (HA).

43. "Staatsabkommen der Länder des Amerikanischen, des Britischen und des Französischen Besatzungsgebietes über die Finanzierung wissenschaftlicher Forschungseinrichtungen," Königstein, 1 Apr 1949, in *50 Jahre Kaiser Wilhelm-Gesellschaft und Max Planck-Gesellschaft 1911–1961* (Bonn: Max Planck-Gesellschaft, 1961), pp. 227–231.

44. [WH], "Denkschrift des DFR über die Betreuung der wissenschaftlicher Forschung im Rahmen der Deutschen Bundesregierung," Göttingen, 1 Sep 1949, in *Abschlussbericht*, note 38, pp. 117–121.

45. WH to J. Zenneck, 17 Dec 1949 (HA).

46. WH to Oberregierungsrat Dr. Rust in the Bundeskanzleramt, 8 Jun 1950 (HA).

47. For example, WH, "Erforschung und wirtschaftliche Nutzbarmachung der Atomenergie im Frieden," attached to memo of 19 Jun 1951 (HA).

48. Dr. E. Lehnartz, Vorsitzender des Hauptausschusses der Notgemeinschaft, to WH, official and unofficial letters of 17 Oct 1949 (HA).

49. Adenauer to WH, 11 May 1951 (HA); WH, "Memorandum des DFR über die Organisation der wissenschaftlichen Forschung in Deutschland," 19 Jun 1951, in Abschlussbericht, note 38, pp. 122–126, without the appendix cited in note 47.

50. Adenauer to WH, 2 Jul and 17 Jul 1951 (HA).

51. In particular, the Program for New Technologies, 1969–1974.

52. WH, "Vorwort," in Abschlussbericht, note 38, p. 9.

53. Heisenberg's papers are reprinted in HCW A3, forthcoming. They are discussed in an introduction by a coworker of that period, H. Koppe, "Über Heisenbergs Arbeiten zur Supraleitung."

54. Heisenberg's papers on turbulence are reprinted in HCW A1, group 1. They are discussed in introductory essays by S. Chandrasekhar and H. Rechenberg.

55. Renormalization, postwar quantum field theory, and their history have been discussed in a number of works, among them Pais, Inward; S. S. Schweber, "Some chapters for a history of quantum field theory: 1938–1952," in Les houches, ed. B. S. DeWitt and R. Stora, session XL (Amsterdam: North-Holland, 1984), pp. 37–220; Schweber; "Particle theory in the 50's: An historical assessment," in L. Brown et al., ed., Pions and quarks: Particle physics in the 1950s (Cambridge: Cambridge University Press, 1989), pp. 668–693. and Andrew Pickering, Constructing quarks: A sociological history of particle physics (Chicago: University of Chicago Press, 1984). Some of the most important papers are reprinted in Selected papers on quantum electrodynamics, ed. Julian Schwinger (New York: Dover, 1958).

56. Heisenberg's postwar papers on field theory are reprinted in HCW A3. He summarized his work through 1957 in "Quantum theory of fields and elementary particles" (1957), HCW B, 552–561. Heisenberg's postwar work has not yet received adequate historical analysis. It is summarized in Hans-Peter Dürr, "Heisenbergs einheitliche Feldtheorie der Elementarteilchen," manuscript for Deutsche Akademie der Naturforscher Leopoldina, 1981, and in his introductory essays to works in HCW A3.

57. WH, "Zur Frage der Kausalität in der Quantentheorie der Elementarteilchen" (1951), HCW A3.

58. WH, "Der mathematische Rahmen der Quantentheorie der Wellenfelder" (1946), HCW A2, 699–713.

59. See works cited in note 56.

60. WH-Pauli correspondence (AHQP 82); recalled in WH, PB, pp. 223–226.

61. WH to Pauli, 16 Dec 1957 and further correspondence in 1958 (AHQP 82).

62. WH and Pauli, "On the isospingroup in the theory of the elementary particles," unpublished mimeographed typescript, 2 editions (HA and AHQP 82).

63. Quoted in Weisskopf to Pauli, 7 Mar 1958 (AHQP 82).

64. The Heisenberg episode is recounted in Hermann, Jahrhundert, and recalled in WH, PB, ch. 19. The New York Times fascination with Einstein is followed in Pais, Subtle; the article appeared in the 27 Dec 1949 issue.

65. Pauli, statement, dated 8 Apr 1958, and WH to Pauli, 13 Apr 1958 (both AHQP 82).

66. Quoted in Hermann, Jahrhundert, p. 236.

67. Heisenberg's paper, "Remarks on the nonlinear spinor theory with indefinite

metric in Hilbert space," and the published version of the conference discussion, are reprinted in HCW B, 563–570.

68. WH, *PB*, p. 235.

69. Heisenberg wrote obituaries for each (HCW C4).

70. For instance, WH to Sommerfeld, 25 Nov 1950 (SN); and WH to his wife, 11 Feb 1956 (HP).

71. Communication from Manfred Schroeder.

72. WH, "Physics and philosophy," the Gifford Lectures (German version in HCW C2, 1–201). Heisenberg's philosophy and its relation to Greek sources have been explored in Patrick A. Heelan, *Quatnum mechanics and objectivity: A study of the physical philosophy of Werner Heisenberg* (The Hague: Martinus Nijhoff, 1965); and Heelan, "Heisenberg and radical theoretic change," *Zeitschrift für allgemeine Wissenschaftstheorie*, 6 (1975), 113–138, with a reply by Heisenberg.

73. WH, "The nature of elementary particles" (1976), HCW B, 917–927, on 924.

74. WH, *PB*, p. 244.

75. Ibid., p. 247.

NAME INDEX

Subject Index

UNCERTAINTY

THE LIFE AND SCIENCE OF
WERNER
HEISENBERG

DAVID C. CASSIDY

Werner Heisenberg (1901–1976) was one of the most brilliant and renowned physicists of his day. A Nobel laureate at the age of 32, he is ranked among the greatest geniuses of this or any century. Heisenberg's extraordinary scientific accomplishments include his seminal work in the development of quantum physics, elementary particles, and nuclear physics. But his masterpiece was his articulation of the Uncertainty Principle, a scientific insight so profound that it has become a metaphor for our understanding of all human phenomena.

Heisenberg's life was one of dramatic controversy as well as spectacular triumph. Unwilling to leave his beloved homeland, he remained in Germany during the rise and rule of the Third Reich—and made the personal and professional compromises such a choice entailed. During World War II, he played a major role in conducting nuclear research under the Nazis, even as Robert Oppenheimer and his colleagues strove to develop the atom bomb at Los Alamos. The actions and motives of Heisenberg and his fellow researchers are to this day a matter of heated debate.

Uncertainty: The Life and Science of Werner Heisenberg is the first full-scale biography of this towering figure—an elegant, finely observed, intensively researched portrait by the noted science historian and Heisenberg scholar David C. Cassidy.

(continued on back flap)